Operator Theory: Advances and Applications
Volume 226

Founded in 1979 by Israel Gohberg

Daniel Alpay
Bernd Kirstein
Editors

Interpolation, Schur Functions and Moment Problems II

L | Linear
O | Operators and
L | Linear
S | Systems

Birkhäuser

Editors
Daniel Alpay
Department of Mathematics
   and Computer Science
Ben-Gurion University of the Negev
Beer Sheva
Israel

Bernd Kirstein
Mathematisches Institut
Universität Leipzig
Leipzig
Germany

ISBN 978-3-0348-0427-1   ISBN 978-3-0348-0428-8 (eBook)
DOI 10.1007/978-3-0348-0428-8
Springer Basel Heidelberg New York Dordrecht London

Library of Congress Control Number: 2012944402

Mathematics Subject Classification: 44A60, 47A57, 30E05

Printed on acid-free paper

Springer Basel AG is part of Springer Science+Business Media (www.birkhauser-science.com)

Contents

Editorial Introduction 1

B. Fritzsche, B. Kirstein, C. Mädler and T. Schwarz
    On the Concept of Invertibility for Sequences
    of Complex $p \times q$-matrices and its Application
    to Holomorphic $p \times q$-matrix-valued Functions 9

B. Fritzsche, B. Kirstein, A. Lasarow and A. Rahn
    On Reciprocal Sequences of Matricial Carathéodory
    Sequences and Associated Matrix Functions 57

B. Fritzsche, B. Kirstein, C. Mädler and T. Schwarz
    On a Schur-type Algorithm for Sequences of Complex
    $p \times q$-matrices and its Interrelations with the Canonical
    Hankel Parametrization 117

A.E. Choque Rivero
    Multiplicative Structure of the Resolvent Matrix for
    the Truncated Hausdorff Matrix Moment Problem 193

B. Fritzsche, B. Kirstein and C. Mädler
    On a Special Parametrization of Matricial α-Stieltjes
    One-sided Non-negative Definite Sequences 211

B. Fritzsche, B. Kirstein and A. Lasarow
    On Maximal Weight Solutions of a Moment Problem
    for Rational Matrix-valued Functions 251

Operator Theory:
Advances and Applications, Vol. 226, 1–7

Editorial Introduction

The present volume, entitled *"Interpolation, Schur functions and moment problems II"*, contains a selection of essays on various topics in Schur Analysis and addresses the goals set in 2006 in Volume 165 *"Interpolation, Schur functions and moment problems"* of the OT Series.

The origins of Schur Analysis lie in I. Schur's remarkable 1917/1918 two-part paper [33] which includes his insightful approach to constructing an algorithm for solving the coefficient problem for functions holomorphic on the unit disk $\mathbb{D} := \{z \in \mathbb{C} : |z| < 1\}$ that are also bounded by 1. The class of all such functions is now referred to as the *Schur class*. The algorithm has come to be known as the *Schur algorithm*. The English translation of Schur's paper [33] and research papers on the Schur algorithm were published in volume 18 of the OT Series (see [19]).

Schur's parametrization ([33]) of a Schur function's Taylor coefficients using a sequence of contractive parameters was the starting point for many applications of Schur analytic methods to various types of block matrices. Schur showed that lower triangular Toeplitz matrices resulting from the Taylor coefficient sequence of a Schur function are contractive. Schur's results represent a particularly noteworthy special case, which opened the gateway to more general cases and led to the development of Schur analytic methods for other block matrices. The theory developed for non-negative Hermitian matrices is a particularly interesting example of these later developments. More specifically, it was shown that non-negative Hermitian matrices can be fully described using their diagonal blocks and a triangular configuration of contractive matrices, often referred to as a *choice triangle*. Every non-negative Hermitian matrix corresponds to a triangular configuration of matrix balls. A contractive parameter taken from the choice triangle then indicates the position of the relevant block within the corresponding matrix ball. Further information on Schur Analysis of block matrices and related topics can be found, for instance, in Bakonyi/Woerdeman [9], Constantinescu [11] as well as Dubovoj/Fritzsche/Kirstein [12] and references therein.

R. Nevanlinna deserves particular mention for being among the first to recognize, adapt and further develop Schur's ideas. In his dissertation (see [27]), R. Nevanlinna arrived at a modification of the Schur algorithm with which he was able to describe all Schur functions satisfying certain interpolation conditions. Nevanlinna was unaware of the fact (likely due to the tumultuous times of World

War I in which he obtained his results) that Pick [30] had already determined necessary and sufficient conditions for the aforementioned interpolation problem in terms of whether or not a particular matrix (later given the name *Pick matrix*) constructed from the given data was non-negative Hermitian. Interpolation problems remained a favourite subject of R. Nevanlinna. His most significant achievements in this field are found in his 1929 article [29] in which he discusses many new ideas on characterizing the connection between limit point and limit circle cases. It should be mentioned that reprints of the papers Schur [33], Nevanlinna [29] and Pick [30] as well as Herglotz [23] and Weyl [36] are all to be found in the collection [17].

R. Nevanlinna is also responsible for introducing a very interesting approach to power moment problems on the real axis (see [28]). Inspired by his work on interpolation problems in [27], R. Nevanlinna found a way to restate the Hamburger moment problem on the real axis as an equivalent problem. The equivalent problem involves finding functions holomorphic in the open upper half-plane $\Pi^+ := \{z \in \mathbb{C} \ : \ \operatorname{Im} z \in (0, +\infty)\}$ with a given asymptotic expansion and having non-negative imaginary parts. This reformulation of the problem makes it possible to apply Schur's method for power moment problems. The, up to this point, groundbreaking work of Stieltjes [34] and Hamburger [20] was based on continued fractions methods. A survey of the theory of classical power moment problems is found in Akhiezer [5].

Bolstered by the advances in operator theory and influenced by the needs of electrical engineering, signal transmission and processing, as well as prediction theory for stationary stochastic processes, Schur's method experienced a renaissance near the end of the 1960s. The publications of Adamjan–Arov–Krein [1–4] laid the groundwork for this resurgence of the Schur method and led to intensive research on matrix and operator versions of classical interpolation and moment problems. The flexibility and variety of Schur analytic methods led to the publishing of an exceptionally large number of books dedicated to this topic: Alpay [6], Alpay/Dijksma/Rovnyak/de Snoo [7], Bakonyi/Constantinescu [8], Bakonyi/-Woerdeman [9], Ball/Gohberg/Rodman [10], Constantinescu [11], Dubovoj/Fritzsche/Kirstein [12], Dym [13], Foias/Frazho [15], Foias/Frazho/Gohberg/Kaashoek [16], Helton [24], Katsnelson [25] (see also [26], which is the English translation of the first part of [25]), Rosenblum/Rovnyak [31], Sakhnovich [32]. These books share the common feature that they all deal with matrix and operator generalizations of Schur's method.

Schur Analysis' rapid growth and evolution is documented by the numerous articles, particularly the many recent articles, published on topics in Schur Analysis. It is to be expected that these developments will continue for quite a while to come. This volume addresses a number of current trends and developments in Schur Analysis. By and large, the focus is on matricial generalizations of Schur's and Nevanlinna's earlier described classical results.

The first central theme of this volume involves matrix versions of power moment problems on partial intervals of the real axis as well as how these problems might be approached. The solvability of such moment problems can be character-ized by regarding a block Hankel matrix constructed from the initial data and con-sidering its Hermitian non-negativeness. This leads to a need to know more about the inner structure of block Hankel matrices. Before going into further detail on this, it is worth recalling a similar topic, very much related to non-negative Hermit-ian block Toeplitz matrices. As part of developing approaches to matrix versions of interpolation problems for the Carathéodory and Schur function classes, much was done in the 1980s to describe the inner structure of non-negative Hermitian and contractive block matrices (see, for instance, Dubovoj/Fritzsche/Kirstein [12]). The focus was on block Toeplitz matrices and it was determined that a non-negative Hermitian block Toeplitz matrix could be uniquely described with a se-quence of contractive parameters. This parameter sequence, often referred to as a *Schur parameter sequence*, contains considerable information on the inner structure of a given block Toeplitz matrix. The next task was to find a similarly useful inner parametrization for non-negative Hermitian block Hankel matrices. The canonical Hankel parametrization, which was introduced in [14] and [18], proved to be a very effective tool. A number of questions concerning the inner structure of spe-cial non-negative Hermitian block Hankel matrices could thus be resolved. Closely related to the search for these answers is the need for a corresponding Schur-type algorithm for finite and infinite sequences of complex matrices.

The second main theme of this volume relates to the theory of matricial Carathéodory functions and deals with various aspects of this theory. The close connection to the theory of orthogonal rational matrix functions on the unit circle is among the most important tools used in discussing these topics.

This volume contains six papers, and we now review their contents.

Reciprocal sequences (of finite and infinite sequences of matrices) and their applications: The paper *"On the concept of invertibility for sequences of complex $p \times q$ matrices and its applications to holomorphic $p \times q$ matrix-valued functions"* by **B. Fritzsche**, **B. Kirstein**, **C. Mädler** and **T. Schwarz** focuses on a type of in-vertibility for finite and infinite sequences of complex $p \times q$ matrices. The problem of characterizing all invertible sequences of complex $p \times q$ matrices naturally leads to what will be referred to as *first term dominant* sequences of complex $p \times q$ ma-trices. Given an invertible sequence of complex $p \times q$ matrices, the question of how to determine its inverse sequence (of complex $q \times p$ matrices) is answered using a recursively defined sequence of $q \times p$ matrices, called its *reciprocal sequence*. This definition opens up many possibilities for analyzing and solving a great number of matricial complex function theory problems. This includes, for example, the prob-lem of characterizing the holomorphicity of a holomorphic $p \times q$ matrix-function F's Moore-Penrose inverse F^+.

The paper *"On reciprocal sequences of matricial Carathéodory sequences and associated matrix functions"* by **B. Fritzsche**, **B. Kirstein**, **A. Lasarow** and **A. Rahn** discusses, among other topics, applications of reciprocal sequences to a particular subclass from matricial complex function theory. The focus is, in particular, on the case $p = q$ and the class $\mathcal{C}_q(\mathbb{D})$ of all $q \times q$ Carathéodory functions on the open unit disk $\mathbb{D} := \{ w \in \mathbb{C} : |w| < 1 \}$. These are all $q \times q$ matrix-functions Ω that are holomorphic in $\mathbb{D}$ and for which $\frac{1}{2}[\Omega(w) + \Omega^*(w)]$ is non-negative Hermitian for all $w \in \mathbb{D}$. As an important result, it is shown that for any function $\Omega \in \mathcal{C}_q(\mathbb{D})$, the Moore-Penrose inverse Ω^+ of this function again belongs to $\mathcal{C}_q(\mathbb{D})$. It is also determined that the Taylor coefficient sequence of Ω^+ is the reciprocal sequence to Ω's Taylor coefficient sequence.

In the paper "*On a Schur-type algorithm for sequences of complex $p \times q$ matrices and its interrelations with the canonical Hankel parametrization*" by **B. Fritzsche, B. Kirstein, C. Mädler** and **T. Schwarz**, a further application of reciprocal sequences is discussed. The central focus is on special classes of sequences of complex $q \times q$ matrices that are closely related to a number of different matricial versions of the Hamburger Moment Problem on the real axis $\mathbb{R}$. Given a sequence $(s_j)_{j=0}^{\infty}$ of complex $p \times q$ matrices, the reciprocal sequence of complex $q \times p$ matrices $(s_j^{\sharp})_{j=0}^{\infty}$ is used to develop a Schur-type algorithm for sequences of $p \times q$ matrices. This makes it possible for an algebraic approach to matricial versions of the Hamburger Moment Problem to be modelled on the basis of this algorithm.

Matricial power moment problems: With "*The multiplicative structure of the resolvent matrix for the truncated Hausdorff matrix moment problem*", **Abdon E. Choque Rivero** offers a continued look into topics from his earlier collaborations with Yu.M. Dyukarev, B. Fritzsche and B. Kirstein. These collaborations yielded a constructive approach to obtaining a special polynomial resolvent matrix. Choque Rivero derives a factorization of this polynomial as a product of linear factors.

The paper "*On a special parametrization of matricial α-Stieltjes one-sided non-negative definite sequences*", by **B. Fritzsche, B. Kirstein** and **C. Mädler**, deals with topics concerning the matrix version of the Stieltjes moment problem for a half-infinite closed interval. The authors obtained characterizations of the solvability for the relevant moments in earlier collaborations with Yu.M. Dyukarev. These solvability criteria say that the initial data sequence must be one-sided α-Stieltjes non-negative definite. The class of all such matrix sequences serves as a starting point for this paper. The objective here is to derive an inner parametrization for sequences of this class. This parametrization should, in a particular manner, reflect the one-sided α-Stieltjes structure of the aforementioned sequences. The authors used a similar approach in their earlier work regarding Hankel non-negative definite sequences. There, they obtained that the structure of these sequences was clearly recognizable in their canonical Hankel parametrizations. The authors, furthermore, constructed a Schur-type algorithm which yielded the Hankel parametrization as part of its result. The approach used here for the α-Stieltjes case is quite similar. The first step involves a detailed look at the constructed α-Stieltjes parametrization. In particular, a relationship between this parametrization and the Hankel parametrization is established.

Orthogonal rational matrix-valued functions: In a series of papers, over the past decade, **B. Fritzsche**, **B. Kirstein** and **A. Lasarow** worked out essential features of a Szegő theory for orthogonal rational matrix-valued functions on the unit circle and applied this theory to corresponding interpolation and moment problems. The paper *"On maximal weight solutions of a moment problem for rational matrix-valued functions"* continues these studies and focuses on special canonical solutions of the matricial moment problem under consideration. These canonical solutions turn out to be molecular non-negative Hermitian measures on the unit circle, which means that they are concentrated on a finite set of points of the unit circle. It is shown that these canonical solutions have a particular extremal property which is connected to a maximal mass condition in a prescribed point of the unit circle.

References

[1] V.M. Adamjan, D.Z. Arov, and M.G. Kreĭn, *Infinite Hankel matrices and generalized problems of Carathéodory-Fejér and F. Riesz*, Funkcional. Anal. i Priložen. **2** (1968), no. 1, 1–19 (Russian). MR0234274 (38 #2591)

[2] ———, *Infinite Hankel matrices and generalized Carathéodory-Fejér and I. Schur problems*, Funkcional. Anal. i Priložen. **2** (1968), no. 4, 1–17 (Russian). MR0636333 (58 #30446)

[3] ———, *Analytic properties of the Schmidt pairs of a Hankel operator and the generalized Schur-Takagi problem*, Mat. Sb. (N.S.) **86(128)** (1971), 34–75 (Russian). MR0298453 (45 #7505)

[4] ———, *Infinite Hankel block matrices and related problems of extension*, Izv. Akad. Nauk Armjan. SSR Ser. Mat. **6** (1971), no. 2-3, 87–112 (Russian, with Armenian and English summaries). MR0298454 (45 #7506)

[5] N.I. Akhiezer, *The classical moment problem and some related questions in analysis*, Translated by N. Kemmer, Hafner Publishing Co., New York, 1965. MR0184042 (32 #1518)

[6] D. Alpay, *The Schur algorithm, reproducing kernel spaces and system theory*, SMF/AMS Texts and Monographs, vol. 5, American Mathematical Society, Providence, RI, 2001. Translated from the 1998 French original by Stephen S. Wilson. MR1839648 (2002b:47144)

[7] D. Alpay, A. Dijksma, J. Rovnyak, and H. de Snoo, *Schur functions, operator colligations, and reproducing kernel Pontryagin spaces*, Operator Theory: Advances and Applications, vol. 96, Birkhäuser Verlag, Basel, 1997. MR1465432 (2000a:47024)

[8] M. Bakonyi and T. Constantinescu, *Schur's algorithm and several applications*, Pitman Research Notes in Mathematics Series, vol. 261, Longman Scientific & Technical, Harlow, 1992. MR1192547 (94b:47022)

[9] M. Bakonyi and H.J. Woerdeman, *Matrix completions, moments, and sums of Hermitian squares*, Princeton University Press, Princeton, NJ, 2011. MR2807419

[10] J.A. Ball, I. Gohberg, and L. Rodman, *Interpolation of rational matrix functions*, Operator Theory: Advances and Applications, vol. 45, Birkhäuser Verlag, Basel, 1990. MR1083145 (92m:47027)

[11] T. Constantinescu, *Schur parameters, factorization and dilation problems*, Operator Theory: Advances and Applications, vol. 82, Birkhäuser Verlag, Basel, 1996. MR1399080 (97g:47012)

[12] V.K. Dubovoj, B. Fritzsche, and B. Kirstein, *Matricial version of the classical Schur problem*, Teubner-Texte zur Mathematik [Teubner Texts in Mathematics], vol. 129, B.G. Teubner Verlagsgesellschaft mbH, Stuttgart, 1992. With German, French and Russian summaries. MR1152328 (93e:47021)

[13] H. Dym, *J contractive matrix functions, reproducing kernel Hilbert spaces and interpolation*, CBMS Regional Conference Series in Mathematics, vol. 71, Published for the Conference Board of the Mathematical Sciences, Washington, DC, 1989. MR1004239 (90g:47003)

[14] Yu.M. Dyukarev, B. Fritzsche, B. Kirstein, C. Mädler, and H.C. Thiele, *On distinguished solutions of truncated matricial Hamburger moment problems*, Complex Anal. Oper. Theory **3** (2009), no. 4, 759–834, DOI 10.1007/s11785-008-0061-2. MR2570113

[15] C. Foias and A.E. Frazho, *The commutant lifting approach to interpolation problems*, Operator Theory: Advances and Applications, vol. 44, Birkhäuser Verlag, Basel, 1990. MR1120546 (92k:47033)

[16] C. Foias, A.E. Frazho, I. Gohberg, and M.A. Kaashoek, *Metric constrained interpolation, commutant lifting and systems*, Operator Theory: Advances and Applications, vol. 100, Birkhäuser Verlag, Basel, 1998. MR1635831 (99i:47027)

[17] B. Fritzsche and B. Kirstein (eds.), *Ausgewählte Arbeiten zu den Ursprüngen der Schur-Analysis*, Teubner – Archiv zur Mathematik, vol. 16, B.G. Teubner Verlagsgesellschaft, Stuttgart, 1991. MR1162976 (93d:01095)

[18] B. Fritzsche, B. Kirstein, and C. Mädler, *On Hankel nonnegative definite sequences, the canonical Hankel parametrization, and orthogonal matrix polynomials*, Complex Anal. Oper. Theory **5** (2011), no. 2, 447–511, DOI 10.1007/s11785-010-0054-9. MR2805417

[19] I. Gohberg (ed.), *I. Schur methods in operator theory and signal processing*, Operator Theory: Advances and Applications, vol. 18, Birkhäuser Verlag, Basel, 1986. MR902600 (88d:00006)

[20] H. Hamburger, *Über eine Erweiterung des Stieltjesschen Momentenproblems*, Math. Ann. **81** (1920), no. 2-4, 235–319, DOI 10.1007/BF01564869 (German). MR1511966

[21] _____, *Über eine Erweiterung des Stieltjesschen Momentenproblems*, Math. Ann. **82** (1920), no. 1-2, 120–164, DOI 10.1007/BF01457982 (German). MR1511978

[22] _____, *Über eine Erweiterung des Stieltjesschen Momentenproblems*, Math. Ann. **82** (1921), no. 3-4, 168–187, DOI 10.1007/BF01498663 (German). MR1511981

[23] G. Herglotz, *Über Potenzreihen mit positivem, reellen Teil im Einheitskreis*, Berichte über die Verhandlungen der Königlich Sächsischen Gesellschaft der Wissenschaften zu Leipzig. Mathematisch-physische Klasse **63** (1911), 501–511 (German).

[24] J.W. Helton, *Operator theory, analytic functions, matrices, and electrical engineering*, CBMS Regional Conference Series in Mathematics, vol. 68, Published for the Conference Board of the Mathematical Sciences, Washington, DC, 1987. MR896034 (89f:47001)

[25] V.È. Katsnel'son, *Methods of J-theory in continuous interpolation problems of analysis*, 1983. Deposited in VINITI. Part 1 translated from the Russian in [26].

[26] _____, *Methods of J-theory in continuous interpolation problems of analysis. Part I*, translated by T. Ando, T. Ando Hokkaido University, Sapporo, 1985. MR777324 (86i:47048)

[27] R. Nevanlinna, *Über beschränkte Funktionen, die in gegebenen Punkten vorgeschriebene Werte annehmen*, Ann. Acad. Sci. Fenn. **A 13** (1919), no. 1, 1–71.

[28] _____, *Asymptotische Entwicklungen beschränkter Funktionen und das Stieltjessche Momentenproblem*, Ann. Acad. Sci. Fenn. **A 18** (1922), no. 5, 1–53.

[29] _____, *Über beschränkte analytische Funktionen*, Ann. Acad. Sci. Fenn. **A 32** (1929), no. 7, 1–75.

[30] G. Pick, *Über die Beschränkungen analytischer Funktionen, welche durch vorgegebene Funktionswerte bewirkt werden*, Math. Ann. **77** (1915), no. 1, 7–23, DOI 10.1007/BF01456817 (German). MR1511844

[31] M. Rosenblum and J. Rovnyak, *Hardy classes and operator theory*, Oxford Mathematical Monographs, The Clarendon Press Oxford University Press, New York, 1985. Oxford Science Publications. MR822228 (87e:47001)

[32] L.A. Sakhnovich, *Interpolation theory and its applications*, Mathematics and its Applications, vol. 428, Kluwer Academic Publishers, Dordrecht, 1997. MR1631843 (99j:47016)

[33] I. Schur, *Über Potenzreihen, die im Innern des Einheitskreises beschränkt sind*. Teil I **147** (1917), 205–232; Teil II **148** (1918), 122–145, J. reine u. angew. Math.

[34] T.-J. Stieltjes, *Recherches sur les fractions continues*, Ann. Fac. Sci. Toulouse Sci. Math. Sci. Phys. **8** (1894), no. 4, J1–J122 (French). MR1508159

[35] _____, *Recherches sur les fractions continues [Suite et fin]*, Ann. Fac. Sci. Toulouse Sci. Math. Sci. Phys. **9** (1895), no. 1, A5–A47 (French). MR1508160

[36] H. Weyl, *Über das Pick–Nevanlinna'sche Interpolationsproblem und sein infinitesimales Analogon*, Annals of Mathematics **36** (1935), no. 2, 230–254 (German). MR1503221

Daniel Alpay, Bernd Kirstein

Operator Theory:
Advances and Applications, Vol. 226, 9–56

On the Concept of Invertibility for Sequences of Complex $p \times q$-matrices and its Application to Holomorphic $p \times q$-matrix-valued Functions

Bernd Fritzsche, Bernd Kirstein, Conrad Mädler and Tilo Schwarz

Abstract. The main topic of this paper is the invertibility of finite and infinite sequences of complex $p \times q$-matrices. This concept was previously considered in the mathematical literature for the special case in which $p = q$, under certain regularity conditions, in the context of matricial power series inversion. The problem of describing all (finite and infinite) invertible sequences of complex $p \times q$-matrices leads directly to the class of "first term dominant" sequences $(s_j)_{j=0}^{\kappa}$ of complex $p \times q$-matrices. These sequences have the property that the null space of s_0 is contained in the null spaces of all s_j, while the range of s_0 encompasses the range of every s_j. The inverse sequence $(s_j^{\ddagger})_{j=0}^{\kappa}$ in $\mathbb{C}^{q \times p}$ for an invertible sequence $(s_j)_{j=0}^{\kappa}$ in $\mathbb{C}^{p \times q}$ is then constructed. This leads, in conjunction with the concept of power series inversion, to a general recursive method for constructing a reciprocal sequence $(s_j^{\sharp})_{j=0}^{\kappa}$ in $\mathbb{C}^{q \times p}$ to any given sequence $(s_j)_{j=0}^{\kappa}$ in $\mathbb{C}^{p \times q}$. It is shown that if $(s_j)_{j=0}^{\kappa}$ is invertible, then its inverse and reciprocal sequences coincide, i.e., $(s_j^{\ddagger})_{j=0}^{\kappa} = (s_j^{\sharp})_{j=0}^{\kappa}$.

Using reciprocal sequences allows for interesting new approaches to a number of fascinating problems in matricial complex analysis. This paper considers the holomorphicity of the Moore-Penrose inverse of a $p \times q$-matrix-function, using an approach based on analyzing the structure of Taylor coefficient sequences. A main result of this paper states that the Moore-Penrose inverse $F^{\dagger}$ of a complex $p \times q$-matrix-function F which is holomorphic in an open disk K of the complex plane $\mathbb{C}$ is holomorphic in K if, and only if, the Taylor-McLaurin coefficient sequence $(s_j)_{j=0}^{\kappa}$ for F at the center z_0 of K is invertible. When this is the case, the reciprocal sequence $(s_j^{\sharp})_{j=0}^{\kappa}$ to $(s_j)_{j=0}^{\kappa}$ is the Taylor-McLaurin coefficient sequence for $F^{\dagger}$ in z_0.

Mathematics Subject Classification (2010). 30E05; 15A09.

Keywords. Inverse sequence corresponding to an invertible sequence of complex $p \times q$-matrices, reciprocal sequence corresponding to an arbitrary sequence of complex matrices, invertible sequences of complex $p \times q$-matrices.

0. Introduction

Let q be a positive integer and let f be a holomorphic $q \times q$-matrix-valued function, defined in some disk, K, around the origin, and satisfying

$$\det\, [f(z)] \neq 0,$$

for all $z \in K$. The $q \times q$-matrix-valued function,

$$f^{-1}\, :\, K \longrightarrow \mathbb{C}^{q \times q}, \qquad \text{defined by} \qquad f^{-1}(z) := [f(z)]^{-1}$$

is then also holomorphic in K. Suppose that the Taylor series of f and f^{-1} around the origin are given by

$$f(z) = \sum_{j=0}^{\infty} s_j z^j \qquad \text{and} \qquad f^{-1}(z) = \sum_{j=0}^{\infty} r_j z^j.$$

Kaluza [13] was the first to study the relationship between the sequences $(r_j)_{j=0}^{\infty}$ and $(s_j)_{j=0}^{\infty}$ for the scalar case, $q = 1$. Jurkat [12], Lamperti [14], Brietzke [3] and Baricz/Vesti/Vuorinen [1] also continued to study this relationship. It should be mentioned that Kaluza's results are useful in the study of renewal sequences (see Horn [11], Hansen/Steutel [10]).

For each non-negative integer n, the sequences $(r_j)_{j=0}^{\infty}$ and $(s_j)_{j=0}^{\infty}$ are then linked by the formulas

$$\begin{pmatrix} s_0 & 0 & \cdots & 0 \\ s_1 & s_0 & \cdots & 0 \\ \vdots & \vdots & \ddots & \vdots \\ s_n & s_{n-1} & \cdots & s_0 \end{pmatrix} \begin{pmatrix} r_0 & 0 & \cdots & 0 \\ r_1 & r_0 & \cdots & 0 \\ \vdots & \vdots & \ddots & \vdots \\ r_n & r_{n-1} & \cdots & r_0 \end{pmatrix} = I_{(n+1)q}, \qquad (0.1)$$

where $I_{(n+1)q}$ is the unit matrix of order $(n+1)q$. Thus, for each non-negative integer n, the matrices of the left-hand side of (0.1) are non-singular and

$$\begin{pmatrix} s_0 & 0 & \cdots & 0 \\ s_1 & s_0 & \cdots & 0 \\ \vdots & \vdots & \ddots & \vdots \\ s_n & s_{n-1} & \cdots & s_0 \end{pmatrix}^{-1} = \begin{pmatrix} r_0 & 0 & \cdots & 0 \\ r_1 & r_0 & \cdots & 0 \\ \vdots & \vdots & \ddots & \vdots \\ r_n & r_{n-1} & \cdots & r_0 \end{pmatrix}.$$

We are interested in a generalization of these results to the case in which p and q are positive integers and f is a holomorphic $p \times q$-matrix-valued function, defined in some disk, K, around the origin.

For each complex $p \times q$-matrix A, we will use $A^{\dagger}$ to denote the Moore-Penrose inverse of A. Given f, we define

$$f^{\dagger}\, :\, K \longrightarrow \mathbb{C}^{q \times p} \qquad \text{as} \qquad f^{\dagger}(z) := [f(z)]^{\dagger}.$$

We want to find suitable characterizations of the holomorphicity of $f^{\dagger}$ in K. Specifically, our goal is to characterize the holomorphicity of $f^{\dagger}$ in terms of the sequence, $(s_j)_{j=0}^{\infty}$, of Taylor coefficients belonging to f.

We are, thus, naturally led to consider the class of $p \times q$-matrix-sequences, $(s_j)_{j=0}^{\infty}$, for which there exists a $q \times p$-matrix-sequence $(r_j)_{j=0}^{\infty}$, such that for each non-negative integer n:

$$
\begin{pmatrix}
s_0 & 0 & \cdots & 0 \\
s_1 & s_0 & \cdots & 0 \\
\vdots & \vdots & \ddots & \vdots \\
s_n & s_{n-1} & \cdots & s_0
\end{pmatrix}^{\dagger}
=
\begin{pmatrix}
r_0 & 0 & \cdots & 0 \\
r_1 & r_0 & \cdots & 0 \\
\vdots & \vdots & \ddots & \vdots \\
r_n & r_{n-1} & \cdots & r_0
\end{pmatrix}.
\tag{0.2}
$$

Sequences of this type will be called *invertible* sequences and will serve as the main focus of the first part of this paper. We will later give a complete description of all invertible sequences of $p \times q$-matrices. If $(s_j)_{j=0}^{\infty}$ is one such invertible sequence, then it will be shown that there exists a unique sequence, $(r_j)_{j=0}^{\infty}$, satisfying (0.2) for each non-negative integer n. We will, furthermore, present a concrete method for constructing $(r_j)_{j=0}^{\infty}$ from $(s_j)_{j=0}^{\infty}$ (see Theorem 4.21).

This paper is organized as follows. Section 1 serves as a quick review and summary of some of the notation used in this paper.

In Section 2 we introduce the concept of invertibility for finite and infinite sequences of complex $p \times q$-matrices. We pay special attention to a particular subclass of invertible sequences. The results we obtain for this subclass then dictate how we proceed further. One of our main objectives for the remainder of this paper then consists of extending the results for this subclass to the greater class of invertible sequences.

We give an overview of important rules for working with invertible sequences in Section 3.

Central to this paper is Section 4, in which we give a characterization of invertible sequences and, furthermore, explicit formulas for constructing the inverse sequence (see Theorem 4.21).

Section 4's results give us a new perspective on the class of invertible sequences and thus serve as our motivation for Section 5. In particular, we take an extensive look at the construction of reciprocal sequences for finite and infinite sequences of complex $p \times q$-matrices. This leads us to a characterization of the invertibility of a sequence in terms of its reciprocal sequence (see Proposition 5.13).

To begin Section 6, we present some additional rules for working with invertible sequences. The main result of Section 6 is Theorem 6.11, which offers us important insight into the structure of a given invertible sequence $(s_j)_{j=0}^{\kappa}$ of complex $p \times q$-matrices. Specifically, we find that every such sequence is generated (in a certain way) by a suitably constructed sequence, $(\widehat{s}_j)_{j=0}^{\kappa}$, of complex $r \times r$-matrices with $r := \operatorname{rank} s_0$ and $\det \widehat{s}_0 \neq 0$.

Section 7 focuses on a special subclass of invertible sequences of complex $q \times q$ matrices, called EP sequences (see Definition 7.1). These are sequences in $\mathbb{C}^{q \times q}$ for which every lower-triangular block Toeplitz matrix of the type defined in Notation 2.1 is an EP matrix.

In Section 8 we examine the properties of the Moore-Penrose inverse $F^\dagger$, for a holomorphic (in a non-empty open and connected subset, $\mathcal{G}$, of $\mathbb{C}$) $p \times q$-matrix-function F. The main result of Section 8 is Theorem 8.9, which says that, if F is a matrix-function, holomorphic in a point z_0 of its domain, then $F^\dagger$ is holomorphic in z_0 if, and only if, the sequence of F's Taylor coefficients, $\left(\frac{F^{(j)}(z_0)}{j!} \right)_{j=0}^\infty$, is invertible. Whenever this is indeed the case, we also recognize $\left(\frac{(F^\dagger)^{(j)}(z_0)}{j!} \right)_{j=0}^\infty$ as the inverse sequence to $\left(\frac{F^{(j)}(z_0)}{j!} \right)_{j=0}^\infty$.

In this paper, we will often draw upon the theory of Moore-Penrose inverses. For this reason, we have summarized some of the relevant information on the Moore-Penrose inverse in Appendix A. In particular, we look at the Moore-Penrose inverse of a product of matrices (see Proposition A.5 and Proposition A.7) and verify special results on the Moore-Penrose inverse of a block-matrix under certain constraints (see Proposition A.10, Corollary A.11 and Corollary A.12).

1. Some notation

Let $\mathbb{C}$, $\mathbb{R}$, $\mathbb{N}_0$, and $\mathbb{N}$ be the set of all complex numbers, the set of all real numbers, the set of all non-negative integers, and the set of all positive integers, respectively. For every choice of $\alpha, \beta \in \mathbb{R} \cup \{-\infty, +\infty\}$, let $\mathbb{Z}_{\alpha,\beta}$ be the set of all integers k for which $\alpha \le k \le \beta$. Whenever there is no ambiguity with regard to the size of a matrix, we will omit the indices.

Throughout this paper, let p and q be positive integers. If $\mathcal{X}$ is a non-empty set, then let $\mathcal{X}^{p \times q}$ be the set of all $p \times q$-matrices with elements in $\mathcal{X}$. We will write $0_{p \times q}$ for the zero-matrix in $\mathbb{C}^{p \times q}$, and will use I_q to denote the unit matrix in $\mathbb{C}^{q \times q}$.

We will write $\mathbb{C}_H^{q \times q}$ and $\mathbb{C}_\ge^{q \times q}$, respectively, for the set of all Hermitian and non-negative Hermitian complex $q \times q$-matrices. For each $A \in \mathbb{C}^{p \times q}$, let $\mathcal{R}(A)$ be the range of A and let $\mathcal{N}(A)$ be the null-space of A. If $n \in \mathbb{N}$ and if $(v_j)_{j=1}^n$ is a sequence of complex $p \times q$-matrices, then let

$$\text{row}(v_k)_{k=1}^n := (v_1, v_2, \ldots, v_n) \quad \text{and} \quad \text{col}(v_j)_{j=1}^n := (v_1^*, v_2^*, \ldots, v_n^*)^*.$$

With $A \otimes B$, we denote the Kronecker product of the matrices $A = (a_{jk})_{\substack{j=1,\ldots,p \\ k=1,\ldots,q}} \in \mathbb{C}^{p \times q}$ and $B \in \mathbb{C}^{r \times s}$: $A \otimes B := (a_{jk}B)_{\substack{j=1,\ldots,p \\ k=1,\ldots,q}}$. We draw attention to the fact that, if $s_0 \in \mathbb{C}^{p \times q}$ and if $m \in \mathbb{N}_0$, then the $(m+1)p \times (m+1)q$-matrix $\text{diag}(s_0, s_0, \ldots, s_0)$ can be expressed as $I_{m+1} \otimes s_0$.

If M is a non-empty subset of $\mathbb{C}^q$, then $M^\perp$ stands for the orthogonal complement of M in $\mathbb{C}^q$ (with respect to the Euclidean inner product). Furthermore, if $\mathcal{U}$ is a linear subspace of $\mathbb{C}^p$ and if $\mathcal{V}$ is a linear subspace of $\mathbb{C}^q$, then $\mathcal{U} \oplus \mathcal{V}$ is the set of all vectors $w \in \mathbb{C}^{p \times q}$ which admit the representation $w = \begin{pmatrix} u \\ v \end{pmatrix}$ for some $u \in \mathcal{U}$ and $v \in \mathcal{V}$.

2. A first look at invertible sequences in $\mathbb{C}^{p \times q}$

In this section we will introduce the concept of invertibility for finite as well as infinite sequences of complex matrices. We begin with some notation, which will help simplify things.

Notation 2.1. Given a $\kappa \in \mathbb{N}_0 \cup \{+\infty\}$ and a sequence $(s_j)_{j=0}^{\kappa}$ in $\mathbb{C}^{p \times q}$, we define, for each $m \in \mathbb{Z}_{0, \kappa}$, the lower-triangular block Toeplitz matrix $\mathbf{S}_m^{(s)}$ as

$$\mathbf{S}_m^{(s)} := \begin{pmatrix} s_0 & 0 & 0 & \cdots & 0 \\ s_1 & s_0 & 0 & \cdots & 0 \\ s_2 & s_1 & s_0 & \cdots & 0 \\ \vdots & \vdots & \vdots & \ddots & \vdots \\ s_m & s_{m-1} & s_{m-2} & \cdots & s_0 \end{pmatrix}. \tag{2.1}$$

Whenever it is clear which sequence is meant, we will simply write $\mathbf{S}_m$ instead of $\mathbf{S}_m^{(s)}$.

The following definition will prove to be, for our purposes, a very important and useful one.

Definition 2.2. Let $m \in \mathbb{N}_0$. A sequence $(s_j)_{j=0}^{m}$ of complex $p \times q$-matrices is called *invertible* if there is a sequence $(r_j)_{j=0}^{m}$ of complex $q \times p$-matrices such that the Moore-Penrose inverse $\left[\mathbf{S}_m^{(s)} \right]^{\dagger}$ of the matrix $\mathbf{S}_m^{(s)}$ coincides with the block Toeplitz matrix $\mathbf{S}_m^{(r)}$.

Remark 2.3. If $m \in \mathbb{N}_0$ and $(s_j)_{j=0}^{m}$ is an invertible sequence of complex $p \times q$-matrices, then there is a unique sequence $(r_j)_{j=0}^{m}$ of complex $q \times p$-matrices such that

$$\left[\mathbf{S}_m^{(s)} \right]^{\dagger} = \mathbf{S}_m^{(r)}. \tag{2.2}$$

Recalling Corollary A.11, we see that this sequence, $(r_j)_{j=0}^{m}$, satisfies $\left[\mathbf{S}_k^{(s)} \right]^{\dagger} = \mathbf{S}_k^{(r)}$ for each $k \in \mathbb{Z}_{0,m}$.

Remark 2.3 leads us to the following definition, which complements Definition 2.2.

Definition 2.4. A sequence $(s_j)_{j=0}^{\infty}$ of complex $p \times q$-matrices is called *invertible* if there is a sequence, $(r_j)_{j=0}^{\infty}$, of complex $q \times p$-matrices such that (2.2) holds true for each non-negative integer m.

14   B. Fritzsche, B. Kirstein, C. Mädler and T. Schwarz

Notation 2.5. Given a $\kappa \in \mathbb{N}_0 \cup \{+\infty\}$, we will use $\mathcal{I}_{p\times q,\kappa}$ to denote the set of all invertible sequences, $(s_j)_{j=0}^{\kappa}$, of complex $p \times q$-matrices.

Definitions 2.2 and 2.4 along with Remark 2.3 lead us to:

Definition 2.6. Let $\kappa \in \mathbb{N}_0 \cup \{+\infty\}$ and $(s_j)_{j=0}^{\kappa} \in \mathcal{I}_{p\times q,\kappa}$. The unique sequence $(r_j)_{j=0}^{\kappa}$ with (2.2) for each $m \in \mathbb{Z}_{0,\kappa}$, is called the *inverse sequence corresponding to* $(s_j)_{j=0}^{\kappa}$ and will be denoted by $(s_j^\dagger)_{j=0}^{\kappa}$.

Remark 2.7. Let $\kappa \in \mathbb{N}_0 \cup \{+\infty\}$ and $(s_j)_{j=0}^{\kappa} \in \mathcal{I}_{p\times q,\kappa}$. From Remark 2.3, we see that the following statements hold:

(a) $(s_j)_{j=0}^{m} \in \mathcal{I}_{p\times q,m}$ for all $m \in \mathbb{Z}_{0,\kappa}$.

(b) For all $m \in \mathbb{Z}_{0,\kappa}$, the inverse sequence corresponding to $(s_j)_{j=0}^{m}$ is $(s_j^\dagger)_{j=0}^{m}$.

(c) For the first element of the inverse sequence, we have $s_0^\dagger = s_0^\dagger$.

Remark 2.8. Let $\kappa \in \mathbb{N}_0 \cup \{+\infty\}$ and $(s_j)_{j=0}^{\kappa} \in \mathcal{I}_{p\times q,\kappa}$. Using (A.1) and recalling Definitions 2.2 and 2.4, we see that $(s_j^\dagger)_{j=0}^{\kappa} \in \mathcal{I}_{q\times p,\kappa}$ and, furthermore, that $(s_j)_{j=0}^{\kappa}$ is the inverse sequence corresponding to $(s_j^\dagger)_{j=0}^{\kappa}$.

Remark 2.9. Let $r, s \in \mathbb{N}$, $U \in \mathbb{C}^{r\times p}$ with $U^*U = I_p$, $\kappa \in \mathbb{N}_0 \cup \{+\infty\}$, $(s_j)_{j=0}^{\kappa} \in \mathcal{I}_{p\times q,\kappa}$, and $V \in \mathbb{C}^{q\times s}$ with $VV^* = I_q$. Using part (c) of Lemma A.3, we see that

$$\left[\operatorname{diag}(U)_{j=0}^{m} \cdot \mathbf{S}_m \cdot \operatorname{diag}(V)_{j=0}^{m}\right]^\dagger = \operatorname{diag}(V^*)_{j=0}^{m} \cdot \mathbf{S}_m^\dagger \cdot \operatorname{diag}(U^*)_{j=0}^{m},$$

for all $m \in \mathbb{Z}_{0,\kappa}$. Consequently, $(Us_jV)_{j=0}^{\kappa}$ belongs to $\mathcal{I}_{r\times s,\kappa}$ and $(V^*s_j^\dagger U^*)_{j=0}^{\kappa}$ is the inverse sequence corresponding to $(Us_jV)_{j=0}^{\kappa}$.

We now consider a generic subclass of invertible sequences. The results we obtain for this subclass (see, for instance, Proposition 2.11 and Proposition 2.14) will later dictate how we proceed in studying the larger class of invertible sequences. One of our goals will be to, in some way, extend these special-case results to the greater class.

Notation 2.10. Given a $\kappa \in \mathbb{N} \cup \{+\infty\}$, we will use $\widetilde{\mathcal{I}}_{q\times q,\kappa}$ to denote the set of all sequences $(s_j)_{j=0}^{\kappa}$ such that s_0 is invertible.

Applying [8, Lemma 5.6] yields the following useful properties for $\widetilde{\mathcal{I}}_{q\times q,\kappa}$.

Proposition 2.11. *If $\kappa \in \mathbb{N}_0 \cup \{+\infty\}$ and $(s_j)_{j=0}^{\kappa} \in \widetilde{\mathcal{I}}_{q\times q,\kappa}$, then*

(a) $(s_j)_{j=0}^{\kappa} \in \mathcal{I}_{q\times q,\kappa}$ *and* $(s_j^\dagger)_{j=0}^{\kappa} \in \widetilde{\mathcal{I}}_{q\times q,\kappa}$.

(b) *The sequence* $(s_j^\dagger)_{j=0}^{\kappa}$ *is given by*

$$s_k^\dagger = \begin{cases} s_0^{-1}, & \text{if } k = 0 \\ -s_0^{-1}\left(\displaystyle\sum_{j=0}^{k-1} s_{k-j}s_j^\dagger\right), & \text{if } k \in \mathbb{Z}_{1,\kappa} \end{cases}$$

or, alternatively, by

$$
s_k^{\ddagger} = \begin{cases} s_0^{-1}, & \text{if } k = 0 \\ -\left(\displaystyle\sum_{j=0}^{k-1} s_{k-j}^{\ddagger} s_j \right) s_0^{-1}, & \text{if } k \in \mathbb{Z}_{1,\kappa}. \end{cases}
$$

(c) If $(s_j)_{j=0}^{\kappa}$ is a sequence of Hermitian matrices, then $(s_j^{\ddagger})_{j=0}^{\kappa}$ is also a sequence of Hermitian matrices. Correspondingly, the inverse sequence for a sequence of symmetric matrices will also be a sequence of symmetric matrices.

(d) $\det \mathbf{S}_m^{(s)} = (\det s_0)^{m+1} \neq 0$ *and* $\left[\mathbf{S}_m^{(s)} \right]^{-1} = \mathbf{S}_m^{(s^{\ddagger})}$, *for each* $m \in \mathbb{Z}_{0,\kappa}$.

Next, we show that $\widetilde{\mathcal{I}}_{q \times q, \kappa}$ is closed with respect to the Cauchy product, which is defined as follows:

Notation 2.12. Let $\kappa \in \mathbb{N}_0 \cup \{+\infty\}$. For sequences $(s_j)_{j=0}^{\kappa}$ and $(t_j)_{j=0}^{\kappa}$ in $\mathbb{C}^{p \times q}$ and $\mathbb{C}^{q \times r}$, respectively, we will denote the *Cauchy product* of these two sequences by $((s \odot t)_j)_{j=0}^{\kappa}$, i.e., for all $j \in \mathbb{Z}_{0,\kappa}$:

$$
(s \odot t)_j = \sum_{l=0}^{j} s_l t_{j-l}.
$$

Remark 2.13. Let $\kappa \in \mathbb{N}_0 \cup \{+\infty\}$. Suppose, furthermore, that $(s_j)_{j=0}^{\kappa}$ and $(t_j)_{j=0}^{\kappa}$ are, respectively, sequences in $\mathbb{C}^{p \times q}$ and $\mathbb{C}^{q \times r}$. If $(u_j)_{j=0}^{\kappa}$ is a sequence in $\mathbb{C}^{p \times r}$, then

$$
(u_j)_{j=0}^{\kappa} = \left((s \odot t)_j \right)_{j=0}^{\kappa} \quad \text{if and only if} \quad \mathbf{S}_m^{(s)} \mathbf{S}_m^{(t)} = \mathbf{S}_m^{(u)}, \text{ for all } m \in \mathbb{Z}_{0,\kappa}.
$$

Proposition 2.14. *Let* $\kappa \in \mathbb{N}_0 \cup \{+\infty\}$. *If* $(s_j)_{j=0}^{\kappa}$ *and* $(t_j)_{j=0}^{\kappa}$ *are sequences of complex* $q \times q$*-matrices, then:*

(a) $\left((s \odot t)_j \right)_{j=0}^{\kappa} \in \widetilde{\mathcal{I}}_{q \times q, \kappa}$ *if and only if the sequences* $(s_j)_{j=0}^{\kappa}$ *and* $(t_j)_{j=0}^{\kappa}$ *both belong to* $\widetilde{\mathcal{I}}_{q \times q, \kappa}$.

(b) *If* $(s_j)_{j=0}^{\kappa} \in \widetilde{\mathcal{I}}_{q \times q, \kappa}$ *and* $(t_j)_{j=0}^{\kappa} \in \widetilde{\mathcal{I}}_{q \times q, \kappa}$, *then* $\left((s \odot t)_j \right)_{j=0}^{\kappa} \in \widetilde{\mathcal{I}}_{q \times q, \kappa}$ *and*

$$
\left((s \odot t)_j^{\ddagger} \right)_{j=0}^{\kappa} = \left((t^{\ddagger} \odot s^{\ddagger})_j \right)_{j=0}^{\kappa}.
$$

Proof. (a) Follows directly from $(s \odot t)_0 = s_0 t_0$.

(b) Let $m \in \mathbb{Z}_{0,\kappa}$. Using part (d) of Proposition 2.11, we have

$$
\left[\mathbf{S}_m^{(s)} \right]^{-1} = \mathbf{S}_m^{(s^{\ddagger})}, \qquad \left[\mathbf{S}_m^{(t)} \right]^{-1} = \mathbf{S}_m^{(t^{\ddagger})} \quad \text{and} \quad \left[\mathbf{S}_m^{(t \odot s)} \right]^{-1} = \mathbf{S}_m^{((t \odot s)^{\ddagger})}.
$$

Combining this with Remark 2.13, we obtain

$$
\mathbf{S}_m^{(s^{\ddagger} \odot t^{\ddagger})} = \left[\mathbf{S}_m^{(s^{\ddagger})} \right] \left[\mathbf{S}_m^{(t^{\ddagger})} \right] = \left[\mathbf{S}_m^{(s)} \right]^{-1} \left[\mathbf{S}_m^{(t)} \right]^{-1}
$$

$$= \left(\begin{bmatrix} \mathbf{S}_m^{(t)} \end{bmatrix} \begin{bmatrix} \mathbf{S}_m^{(s)} \end{bmatrix} \right)^{-1} = \left[\mathbf{S}_m^{(t \odot s)} \right]^{-1} = \mathbf{S}_m^{((t \odot s)^\ddagger)},$$

from which we get (b). $\qquad\square$

Now that we have covered the class $\widetilde{\mathcal{I}}_{q \times q, \kappa}$, we again turn our attention to the general class $\mathcal{I}_{p \times q, \kappa}$ of invertible matrix-sequences. We next show that our definition for invertibility could easily have been formulated using upper-triangular instead of lower-triangular block Toeplitz matrices.

Notation 2.15. Given a $\kappa \in \mathbb{N}_0 \cup \{+\infty\}$ and a sequence $(s_j)_{j=0}^\kappa$ in $\mathbb{C}^{p \times q}$, we define, for each $m \in \mathbb{Z}_{0, \kappa}$, the upper-triangular block Toeplitz matrix $\mathbb{S}_m^{(s)}$ as

$$\mathbb{S}_m^{(s)} := \begin{pmatrix} s_0 & s_1 & s_2 & \cdots & s_m \\ 0 & s_0 & s_1 & \cdots & s_{m-1} \\ 0 & 0 & s_0 & \cdots & s_{m-2} \\ \vdots & \vdots & \vdots & \ddots & \vdots \\ 0 & 0 & 0 & \cdots & s_0 \end{pmatrix}. \tag{2.3}$$

Whenever it is clear which sequence is meant, we will simply write $\mathbb{S}_m$ for $\mathbb{S}_m^{(s)}$.

Remark 2.16. It should be noted that, for each $\kappa \in \mathbb{N}_0 \cup \{+\infty\}$, for each sequence $(s_j)_{j=0}^\kappa$ of complex $p \times q$-matrices, and for each $m \in \mathbb{Z}_{0,\kappa}$, the block Toeplitz matrices $\mathbf{S}_m$ and $\mathbb{S}_m$, are connected via the equality

$$\mathbb{S}_m = U_{p,m} \mathbf{S}_m U_{q,m}, \tag{2.4}$$

where $U_{q,m} := (\delta_{j,m-k} I_q)_{j,k=0}^m$ and δ_{ln} is the Kronecker symbol, i.e., $\delta_{ln} = 1$ for $l = n$ and $\delta_{ln} = 0$ for $l \neq n$. Since the matrices $U_{p,m}$ and $U_{q,m}$ are unitary and Hermitian, we obtain the following identity from (2.4) and part (c) of Lemma A.3:

$$\mathbb{S}_m^\dagger = U_{q,m}^* \mathbf{S}_m^\dagger U_{p,m}^* = U_{q,m} \mathbf{S}_m^\dagger U_{p,m}. \tag{2.5}$$

Remark 2.17. Let $m \in \mathbb{N}_0$. Furthermore, suppose that $(s_j)_{j=0}^m$ and $(r_j)_{j=0}^m$ are, respectively, sequences in $\mathbb{C}^{p \times q}$ and $\mathbb{C}^{q \times p}$. Using (2.1) and (2.3), we see that

$$\left[\mathbb{S}_m^{(s)} \right]^* = \mathbf{S}_m^{(r)} \qquad \text{if and only if} \qquad (s_j)_{j=0}^m = (r_j^*)_{j=0}^m.$$

Proposition 2.18. *If $\kappa \in \mathbb{N}_0 \cup \{+\infty\}$ and $(s_j)_{j=0}^\kappa$ is a sequence of complex $p \times q$-matrices, then the following statements are all equivalent:*

(i) *$(s_j)_{j=0}^\kappa \in \mathcal{I}_{p \times q, \kappa}$.*

(ii) *There is a sequence $(r_j)_{j=0}^\kappa$ of complex $q \times p$-matrices such that $\left[\mathbb{S}_m^{(s)} \right]^\dagger = \mathbb{S}_m^{(r)}$ for each $m \in \mathbb{Z}_{0,\kappa}$.*

When condition (i) is met, the sequence $(r_j)_{j=0}^\kappa$ is uniquely determined and $r_j = s_j^\dagger$ for each $j \in \mathbb{Z}_{0,m}$.

Proof. Combine Remark 2.16 and Remark 2.3. $\qquad\square$

From Proposition 2.18 we see that, as announced, the concept of invertibility could also have been based on upper-triangular block Toeplitz matrices.

3. The arithmetic of invertible sequences

In this section, we derive rules for working with invertible sequences.

Remark 3.1. If $m \in \mathbb{N}_0$ and $(s_j)_{j=0}^m$ is a sequence in $\mathbb{C}^{p \times q}$, then

$$\mathbb{S}_m^{(s)} = \left[\mathbf{S}_m^{(s^*)} \right]^* = \left[\mathbf{S}_m^{(s^{\mathrm{T}})} \right]^{\mathrm{T}},$$

where $\mathbf{S}_m^{(s^*)}$ and $\mathbf{S}_m^{(s^{\mathrm{T}})}$ are, respectively, the lower-triangular block-Toeplitz-matrices (see (2.1)) corresponding to the sequences $(s_j^*)_{j=0}^m$ and $(s_j^{\mathrm{T}})_{j=0}^m$.

Lemma 3.2. *Suppose* $\kappa \in \mathbb{N}_0 \cup \{+\infty\}$ *and that* $(s_j)_{j=0}^\kappa$ *is a sequence from* $\mathbb{C}^{p \times q}$. *Then*

$$(s_j^*)_{j=0}^\kappa \in \mathcal{I}_{q \times p, \kappa} \qquad \text{if and only if} \qquad (s_j)_{j=0}^\kappa \in \mathcal{I}_{p \times q, \kappa}.$$

In this case, $\left((s_j^\dagger)^* \right)_{j=0}^\kappa$ *is the inverse sequence corresponding to* $(s_j^*)_{j=0}^\kappa$.

Proof. Combine Proposition 2.18 and Remark 3.1.   □

Remark 3.3. Let $\kappa \in \mathbb{N}_0 \cup \{+\infty\}$ and let $(s_j)_{j=0}^\kappa \in \mathcal{I}_{q \times q, \kappa}$. If $s_j^* = s_j$ for each $j \in \mathbb{Z}_{0,\kappa}$, then, from Lemma 3.2, we see that the inverse sequence $(s_j^\dagger)_{j=0}^\kappa$ corresponding to $(s_j)_{j=0}^\kappa$ is also a sequence of Hermitian matrices.

Proof. Use Lemma 3.2.   □

Lemma 3.4. *Suppose* $\kappa \in \mathbb{N}_0 \cup \{+\infty\}$ *and that* $(s_j)_{j=0}^\kappa$ *is a sequence from* $\mathbb{C}^{p \times q}$. *It then holds that*

$$(s_j^{\mathrm{T}})_{j=0}^\kappa \in \mathcal{I}_{q \times p, \kappa} \qquad \text{if and only if} \qquad (s_j)_{j=0}^\kappa \in \mathcal{I}_{p \times q, \kappa}.$$

In this case, $\left((s_j^\dagger)^{\mathrm{T}} \right)_{j=0}^\kappa$ *is the inverse sequence corresponding to* $(s_j^{\mathrm{T}})_{j=0}^\kappa$.

Proof. Apply Proposition 2.18 and Remark 3.1.   □

Remark 3.5. Let $\kappa \in \mathbb{N}_0 \cup \{+\infty\}$ and let $(s_j)_{j=0}^\kappa \in \mathcal{I}_{q \times q, \kappa}$. If $s_j^{\mathrm{T}} = s_j$, for each $j \in \mathbb{Z}_{0,\kappa}$, then we see from Lemma 3.4 and Remark 3.1 that the inverse sequence, $(s_j^\dagger)_{j=0}^\kappa$, of $(s_j)_{j=0}^\kappa$ satisfies $(s_j^\dagger)^{\mathrm{T}} = s_j^\dagger$, for each $j \in \mathbb{Z}_{0,\kappa}$.

Lemma 3.6. *If* $\kappa \in \mathbb{N}_0 \cup \{+\infty\}$ *and* $(s_j)_{j=0}^\kappa \in \mathcal{I}_{p \times q, \kappa}$, *then, for each* $m \in \mathbb{Z}_{0,\kappa}$:

$$\mathbf{S}_m \mathbf{S}_m^\dagger = I_{m+1} \otimes (s_0 s_0^\dagger), \qquad \mathbf{S}_m^\dagger \mathbf{S}_m = I_{m+1} \otimes (s_0^\dagger s_0) \qquad (3.1)$$

and

$$\mathbb{S}_m \mathbb{S}_m^\dagger = I_{m+1} \otimes (s_0 s_0^\dagger), \qquad\qquad \mathbb{S}_m^\dagger \mathbb{S}_m = I_{m+1} \otimes (s_0^\dagger s_0). \qquad (3.2)$$

If, furthermore, $p = q$ and if $\mathcal{R}(s_0) = \mathcal{R}(s_0^*)$, then

$$\mathbf{S}_m \mathbf{S}_m^\dagger = \mathbf{S}_m^\dagger \mathbf{S}_m \qquad\qquad and \qquad\qquad \mathbb{S}_m \mathbb{S}_m^\dagger = \mathbb{S}_m^\dagger \mathbb{S}_m,$$

for every $m \in \mathbb{Z}_{0,\kappa}$.

Proof. We start with the proof of the first identity in (3.1). Let $m \in \mathbb{Z}_{0,\kappa}$. We see from the construction of the sequence $(s_j^\dagger)_{j=0}^\kappa$ that

$$\mathbf{S}_m^\dagger = \mathbf{S}_m^{(s^\dagger)}. \qquad (3.3)$$

Let $(C_j)_{j=0}^\kappa := \left((s \odot s^\dagger)_j \right)_{j=0}^\kappa$ be the Cauchy product of $(s_j)_{j=0}^\kappa$ and $(s_j^\dagger)_{j=0}^\kappa$. For each $j \in \mathbb{Z}_{0,m}$, we thus have

$$C_j = \sum_{k=0}^j s_j s_{k-j}^\dagger. \qquad (3.4)$$

From (3.4) and part (c) of Remark 2.7, it follows that $C_0 = s_0 s_0^\dagger$. Using Remark 2.13 and (3.3), we obtain

$$\mathbf{S}_m^{(C)} = \mathbf{S}_m \mathbf{S}_m^{(s^\dagger)} = \mathbf{S}_m \mathbf{S}_m^\dagger. \qquad (3.5)$$

Since $\left(\mathbf{S}_m \mathbf{S}_m^\dagger \right)^* = \mathbf{S}_m \mathbf{S}_m^\dagger$, we see from (3.5) that

$$\left(\mathbf{S}_m^{(C)} \right)^* = \mathbf{S}_m^{(C)}. \qquad (3.6)$$

Using (3.6), $C_0 = s_0 s_0^\dagger$ and the definition of $\mathbf{S}_m^{(C)}$ (see (2.1)), we conclude that $\mathbf{S}_m^{(C)} = I_{m+1} \otimes (s_0 s_0^\dagger)$. Thus, equation (3.5) implies the first equation in (3.1).

The second identity in (3.1) can be similarly verified.

Having just proved (3.1), we can now use these identities to obtain (3.2), since, using (2.4) and (2.5), the identities in (3.2) follow from (3.1).

If $p = q$ and $\mathcal{N}(s_0) = \mathcal{N}(s_0^*)$, then from Proposition A.4 we see that $s_0 s_0^\dagger = s_0^\dagger s_0$. Thus, combining (3.1) and (3.2) yields the remaining assertions. $\square$

Remark 3.7. If $\kappa \in \mathbb{N}_0 \cup \{+\infty\}$ and $(s_j)_{j=0}^\kappa \in \mathcal{I}_{p \times q, \kappa}$, then the proof of Lemma 3.6 shows that $(s \odot s^\dagger)_j = \delta_{0,j} \cdot s_0 s_0^\dagger$ and $(s^\dagger \odot s)_j = \delta_{0,j} \cdot s_0^\dagger s_0$ for each $j \in \mathbb{Z}_{0,\kappa}$.

Now we turn our attention to the Cauchy product of invertible sequences of complex matrices. The following result generalizes Proposition 2.14.

Proposition 3.8. *Let $\kappa \in \mathbb{N}_0 \cup \{+\infty\}$ and $(s_j)_{j=0}^\kappa \in \mathcal{I}_{p \times q, \kappa}$. If $(t_j)_{j=0}^\kappa \in \mathcal{I}_{q \times r, \kappa}$ with $\mathcal{R}(s_0^*) = \mathcal{R}(t_0)$, then*

$$\left((s \odot t)_j \right)_{j=0}^\kappa \in \mathcal{I}_{p \times r, \kappa} \qquad\qquad and \qquad\qquad \left((s \odot t)_j^\dagger \right)_{j=0}^\kappa = \left((t^\dagger \odot s^\dagger)_j \right)_{j=0}^\kappa.$$

Proof. Let $m \in \mathbb{Z}_{0,\kappa}$. By assumption, we have

$$\left[\mathbf{S}_m^{(s)} \right]^\dagger = \mathbf{S}_m^{\left(s^\dagger \right)} \qquad \text{and} \qquad \left[\mathbf{S}_m^{(t)} \right]^\dagger = \mathbf{S}_m^{\left(t^\dagger \right)}. \qquad (3.7)$$

Remark 2.13 yields

$$\mathbf{S}_m^{(s \odot t)} = \mathbf{S}_m^{(s)} \cdot \mathbf{S}_m^{(t)} \qquad \text{and} \qquad \mathbf{S}_m^{\left(t^\dagger \odot s^\dagger \right)} = \mathbf{S}_m^{\left(t^\dagger \right)} \cdot \mathbf{S}_m^{\left(s^\dagger \right)}. \qquad (3.8)$$

Formula (A.4) yields

$$s_0^* \left(s_0^* \right)^\dagger = s_0^\dagger s_0 \qquad (3.9)$$

and

$$\left[\mathbf{S}_m^{(s)} \right]^* \left(\left[\mathbf{S}_m^{(s)} \right]^* \right)^\dagger = \left(\mathbf{S}_m^{(s)} \right)^\dagger \mathbf{S}_m^{(s)}. \qquad (3.10)$$

Since $\mathcal{R}(s_0^*) = \mathcal{R}(t_0)$, part (b) of Lemma A.3 yields $s_0^* (s_0^*)^\dagger = t_0 t_0^\dagger$. Thus, from (3.9), we see that

$$s_0^\dagger s_0 = t_0 t_0^\dagger. \qquad (3.11)$$

Applying Lemma 3.6 gives us

$$\left[\mathbf{S}_m^{(s)} \right]^\dagger \mathbf{S}_m^{(s)} = I_{m+1} \otimes \left(s_0^\dagger s_0 \right) \qquad (3.12)$$

and

$$\mathbf{S}_m^{(t)} \left[\mathbf{S}_m^{(t)} \right]^\dagger = I_{m+1} \otimes \left(t_0 t_0^\dagger \right). \qquad (3.13)$$

Using (3.10), (3.12), (3.11) and (3.13), we get

$$\left[\mathbf{S}_m^{(s)} \right]^* \left(\left[\mathbf{S}_m^{(s)} \right]^* \right)^\dagger = \left[\mathbf{S}_m^{(s)} \right]^\dagger \mathbf{S}_m^{(s)} = I_{m+1} \otimes \left(s_0^\dagger s_0 \right)$$
$$= I_{m+1} \otimes \left(t_0 t_0^\dagger \right) = \mathbf{S}_m^{(t)} \left[\mathbf{S}_m^{(t)} \right]^\dagger.$$

Thus, part (b) of Lemma A.3 yields

$$\mathcal{R}\left(\left[\mathbf{S}_m^{(s)} \right]^* \right) = \mathcal{R}\left(\mathbf{S}_m^{(t)} \right).$$

Hence, from Proposition A.7, we get

$$\left[\mathbf{S}_m^{(s)} \mathbf{S}_m^{(t)} \right]^\dagger = \left[\mathbf{S}_m^{(t)} \right]^\dagger \left[\mathbf{S}_m^{(s)} \right]^\dagger. \qquad (3.14)$$

Finally, applying (3.8), (3.14) and (3.7) yields

$$\left[\mathbf{S}_m^{(s \odot t)} \right]^\dagger = \left[\mathbf{S}_m^{(s)} \mathbf{S}_m^{(t)} \right]^\dagger = \left[\mathbf{S}_m^{(t)} \right]^\dagger \left[\mathbf{S}_m^{(s)} \right]^\dagger = \mathbf{S}_m^{\left(t^\dagger \right)} \cdot \mathbf{S}_m^{\left(s^\dagger \right)} = \mathbf{S}_m^{\left(t^\dagger \odot s^\dagger \right)},$$

which completes the proof. $\qquad \square$

If $l \in \mathbb{Z}_{2,+\infty}$, $\kappa \in \mathbb{N}_0 \cup \{+\infty\}$, and $(s_j)_{j=0}^{\kappa}$ is a sequence in $\mathbb{C}^{q \times q}$ with non-singular matrix s_0, then Proposition 2.14 yields $(s_j)_{j=0}^{\kappa} \in \mathcal{I}_{q \times q, \kappa}$. Furthermore, we know that the l-th Cauchy product of $(s_j)_{j=0}^{\kappa}$ is in $\mathcal{I}_{q \times q, \kappa}$, and that

$$\left(\left(\bigodot_{k=1}^{l} s^{\ddagger} \right)_j \right)_{j=0}^{\kappa} \quad \text{is the inverse sequence to} \quad \left(\left(\bigodot_{k=1}^{l} s \right)_j \right)_{j=0}^{\kappa}, \quad \text{i.e.:}$$

$$\left(\left(\bigodot_{k=1}^{l} s \right)_j \right)_{j=0}^{\kappa} \in \mathcal{I}_{q \times q, \kappa} \quad \text{and} \quad \left(\left(\bigodot_{k=1}^{l} s \right)_j^{\ddagger} \right)_{j=0}^{\kappa} = \left(\left(\bigodot_{k=1}^{l} s^{\ddagger} \right)_j \right)_{j=0}^{\kappa}.$$

If s_0 satisfies $\mathcal{R}(s_0) = \mathcal{R}(s_0^*)$, then this result can be extended to the degenerate case:

Proposition 3.9. *If $l \in \mathbb{Z}_{2,+\infty}$, $\kappa \in \mathbb{N}_0 \cup \{+\infty\}$, and $(s_j)_{j=0}^{\kappa} \in \mathcal{I}_{q \times q, \kappa}$ are such that s_0 satisfies $\mathcal{R}(s_0) = \mathcal{R}(s_0^*)$, then*

$$\left(\left(\bigodot_{k=1}^{l} s \right)_j \right)_{j=0}^{\kappa} \in \mathcal{I}_{q \times q, \kappa} \quad \text{and} \quad \left(\left(\bigodot_{k=1}^{l} s \right)_j^{\ddagger} \right)_{j=0}^{\kappa} = \left(\left(\bigodot_{k=1}^{l} s^{\ddagger} \right)_j \right)_{j=0}^{\kappa}.$$

Proof. From Corollary A.8 for each $k \in \mathbb{N}$, we get $\mathcal{R}(s_0^k) = \mathcal{R}(s_0)$ and consequently $\mathcal{R}(s_0^k) = \mathcal{R}(s_0^*)$. Now the assertion follows by induction from Proposition 3.8. $\qquad\square$

From Remark A.9 we see that Proposition 3.9 works, in particular, for the case in which s_0 is a normal matrix.

4. Constructing the inverse sequence

The main focus of this section will, firstly, be on coming to an explicit description of the set $\mathcal{I}_{p \times q, \kappa}$, of all invertible sequences, $(s_j)_{j=0}^{\kappa}$, in $\mathbb{C}^{p \times q}$. We will, secondly, focus on finding an effective method for constructing the inverse sequence associated with a sequence, $(s_j)_{j=0}^{\kappa} \in \mathcal{I}_{p \times q, \kappa}$. Our next task will be to find a suitable approach to solving these problems. The following result will help point us in the right direction.

Proposition 4.1. *If $m \in \mathbb{N}$ and $(s_j)_{j=0}^{m} \in \mathcal{I}_{p \times q, m}$, then:*

(a) $\mathcal{R}(s_j) \subseteq \mathcal{R}(s_0)$ *for each $j \in \mathbb{Z}_{0,m}$.*

(b) $\mathcal{N}(s_0) \subseteq \mathcal{N}(s_j)$ *for each $j \in \mathbb{Z}_{0,m}$.*

(c) $s_m^{\ddagger} = -s_0^{\ddagger} \left(\displaystyle\sum_{k=0}^{m-1} s_{m-k} s_k^{\ddagger} \right).$

(d) $s_m^{\ddagger} = -\left(\displaystyle\sum_{k=1}^{m} s_{m-k}^{\ddagger} s_k \right) s_0^{\ddagger}.$

Proof. The proof relies on applying Corollary A.11, which requires taking corresponding block partitions of the matrices $\mathbf{S}_m^{(s)}$ and $\mathbf{S}_m^{(s^\dagger)}$ into account. From the definition of the sequence $(s_j^\dagger)_{j=0}^m$, we see that

$$
\begin{pmatrix} \mathbf{S}_{m-1}^{(s^\dagger)} & 0_{mp\times q} \\ (s_m^\dagger \ s_{m-1}^\dagger \ \cdots \ s_1^\dagger) & s_0^\dagger \end{pmatrix} = \mathbf{S}_m^{(s^\dagger)} = \left[\mathbf{S}_m^{(s)} \right]^\dagger = \begin{pmatrix} \mathbf{S}_{m-1}^{(s)} & 0_{mp\times q} \\ (s_m \ s_{m-1} \ \cdots \ s_1) & s_0 \end{pmatrix}^\dagger.
$$
(4.1)

(a) From (4.1) and Corollary A.11 we get $\mathcal{R}\left((s_m \ \cdots \ s_1) \right) \subseteq \mathcal{R}(s_0)$, from which (a) follows directly.

(b) Because of $(s_j)_{j=0}^m \in \mathcal{I}_{p\times q,m}$, Lemma 3.2 yields $(s_j^*)_{j=0}^m \in \mathcal{I}_{q\times p,m}$. Thus, (a) implies $\mathcal{R}(s_j^*) \subseteq \mathcal{R}(s_0^*)$ for each $j \in \mathbb{Z}_{0,m}$. By going over to the orthogonal complements, we obtain (b).

(c) From (4.1) and Corollary A.11 we get

$$
\begin{pmatrix} s_m^\dagger & s_{m-1}^\dagger & \cdots & s_1^\dagger \end{pmatrix} = -s_0^\dagger \begin{pmatrix} s_m & s_{m-1} & \cdots & s_1 \end{pmatrix} \left[\mathbf{S}_{m-1}^{(s)} \right]^\dagger. \tag{4.2}
$$

Remark 2.3 yields

$$
\left[\mathbf{S}_{m-1}^{(s)} \right]^\dagger = \mathbf{S}_{m-1}^{(s^\dagger)}. \tag{4.3}
$$

Combining (4.2) and (4.3) gives us

$$
s_m^\dagger = -s_0^\dagger \begin{pmatrix} s_m & s_{m-1} & \cdots & s_1 \end{pmatrix} \begin{pmatrix} s_0^\dagger \\ s_1^\dagger \\ \vdots \\ s_{m-2}^\dagger \end{pmatrix} = -s_0^\dagger \left(\sum_{k=0}^{m-1} s_{m-k} s_k^\dagger \right).
$$

(d) Applying (c) to the sequence $(s_j^*)_{j=0}^m \in \mathcal{I}_{q\times p,m}$, recalling Lemma 3.2 and the identity $(s_0^*)^\dagger = (s_0^\dagger)^*$ (see (A.2)), the assertion follows directly. □

It should be mentioned that, for the special case in which $p = q$ and $\det s_0 \neq 0$, the results of parts (c) and (d) were obtained as part (b) of Proposition 2.11.

Remark 4.2. Let $\kappa \in \mathbb{N}_0 \cup \{+\infty\}$. If $(s_j)_{j=0}^\kappa \in \mathcal{I}_{p\times q,\kappa}$ with $s_0 = 0_{p\times q}$, then part (a) of Proposition 4.1 yields that $s_j = 0_{p\times q}$ for each $j \in \mathbb{Z}_{0,\kappa}$.

Proposition 4.1 gives us a good starting-point and will play a large role in the next few sections. Keeping the results of parts (a) and (b) of Proposition 4.1 in mind, we next introduce a class of sequences of complex matrices which, as we will see, will continue to play a major role throughout the rest of this paper.

Definition 4.3. Let $\kappa \in \mathbb{N}_0 \cup \{+\infty\}$ and $(s_j)_{j=0}^\infty$ be a sequence in in $\mathbb{C}^{p\times q}$. We then say that $(s_j)_{j=0}^\kappa$ is *dominated by its first term* (or, simply, that it is *first* *term dominant*) when

$$\mathcal{N}(s_0) \subseteq \bigcap_{j=0}^{\kappa} \mathcal{N}(s_j) \qquad \text{and} \qquad \bigcup_{j=0}^{\kappa} \mathcal{R}(s_j) \subseteq \mathcal{R}(s_0). \qquad (4.4)$$

The set of all first term dominant sequences $(s_j)_{j=0}^{\kappa}$ in $\mathbb{C}^{p \times q}$ will be denoted by $\mathcal{D}_{p \times q, \kappa}$.

We can now rewrite the results of parts (a) and (b) of Proposition 4.1.

Proposition 4.4. *If $\kappa \in \mathbb{N}_0 \cup \{+\infty\}$, then $\mathcal{I}_{p \times q, \kappa} \subseteq \mathcal{D}_{p \times q, \kappa}$.*

Proof. From the definition of the set $\mathcal{D}_{p \times q, \kappa}$ it is clear that the assertion follows from parts (a) and (b) of Proposition 4.1. $\qquad\qquad\qquad\square$

We now take a closer look at the class $\mathcal{D}_{p \times q, \kappa}$ and will obtain results, which will later help us to show that $\mathcal{D}_{p \times q, \kappa} \subseteq \mathcal{I}_{p \times q, \kappa}$.

Remark 4.5. If $\kappa \in \mathbb{N}_0 \cup \{+\infty\}$ and if $(\alpha_j)_{j=0}^{\kappa}$ is a sequence in $\mathbb{C}$, then $(\alpha_j)_{j=0}^{\kappa} \in \mathcal{D}_{1 \times 1, \kappa}$ if and only if $\alpha_0 \neq 0$ or $(\alpha_j)_{j=0}^{\kappa}$ is the constant sequence with value 0.

It should be noted that, for each $\kappa \in \mathbb{N}_0 \cup \{+\infty\}$, the set $\mathcal{D}_{q \times q, \kappa}$ contains all sequences $(s_j)_{j=0}^{\kappa}$ of complex $q \times q$-matrices having non-singular matrix s_0. Additionally, we make the following simple observations:

Remark 4.6. Let $\kappa \in \mathbb{N}_0 \cup \{+\infty\}$ and $(s_j)_{j=0}^{\kappa}$ be a sequence of complex $p \times q$-matrices. Then

$$(s_j)_{j=0}^{\kappa} \in \mathcal{D}_{p \times q, \kappa} \qquad \text{if and only if} \qquad (s_j)_{j=0}^{m} \in \mathcal{D}_{p \times q, m}, \text{ for all } m \in \mathbb{Z}_{0, \kappa}.$$

Remark 4.7. Let $\kappa \in \mathbb{N}_0 \cup \{+\infty\}$ and $(s_j)_{j=0}^{\kappa}$ be a sequence of complex $p \times q$-matrices. If, for each $j \in \mathbb{Z}_{0, \kappa}$, we set

$$\tilde{s}_j := s_0 s_0^{\dagger} s_j s_0^{\dagger} s_0,$$

then $\tilde{s}_0 = s_0$ and $(\tilde{s}_j)_{j=0}^{\kappa} \in \mathcal{D}_{p \times q, \kappa}$.

Remark 4.8. Suppose $m \in \mathbb{N}_0$ and that $(s_j)_{j=0}^{m} \in \mathcal{D}_{p \times q, m}$. Furthermore, suppose that $\kappa \in \mathbb{Z}_{m+1, +\infty} \cup \{+\infty\}$ and also that $(A_j)_{j=m+1}^{\kappa}$ is a sequence of complex $q \times p$-matrices. If we set $s_j := s_0 A_j s_0$, for each $j \in \mathbb{Z}_{m+1, \kappa}$, then, clearly, we get a sequence $(s_j)_{j=0}^{\kappa}$ belonging to $\mathcal{D}_{p \times q, \kappa}$.

Our focus now will be on matters related to the column-space of the matrix $\mathbf{S}_m^{(s)}$ given in (2.1).

Lemma 4.9. *If $m \in \mathbb{N}_0$ and $(s_j)_{j=0}^{m}$ is a sequence of complex $p \times q$-matrices, then the following statements are all equivalent:*

(i) $\mathcal{R}(s_j) \subseteq \mathcal{R}(s_0)$ for each $j \in \mathbb{Z}_{0, m}$.
(ii) $\mathcal{R}(\mathbf{S}_j) = \mathcal{R}(I_{j+1} \otimes s_0)$ for each $j \in \mathbb{Z}_{0, m}$.
(iii) $\mathcal{R}(\mathbf{S}_m) = \mathcal{R}(I_{m+1} \otimes s_0)$.

Proof. The case $m = 0$ is trivial. Suppose $m \geq 1$.

"(i)$\Rightarrow$(ii)": We prove (ii) by induction. There is an $n \in \mathbb{Z}_{1,m}$ such that

$$\mathcal{R}(\mathbf{S}_j) = \mathcal{R}(I_{j+1} \otimes s_0) \tag{4.5}$$

for each $j \in \mathbb{Z}_{0,n-1}$. From (2.1), we obtain the block-partition

$$\mathbf{S}_n = \begin{pmatrix} \mathbf{S}_{n-1} & 0 \\ \widehat{z}_{1,n} & s_0 \end{pmatrix},$$

where, in view of (i), it is obvious that the block $\widehat{z}_{1,n} := \begin{pmatrix} s_n & s_{n-1} & \cdots & s_1 \end{pmatrix}$ satisfies $\mathcal{R}(\widehat{z}_{1,n}) \subseteq \mathcal{R}(s_0)$. Hence, part (a) of Lemma A.13 yields

$$\mathcal{R}(\mathbf{S}_n) = \mathcal{R}\left(\mathrm{diag}(\mathbf{S}_{n-1}, s_0)\right). \tag{4.6}$$

On the other hand, if we consider (4.5) for $j = n - 1$, we get

$$\mathcal{R}\left(\mathrm{diag}(\mathbf{S}_{n-1}, s_0)\right) = \mathcal{R}(\mathbf{S}_{n-1}) \oplus \mathcal{R}(s_0) = \mathcal{R}(I_n \otimes s_0) \oplus \mathcal{R}(s_0) = \mathcal{R}(I_{n+1} \otimes s_0). \tag{4.7}$$

Thus, in view of (4.6) and (4.7), condition (ii) is proved by induction.

"(ii)$\Rightarrow$(iii)": This implication is obvious.

"(iii)$\Rightarrow$(i)": Let $j \in \mathbb{Z}_{0,m}$. Suppose that $y \in \mathcal{R}(s_j)$. There is, then, an $x \in \mathbb{C}^q$ such that $y = s_j x$. Obviously, $w := \mathrm{col}(s_k x)_{k=0}^m$ can be expressed as

$$w = \mathbf{S}_m \cdot \mathrm{col}(\delta_{j0} x)_{j=0}^m.$$

Hence, from (iii) we see that $w \in \mathcal{R}(I_{m+1} \otimes s_0)$. Thus, there exists a $v \in \mathbb{C}^{(m+1)q}$ such that $w = (I_{m+1} \otimes s_0)v$. Consequently,

$$\begin{aligned} y = s_j x &= (\delta_{0j} I_q, \delta_{1j} I_q, \ldots, \delta_{mj} I_q) w \\ &= (\delta_{0j} I_q, \delta_{1j} I_q, \ldots, \delta_{mj} I_q)(I_{m+1} \otimes s_0)v \\ &= (\delta_{0j} s_0, \delta_{1j} s_0, \ldots, \delta_{mj} s_0)v \\ &= s_0(\delta_{0j} I_q, \delta_{1j} I_q, \ldots, \delta_{mj} I_q)v. \end{aligned}$$

In particular, $y \in \mathcal{R}(s_0)$. This means $\mathcal{R}(s_j) \subseteq \mathcal{R}(s_0)$. Therefore, (i) holds true. $\square$

Using part (a) of Lemma A.14 instead of part (a) of Lemma A.13, the following result can be proved in much the same way as Lemma 4.9.

Lemma 4.10. *If $m \in \mathbb{N}_0$ and $(s_j)_{j=0}^m$ is a sequence of complex $p \times q$-matrices, then the following statements are all equivalent:*

(i) $\mathcal{R}(s_j) \subseteq \mathcal{R}(s_0)$ *for each* $j \in \mathbb{Z}_{0,m}$.
(ii) $\mathcal{R}(\mathbb{S}_j) = \mathcal{R}(I_{j+1} \otimes s_0)$ *for each* $j \in \mathbb{Z}_{0,m}$.
(iii) $\mathcal{R}(\mathbb{S}_m) = \mathcal{R}(I_{m+1} \otimes s_0)$.

Having obtained Lemmas 4.9 and 4.10, we now present corresponding results for null-spaces.

Lemma 4.11. *Let $m \in \mathbb{N}_0$ and let $(s_j)_{j=0}^m$ be a sequence of complex $p \times q$-matrices. The following statements are all equivalent:*

(i) $\mathcal{N}(s_0) \subseteq \mathcal{N}(s_j)$ *for each* $j \in \mathbb{Z}_{0,m}$.
(ii) $\mathcal{N}(\mathbf{S}_j) = \mathcal{N}(I_{j+1} \otimes s_0)$ *for each* $j \in \mathbb{Z}_{0,m}$.

(iii) $\mathcal{N}(\mathbf{S}_m) = \mathcal{N}(I_{m+1} \otimes s_0)$.

(iv) $\mathcal{N}(\mathbb{S}_j) = \mathcal{N}(I_{j+1} \otimes s_0)$ *for each $j \in \mathbb{Z}_{0,m}$*.

(v) $\mathcal{N}(\mathbb{S}_m) = \mathcal{N}(I_{m+1} \otimes s_0)$.

Proof. Consider the matrices $\mathbf{S}_m^*$ and $\mathbb{S}_m^*$, apply Lemma 4.9 and Lemma 4.10 to the sequence $(s_j^*)_{j=0}^m$, and form orthogonal complements. $\qquad\square$

Remark 4.12. Let $\kappa \in \mathbb{N} \cup \{+\infty\}$ and let $(s_j)_{j=0}^\kappa$ be a sequence of Hermitian complex $q \times q$-matrices. Using Remark 2.17, Lemma 4.11, Lemma 4.10, Lemma 4.9 and Lemma A.3 (specifically, parts (a) and (b)), we see that the following statements hold:

(a) $\mathbb{S}_j = \mathbf{S}_j^*$ for each $j \in \mathbb{Z}_{0,\kappa}$.

(b) The following statements are all equivalent:

    (i) $(s_j)_{j=0}^\kappa \in \mathcal{D}_{q \times q, \kappa}$.

    (ii) $\mathcal{N}(s_0) \subseteq \mathcal{N}(s_j)$ for all $j \in \mathbb{Z}_{0,\kappa}$.

    (iii) $\mathcal{R}(s_j) \subseteq \mathcal{R}(s_0)$ for all $j \in \mathbb{Z}_{0,\kappa}$.

    (iv) $s_j s_0^\dagger s_0 = s_j$ for all $j \in \mathbb{Z}_{0,\kappa}$.

    (v) $s_0 s_0^\dagger s_j = s_j$ for all $j \in \mathbb{Z}_{0,\kappa}$.

(c) If condition (i) is met, then $\mathcal{N}(\mathbf{S}_j^*) = \mathcal{N}(\mathbb{S}_j)$ and $\mathcal{R}(\mathbf{S}_j^*) = \mathcal{R}(\mathbb{S}_j)$ for all $j \in \mathbb{Z}_{0,\kappa}$.

Our next objective can be described as follows: If $m \in \mathbb{N}$ and $(s_j)_{j=0}^m \in \mathcal{D}_{p \times q, m}$, then we want to show that $(s_j)_{j=0}^m \in \mathcal{I}_{p \times q, m}$ by explicitly constructing a sequence $(r_j)_{j=0}^m$ in $\mathbb{C}^{q \times p}$ such that $\left[\mathbf{S}_m^{(s)}\right]^\dagger = \mathbf{S}_m^{(r)}$. With this in mind and recalling part (c) of Proposition 4.1, we proceed with the following definition:

Definition 4.13. Let $\kappa \in \mathbb{N}_0 \cup \{+\infty\}$ and let $(s_j)_{j=0}^\kappa$ be a sequence of complex $p \times q$-matrices. The sequence $(s_j^\sharp)_{j=0}^\kappa$ defined by

$$s_k^\sharp := \begin{cases} s_0^\dagger, & \text{if } k = 0 \\ -s_0^\dagger \sum_{l=0}^{k-1} s_{k-l} s_l^\sharp, & \text{if } k \in \mathbb{Z}_{1,\kappa} \end{cases} \tag{4.8}$$

is called the *reciprocal sequence* corresponding to $(s_j)_{j=0}^\kappa$.

Notation 4.14. Given a $\kappa \in \mathbb{N}_0 \cup \{+\infty\}$ and a sequence $(s_j)_{j=0}^\kappa$ of complex $p \times q$-matrices, we will always use $(s_j^\sharp)_{j=0}^\kappa$ to denote the reciprocal sequence corresponding to $(s_j)_{j=0}^\kappa$. Also, for each $m \in \mathbb{Z}_{0,\kappa}$, we set (using Notation 2.1, with $(s_j^\sharp)_{j=0}^\kappa$ instead of $(s_j)_{j=0}^\kappa$):

$$\mathbf{S}_m^\sharp := \mathbf{S}_m^{(s^\sharp)}. \tag{4.9}$$

Remark 4.15. If $\kappa \in \mathbb{N} \cup \{+\infty\}$ and $(s_j)_{j=0}^\kappa$ is a sequence in $\mathbb{C}^{p \times q}$, then

$$s_1^\sharp = -s_0^\dagger s_1 s_0^\dagger.$$

Example 4.16. Let $s_0 \in \mathbb{C}^{p \times q}$ and $\kappa \in \mathbb{N}_0 \cup \{+\infty\}$. From Definition 4.13, we see that if $s_j = \delta_{0,j} \cdot s_0$ for each $j \in \mathbb{Z}_{0,\kappa}$, then $s_j^\sharp = \delta_{0,j} \cdot s_0^\sharp$ for each $j \in \mathbb{Z}_{0,\kappa}$.

Remark 4.17. Let $\kappa \in \mathbb{N}_0 \cup \{+\infty\}$. If $(s_j)_{j=0}^\kappa$ is a sequence in $\mathbb{C}^{p \times q}$, then it immediately follows from Definition 4.13 that, for each $m \in \mathbb{Z}_{0,\kappa}$, the reciprocal sequence for $(s_j)_{j=0}^m$ is $\left(s_j^\sharp\right)_{j=0}^m$.

To prove that $\mathbf{S}_m^\dagger = \mathbf{S}_m^\sharp$ for any sequence $(s_j)_{j=0}^m \in \mathcal{D}_{p \times q, m}$, we will need the following result on the Moore-Penrose inverse of the block Toeplitz matrix $\mathbf{S}_m$.

Lemma 4.18. *Let* $\kappa \in \mathbb{N} \cup \{+\infty\}$ *and let* $(s_j)_{j=0}^\kappa \in \mathcal{D}_{p \times q, \kappa}$. *For each* $m \in \mathbb{Z}_{1,\kappa}$, *the Moore-Penrose inverse* $\mathbf{S}_m^\dagger$ *of the block Toeplitz matrix* $\mathbf{S}_m$ *admits the block representations*

$$\mathbf{S}_m^\dagger = \begin{pmatrix} \mathbf{S}_{m-1}^\dagger & 0_{mq \times q} \\ -s_0^\dagger \widehat{z}_{1,m} \mathbf{S}_{m-1}^\dagger & s_0^\dagger \end{pmatrix} \quad and \quad \mathbf{S}_m^\dagger = \begin{pmatrix} s_0^\dagger & 0_{q \times mq} \\ -\mathbf{S}_{m-1}^\dagger y_{1,m} s_0^\dagger & \mathbf{S}_{m-1}^\dagger \end{pmatrix}, \quad (4.10)$$

where

$$\widehat{z}_{1,m} := \begin{pmatrix} s_m & s_{m-1} & \cdots & s_1 \end{pmatrix} \quad and \quad y_{1,m} := \operatorname{col}(s_j)_{j=1}^m. \quad (4.11)$$

Proof. Let $m \in \mathbb{Z}_{1,\kappa}$. We will apply Corollary A.11. From (2.1) we see that $\mathbf{S}_m$ admits the block partitions

$$\mathbf{S}_m = \begin{pmatrix} \mathbf{S}_{m-1} & 0 \\ \widehat{z}_{1,m} & s_0 \end{pmatrix} \quad and \quad \mathbf{S}_m = \begin{pmatrix} s_0 & 0 \\ y_{1,m} & \mathbf{S}_{m-1} \end{pmatrix}. \quad (4.12)$$

Since $(s_j)_{j=0}^m \in \mathcal{D}_{p \times q, m}$, we have

$$\mathcal{N}(s_0) \subseteq \bigcap_{j=1}^m \mathcal{N}(s_j) \quad and \quad \bigcup_{j=1}^m \mathcal{R}(s_j) \subseteq \mathcal{R}(s_0). \quad (4.13)$$

Clearly,

$$\mathcal{R}(\widehat{z}_{1,m}) \subseteq \mathcal{R}(s_0), \quad (4.14)$$

Lemma 4.9 yields $\mathcal{R}(\mathbf{S}_{m-1}) = \mathcal{R}(I_m \otimes s_0)$. Thus, taking part (b) of Lemma A.3 into account, we obtain

$$\mathbf{S}_{m-1}\mathbf{S}_{m-1}^\dagger = (I_m \otimes s_0)(I_m \otimes s_0)^\dagger = (I_m \otimes s_0)(I_m \otimes s_0^\dagger) = I_m \otimes (s_0 s_0^\dagger). \quad (4.15)$$

Formulas (4.13) and part (b) of Lemma A.3, together, yield $s_0 s_0^\dagger s_j = s_j$ for each $j \in \mathbb{Z}_{0,m}$. Using (4.15), this gives us

$$\mathbf{S}_{m-1}\mathbf{S}_{m-1}^\dagger y_{1,m} = \left[I_m \otimes (s_0 s_0^\dagger)\right] \cdot \operatorname{col}(s_j)_{j=1}^m = \operatorname{col}(s_0 s_0^\dagger s_j)_{j=1}^m = \operatorname{col}(s_j)_{j=1}^m = y_{1,m}.$$

Thus, using Lemma A.3, part (b), we obtain

$$\mathcal{R}(y_{1,m}) \subseteq \mathcal{R}(\mathbf{S}_{m-1}). \quad (4.16)$$

The first inclusion in (4.13) gives us

$$\mathcal{N}(s_0) \subseteq \mathcal{N}(y_{1,m}). \quad (4.17)$$

Lemma 4.11 yields $\mathcal{N}(\mathbf{S}_{m-1}) = \mathcal{N}(I_m \otimes s_0)$. Since the first inclusion in (4.13) implies $\mathcal{N}(I_m \otimes s_0) \subseteq \mathcal{N}(\widehat{z}_{1,m})$, we also get

$$\mathcal{N}(\mathbf{S}_{m-1}) \subseteq \mathcal{N}(\widehat{z}_{1,m}). \tag{4.18}$$

Combining (4.12) with (4.14) and (4.18), and applying Corollary A.11 yields the first block-representation of $\mathbf{S}_m^\dagger$ in (4.10). Similarly, to obtain the second block-representation in (4.10), we combine (4.12) with (4.16) and (4.17). $\qquad\square$

It should be noted that, for Lemma 4.18, a corresponding result can be shown for the upper block Toeplitz matrices $\mathbb{S}_m$, introduced via $\mathbb{S}_m := \mathbb{S}_m^{(s)}$ and (2.3).

Notation 4.19. In the following, for each $\kappa \in \mathbb{N}_0 \cup \{+\infty\}$, for each sequence $(s_j)_{j=0}^\kappa$ of complex $p \times q$-matrices, and for each $m \in \mathbb{Z}_{0,\kappa}$, let the block Toeplitz matrices $\mathbf{S}_m^\sharp$ and $\mathbb{S}_m^\sharp$ be given, respectively, by formula (4.9) and (using Notation 2.15, with $(s_j^\sharp)_{j=0}^\kappa$ instead of $(s_j)_{j=0}^\kappa$):

$$\mathbb{S}_m^\sharp := \mathbb{S}_m^{(s^\sharp)}.$$

Given a sequence $(s_j)_{j=0}^m \in \mathcal{D}_{p\times q,m}$, we now have a new way of looking at the Moore-Penrose inverses of the matrices $\mathbf{S}_m$ and $\mathbb{S}_m$.

Proposition 4.20. If $\kappa \in \mathbb{N}_0 \cup \{+\infty\}$ and $(s_j)_{j=0}^\kappa \in \mathcal{D}_{p\times q,\kappa}$, then, for each $m \in \mathbb{Z}_{0,\kappa}$:

$$\mathbf{S}_m^\dagger = \mathbf{S}_m^\sharp \qquad \text{and} \qquad \mathbb{S}_m^\dagger = \mathbb{S}_m^\sharp.$$

Proof. The case $\kappa = 0$ is trivial. Suppose, therefore, that $\kappa \geq 1$. Because $\mathbf{S}_0^\dagger = \mathbf{S}_0^\sharp$, there exists an $m \in \mathbb{Z}_{1,\kappa}$ such that $\mathbf{S}_k^\dagger = \mathbf{S}_k^\sharp$ for each $k \in \mathbb{Z}_{0,m-1}$. From Lemma 4.18, we then get

$$\mathbf{S}_m^\dagger = \begin{pmatrix} \mathbf{S}_{m-1}^\sharp & 0 \\ -s_0^\dagger \widehat{z}_{1,m} \mathbf{S}_{m-1}^\sharp & s_0^\sharp \end{pmatrix}. \tag{4.19}$$

Hence, Definition 4.13 yields

$$-s_0^\dagger \widehat{z}_{1,m} \mathbf{S}_{m-1}^\sharp = \left(-\sum_{j=0}^{m-1} s_0^\dagger s_{m-j} s_j^\sharp, -\sum_{j=0}^{m-2} s_0^\dagger s_{m-1-j} s_j^\sharp, \dots, -\sum_{j=0}^{1-1} s_0^\dagger s_{1-j} s_j^\sharp \right)$$

$$= (s_m^\sharp, s_{m-1}^\sharp, \dots, s_1^\sharp). \tag{4.20}$$

Combining (4.19) and (4.20), we obtain $\mathbf{S}_m^\dagger = \mathbf{S}_m^\sharp$. Thus, by induction, we have shown that $\mathbf{S}_m^\dagger = \mathbf{S}_m^\sharp$ for each $m \in \mathbb{Z}_{0,\kappa}$. Furthermore, since the matrix U in (2.5) is unitary, we see from part (c) of Lemma A.3 that $\mathbb{S}_m^\dagger = U_{q,m} \mathbf{S}_m^\sharp U_{p,m} = \mathbb{S}_m^\sharp$ for each $m \in \mathbb{Z}_{0,\kappa}$ $\qquad\square$

We next come to the main result of this section.

Theorem 4.21. If $\kappa \in \mathbb{N}_0 \cup \{+\infty\}$, then

(a) $\mathcal{I}_{p\times q,\kappa} = \mathcal{D}_{p\times q,\kappa}$.
(b) If $(s_j)_{j=0}^\kappa \in \mathcal{I}_{p\times q,\kappa}$, then the inverse sequence corresponding to $(s_j)_{j=0}^\kappa$ coincides with the reciprocal sequence corresponding to $(s_j)_{j=0}^\kappa$.

Proof. (a) From Proposition 4.4, we get $\mathcal{I}_{p \times q, \kappa} \subseteq \mathcal{D}_{p \times q, \kappa}$. To prove (a) we therefore only need to show that $\mathcal{I}_{p \times q, \kappa} \supseteq \mathcal{D}_{p \times q, \kappa}$. To this end, let $(s_j)_{j=0}^{\kappa} \in \mathcal{D}_{p \times q, \kappa}$. From Proposition 4.20 it follows that $\mathbf{S}_m^{\dagger} = \mathbf{S}_m^{\sharp}$ for all $m \in \mathbb{Z}_{0, \kappa}$. From (4.9) and Definition 2.2 we then see that $(s_j)_{j=0}^{\kappa} \in \mathcal{I}_{p \times q, \kappa}$, and thus, $\mathcal{I}_{p \times q, \kappa} \supseteq \mathcal{D}_{p \times q, \kappa}$.

(b) Suppose $(s_j)_{j=0}^{\kappa} \in \mathcal{I}_{p \times q, \kappa}$. From Proposition 4.4, we see that $(s_j)_{j=0}^{\kappa} \in \mathcal{D}_{p \times q, \kappa}$. Using Proposition 4.20, we then obtain $\mathbf{S}_m^{\dagger} = \mathbf{S}_m^{\sharp}$ for all $m \in \mathbb{Z}_{0, \kappa}$. From Definition 2.6, we now see that the inverse sequence corresponding to $(s_j)_{j=0}^{\kappa}$ coincides with the reciprocal sequence $(s_j^{\sharp})_{j=0}^{\kappa}$ corresponding to $(s_j)_{j=0}^{\kappa}$. $\quad\square$

Corollary 4.22. *If* $\kappa \in \mathbb{N}_0 \cup \{+\infty\}$ *and* $(s_j)_{j=0}^{\kappa} \in \mathcal{D}_{p \times q, \kappa}$, *then the reciprocal sequence corresponding to* $\left(s_j^{\sharp}\right)_{j=0}^{\kappa}$ *is* $(s_j)_{j=0}^{\kappa}$.

Proof. Combine Theorem 4.21 and Remark 2.8. $\quad\square$

Corollary 4.23. *Let* $\kappa \in \mathbb{N} \cup \{+\infty\}$ *and* $(s_j)_{j=0}^{\kappa} \in \mathcal{D}_{p \times q, \kappa}$. *For each* $m \in \mathbb{Z}_{1, \kappa}$, *then*

$$\begin{pmatrix} s_m^{\sharp} & s_{m-1}^{\sharp} & \cdots & s_1^{\sharp} \end{pmatrix} = -s_0^{\dagger} \widehat{z}_{1, m} \mathbf{S}_{m-1}^{\dagger} \qquad \text{and} \qquad \mathrm{col}\left(s_j^{\sharp} \right)_{j=1}^{m} = -\mathbf{S}_{m-1}^{\dagger} y_{1, m} s_0^{\dagger},$$

where $\widehat{z}_{1, m}$ *and* $y_{1, m}$ *are defined as in* (4.11).

Proof. Consider the left lower blocks in the block partitions (4.10) and take into account that Theorem 4.21 implies $\mathbf{S}_m^{\dagger} = \mathbf{S}_m^{\sharp}$. $\quad\square$

Corollary 4.24. *Let* $\kappa \in \mathbb{Z}_{2, +\infty} \cup \{+\infty\}$ *and* $(s_j)_{j=0}^{\kappa} \in \mathcal{D}_{p \times q, \kappa}$. *If* $k \in \mathbb{Z}_{2, \kappa}$, *then*

$$s_k^{\sharp} = s_0^{\dagger} \left(-s_k + \widehat{z}_{1, k-1} \mathbf{S}_{k-2}^{\dagger} y_{1, k-1} \right) s_0^{\dagger},$$

where $\widehat{z}_{1, m}$ *and* $y_{1, m}$ *are defined as in* (4.11).

Proof. By Corollary 4.23, it follows that

$$\begin{pmatrix} s_k^{\sharp} & s_{k-1}^{\sharp} & \cdots & s_1^{\sharp} \end{pmatrix} = -s_0^{\dagger} \widehat{z}_{1, k} \mathbf{S}_{k-1}^{\dagger} = -s_0^{\dagger} \begin{pmatrix} s_k & s_{k-1} & \cdots & s_1 \end{pmatrix} \mathbf{S}_{k-1}^{\dagger}.$$

Thus,

$$\begin{aligned} s_k^{\sharp} &= \begin{pmatrix} s_m^{\sharp} & s_{m-1}^{\sharp} & \cdots & s_1^{\sharp} \end{pmatrix} \begin{pmatrix} I_p \\ 0_{(k-1)p \times p} \end{pmatrix} \\ &= -s_0^{\dagger} \begin{pmatrix} s_k & s_{k-1} & \cdots & s_1 \end{pmatrix} \mathbf{S}_{k-1}^{\dagger} \begin{pmatrix} I_p \\ 0_{(k-1)p \times p} \end{pmatrix}. \end{aligned} \tag{4.21}$$

Lemma 4.18 gives us the block partition

$$\mathbf{S}_{k-1}^{\dagger} = \begin{pmatrix} s_0^{\dagger} & 0_{q \times (k-1)q} \\ -\mathbf{S}_{k-2}^{\dagger} y_{1,k-1} s_0^{\dagger} & \mathbf{S}_{k-2}^{\dagger} \end{pmatrix}.$$

Thus, it follows that

$$\mathbf{S}_{k-1}^{\dagger} \begin{pmatrix} I_p \\ 0_{(k-1)p \times p} \end{pmatrix} = \begin{pmatrix} s_0^{\dagger} \\ -\mathbf{S}_{k-2}^{\dagger} y_{1, k-1} s_0^{\dagger} \end{pmatrix}. \tag{4.22}$$

Furthermore,

$$\begin{pmatrix} s_k & s_{k-1} & \cdots & s_1 \end{pmatrix} = \begin{pmatrix} s_k & \widehat{z}_{1,\,k-1} \end{pmatrix}. \tag{4.23}$$

Using (4.21)–(4.23), we obtain

$$\begin{aligned} s_k^\sharp &= -s_0^\dagger \begin{pmatrix} s_k & \widehat{z}_{1,\,k-1} \end{pmatrix} \begin{pmatrix} s_0^\dagger \\ -\mathbf{S}_{k-2}^\dagger y_{1,\,k-1} s_0^\dagger \end{pmatrix} \\ &= -s_0^\dagger s_k s_0^\dagger + s_0^\dagger \widehat{z}_{1,\,k-1} \mathbf{S}_{k-2}^\dagger y_{1,\,k-1} s_0^\dagger \\ &= s_0^\dagger \left(-s_k + \widehat{z}_{1,\,k-1} \mathbf{S}_{k-2}^\dagger y_{1,\,k-1} \right) s_0^\dagger, \end{aligned}$$

which completes the proof. $\qquad\square$

5. A closer look at the set $\mathcal{D}_{p\times q,\kappa}$ and reciprocal sequences

Theorem 4.21 shows us that the sets $\mathcal{I}_{p\times q,\kappa}$ and $\mathcal{D}_{p\times q,\kappa}$ are equal to one another. The same theorem also tells us that the problem of constructing a sequence's inverse amounts to constructing its reciprocal. To find out more about $\mathcal{I}_{p\times q,\kappa}$, we can therefore look at $\mathcal{D}_{p\times q,\kappa}$. By examining the properties of reciprocal sequences and determining the rules that apply when working with these sequences, we can learn more about these same aspects as they apply to inverse sequences. This section will serve that very purpose. We now take a closer look at the set $\mathcal{D}_{p\times q,\kappa}$ and its elements.

Proposition 5.1. *If $\kappa \in \mathbb{N}_0 \cup \{+\infty\}$ and $(s_j)_{j=0}^\kappa$ is a sequence in $\mathbb{C}^{p\times q}$, then the following statements are all equivalent:*

(i) $(s_j)_{j=0}^\kappa \in \mathcal{D}_{p\times q,\kappa}$.

(ii) $\mathcal{N}(s_0) \subseteq \bigcap_{j=0}^\kappa \mathcal{N}(s_j)$ and $\mathcal{N}(s_0^*) \subseteq \bigcap_{j=0}^\kappa \mathcal{N}(s_j^*)$.

(iii) $\bigcup_{j=0}^\kappa \mathcal{R}(s_j) \subseteq \mathcal{R}(s_0)$ and $\bigcup_{j=0}^\kappa \mathcal{R}(s_j^*) \subseteq \mathcal{R}(s_0^*)$.

(iv) $(s_j^*)_{j=0}^\kappa \in \mathcal{D}_{q\times p,\kappa}$.

(v) $s_j s_0^\dagger s_0 = s_j$ and $s_0 s_0^\dagger s_j = s_j$ for all $j \in \mathbb{Z}_{0,\,\kappa}$.

(vi) $s_0 s_0^\dagger s_j s_0^\dagger s_0 = s_j$ for all $j \in \mathbb{Z}_{0,\,\kappa}$.

Proof. Use parts (a) and (b) of Lemma A.3 and consider orthogonal complements. $\qquad\square$

Remark 5.2. If $\kappa \in \mathbb{N}_0 \cup \{+\infty\}$ and $(s_j)_{j=0}^\kappa \in \mathcal{D}_{p\times q,\kappa}$, then Remark 4.5 yields

$$(\operatorname{rank} s_j)_{j=0}^\kappa \in \mathcal{D}_{1\times 1,\kappa} \qquad\text{and}\qquad (\det s_j)_{j=0}^\kappa \in \mathcal{D}_{1\times 1,\kappa}.$$

Remark 5.3. If $\kappa \in \mathbb{N}_0 \cup \{+\infty\}$, $(\alpha_j)_{j=0}^\kappa \in \mathcal{D}_{1\times 1,\kappa}$ and $(s_j)_{j=0}^\kappa \in \mathcal{D}_{p\times q,\kappa}$, then, taking Remark 4.5 into account, we see that $(\alpha_j s_j)_{j=0}^\kappa \in \mathcal{D}_{p\times q,\kappa}$.

Lemma 5.4. *Suppose $\kappa \in \mathbb{N}_0 \cup \{+\infty\}$ and $(s_j)_{j=0}^{\kappa} \in \mathcal{D}_{p \times q, \kappa}$. Let $L \in \mathbb{C}^{k \times p}$ and $R \in \mathbb{C}^{q \times n}$ be such that $\mathcal{R}(s_0) \subseteq \mathcal{R}(L^*)$ and $\mathcal{R}(s_0^*) \subseteq \mathcal{R}(R)$. Then $(Ls_j R)_{j=0}^{\kappa} \in \mathcal{D}_{k \times n, \kappa}$, $(Ls_j)_{j=0}^{\kappa} \in \mathcal{D}_{k \times q, \kappa}$ and $(s_j R)_{j=0}^{\kappa} \in \mathcal{D}_{p \times n, \kappa}$.*

Proof. Because of (A.8), we have $\mathcal{R}(s_0^*) = \mathcal{R}(s_0^{\dagger})$, and thus $\mathcal{R}(s_0^{\dagger}) \subseteq \mathcal{R}(R)$. Hence, part (b) of Lemma A.3 implies

$$RR^{\dagger}s_0^{\dagger} = s_0^{\dagger}. \tag{5.1}$$

Because of $\mathcal{R}(s_0) \subseteq \mathcal{R}(L^*)$, part (b) of Lemma A.3 yields $L^*(L^*)^{\dagger}s_0 = s_0$. Using (A.4), we now get

$$L^{\dagger}Ls_0 = s_0. \tag{5.2}$$

Let $j \in \mathbb{Z}_{0,\kappa}$. Because of Proposition 5.1, we have

$$s_j = s_0 s_0^{\dagger} s_j s_0^{\dagger} s_0.$$

Thus, taking both (5.1) and (5.2) into account, we get

$$
\begin{aligned}
Ls_j R &= Ls_0 s_0^{\dagger} s_j s_0^{\dagger} s_0 R \\
&= Ls_0 (RR^{\dagger}s_0^{\dagger}) s_j s_0^{\dagger} (L^{\dagger}Ls_0) R \\
&= (Ls_0 R)(R^{\dagger}s_0^{\dagger} s_j s_0^{\dagger} L^{\dagger})(Ls_0 R).
\end{aligned}
$$

This implies

$$\mathcal{N}(Ls_j R) \subseteq \mathcal{N}(Ls_0 R) \qquad \text{and} \qquad \mathcal{R}(Ls_j R) \subseteq \mathcal{R}(Ls_0 R).$$

Hence, $(Ls_j R)_{j=0}^{\kappa}$ belongs to $\mathcal{D}_{k \times n, \kappa}$. The two remaining assertions can be similarly verified. $\qquad \square$

Remark 5.5. Let $L \in \mathbb{C}^{k \times p}$, $R \in \mathbb{C}^{q \times n}$, $\kappa \in \mathbb{N}_0 \cup \{+\infty\}$, and $(s_j)_{j=0}^{\kappa} \in \mathcal{D}_{p \times q, \kappa}$. Suppose, furthermore, that rank $L = p$ and rank $R = q$. It follows then, that $\mathcal{R}(L^*) = \mathbb{C}^p$ and $\mathcal{R}(R) = \mathbb{C}^q$. Thus, from Lemma 5.4 we then see that $(Ls_j R)_{j=0}^{\kappa} \in \mathcal{D}_{k \times n, \kappa}$.

Remark 5.6. Let $(p_l)_{l=1}^{k}$ and $(q_l)_{l=1}^{k}$ be sequences in $\mathbb{N}$ and let $\kappa \in \mathbb{N}_0 \cup \{+\infty\}$. For each $l \in \mathbb{Z}_{1,k}$, let $(s_j^{(l)})_{j=0}^{\kappa}$ be a sequence in $\mathbb{C}^{p_l \times q_l}$. Setting $\tilde{p} := \sum_{l=1}^{k} p_l$ and $\tilde{q} := \sum_{l=1}^{k} q_l$, we then see from (4.4) that

$$\left(\operatorname{diag}(s_j^{(l)})_{l=1}^{k} \right)_{j=0}^{\kappa} \in \mathcal{D}_{\tilde{p} \times \tilde{q}, \kappa} \qquad \text{iff} \qquad (s_j^{(l)})_{j=0}^{\kappa} \in \mathcal{D}_{p_l \times q_l, \kappa}, \quad \text{for all } l \in \mathbb{Z}_{1,k}.$$

Remark 5.7. Let $\kappa \in \mathbb{N}_0 \cup \{+\infty\}$ and $(s_j)_{j=0}^{\kappa}$ be a sequence from $\mathbb{C}^{p \times q}$ such that

$$s_{2l+1} = 0_{p \times q} \qquad \qquad \text{for all } l \in \mathbb{N}_0 \text{ with } 2l + 1 \leq \kappa.$$

Recalling Definition 4.13, it can be shown by induction that

$$s_{2l+1}^{\sharp} = 0_{q \times p} \qquad \qquad \text{for all } l \in \mathbb{N}_0 \text{ with } 2l + 1 \leq \kappa.$$

Remark 5.8. Let $\alpha \in \mathbb{C}$, $\kappa \in \mathbb{N}_0 \cup \{+\infty\}$, and $(s_j)_{j=0}^{\kappa}$ be a sequence in $\mathbb{C}^{p \times q}$. Recalling Definition 4.13, it can be shown by induction that

$$\left((\alpha s_j)^{\sharp} \right)_{j=0}^{\kappa} = \left(\alpha^{\dagger} s_j^{\sharp} \right)_{j=0}^{\kappa},$$

i.e., that $(\alpha^{\dagger} s_j^{\sharp})_{j=0}^{\kappa}$ is the reciprocal sequence corresponding to $(\alpha s_j)_{j=0}^{\kappa}$.

Remark 5.9. Let $\alpha \in \mathbb{C}$, $\kappa \in \mathbb{N}_0 \cup \{+\infty\}$ and let $(s_j)_{j=0}^{\kappa}$ be a sequence in $\mathbb{C}^{p \times q}$. Recalling Definition 4.13, it can be verified by induction that

$$\left((\alpha^j s_j)^{\sharp} \right)_{j=0}^{\kappa} = \left(\alpha^j s_j^{\sharp} \right)_{j=0}^{\kappa}.$$

We will next recognize that constructing the sequence reciprocal to a sequence $(s_j)_{j=0}^{\kappa}$ in $\mathbb{C}^{p \times q}$ always yields an element of $\mathcal{D}_{q \times p, \kappa}$.

Proposition 5.10. *If $\kappa \in \mathbb{N}_0 \cup \{+\infty\}$ and $(s_j)_{j=0}^{\kappa}$ is a sequence in $\mathbb{C}^{p \times q}$, then:*

(a) $\left(s_j^{\sharp} \right)_{j=0}^{\kappa} \in \mathcal{D}_{q \times p, \kappa}$.

(b) *For each $j \in \mathbb{Z}_{0, \kappa}$, both of the following identities hold:*

$$s_0^{\dagger} s_0 s_j^{\sharp} = s_j^{\sharp} \qquad and \qquad s_j^{\sharp} s_0 s_0^{\dagger} = s_j^{\sharp}.$$

(c) $s_0^{\sharp} \left[\sum\limits_{l=0}^{k} s_{k-l} s_l^{\sharp} \right] = 0_{p \times q}$ *for each $k \in \mathbb{Z}_{1, \kappa}$.*

(d) *If $\bigcup\limits_{j=1}^{\kappa} \mathcal{R}(s_j) \subseteq \mathcal{R}(s_0)$, then $\sum\limits_{l=0}^{k} s_{l-k} s_l^{\sharp} = 0_{p \times p}$ for each $k \in \mathbb{Z}_{1, \kappa}$.*

Proof. (a) From Definition 4.13 it is immediately clear that $\mathcal{R}(s_j^{\sharp}) \subseteq \mathcal{R}(s_0^{\sharp})$, for each $j \in \mathbb{Z}_{0, \kappa}$. Using Definition 4.13, it follows by induction that $\mathcal{N}(s_0^{\sharp}) \subseteq \mathcal{N}(s_j^{\sharp})$, for each $j \in \mathbb{Z}_{0, \kappa}$.

(b) Taking (A.1) and Definition 4.13 into account, we obtain $s_0 = (s_0^{\dagger})^{\dagger} = (s_0^{\sharp})^{\dagger}$. Thus, combining (a) and (4.4) with parts (b) and (a) of Lemma A.3, we conclude that

$$s_0^{\dagger} s_0 s_j^{\sharp} = s_0^{\sharp} (s_0^{\sharp})^{\dagger} s_j^{\sharp} = s_j^{\sharp} \qquad and \qquad s_j^{\sharp} s_0 s_0^{\dagger} = s_j^{\sharp} (s_0^{\sharp})^{\dagger} s_0^{\sharp} = s_j^{\sharp},$$

for each $j \in \mathbb{Z}_{0, \kappa}$.

(c) Using (b) and Definition 4.13, we get

$$s_0^{\sharp} \left[\sum_{l=0}^{k} s_{k-l} s_l^{\sharp} \right] = s_0^{\dagger} s_0 s_k^{\sharp} + s_0^{\sharp} \left[\sum_{l=1}^{k} s_{k-l} s_l^{\sharp} \right] = s_k^{\sharp} - s_k^{\sharp} = 0_{p \times q}.$$

(d) Using part (b) of Lemma A.3 and (c), we see that

$$\sum_{l=0}^{k} s_{k-l} s_l^{\sharp} = \sum_{l=0}^{k} s_0 s_0^{\dagger} s_{k-l} s_l^{\sharp} = s_0 \left(s_0^{\sharp} \left[\sum_{l=0}^{k} s_{k-l} s_l^{\sharp} \right] \right) = s_0 \cdot 0_{p \times q} = 0_{p \times p}. \qquad \square$$

We now characterize the case in which two sequences of complex $p \times q$-matrices have the same reciprocal sequence.

Proposition 5.11. *Let $\kappa \in \mathbb{N}_0 \cup \{+\infty\}$. If $(s_j)_{j=0}^\kappa$ and $(t_j)_{j=0}^\kappa$ are sequences in $\mathbb{C}^{p \times q}$, then the following two statements are equivalent to one another:*

(i) *For every $j \in \mathbb{Z}_{0,\kappa}$:*

$$s_0 s_0^\dagger s_j s_0^\dagger s_0 = t_0 t_0^\dagger t_j t_0^\dagger t_0.$$

(ii) *For every $j \in \mathbb{Z}_{0,\kappa}$:*

$$s_j^\sharp = t_j^\sharp.$$

Proof. From part (b) of Proposition 5.10 we obtain

$$s_0^\dagger s_0 s_j^\sharp = s_j^\sharp \qquad \text{and} \qquad t_0^\dagger t_0 t_j^\sharp = t_j^\sharp, \qquad \text{for every } j \in \mathbb{Z}_{0,\kappa}. \tag{5.3}$$

"(i)$\Rightarrow$(ii)": From Definition 4.13, Remark 4.7 and (i) we see that

$$s_0^\sharp = s_0^\dagger = s_0 s_0^\dagger s_0 s_0^\dagger s_0 = t_0 t_0^\dagger t_0 t_0^\dagger t_0 = t_0^\dagger = t_0^\sharp. \tag{5.4}$$

For $\kappa = 0$, we have thus proved (ii).

Suppose now that $\kappa \geq 1$ and that for some $m \in \mathbb{Z}_{1,\kappa}$:

$$s_j^\sharp = t_j^\sharp, \qquad \text{for every } j \in \mathbb{Z}_{0,m-1}.$$

From Definition 4.13, (5.3), and (i) it then follows that

$$s_m^\sharp = -s_0^\dagger \sum_{j=0}^{m-1} s_{m-j} s_j^\sharp = -s_0^\dagger \sum_{j=0}^{m-1} s_0 s_0^\dagger s_{m-j} s_0^\dagger s_0 s_j^\sharp$$

$$= -t_0^\dagger \sum_{j=0}^{m-1} t_0 t_0^\dagger t_{m-j} t_0^\dagger t_0 t_j^\sharp = -t_0^\dagger \sum_{j=0}^{m-1} t_{m-j} t_j^\sharp = t_m^\sharp.$$

Thus we have proved (ii) by induction.

"(ii)$\Rightarrow$(i)": From (A.1), Definition 4.13, and (ii), we obtain

$$s_0 = (s_0^\dagger)^\dagger = (s_0^\sharp)^\dagger = (t_0^\sharp)^\dagger = (t_0^\dagger)^\dagger = t_0. \tag{5.5}$$

Taking (5.5) into account, we see that $s_0 s_0^\dagger s_0 s_0^\dagger s_0 = t_0 t_0^\dagger t_0 t_0^\dagger t_0$. For $\kappa = 0$ we have therefore proved (i).

Suppose now that $\kappa \geq 1$. From Remark 2.17, (ii) and (5.5), we get

$$s_0 s_0^\dagger s_1 s_0^\dagger s_0 = s_0 \left(-s_1^\sharp \right) s_0 = t_0 \left(-t_1^\sharp \right) t_0 = t_0 t_0^\dagger t_1 t_0^\dagger t_0. \tag{5.6}$$

We have thus proved (i) for $\kappa = 1$.

Now suppose that $\kappa \geq 2$ and that for some $m \in \mathbb{Z}_{2,\kappa}$:

$$s_0 s_0^\dagger s_j s_0^\dagger s_0 = t_0 t_0^\dagger t_j t_0^\dagger t_0, \qquad \text{for every } j \in \mathbb{Z}_{1,m-1}.$$

From Definition 4.13, (5.3), (ii) and (5.5), we obtain

$$-s_0^\dagger s_m s_0^\dagger = -s_0^\dagger s_m s_0^\sharp$$

$$= -s_0^\dagger \sum_{j=0}^{m-1} s_{m-j} s_j^\sharp + s_0^\dagger \sum_{j=1}^{m-1} s_{m-j} s_j^\sharp$$

$$= s_m^\sharp + s_0^\dagger \sum_{j=1}^{m-1} s_0 s_0^\dagger s_{m-j} s_0^\dagger s_0 s_j^\sharp$$

$$= t_m^\sharp + t_0^\dagger \sum_{j=1}^{m-1} t_0 t_0^\dagger t_{m-j} t_0^\dagger t_0 t_j^\sharp$$

$$= -t_0^\dagger \sum_{j=0}^{m-1} t_{m-j} t_j^\sharp + t_0^\dagger \sum_{j=1}^{m-1} t_{m-j} t_j^\sharp$$

$$= -t_0^\dagger t_m t_0^\sharp = -t_0^\dagger t_m t_0^\dagger.$$

Taking (5.5) into account, we see that $s_0 s_0^\dagger s_m s_0^\dagger s_0 = t_0 t_0^\dagger t_m t_0^\dagger t_0$. By induction, we have thus proved (i). □

Corollary 5.12. *Let $m \in \mathbb{N}_0$ and $(s_j)_{j=0}^m$ be a sequence in $\mathbb{C}^{p \times q}$. Furthermore, let*

$$t_j := \begin{cases} s_j, & \text{if } m > 0 \text{ and } j \in \mathbb{Z}_{0,m-1} \\ s_0 s_0^\dagger s_j s_0^\dagger s_0, & \text{if } j = m. \end{cases}$$

Then both $(s_j)_{j=0}^m$ and $(t_j)_{j=0}^m$ have the same reciprocal sequence, i.e., $t_j^\sharp = s_j^\sharp$ holds true for all $j \in \mathbb{Z}_{0,m}$.

Proof. Use Proposition 5.11. □

Now we present a characterization of the invertibility of a sequence in terms of the reciprocal sequence.

Proposition 5.13. *Let $\kappa \in \mathbb{N}_0 \cup \{+\infty\}$. If $(s_j)_{j=0}^\kappa$ is a sequence in $\mathbb{C}^{p \times q}$ and if we then define the sequence $(\tilde{s}_j)_{j=0}^\kappa$ via $\tilde{s}_j := s_0 s_0^\dagger s_j s_0^\dagger s_0$, for each $j \in \mathbb{Z}_{0,\kappa}$, then:*

(a) *Both of these sequences have the same reciprocal sequence, i.e.:*

$$(s_j^\sharp)_{j=0}^\kappa = (\tilde{s}_j^\sharp)_{j=0}^\kappa. \tag{5.7}$$

(b) *The sequence $(\tilde{s}_j)_{j=0}^\kappa$ is invertible and its inverse sequence $(\tilde{s}_j^\dagger)_{j=0}^\kappa$ is the sequence reciprocal to $(s_j)_{j=0}^\kappa$, i.e.:*

$$(\tilde{s}_j)_{j=0}^\kappa \in \mathcal{I}_{p \times q, \kappa} \qquad \text{and} \qquad (\tilde{s}_j^\dagger)_{j=0}^\kappa = (s_j^\sharp)_{j=0}^\kappa.$$

(c) *The reciprocal sequence* $\left((s_j^\sharp)^\sharp \right)_{j=0}^{\kappa}$ *for the sequence* $(s_j^\sharp)_{j=0}^{\kappa}$ *satisfies*

$$\left((s_j^\sharp)^\sharp \right)_{j=0}^{\kappa} = (\tilde{s}_j)_{j=0}^{\kappa}. \tag{5.8}$$

(d) *The following three statements are all equivalent:*

(i) $(s_j)_{j=0}^{\kappa}$ *is the reciprocal sequence for* $(s_j^\sharp)_{j=0}^{\kappa}$.

(ii) $(s_j)_{j=0}^{\kappa} = (\tilde{s}_j)_{j=0}^{\kappa}$.

(iii) $(s_j)_{j=0}^{\kappa} \in \mathcal{I}_{p\times q,\kappa}$.

Proof. (a) Remark 4.7 gives us $s_0 = \tilde{s}_0$. Thus, for each $j \in \mathbb{Z}_{0,\kappa}$, we obtain

$$\tilde{s}_0 \tilde{s}_0^\dagger \tilde{s}_j \tilde{s}_0^\dagger \tilde{s}_0 = s_0 s_0^\dagger \tilde{s}_j s_0^\dagger s_0 = s_0 s_0^\dagger \left(s_0 s_0^\dagger s_j s_0^\dagger s_0 \right) s_0^\dagger s_0 = s_0 s_0^\dagger s_j s_0^\dagger s_0.$$

Hence, Proposition 5.11 yields (5.7).

(b) Recalling Remark 4.7, we see that

$$(\tilde{s}_j)_{j=0}^{\kappa} \in \mathcal{D}_{p\times q,\kappa}. \tag{5.9}$$

From (5.9) and part (a) of Theorem 4.21 we get $(\tilde{s}_j)_{j=0}^{\kappa} \in \mathcal{I}_{p\times q,\kappa}$. From part (b) of Theorem 4.21 we get and (a) we conclude that

$$(\tilde{s}_j^\dagger)_{j=0}^{\kappa} = (\tilde{s}_j^\sharp)_{j=0}^{\kappa} = (s_j^\sharp)_{j=0}^{\kappa}.$$

(c) From (5.9) and Corollary 4.22, it follows that

$$\left((\tilde{s}_j^\sharp)^\sharp \right)_{j=0}^{\kappa} = (\tilde{s}_j)_{j=0}^{\kappa}.$$

Thus, (a) implies (5.8).

(d) "(i) $\Longleftrightarrow$ (ii)" This equivalence follows from (c).

"(ii) $\Longleftrightarrow$ (iii)" This follows from Proposition 5.1 and part (a) of Theorem 4.21. $\qquad\square$

Corollary 5.14. *Let* $\kappa \in \mathbb{N} \cup \{ +\infty \}$ *and let* $(s_j)_{j=0}^{\kappa} \in \mathcal{D}_{p\times q,\kappa}$. *For each* $k \in \mathbb{Z}_{1,\kappa}$,

then $\sum_{l=0}^{k} s_{l-k}^\sharp s_l = 0_{q\times q}$.

Proof. Let $k \in \mathbb{Z}_{1,\kappa}$. It follows from part (a) of Proposition 5.10 and Definition 4.13 that $\bigcup_{l=0}^{k} \mathcal{R}(s_{l-k}^\sharp) \subseteq \mathcal{R}(s_0^\sharp)$. Thus, part (d) of Proposition 5.10 yields

$$\sum_{l=0}^{k} s_{l-k}^\sharp (s_l^\sharp)^\sharp = 0_{q\times q}. \tag{5.10}$$

Because $(s_j)_{j=0}^{\kappa} \in \mathcal{D}_{p\times q,\kappa}$, we see from part (d) of Proposition 5.13 that

$$\left((s_j^\sharp)^\sharp \right)_{j=0}^{\kappa} = (s_j)_{j=0}^{\kappa} .$$

Combining this with (5.10) completes the proof. $\qquad\square$

We next look to find generalizations of some parts of Lemma 3.2. We will first need the following results.

Remark 5.15. Suppose that $k \in \mathbb{N}_0 \cup \{+\infty\}$ and that $(s_j)_{j=0}^{\kappa}$ is a sequence in $\mathbb{C}^{p \times q}$. For each $j \in \mathbb{Z}_{0,\kappa}$, we set $\widetilde{s}_j := s_0 s_0^\dagger s_j s_0^\dagger s_0$, $t_j := s_j^\dagger$ and $\widetilde{t}_j := t_0 t_0^\dagger t_j t_0^\dagger t_0$. Then, using (A.2), we get

$$\widetilde{t}_j = s_0^*(s_0^*)^\dagger s_j^*(s_0^*)^\dagger s_0^* = s_0^*(s_0^\dagger)^* s_j^*(s_0^\dagger)^* s_0^* = \left(s_0 s_0^\dagger s_j s_0^\dagger s_0\right)^* = \widetilde{s}_j^*.$$

Proposition 5.16. *Suppose that* $k \in \mathbb{N}_0 \cup \{+\infty\}$ *and that* $(s_j)_{j=0}^{\kappa}$ *is a sequence in* $\mathbb{C}^{p \times q}$. *Then*

$$\left((s_j^*)^\sharp\right)_{j=0}^{\kappa} = \left((s_j^\sharp)^*\right)_{j=0}^{\kappa}.$$

Proof. For each $j \in \mathbb{Z}_{0,\kappa}$ we set $\widetilde{s}_j := s_0 s_0^\dagger s_j s_0^\dagger s_0$, $t_j := s_j^*$ and $\widetilde{t}_j := t_0 t_0^\dagger t_j t_0^\dagger t_0$. Then Remark 5.15 yields

$$\left(\widetilde{t}_j\right)_{j=0}^{\kappa} = \left(\widetilde{s}_j^*\right)_{j=0}^{\kappa}. \tag{5.11}$$

Part (a) of Proposition 5.13 yields (5.7) and

$$\left(t_j^\sharp\right)_{j=0}^{\kappa} = \left(\widetilde{t}_j^\sharp\right)_{j=0}^{\kappa}. \tag{5.12}$$

Combining Remark 4.7 and part (a) of Theorem 4.21 we see that

$$\left(\widetilde{s}_j\right)_{j=0}^{\kappa} \in \mathcal{I}_{p \times q, \kappa}. \tag{5.13}$$

Because of (5.13), applying Lemma 3.2 yields $\left(\widetilde{s}_j^*\right)_{j=0}^{\kappa} \in \mathcal{I}_{q \times p, \kappa}$ and

$$\left((\widetilde{s}_j^*)^\ddagger\right)_{j=0}^{\kappa} = \left((\widetilde{s}_j^\dagger)^*\right)_{j=0}^{\kappa}. \tag{5.14}$$

In view of part (b) of Theorem 4.21, formula (5.14) can be rewritten as

$$\left((\widetilde{s}_j^*)^\sharp\right)_{j=0}^{\kappa} = \left((\widetilde{s}_j^\sharp)^*\right)_{j=0}^{\kappa}. \tag{5.15}$$

Using (5.7), (5.15), (5.11), (5.12) and (5.11) again, we obtain

$$\left((s_j^\sharp)^*\right)_{j=0}^{\kappa} = \left((\widetilde{s}_j^\sharp)^*\right)_{j=0}^{\kappa} = \left((\widetilde{s}_j^*)^\sharp\right)_{j=0}^{\kappa} = \left(\widetilde{t}_j^\sharp\right)_{j=0}^{\kappa} = \left(t_j^\sharp\right)_{j=0}^{\kappa} = \left((s_j^*)^\sharp\right)_{j=0}^{\kappa}.$$

Thus, the proof is complete. $\qquad\square$

Corollary 5.17. *If* $k \in \mathbb{N}_0 \cup \{+\infty\}$ *and* $(s_j)_{j=0}^{\kappa}$ *is a sequence in* $\mathbb{C}_{\mathrm{H}}^{q \times q}$, *then* $\left(s_j^\sharp\right)_{j=0}^{\kappa}$ *is also a sequence in* $\mathbb{C}_{\mathrm{H}}^{q \times q}$.

Taking part (b) of Theorem 4.21 into account, we see that Proposition 5.16 and Corollary 5.17 are extensions of Lemma 3.2 and Remark 3.3, respectively.

Lemma 5.18. *Let $\kappa \in \mathbb{N}_0 \cup \{+\infty\}$ and $(s_j)_{j=0}^{\kappa}$ be a sequence in $\mathbb{C}^{p \times q}$. Let $L \in \mathbb{C}^{k \times p}$ and $R \in \mathbb{C}^{q \times n}$ be such that $\mathcal{R}(L^*) = \mathcal{R}(s_0)$ and $\mathcal{R}(s_0^*) = \mathcal{R}(R)$. Then*

$$\left((Ls_j R)^{\sharp} \right)_{j=0}^{\kappa} = \left(R^{\dagger} s_j^{\sharp} L^{\dagger} \right)_{j=0}^{\kappa}, \qquad \left((Ls_j)^{\sharp} \right)_{j=0}^{\kappa} = \left(s_j^{\sharp} L^{\dagger} \right)_{j=0}^{\kappa},$$

and

$$\left((s_j R)^{\sharp} \right)_{j=0}^{\kappa} = \left(R^{\dagger} s_j^{\sharp} \right)_{j=0}^{\kappa}.$$

Proof. From Proposition A.7 we see that

$$(Ls_0 R)^{\dagger} = R^{\dagger} s_0^{\dagger} L^{\dagger}. \tag{5.16}$$

Combining (5.16) with Definition 4.13 gives us $(Ls_0 R)^{\sharp} = R^{\dagger} s_0^{\sharp} L^{\dagger}$. Hence, there is an $m \in \mathbb{Z}_{1,\kappa}$ such that

$$(Ls_l R)^{\sharp} = R^{\dagger} s_l^{\sharp} L^{\dagger}, \qquad \text{for all } l \in \mathbb{Z}_{0,m-1}.$$

Since $\mathcal{R}(L^*) = \mathcal{R}(s_0)$, considering orthogonal complements, we obtain $\mathcal{N}(s_0^*) = \mathcal{N}(L)$, and, because of (A.9), thus, $\mathcal{N}(s_0^{\dagger}) = \mathcal{N}(L)$. Now part (a) of Lemma A.3 yields

$$s_0^{\dagger} L^{\dagger} L = s_0^{\dagger}. \tag{5.17}$$

Since we assumed that $\mathcal{R}(s_0^*) = \mathcal{R}(R)$, from (A.8), we get that $\mathcal{R}(s_0^{\dagger}) = \mathcal{R}(R)$. Thus, Remark A.6 implies

$$RR^{\dagger} s_0^{\dagger} = s_0^{\dagger}. \tag{5.18}$$

Consequently, if $\kappa \geq 1$, then from Definition 4.13, (5.16), (5.17), (5.18) and part (b) of Proposition 5.10 we get

$$(Ls_m R)^{\sharp} = -(Ls_0 R)^{\dagger} \sum_{l=0}^{m-1} (Ls_{m-l} R)(Ls_l R)^{\sharp}$$

$$- -R^{\dagger} s_0^{\dagger} L^{\dagger} \sum_{l=0}^{m-1} (Ls_{m-l} R) \left(R^{\dagger} s_l^{\sharp} L^{\dagger} \right)$$

$$= -R^{\dagger} \left(s_0^{\dagger} L^{\dagger} L \right) \sum_{l=0}^{m-1} s_{m-l} \left(RR^{\dagger} s_0^{\dagger} \right) s_0 s_l^{\sharp} L^{\dagger}$$

$$= -R^{\dagger} s_0^{\dagger} \sum_{l=0}^{m-1} s_{m-l} \left(s_0^{\dagger} s_0 s_l^{\sharp} \right) L^{\dagger} = -R^{\dagger} s_0^{\dagger} \sum_{l=0}^{m-1} s_{m-l} s_l^{\sharp} L^{\dagger}$$

$$= R^{\dagger} \left(-s_0^{\dagger} \sum_{l=0}^{m-1} s_{m-l} s_l^{\sharp} \right) L^{\dagger} = R^{\dagger} s_m^{\sharp} L^{\dagger}.$$

Thus, by induction, the first assertion is proved. The remaining two assertions can be similarly verified. $\qquad \square$

Lemma 5.19. *Let $\kappa \in \mathbb{N}_0 \cup \{+\infty\}$ and $(s_j)_{j=0}^{\kappa}$ be a sequence in $\mathbb{C}^{p \times q}$. If $U \in \mathbb{C}^{k \times p}$ with $U^* U = I_p$ and $V \in \mathbb{C}^{q \times n}$ with $V V^* = I_q$, then*

$$(U s_j V)^{\sharp} = V^* s_j^{\sharp} U^* \qquad \text{for each } j \in \mathbb{Z}_{0,\kappa}.$$

Proof. Using Definition 4.13 and part (c) of Lemma A.3, we get

$$(U s_0 V)^{\sharp} = (U s_0 V)^{\dagger} = V^* s_0^{\dagger} U^* = V^* s_0^{\sharp} U^*. \qquad (5.19)$$

Now suppose that $m \in \mathbb{Z}_{1,\kappa}$ and that the equation $(U s_j V)^{\sharp} = V^* s_j^{\sharp} U^*$ holds for each $j \in \mathbb{Z}_{0,m-1}$. Then, using Definition 4.13 and (5.19), we conclude that

$$(U s_m V)^{\sharp} = -(U s_0 V)^{\dagger} \sum_{l=0}^{m-1} (U s_{m-l} V)(U s_l V)^{\sharp}$$

$$= -V^* s_0^{\dagger} U^* \sum_{l=0}^{m-1} U s_{m-l} V V^* s_l^{\sharp} U^*$$

$$= V^* \left(-s_0^{\dagger} \sum_{l=0}^{m-1} s_{m-l} s_l^{\sharp} \right) U^* = V^* s_m^{\sharp} U^*.$$

Thus, the proof is complete. □

Remark 5.20. Let $(p_l)_{l=1}^n$ and $(q_l)_{l=1}^n$ be sequences of positive integers and let $\kappa \in \mathbb{N}_0 \cup \{+\infty\}$. Furthermore, for each $l \in \mathbb{Z}_{1,n}$, let $\left(s_j^{(l)} \right)_{j=0}^{\kappa}$ be a sequence in $\mathbb{C}^{p_l \times q_l}$. It is then not difficult to show, by induction, that

$$\left[\operatorname{diag} \left(s_j^{(l)} \right)_{l=1}^n \right]^{\sharp} = \operatorname{diag} \left(\left(s_j^{(l)} \right)^{\sharp} \right)_{l=1}^n \qquad \text{for all } j \in \mathbb{Z}_{0,\kappa}.$$

As our final topic in this section, we now focus on a subclass of the class $\mathcal{D}_{p \times q, m}$, introduced in Definition 4.3.

Notation 5.21. For each $m \in \mathbb{N}$ we define $\widetilde{\mathcal{D}}_{p \times q, m}$ as the set of all sequences $(s_j)_{j=0}^m$ in $\mathbb{C}^{p \times q}$, which satisfy $(s_j)_{j=0}^{m-1} \in \mathcal{D}_{p \times q, m-1}$.

Remark 5.22. Let $m \in \mathbb{N}$ and $(s_j)_{j=0}^m$ be a sequence in $\mathbb{C}^{p \times q}$. For $j \in \mathbb{Z}_{0,m}$, let $\tilde{s}_j := s_0 s_0^{\dagger} s_j s_0^{\dagger} s_0$. Then Proposition 5.1 shows that

$$(s_j)_{j=0}^m \in \widetilde{\mathcal{D}}_{p \times q, m} \qquad \text{if and only if} \qquad (s_j)_{j=0}^{m-1} = (\tilde{s}_j)_{j=0}^{m-1}.$$

Remark 5.23. Let $m \in \mathbb{N}$. Remark 4.6 then yields $\mathcal{D}_{p \times q, m} \subseteq \widetilde{\mathcal{D}}_{p \times q, m}$.

Because of part (b) of Theorem 4.21, the next result can be considered an extension of part (d) of Proposition 4.1.

Proposition 5.24. *Let $m \in \mathbb{N}$ and $(s_j)_{j=0}^m \in \widetilde{\mathcal{D}}_{p \times q, m}$. Then*

$$s_m^{\sharp} = - \left(\sum_{k=1}^m s_{m-k}^{\sharp} s_k \right) s_0^{\dagger}.$$

Proof. In view of Definition 4.13, we have $s_0^\sharp = s_0^\dagger$. Using this and Remark 2.17, we get

$$-\left(\sum_{k=1}^{1} s_{m-k}^\sharp s_k\right) s_0^\dagger = -s_0^\sharp s_1 s_0^\dagger = -s_0^\dagger s_1 s_0^\dagger = s_1^\sharp.$$

Thus, the assertion is proved for $m = 1$. Now let $m \in \{2, 3, \ldots\}$. For each $j \in \mathbb{Z}_{0,m}$, let

$$\widetilde{s}_j := s_0 s_0^\dagger s_j s_0^\dagger s_0.$$

Then Remark 5.22 yields

$$(s_j)_{j=0}^{m-1} = (\widetilde{s}_j)_{j=0}^{m-1}. \tag{5.20}$$

Combining Remark 4.7 and part (a) of Theorem 4.21 we see that

$$(\widetilde{s}_j)_{j=0}^{m} \in \mathcal{I}_{p\times q,m}. \tag{5.21}$$

Furthermore, part (a) of Proposition 5.13 gives us

$$(s_j^\sharp)_{j=0}^{m} = (\widetilde{s}_j^\sharp)_{j=0}^{m}. \tag{5.22}$$

In view of (5.21), we see from Theorem 4.21 and part (d) of Proposition 4.1 that

$$s_m^\sharp = -\left(\sum_{k=1}^{m} \widetilde{s}_{m-k}^\sharp \widetilde{s}_k\right) s_0^\dagger = -\left(\widetilde{s}_0^\sharp \widetilde{s}_m s_0^\dagger + \sum_{k=1}^{m} \widetilde{s}_{m-k}^\sharp \widetilde{s}_k s_0^\dagger\right). \tag{5.23}$$

From (5.20) and (5.22) we see that

$$\sum_{k=1}^{m-1} \widetilde{s}_{m-k}^\sharp \widetilde{s}_k s_0^\dagger = \sum_{k=1}^{m-1} s_{m-k}^\sharp s_k s_0^\dagger. \tag{5.24}$$

Because of Definition 4.13 and Remark 4.7, we obtain $\widetilde{s}_0^\sharp = \widetilde{s}_0^\dagger = s_0^\dagger$. Thus, taking $s_0^\sharp = s_0^\dagger$ into account, we get

$$\widetilde{s}_0^\sharp \widetilde{s}_m s_0^\dagger = s_0^\dagger\left(s_0 s_0^\dagger s_m s_0^\dagger s_0\right) = s_0^\dagger s_m s_0^\dagger = s_0^\sharp s_m s_0^\dagger. \tag{5.25}$$

Combining (5.22), (5.23), (5.24) and (5.25) we conclude that

$$s_m^\sharp = \widetilde{s}_m^\sharp = -\left(\sum_{k=1}^{m-1} s_{m-k}^\sharp s_k\right) s_0^\dagger.$$

This completes the proof. $\qquad\qquad\square$

6. Further observations on invertible sequences

We again focus on the class of invertible sequences of complex $p \times q$-matrices and here begin with some simple observations. Throughout this section, let m and n be positive integers.

Remark 6.1. If $\kappa \in \mathbb{N}_0 \cup \{+\infty\}$ and if $(\alpha_j)_{j=0}^{\kappa}$ is a sequence in $\mathbb{C}$, then we see from Remark 4.5 and part (a) of Theorem 4.21, that $(\alpha_j)_{j=0}^{\kappa} \in \mathcal{I}_{1\times 1,\kappa}$ if and only if $\alpha_0 \neq 0$ or $(\alpha_j)_{j=0}^{\kappa}$ is the constant sequence with value 0.

Remark 6.2. Let $\kappa \in \mathbb{N}_0 \cup \{+\infty\}$. If $(s_j)_{j=0}^{\kappa} \in \mathcal{I}_{p\times q,\kappa}$, then we see from part (a) of Theorem 4.21 as well as from Remark 5.2 that

$$(\operatorname{rank} s_j)_{j=0}^{\kappa} \in \mathcal{I}_{1\times 1,\kappa} \qquad \text{and} \qquad (\det s_j)_{j=0}^{\kappa} \in \mathcal{I}_{1\times 1,\kappa}.$$

Remark 6.3. Let $\kappa \in \mathbb{N}_0 \cup \{+\infty\}$. If $(s_j)_{j=0}^{\kappa} \in \mathcal{I}_{p\times q,\kappa}$, then we see from Remark 2.8, part (b) of Theorem 4.21 and Remark 5.7, that

$$s_{2k+1}^{\dagger} = 0_{q\times p}, \qquad \text{for all } k \in \mathbb{N}_0 \text{ with } 2k+1 \leq \kappa,$$

if and only if

$$s_{2k+1} = 0_{p\times q}, \qquad \text{for all } k \in \mathbb{N}_0 \text{ with } 2k+1 \leq \kappa.$$

Remark 6.4. Let $\kappa \in \mathbb{N}_0 \cup \{+\infty\}$. If $(\alpha_j)_{j=0}^{\kappa} \in \mathcal{I}_{1\times 1,\kappa}$ and $(s_j)_{j=0}^{\kappa} \in \mathcal{I}_{p\times q,\kappa}$, then we see from part (a) of Theorem 4.21 and Remark 5.3 that $(\alpha_j s_j)_{j=0}^{\kappa} \in \mathcal{I}_{p\times q,\kappa}$.

Remark 6.5. Let $\alpha \in \mathbb{C}$ and $\kappa \in \mathbb{N}_0 \cup \{+\infty\}$. If $(s_j)_{j=0}^{\kappa} \in \mathcal{I}_{p\times q,\kappa}$, then Remark 6.4, part (b) of Theorem 4.21 and Remark 5.9 give us that

$$(\alpha^j s_j)_{j=0}^{\kappa} \in \mathcal{I}_{p\times q,\kappa} \qquad \text{and} \qquad \left((\alpha^j s_j)^{\dagger} \right)_{j=0}^{\kappa} = (\alpha^j s_j^{\dagger})_{j=0}^{\kappa}.$$

Lemma 6.6. *Let* $\kappa \in \mathbb{N}_0 \cup \{+\infty\}$. *If* $(s_j)_{j=0}^{\kappa} \in \mathcal{I}_{p\times q,\kappa}$, $L \in \mathbb{C}^{m\times p}$ *and* $R \in \mathbb{C}^{q\times n}$ *such that* $\mathcal{R}(s_0) \subseteq \mathcal{R}(L^*)$ *and* $\mathcal{R}(s_0^*) \subseteq \mathcal{R}(R)$, *then* $(Ls_j R)_{j=0}^{\kappa} \in \mathcal{I}_{m\times n,\kappa}$.

Proof. Use part (a) of Theorem 4.21 and Lemma 5.4. $\qquad\square$

Remark 6.7. Let $\kappa \in \mathbb{N}_0 \cup \{+\infty\}$. If $(s_j)_{j=0}^{\kappa} \in \mathcal{I}_{p\times q,\kappa}$, $L \in \mathbb{C}^{m\times p}$ and $R \in \mathbb{C}^{q\times n}$ with $\operatorname{rank} L = p$ and $\operatorname{rank} R = q$, then, $\mathcal{R}(L^*) = \mathbb{C}^p$ and $\mathcal{R}(R) = \mathbb{C}^q$. Thus, Lemma 6.6 gives us that $(Ls_j R)_{j=0}^{\kappa} \in \mathcal{I}_{m\times n,\kappa}$.

Lemma 6.8. *Let* $\kappa \in \mathbb{N}_0 \cup \{+\infty\}$. *If* $(s_j)_{j=0}^{\kappa} \in \mathcal{I}_{p\times q,\kappa}$, $L \in \mathbb{C}^{m\times p}$ *and* $R \in \mathbb{C}^{q\times n}$ *with* $\mathcal{R}(s_0) = \mathcal{R}(L^*)$ *and* $\mathcal{R}(s_0^*) = \mathcal{R}(R)$, *then:*

(a) $(Ls_j R)_{j=0}^{\kappa} \in \mathcal{I}_{m\times n,\kappa}$ *and* $\left((Ls_j R)^{\dagger} \right)_{j=0}^{\kappa} = (R^{\dagger} s_j^{\dagger} L^{\dagger})_{j=0}^{\kappa}$.

(b) $\left((Ls_j)^{\dagger} \right)_{j=0}^{\kappa} \in \mathcal{I}_{m\times q,\kappa}$ *and* $\left((Ls_j)^{\dagger} \right)_{j=0}^{\kappa} = (s_j^{\dagger} L^{\dagger})_{j=0}^{\kappa}$.

(c) $\left((s_j R)^{\dagger} \right)_{j=0}^{\kappa} \in \mathcal{I}_{p\times n,\kappa}$ *and* $\left((s_j R)^{\dagger} \right)_{j=0}^{\kappa} = (R^{\dagger} s_j^{\dagger})_{j=0}^{\kappa}$.

Proof. Combine Lemma 6.6, part (b) of Theorem 4.21 and Lemma 5.18. $\qquad\square$

Lemma 6.9. *Let* $(p_l)_{l=1}^{m}$ *and* $(q_l)_{l=1}^{m}$ *be sequences of positive integers and let* $\kappa \in \mathbb{N}_0 \cup \{+\infty\}$. *Suppose, for every* $l \in \mathbb{Z}_{1,m}$, *that* $(s_j^{(l)})_{j=0}^{\kappa}$ *is a sequence in* $\mathbb{C}^{p_l \times q_l}$.

If we set $\widetilde{p} := \sum\limits_{l=1}^{m} p_l$ and $\widetilde{q} := \sum\limits_{l=1}^{m} q_l$, then:

$$\left(\operatorname{diag} \left(s_j^{(l)} \right)_{l=1}^{m} \right)_{j=0}^{\kappa} \in \mathcal{I}_{\widetilde{p} \times \widetilde{q}, \kappa} \qquad iff \qquad \left(s_j^{(l)} \right)_{j=0}^{\kappa} \in \mathcal{I}_{p_l \times q_l, \kappa}, \quad for \ all \ l \in \mathbb{Z}_{1,m},$$

and if $\left(\operatorname{diag} \left(s_j^{(l)} \right)_{l=1}^{m} \right)_{j=0}^{\kappa}$ is indeed invertible, then it has the inverse sequence $\left(\operatorname{diag} \left(\left(s_j^{(l)} \right)^{\ddagger} \right)_{l=1}^{m} \right)_{j=0}^{\kappa}$.

Proof. Use Theorem 4.21, Remark 5.6 and Remark 5.20. $\square$

Notation 6.10. Given an integer r with $1 \leq r \leq \min(p, q)$ and a matrix $A \in \mathbb{C}^{r \times r}$, we will use the notation

$$A^{[p,q]} := \begin{cases} A, & if \ p = q = r \\ \begin{pmatrix} A \\ 0_{(p-r) \times q} \end{pmatrix}, & if \ p > q = r \\ \begin{pmatrix} A & 0_{p \times (q-r)} \end{pmatrix}, & if \ q > p = r \\ \begin{pmatrix} A & 0_{r \times (q-r)} \\ 0_{(p-r) \times r} & 0_{(p-r) \times (q-r)} \end{pmatrix}, & if \ \min\{p, q\} > r. \end{cases}$$

The following proposition offers us some important insight into the structure of a given sequence $(s_j)_{j=0}^{\kappa} \in \mathcal{I}_{p \times q, \kappa}$. We will see that when the first element of this sequence is non-zero, i.e., $s_0 \neq 0_{p \times q}$, it is naturally generated by a sequence, $(\widehat{s}_j)_{j=0}^{\kappa} \in \widetilde{\mathcal{I}}_{r \times r, \kappa}$, where $r := \operatorname{rank} s_0$.

Proposition 6.11. *Let $\kappa \in \mathbb{N}_0 \cup \{+\infty\}$ and let $(s_j)_{j=0}^{\kappa} \in \mathcal{I}_{p \times q, \kappa}$ be such that $r := \operatorname{rank} s_0$ is positive. Let $\sigma_1 \geq \sigma_2 \geq \cdots \geq \sigma_r > 0$ be the singular values of s_0 and let $\Delta := \operatorname{diag}(\sigma_1, \sigma_2, \ldots, \sigma_r)$. Furthermore, let U and V be, respectively, unitary complex $q \times q$- and $p \times p$-matrices such that*

$$s_0 = V \Delta^{[p,q]} U^*. \tag{6.1}$$

Let $\widehat{U}$ be the left $q \times r$-block of U and let $\widehat{V}$ be the left $p \times r$-block of V. Then:

(a) *The sequence $(t_j)_{j=0}^{\kappa}$ given by*

$$t_j := \widehat{V}^* s_j \widehat{U} \qquad for \ each \ j \in \mathbb{Z}_{0, \kappa} \tag{6.2}$$

 belongs to $\widetilde{\mathcal{I}}_{r \times r, \kappa}$.

(b) *$s_j = \widehat{V} t_j \widehat{U}^*$ and $s_j = V t_j^{[p,q]} U^*$ for each $j \in \mathbb{Z}_{0, \kappa}$.*

(c) *Let $(s_j^{\ddagger})_{j=0}^{\kappa}$ and $(t_j^{\ddagger})_{j=0}^{\kappa}$ be, respectively, the inverse sequences for $(s_j)_{j=0}^{\kappa}$ and $(t_j)_{j=0}^{\kappa}$. Then $s_j^{\ddagger} = \widehat{U} t_j^{\ddagger} \widehat{V}^*$ for each $j \in \mathbb{Z}_{0, \kappa}$.*

Proof. From (6.1) we see that $t_0 = \Delta$. In particular, we see that the matrix t_0 is non-singular. Hence, Notation 2.10 and Proposition 2.11 yield $(t_j)_{j=0}^{\kappa} \in \tilde{\mathcal{I}}_{r \times r, \kappa}$.

Proposition 4.4 shows that $(s_j)_{j=0}^{\kappa} \in \mathcal{D}_{p \times q, \kappa}$. Thus, from Proposition 5.1 we conclude that

$$\mathcal{N}(s_0) \subseteq \bigcap_{j=0}^{\kappa} \mathcal{N}(s_j) \qquad \text{and} \qquad \mathcal{N}(s_0^*) \subseteq \bigcap_{j=0}^{\kappa} \mathcal{N}(s_j^*) \qquad (6.3)$$

hold true. Because of (6.1), if $r < q$, then the right $q \times (q-r)$-block $\tilde{U}$ of U satisfies the inclusion $\mathcal{R}(\tilde{U}) \subseteq \mathcal{N}(s_0)$ which, in view of (6.3), implies

$$s_j \tilde{U} = 0_{p \times q - r} \qquad \text{for each } j \in \mathbb{Z}_{0, \kappa}. \qquad (6.4)$$

Similarly, because of (6.1) and (6.3), if $r < p$, then the right $p \times (p-r)$-block $\tilde{V}$ of V satisfies

$$s_j^* \tilde{V} = 0_{q \times p - r} \qquad \text{for each } j \in \mathbb{Z}_{0, \kappa}. \qquad (6.5)$$

Using (6.2), (6.4), (6.5) and Notation 6.10, we then obtain $V^* s_j U = t_j^{[p, q]}$ for each $j \in \mathbb{Z}_{0, \kappa}$ and consequently,

$$s_j = V t_j^{[p, q]} U^* = \hat{V} t_j \hat{U}^* \qquad \text{for each } j \in \mathbb{Z}_{0, \kappa}. \qquad (6.6)$$

Using Remark 2.9, we get from (6.6) that $s_j^{\ddagger} = \hat{U} t_j^{\ddagger} \hat{V}^*$ holds true for each $j \in \mathbb{Z}_{0, \kappa}$. $\square$

7. Some considerations on EP sequences

A complex $q \times q$ matrix A is called *range-Hermitian* (also *EP matrix*) if $\mathcal{R}(A) = \mathcal{R}(A^*)$. The class of range-Hermitian matrices belonging to $\mathbb{C}^{q \times q}$ is denoted by $\mathbb{C}_{\mathrm{EP}}^{q \times q}$. The class $\mathbb{C}_{\mathrm{EP}}^{q \times q}$ was introduced in Schwerdtfeger [18]. It is important in the theory of generalized inverses (see, e.g., the monographs Campbell/Meyer [4, Chapter 4, Section 3] and Ben-Israel/Greville [2, Chapter 4, Section 4], and also the papers Pearl [16] and Meyer [15]).

Definition 7.1. Let $n \in \mathbb{N}_0$ and let $(s_j)_{j=0}^n$ be a sequence from $\mathbb{C}^{q \times q}$. Then $(s_j)_{j=0}^n$ is called an *EP sequence* (EP stands for Equal Projectors) if the matrix $\mathbf{S}_n^{(s)}$ defined by (2.1) belongs to $\mathbb{C}_{\mathrm{EP}}^{(n+1)q \times (n+1)q}$.

Remark 7.2. Let $q \in \mathbb{N}$, $n \in \mathbb{N}$ and let $(s_j)_{j=0}^n$ be an EP sequence from $\mathbb{C}^{q \times q}$. For $k \in \mathbb{N}_0$, then $(s_j)_{j=0}^k$ is an EP sequence. Indeed, this is a direct consequence of the block decomposition

$$\mathbf{S}_n = \begin{pmatrix} \mathbf{S}_k & 0_{(k+1)q \times (n-1)q} \\ * & \mathbf{S}_{n-(k+1)} \end{pmatrix}$$

of the matrix $\mathbf{S}_n^{(s)}$ and part (a) of Proposition A.15.

Remark 7.2 leads us to the following definition, which complements Definition 7.1.

Definition 7.3. Let $(s_j)_{j=0}^n$ be a sequence of complex $q \times q$ matrices. Then $(s_j)_{j=0}^\infty$ is called an *EP sequence* if, for each $n \in \mathbb{N}_0$, the sequence $(s_j)_{j=0}^n$ is an EP sequence. Let $q \in \mathbb{N}$ and $\kappa \in \mathbb{N}_0 \cup \{+\infty\}$. Then $\mathcal{F}_{q,\kappa}^{\mathrm{EP}}$ will stand for the set of all EP sequences $(s_j)_{j=0}^\kappa$ from $\mathbb{C}^{q \times q}$.

The following results show that EP sequences are particularly invertible.

Proposition 7.4. *Let $\kappa \in \mathbb{N}_0 \cup \{+\infty\}$ and let $(s_j)_{j=0}^\kappa$ be a sequence from $\mathbb{C}^{q \times q}$.*

(a) *The following statements are equivalent:*
  (i) $(s_j)_{j=0}^\kappa \in \mathcal{F}_{q,\kappa}^{\mathrm{EP}}$.
  (ii) $s_0 \in \mathbb{C}_{\mathrm{EP}}^{q \times q}$ *and* $(s_j)_{j=0}^\kappa \in \mathcal{D}_{q \times q, \kappa}$.
  (iii) $s_0 \in \mathbb{C}_{\mathrm{EP}}^{q \times q}$ *and* $(s_j)_{j=0}^\kappa \in \mathcal{I}_{q \times q, \kappa}$.
(b) *Let (i) be satisfied. Then $s_0 s_0^\dagger = s_0^\dagger s_0$. For each $n \in \mathbb{N}$, furthermore,*

$$\mathbf{S}_n \mathbf{S}_n^\dagger = I_{n+1} \otimes (s_0 s_0^\dagger) \qquad and \qquad \mathbf{S}_n^\dagger \mathbf{S}_n = I_{n+1} \otimes (s_0^\dagger s_0).$$

Proof. (a) 1.) "(i) $\Rightarrow$ (ii)". In view of (i) and Remark 7.2, we have $s_0 \in \mathbb{C}_{\mathrm{EP}}^{q \times q}$. Let $m \in \mathbb{Z}_{1,\kappa}$. Then, from (2.1) and (4.11), we get the block partitions in (4.12). Because of (i), we obtain

$$\mathbf{S}_m \in \mathbb{C}_{\mathrm{EP}}^{(m+1)q \times (m+1)q}. \tag{7.1}$$

In view of (4.11), (4.12) and (7.1), applying part (a) of Proposition A.15 yields

$$\begin{pmatrix} s_m & s_{m-1} & \cdots & s_1 \end{pmatrix} = \widehat{z}_{1,m} = s_0 s_0^\dagger \widehat{z}_{1,m} = \begin{pmatrix} s_0 s_0^\dagger s_m & s_0 s_0^\dagger s_{m-1} & \cdots & s_0 s_0^\dagger s_1 \end{pmatrix}$$

and

$$\mathrm{col}\,(s_j)_{j=1}^m = y_{1,m} = y_{1,m} s_0^\dagger s_0 = \mathrm{col}\,\left(s_j s_0^\dagger s_0\right)_{j=1}^m.$$

Consequently, $s_0 s_0^\dagger s_j = s_j$ and $s_j s_0^\dagger s_0 = s_j$, for each $j \in \mathbb{Z}_{1,m}$. Thus, Proposition 5.1 yields

$$(s_j)_{j=0}^m \in \mathcal{D}_{q \times q, m}. \tag{7.2}$$

Since $(s_j)_{j=0}^0 \in \mathcal{D}_{q \times q, 0}$ obviously holds true, we get from (7.2) and Remark 4.6 that $(s_j)_{j=0}^\kappa \in \mathcal{D}_{q \times q, \kappa}$. Thus, we obtain (ii).

2.) "(ii) $\Rightarrow$ (iii)". This follows from part (a) of Theorem 4.21.

3.) "(iii) $\Rightarrow$ (i)". In view of (iii), we have $s_0 \in \mathbb{C}_{\mathrm{EP}}^{q \times q}$. Thus, Proposition A.4 yields $s_0 s_0^\dagger = s_0^\dagger s_0$. Let $m \in \mathbb{Z}_{0,\kappa}$. In view of (iii), we have $(s_j)_{j=0}^\kappa \in \mathcal{I}_{q \times q, \kappa}$. Thus, Lemma 3.6 implies

$$\mathbf{S}_m \mathbf{S}_m^\dagger = I_{m+1} \otimes (s_0 s_0^\dagger) \qquad and \qquad \mathbf{S}_m^\dagger \mathbf{S}_m = I_{m+1} \otimes (s_0^\dagger s_0).$$

Thus, $\mathbf{S}_m \mathbf{S}_m^\dagger = \mathbf{S}_m^\dagger \mathbf{S}_m$. Hence, Proposition A.4 yields $\mathbf{S}_m \in \mathbb{C}_{\mathrm{EP}}^{(m+1)q \times (m+1)q}$. Hence, (i) holds.

(b) All assertions of (b) were already obtained in the proof of the implication "(iii) $\Rightarrow$ (i)". □

Corollary 7.5. *Let* $\kappa \in \mathbb{N}_0 \cup \{+\infty\}$ *and let* $(s_j)_{j=0}^{\kappa}$ *be a sequence from* $\mathbb{C}^{q\times q}$ *such that* $s_0 \in \mathbb{C}_{\mathrm{EP}}^{q\times q}$. *For each* $j \in \mathbb{Z}_{0,\kappa}$, *let* $\widetilde{s}_j := s_0 s_0^\dagger s_j s_0^\dagger s_0$. *Then* $(\widetilde{s}_j)_{j=0}^{\kappa} \in \mathcal{F}_{q,\kappa}^{\mathrm{EP}}$.

Proof. In view of Remark 4.8, we have $\widetilde{s}_0 = s_0$ and $(\widetilde{s}_j)_{j=0}^{\kappa} \in \mathcal{D}_{q\times q,\kappa}$. Thus, taking into account that $s_0 \in \mathbb{C}_{\mathrm{EP}}^{q\times q}$ and applying part (a) of Proposition 7.4 gives us the assertion. □

Proposition 7.6. *Let* $\kappa \in \mathbb{N}_0 \cup \{+\infty\}$ *and let* $(s_j)_{j=0}^{\kappa}$ *be a sequence from* $\mathbb{C}^{q\times q}$. *Then* $(s_j^\sharp)_{j=0}^{\kappa} \in \mathcal{F}_{q,\kappa}^{\mathrm{EP}}$ *if and only if* $s_0 \in \mathbb{C}_{\mathrm{EP}}^{q\times q}$.

Proof. In view of Definition 4.13, we have $s_0^\sharp = s_0^\dagger$. Let $(s_j^\sharp)_{j=0}^{\kappa} \in \mathcal{F}_{q,\kappa}^{\mathrm{EP}}$. Then part (a) of Proposition 7.4 yields $s_0^\sharp \in \mathbb{C}_{\mathrm{EP}}^{q\times q}$. Hence, $s_0^\sharp = s_0^\dagger$ implies $s_0^\dagger \in \mathbb{C}_{\mathrm{EP}}^{q\times q}$. Consequently, Proposition A.4 gives us $s_0 \in \mathbb{C}_{\mathrm{EP}}^{q\times q}$.

Conversely, assume $s_0 \in \mathbb{C}_{\mathrm{EP}}^{q\times q}$. Then Proposition A.4 implies $s_0^\dagger \in \mathbb{C}_{\mathrm{EP}}^{q\times q}$. Combining this with $s_0^\sharp = s_0^\dagger$, we get $s_0^\sharp \in \mathbb{C}_{\mathrm{EP}}^{q\times q}$. Part (a) of Proposition 5.10 yields $(s_j^\sharp)_{j=0}^{\kappa} \in \mathcal{D}_{q\times q,\kappa}$. Using this and $s_0^\sharp \in \mathbb{C}_{\mathrm{EP}}^{q\times q}$, we see from part (a) of Proposition 7.4 that $(s_j^\sharp)_{j=0}^{\kappa} \in \mathcal{F}_{q,\kappa}^{\mathrm{EP}}$. □

Corollary 7.7. *Let* $\kappa \in \mathbb{N}_0 \cup \{+\infty\}$ *and let* $(s_j)_{j=0}^{\kappa} \in \mathcal{F}_{q,\kappa}^{\mathrm{EP}}$. *Then* $(s_j^\sharp)_{j=0}^{\kappa} \in \mathcal{F}_{q,\kappa}^{\mathrm{EP}}$.

Proof. Part (a) of Proposition 7.4 yields $s_0 \in \mathbb{C}_{\mathrm{EP}}^{q\times q}$. Thus, applying Proposition 7.6 completes the proof. □

8. Moore-Penrose pseudoinverses of holomorphic matrix-malued functions

In the first part of this section, we study the holomorphicity of a $p \times q$-matrix-function's Moore-Penrose inverse.

Remark 8.1. Let $\mathcal{G}$ be a non-empty open and connected subset of $\mathbb{C}$. Since every function $f : \mathcal{G} \longrightarrow \mathbb{C}$ for which both f and $\bar{f}$ are holomorphic in $\mathcal{G}$ is necessarily constant in $\mathcal{G}$ (see, e.g., [6, Ch. II, §2, Proposition 2.10]), the following statements hold:

(a) If $F : \mathcal{G} \longrightarrow \mathbb{C}^{p\times q}$ is a function for which both F and F^* are holomorphic in $\mathcal{G}$, then F is constant in $\mathcal{G}$.
(b) If $H : \mathcal{G} \longrightarrow \mathbb{C}^{q\times q}$ is a holomorphic function in $\mathcal{G}$ with Hermitian values, then F is constant in $\mathcal{G}$.

Lemma 8.2. *Let* $\mathcal{G}$ *be a non-empty open and connected subset of* $\mathbb{C}$. *If* $F : \mathcal{G} \longrightarrow \mathbb{C}^{p\times q}$ *is such that both* F *and* $F^\dagger$ *are holomorphic in* $\mathcal{G}$, *then both* $FF^\dagger$ *and* $F^\dagger F$ *are constant in* $\mathcal{G}$.

Proof. Since $FF^\dagger$ and $F^\dagger F$ are holomorphic functions in $\mathcal{G}$ with Hermitian values, the assertion follows from part (b) of Remark 8.1. □

The following result (which can be found in Campbell/Meyer [4, Theorem 10.5.4]) is, in a sense, converse to Lemma 8.2.

Proposition 8.3. *Let $\mathcal{G}$ be a non-empty open and connected subset of $\mathbb{C}$. If $F : \mathcal{G} \longrightarrow \mathbb{C}^{p \times q}$ is a function, holomorphic in $\mathcal{G}$, such that both $FF^\dagger$ and $F^\dagger F$ are constant in $\mathcal{G}$, then $F^\dagger$ is holomorphic in $\mathcal{G}$.*

These results lead us to a global characterization for the case of holomorphicity for the (pointwise) Moore-Penrose pseudoinverse of a holomorphic matrix-function in a non-empty open and connected subset of $\mathbb{C}$.

Proposition 8.4. *If $\mathcal{G}$ is a non-empty open and connected subset of $\mathbb{C}$, $F \colon \mathcal{G} \longrightarrow \mathbb{C}^{p \times q}$ is holomorphic in $\mathcal{G}$, and $z_0 \in \mathcal{G}$, then all of the following statements are equivalent:*

(i) *$F^\dagger$ is holomorphic in $\mathcal{G}$.*
(ii) *For each $z \in \mathcal{G}$:*

$$(FF^\dagger)(z) = (FF^\dagger)(z_0) \qquad and \qquad (F^\dagger F)(z) = (F^\dagger F)(z_0).$$

(iii) *For each $z \in \mathcal{G}$:*

$$\mathcal{N}(F(z)) = \mathcal{N}(F(z_0)) \qquad and \qquad \mathcal{R}(F(z)) = \mathcal{R}(F(z_0)).$$

Proof. "(i)⇒(ii)": follows from Lemma 8.2.

"(ii)⇒(i)": follows from Proposition 8.3.

"(ii) ⇔ (iii)": Recalling Proposition A.1 and considering orthogonal complements, we see that $F(z)F^\dagger(z)$ and $I_q - F^\dagger(z)F(z)$ are, for each $z \in \mathcal{G}$, the matrices associated with the orthogonal projection from $\mathbb{C}^p$ onto $\mathcal{R}(F(z))$ and from $\mathbb{C}^q$ onto $\mathcal{N}(F(z))$, respectively. Consequently, (ii) and (iii) are equivalent. □

Corollary 8.5. *If $\mathcal{G}$ is a non-empty open and connected subset of $\mathbb{C}$ and $F \colon \mathcal{G} \to \mathbb{C}^{p \times q}$, then the following statements are all equivalent:*

(i) *For each $w \in \mathcal{G}$ and each $z \in \mathcal{G}$:*

$$\mathcal{N}(F^\dagger(w)) = \mathcal{N}(F^\dagger(z)) \qquad and \qquad \mathcal{R}(F^\dagger(w)) = \mathcal{R}(F^\dagger(z)).$$

(ii) *For each $w \in \mathcal{G}$ and each $z \in \mathcal{G}$:*

$$\mathcal{N}(F(w)) = \mathcal{N}(F(z)) \qquad and \qquad \mathcal{R}(F(w)) = \mathcal{R}(F(z)).$$

If condition (i) is met, then $F^\dagger$ is holomorphic if and only if F is holomorphic.

Proof. From (A.8) it follows that

$$\mathcal{N}\left(F^\dagger(z)\right) = \mathcal{N}\left([F(z)]^\dagger\right) = \mathcal{R}\left(\left([F(z)]^\dagger\right)^*\right)^\perp = \mathcal{R}\left(\left([F(z)]^\dagger\right)^\dagger\right)^\perp = \mathcal{R}\left(F(z)\right)^\perp$$

and

$$\mathcal{R}\left(F^\dagger(z)\right) = \mathcal{R}\left([F(z)]^\dagger\right) = \mathcal{R}\left([F(z)]^*\right) = \left[\mathcal{R}\left([F(z)]^*\right)^\perp\right]^\perp = \mathcal{N}\left(F(z)\right)^\perp$$

hold true for all $z \in \mathcal{G}$, hence we see that (i) and (ii) are equivalent.

Now suppose (i). Then (ii) is also true. Since $(F^\dagger)^\dagger = F$, we see from Proposition 8.4 that $F^\dagger$ is holomorphic if and only if F is holomorphic. $\qquad\square$

We now consider the case in which, for a holomorphic $p \times q$-matrix-function F and a z_0 in its domain, $\mathcal{G}$:

$$\left(\frac{F^{(j)}(z_0)}{j!} \right)_{j=0}^{\infty} \in \mathcal{I}_{p\times q, \infty}.$$

We will see that the Moore-Penrose inverse $F^\dagger$ of F is holomorphic in a particular neighbourhood of z_0. We will, furthermore, determine the Taylor series for $F^\dagger$.

For $z_0 \in \mathbb{C}$ and $R \in (0, +\infty)$, we will use

$$K(z_0; R) := \{ z \in \mathbb{C} : |z - z_0| < R \}$$

to denote the open disk with radius R, centered at z_0.

Notation 8.6. Let r be an integer with $1 \leq r \leq \min(p, q)$. If $\mathcal{G}$ is a non-empty subset of $\mathbb{C}$ and if $g : \mathcal{G} \longrightarrow \mathbb{C}^{r\times r}$ is a matrix-valued function, then, in view of Notation 6.10, let $g^{[p,q]} : \mathcal{G} \longrightarrow \mathbb{C}^{p\times q}$ be defined by

$$g^{[p,q]}(z) := [g(z)]^{[p,q]}.$$

Proposition 8.7. *Let $\mathcal{G}$ be a non-empty open and connected subset of $\mathbb{C}$ and let $F : \mathcal{G} \longrightarrow \mathbb{C}^{p\times q}$ be holomorphic in $\mathcal{G}$. Suppose that $z_0 \in \mathcal{G}$ is such that*

$$\left(\frac{F^{(j)}(z_0)}{j!} \right)_{j=0}^{\infty} \in \mathcal{I}_{p\times q, \infty}.$$

Then:

(a) *If $F(z_0) = 0_{p\times q}$, then there exists an $R \in (0, +\infty)$ such that $K(z_0; R) \subseteq \mathcal{G}$ and $F(z) = 0_{p\times q}$ for all $z \in K(z_0; R)$.*

(b) *Let $F(z_0) \neq 0_{p\times q}$ and $r := \operatorname{rank} F(z)$. Let $\sigma_1 \geq \sigma_2 \geq \cdots \geq \sigma_r > 0$ be the singular values of $F(z_0)$ and let $\Delta := \operatorname{diag}(\sigma_1, \sigma_2, \ldots, \sigma_r)$. Furthermore, let U and V, respectively, be unitary complex $q \times q$- and $p \times p$-matrices such that*

$$F(z_0) = V\Delta^{[p,q]}U^*.$$

Let $\widehat{U}$ be the left $q \times r$-block of U and $\widehat{V}$ be the left $p \times r$-block of V. Then:

(b1) *The function*

$$f := \widehat{V}^* F \widehat{U} \tag{8.1}$$

is holomorphic in $\mathcal{G}$.

(b2) *There exists an $R_0 \in (0, +\infty)$ such that $K(z_0; R_0) \subseteq \mathcal{G}$ and $\det f(z) \neq 0$ for all $z \in K(z_0; R_0)$.*

(b3) *If $z \in K(z_0; R_0)$, then*

$$F^\dagger(z) = \widehat{U}[f(z)]^{-1}\widehat{V}^*. \tag{8.2}$$

(b4) *The function $F^\dagger$ is holomorphic in $K(z_0; R_0)$.*

Proof. (a) Remark 4.2 gives us that $\left(\frac{F^{(j)}(z_0)}{j!} \right)_{j=0}^{\infty}$ is the constant sequence with value $0_{p \times q}$. This implies (a).

(b1) This is obvious from the definition of f.

(b2) By construction, we have $\operatorname{rank} f(z_0) = \operatorname{rank} F(z_0) = r$. Thus, $\det f(z_0) \neq 0$. Since the function $\det f$ is continuous, it follows that (b2) holds.

(b3) For each $j \in \mathbb{N}_0$, we set

$$s_j := \frac{F^{(j)}(z_0)}{j!} \tag{8.3}$$

and

$$t_j := \widehat{V}^* s_j \widehat{U}. \tag{8.4}$$

From (8.3), (8.4) and the definition of f, we see that

$$t_j = \frac{f^{(j)}(z_0)}{j!} \qquad \text{for each } j \in \mathbb{N}_0. \tag{8.5}$$

Taking (8.4) into account, we see from part (b) of Proposition 6.11 that

$$s_j = V t_j^{[p,q]} U^* \qquad \text{for each } j \in \mathbb{N}_0. \tag{8.6}$$

From (8.3), (8.5) and (8.6) we obtain

$$F(z) = V f^{[p,q]}(z) U^*. \tag{8.7}$$

Using (8.7) and the fact that the matrices U and V are unitary, the formula $F^\dagger(z) = U \left[f^\dagger(z) \right]^{[p,q]} V^*$ follows by applying part (c) of Lemma A.3. This implies (8.2).

(b4) follows immediately from (b1), (b2) and (8.2). $\qquad \square$

Remark 8.8. Let $\mathcal{F}$ and $\mathcal{G}$ be subsets of $\mathbb{C}$ such that the interior of $\mathcal{F} \cap \mathcal{G}$ is non-empty and let z_0 be an interior point of $\mathcal{F} \cap \mathcal{G}$. If the matrix-valued functions $F: \mathcal{F} \to \mathbb{C}^{p \times q}$ and $G: \mathcal{G} \to \mathbb{C}^{q \times r}$ are holomorphic in z_0, then

$$\mathbf{S}_m^{[F]} \mathbf{S}_m^{[G]} = \mathbf{S}_m^{[FG]}, \qquad \text{for all } m \in \mathbb{N}_0,$$

where $\mathbf{S}_m^{[F]}$, $\mathbf{S}_m^{[G]}$ and $\mathbf{S}_m^{[FG]}$ denote the matrices given by (2.1), substituting, respectively, $\left(\frac{1}{j!} F^{(j)}(z_0) \right)_{j=0}^m$, $\left(\frac{1}{j!} G^{(j)}(z_0) \right)_{j=0}^m$ and $\left(\frac{1}{j!} (FG)^{(j)}(z_0) \right)_{j=0}^m$ for $(s_j)_{j=0}^m$.

Now we come to the main result of this section.

Theorem 8.9. *Let $\mathcal{G}$ be a non-empty and connected subset of $\mathbb{C}$. If z_0 is an interior point of $\mathcal{G}$ and $F: \mathcal{G} \to \mathbb{C}^{p \times q}$ is holomorphic in z_0, then*

$$F^\dagger \text{ is holomorphic in } z_0 \qquad \text{if and only if} \qquad \left(\frac{1}{j!} F^{(j)}(z_0) \right)_{j=0}^{+\infty} \in \mathcal{I}_{p \times q, +\infty}.$$

Whenever this is the case, $\left(\frac{1}{j!}(F^{\dagger})^{(j)}(z_0) \right)_{j=0}^{+\infty}$ *is the inverse sequence correspond-ing to* $\left(\frac{1}{j!}F^{(j)}(z_0) \right)_{j=0}^{+\infty}$.

Proof. First suppose that $F^{\dagger}$ is holomorphic in z_0. From Remark 8.8 we then get

$$\mathbf{S}_m^{[F]}\mathbf{S}_m^{[F^{\dagger}]} = \mathbf{S}_m^{[FF^{\dagger}]} \qquad \text{and} \qquad \mathbf{S}_m^{[F^{\dagger}]}\mathbf{S}_m^{[F]} = \mathbf{S}_m^{[F^{\dagger}F]}$$

for all $m \in \mathbb{N}_0$, where $\mathbf{S}_m^{[F]}$, $\mathbf{S}_m^{[F^{\dagger}]}$, $\mathbf{S}_m^{[FF^{\dagger}]}$ and $\mathbf{S}_m^{[F^{\dagger}F]}$ are given by (2.1) of No-tation 2.1, using $\left(\frac{1}{j!}F^{(j)}(z_0) \right)_{j=0}^{m}$, $\left(\frac{1}{j!}[F^{\dagger}]^{(j)}(z_0) \right)_{j=0}^{m}$, $\left(\frac{1}{j!}(FF^{\dagger})^{(j)}(z_0) \right)_{j=0}^{m}$ and $\left(\frac{1}{j!}(F^{\dagger}F)^{(j)}(z_0) \right)_{j=0}^{m}$, respectively, instead of $(s_j)_{j=0}^{m}$. From the holomorphicity of F and $F^{\dagger}$ in z_0, we see that $FF^{\dagger}$ is holomorphic in z_0. From Remark 8.8 we obtain, for all $m \in \mathbb{N}_0$:

$$\mathbf{S}_m^{[FF^{\dagger}]}\mathbf{S}_m^{[F]} = \mathbf{S}_m^{[FF^{\dagger}F]} \qquad \text{and} \qquad \mathbf{S}_m^{[F^{\dagger}]}\mathbf{S}_m^{[FF^{\dagger}]} = \mathbf{S}_m^{[F^{\dagger}FF^{\dagger}]},$$

where $\mathbf{S}_m^{[FF^{\dagger}F]}$ and $\mathbf{S}_m^{[F^{\dagger}FF^{\dagger}]}$ are given by (2.1) using $\left(\frac{1}{j!}(FF^{\dagger}F)^{(j)}(z_0) \right)_{j=0}^{m}$ and $\left(\frac{1}{j!}(F^{\dagger}FF^{\dagger})^{(j)}(z_0) \right)_{j=0}^{m}$, respectively, instead of $(s_j)_{j=0}^{m}$. Hence, we have

$$\mathbf{S}_m^{[F]}\mathbf{S}_m^{[F^{\dagger}]}\mathbf{S}_m^{[F]} = \mathbf{S}_m^{[FF^{\dagger}]}\mathbf{S}_m^{[F]} = \mathbf{S}_m^{[FF^{\dagger}F]} = \mathbf{S}_m^{[F]} \tag{8.8}$$

and

$$\mathbf{S}_m^{[F^{\dagger}]}\mathbf{S}_m^{[F]}\mathbf{S}_m^{[F^{\dagger}]} = \mathbf{S}_m^{[F^{\dagger}]}\mathbf{S}_m^{[FF^{\dagger}]} = \mathbf{S}_m^{[F^{\dagger}FF^{\dagger}]} = \mathbf{S}_m^{[F^{\dagger}]} \tag{8.9}$$

for all $m \in \mathbb{N}_0$. Since F and $F^{\dagger}$ are holomorphic in z_0, there exists a positive real number R_0 such that F and $F^{\dagger}$ are holomorphic in $K(z_0; R_0)$. Then Proposi-tion 8.4 yields

$$(FF^{\dagger})(z) = (FF^{\dagger})(z_0) \qquad \text{and} \qquad (F^{\dagger}F)(z) = (F^{\dagger}F)(z_0),$$

for all $z \in K(z_0; R_0)$, which implies, for all $j \in \mathbb{N}$:

$$(FF^{\dagger})^{(j)}(z_0) = 0_{p \times p} \qquad \text{and} \qquad (F^{\dagger}F)^{(j)}(z_0) = 0_{q \times q}.$$

It thus follows, for all $m \in \mathbb{N}_0$, that

$$\mathbf{S}_m^{[FF^{\dagger}]} = I_{m+1} \otimes [\frac{1}{0!}(FF^{\dagger})^{(0)}(z_0)] \qquad \text{and} \qquad \mathbf{S}_m^{[F^{\dagger}F]} = I_{m+1} \otimes [\frac{1}{0!}(F^{\dagger}F)^{(0)}(z_0)].$$

Hence, we can conclude, for all $m \in \mathbb{N}_0$, that

$$\mathbf{S}_m^{[F]}\mathbf{S}_m^{[F^{\dagger}]} = \mathbf{S}_m^{[FF^{\dagger}]} = I_{m+1} \otimes \left[(FF^{\dagger})(z_0) \right]$$
$$= I_{m+1} \otimes \left(F(z_0)\,[F(z_0)]^{\dagger} \right) \in \mathbb{C}_{\mathrm{H}}^{(m+1)p \times (m+1)p} \tag{8.10}$$

and, similarly,

$$\mathbf{S}_m^{[F^{\dagger}]}\mathbf{S}_m^{[F]} = \mathbf{S}_m^{[F^{\dagger}F]} = I_{m+1} \otimes \left([F(z_0)]^{\dagger}\,F(z_0) \right) \in \mathbb{C}_{\mathrm{H}}^{(m+1)q \times (m+1)q}. \tag{8.11}$$

Thus, from (8.8), (8.9), (8.10) and (8.11) it follows that

$$\left(\mathbf{S}_m^{[F]} \right)^\dagger = \mathbf{S}_m^{[F^\dagger]} \qquad \text{for all } m \in \mathbb{N}_0.$$

Recalling Definition 2.6, we see that this implies

$$\left(\frac{1}{j!} F^{(j)}(z_0) \right)_{j=0}^{+\infty} \in \mathcal{I}_{p \times q, +\infty} \quad \text{and} \quad \left(\left(\frac{F^{(j)}(z_0)}{j!} \right)^\ddagger \right)_{j=0}^{+\infty} = \left(\frac{[F^\dagger]^{(j)}(z_0)}{j!} \right)_{j=0}^{+\infty}.$$

Now suppose $\left(\frac{1}{j!} F^{(j)}(z_0) \right)_{j=0}^{+\infty} \in \mathcal{I}_{p \times q, +\infty}$. Then, parts (a) and (b4) of Proposition 8.7 imply that $F^\dagger$ is holomorphic in z_0. □

Appendix A. The Moore-Penrose inverse of a complex matrix

Here, we present some facts on the Moore-Penrose inverse. For a comprehensive exposition of the theory of generalized inverses, we refer the reader, for instance, to the monographs Ben-Israel/Greville [2], Campbell/Meyer [4], and Rao/Mitra [17] or [7, Section 1.1].

If $p, q \in \mathbb{N}$ and $A \in \mathbb{C}^{p \times q}$, then (by definition) the Moore-Penrose inverse, $A^\dagger$, of A is the unique matrix, $A^\dagger \in \mathbb{C}^{q \times p}$, which satisfies the conditions

$$AA^\dagger A = A, \qquad A^\dagger A A^\dagger = A^\dagger, \qquad (AA^\dagger)^* = AA^\dagger \qquad \text{and} \qquad (A^\dagger A)^* = A^\dagger A.$$

With the next proposition, we present a well-known characterization of the Moore-Penrose inverse (see, for instance, [7, Theorem 1.1.1]).

Proposition A.1. *If $A \in \mathbb{C}^{p \times q}$, then a matrix $G \in \mathbb{C}^{q \times p}$ is the Moore-Penrose inverse of A if and only if $AG = P_A$ and $GA = P_G$, where P_A and P_G are, respectively, the matrices associated with the orthogonal projection in $\mathbb{C}^p$ onto $\mathcal{R}(A)$ and the orthogonal projection in $\mathbb{C}^q$ onto $\mathcal{R}(G)$.*

For the reader's convenience, we provide (in the next two results, for which we omit the proofs) a short review of basic properties of the Moore-Penrose inverse used in this paper (see, e.g., [2], [4], [17])

Remark A.2. If $A \in \mathbb{C}^{p \times q}$, then from the definition of the Moore-Penrose inverse, the following are obvious:

$$(A^\dagger)^\dagger = A, \tag{A.1}$$
$$(A^\dagger)^* = (A^*)^\dagger, \tag{A.2}$$
$$(A^\dagger)^T = (A^T)^\dagger, \tag{A.3}$$
$$A^\dagger A = A^* (A^*)^\dagger, \tag{A.4}$$
$$AA^\dagger = (A^*)^\dagger A^*, \tag{A.5}$$
$$A^\dagger = A^* (AA^*)^\dagger, \tag{A.6}$$
$$A^\dagger = (A^*A)^\dagger A^*, \tag{A.7}$$
$$\mathcal{R}(A^\dagger) = \mathcal{R}(A^*), \tag{A.8}$$

$$\mathcal{N}(A^\dagger) = \mathcal{N}(A^*), \tag{A.9}$$

$$\operatorname{rank} A = \operatorname{rank}\left(A^\dagger\right), \tag{A.10}$$

$$\operatorname{rank}\left(AA^\dagger\right) = \operatorname{rank} A, \tag{A.11}$$

and

$$\operatorname{rank}\left(AA^\dagger\right) = \operatorname{rank}\left(A^\dagger A\right). \tag{A.12}$$

Lemma A.3. *If $A \in \mathbb{C}^{p\times q}$, then the following statements hold:*

(a) *If $r \in \mathbb{N}$ and $B \in \mathbb{C}^{r\times q}$, then*

$$\mathcal{N}(A) \subseteq \mathcal{N}(B) \qquad \text{if and only if} \qquad BA^\dagger A = B,$$

and

$$\mathcal{N}(A) = \mathcal{N}(B) \qquad \text{if and only if} \qquad A^\dagger A = B^\dagger B.$$

(b) *Let $s \in \mathbb{N}$ and $C \in \mathbb{C}^{p\times s}$, then*

$$\mathcal{R}(C) \subseteq \mathcal{R}(A) \qquad \text{if and only if} \qquad AA^\dagger C = C,$$

and

$$\mathcal{R}(A) = \mathcal{R}(C) \qquad \text{if and only if} \qquad AA^\dagger = CC^\dagger.$$

(c) *Let $r, s \in \mathbb{N}$. For each $U \in \mathbb{C}^{r\times p}$ with $U^*U = I_p$ and each $V \in \mathbb{C}^{q\times s}$ with $VV^* = I_q$, it holds that*

$$(UAV)^\dagger = V^*A^\dagger U^*.$$

We use the following characterizations of the class $\mathbb{C}^{q\times q}_{\mathrm{EP}}$ of all complex $q \times q$ range-Hermitian matrices, which can be found in the papers [5] and [19].

Proposition A.4. *If $A \in \mathbb{C}^{q\times q}$, then all of the following conditions are equivalent:*

(i) $A \in \mathbb{C}^{q\times q}_{\mathrm{EP}}$.
(ii) $\mathcal{N}(A) = \mathcal{N}(A^*)$.
(iii) $AA^\dagger = A^\dagger A$.
(iv) $A^\dagger \in \mathbb{C}^{q\times q}_{\mathrm{EP}}$.
(v) *There exists an $X \in \mathbb{C}^{q\times q}$ such that $AX = A^*$.*
(vi) *There exists a $Y \in \mathbb{C}^{q\times q}$ such that $A^*Y = A$.*

The formula $(AB)^{-1} = B^{-1}A^{-1}$, which holds for all non-singular complex $q \times q$-matrices, can not be generalized to Moore-Penrose inverses without additional constraints. Since a particular sufficient condition we need for

$$\left(\overrightarrow{\prod_{j=1}^{n}}A_j\right)^\dagger = \overleftarrow{\prod_{j=1}^{n}}A_j^\dagger \tag{A.13}$$

can be easily proved, we will state the proofs.

Proposition A.5. *Let $n \in \mathbb{N}$ and let $(p_j)_{j=1}^{n+1}$ be a sequence of positive integers. For each $j \in \mathbb{Z}_{1,n}$, let the matrices $A_j \in \mathbb{C}^{p_j \times p_{j+1}}$ satisfy the conditions:*

$$\left(\overrightarrow{\prod_{j=1}^{n}} A_j \right) \left(\overleftarrow{\prod_{j=1}^{n}} A_j^\dagger \right) = A_1 A_1^\dagger \qquad \text{and} \qquad \left(\overleftarrow{\prod_{j=1}^{n}} A_j^\dagger \right) \left(\overrightarrow{\prod_{j=1}^{n}} A_j \right) = A_n^\dagger A_n. \quad (A.14)$$

Then (A.13) as well as both of the following equations hold true:

$$\mathcal{R}\left(\overrightarrow{\prod_{j=1}^{n}} A_j \right) = \mathcal{R}(A_1) \qquad \text{and} \qquad \mathcal{N}\left(\overrightarrow{\prod_{j=1}^{n}} A_j \right) = \mathcal{N}(A_n). \quad (A.15)$$

Proof. The case $n = 1$ is trivial. Suppose $n \geq 2$. Let

$$B_n := \overrightarrow{\prod_{j=1}^{n}} A_j \qquad \text{and} \qquad C_n := \overleftarrow{\prod_{j=1}^{n}} A_j^\dagger. \quad (A.16)$$

From (A.14) we get that

$$B_n C_n B_n = A_1 A_1^\dagger B_n = A_1 A_1^\dagger A_1 \overrightarrow{\prod_{j=2}^{n}} A_j = A_1 \overrightarrow{\prod_{j=2}^{n}} A_j = B_n$$

and, similarly, $C_n B_n C_n = C_n$. Obviously, (A.14) implies

$$(B_n C_n)^* = (A_1 A_1^\dagger)^* = A_1 A_1^\dagger = B_n C_n$$

and

$$(C_n B_n)^* = (A_n^\dagger A_n)^* = A_n^\dagger A_n = C_n B_n.$$

Consequently, the definition of the Moore-Penrose inverse provides us with $B_n^\dagger = C_n$. The proof of (A.13) is thus complete. Using (A.13), formulas (A.15) follow, respectively, from (A.14) and parts (a) and (b) of Lemma A.3. $\qquad \square$

Remark A.6. Let $p, q, r \in \mathbb{N}$. If $A \in \mathbb{C}^{p \times q}$ and $B \in \mathbb{C}^{q \times r}$, such that $\mathcal{R}(A^*) = \mathcal{R}(B)$, then, using (A.6) and part (b) of Lemma A.3, we see that

$$BB^\dagger A^\dagger = BB^\dagger A^* (AA^*)^\dagger = A^* (AA^*)^\dagger = A^\dagger.$$

Furthermore, applying (A.4) and part (b) of Lemma A.3 yields

$$A^\dagger AB = A^* (A^*)^\dagger B = B.$$

Proposition A.7. *Let $n \in \mathbb{N}$ and let $(p_j)_{j=1}^{n+1}$ be a sequence of positive integers. For each $j \in \mathbb{Z}_{1,n}$, let $A_j \in \mathbb{C}^{p_j \times p_{j+1}}$ such that $\mathcal{R}(A_j^*) = \mathcal{R}(A_{j+1})$. Then both of the equations in (A.14) hold true. Furthermore, (A.13) and (A.15) are fulfilled.*

Proof. The case $n = 1$ is trivial. If $n = 2$, then the first equation in (A.14) follows immediately from Remark A.6. Suppose now that $n \geq 3$. We proceed by induction and we assume that

$$B_{n-1} C_{n-1} = A_1 A_1^\dagger, \quad (A.17)$$

where B_{n-1} and C_{n-1} are defined as in (A.16). Because of Remark A.6 and $\mathcal{R}\left(A_{n-1}^*\right) = \mathcal{R}\left(A_n\right)$, we have

$$A_n A_n^\dagger A_{n-1}^\dagger = A_{n-1}^\dagger \tag{A.18}$$

and, using (A.17) and (A.18), we then get

$$B_n C_n = B_{n-1} A_n A_n^\dagger A_{n-1}^\dagger C_{n-2} = B_{n-1} A_{n-1}^\dagger C_{n-2} = B_{n-1} C_{n-1} = A_1 A_1^\dagger.$$

Thus, the first equation in (A.14) is proved. The second equation can be checked analogously. The rest follows from Proposition A.5. $\qquad\square$

Corollary A.8. *If* $A \in \mathbb{C}_{\mathrm{EP}}^{q \times q}$, *then*

$$A^n\left(A^\dagger\right)^n = AA^\dagger, \qquad \left(A^\dagger\right)^n A^n = A^\dagger A, \qquad \left(A^n\right)^\dagger = \left(A^\dagger\right)^n,$$

$$\mathcal{R}\left(A^n\right) = \mathcal{R}\left(A\right) \qquad and \qquad \mathcal{N}\left(A^n\right) = \mathcal{N}\left(A\right).$$

Remark A.9. If $A \in \mathbb{C}^{q \times q}$ is a normal matrix, then $\mathcal{R}\left(A\right) = \mathcal{R}\left(AA^*\right) = \mathcal{R}\left(A^*A\right) = \mathcal{R}\left(A^*\right)$. Thus we see that $A \in \mathbb{C}_{\mathrm{EP}}^{q \times q}$ and therefore that Corollary A.8 can, in particular, be applied to normal matrices.

Now we derive block representations for the Moore-Penrose inverse of a block matrix with specified constraints on its blocks. Parts of the following result can be found in Campbell/Meyer [4, p. 49, Exercise 7.3].

Proposition A.10. *Let* r, $s \in \mathbb{N}$ *and* $E \in \mathbb{C}^{(p+r) \times (q+s)}$. *Furthermore, let*

$$E = \begin{pmatrix} A & B \\ C & D \end{pmatrix} \tag{A.19}$$

be the block partition of E *with* $p \times q$*-block* A. *Let* $L := D - CA^\dagger B$ *and let*

$$F := \begin{pmatrix} A^\dagger + A^\dagger BL^\dagger CA^\dagger & -A^\dagger BL^\dagger \\ -L^\dagger CA^\dagger & L^\dagger \end{pmatrix}. \tag{A.20}$$

Then the following statements are equivalent:

(i) $E^\dagger = F$.
(ii) *The equations*

$$AA^\dagger B = B, \qquad LL^\dagger C = C, \qquad BL^\dagger L = B, \qquad CA^\dagger A = C,$$

$$DL^\dagger L = D \qquad and \qquad LL^\dagger D = D$$

    hold.
(iii) *The first four of the six equations in* (ii) *hold.*
(iv) *The inclusions*

$$\mathcal{R}(B) \subseteq \mathcal{R}(A), \qquad \mathcal{R}(C) \subseteq \mathcal{R}(L), \qquad \mathcal{N}(L) \subseteq \mathcal{N}(B), \qquad \mathcal{N}(A) \subseteq \mathcal{N}(C),$$

$$\mathcal{N}\left(L\right) \subseteq \mathcal{N}\left(D\right) \text{ and } \mathcal{R}(D) \subseteq \mathcal{R}(L) \text{ hold true.}$$

(v) *The first four of the six inclusions in* (iv) *hold.*

Proof. Let $G := EF$, let $H := FE$, and let $G = (G_{jk})^2_{j,k=1}$ and $H = (H_{jk})^2_{j,k=1}$ be the block partitions of G with $p \times p$-block G_{11} and H with $q \times q$-block H_{11}. Then $G_{11} = AA^\dagger - (I_p - AA^\dagger)BL^\dagger CA^\dagger$, $G_{12} = (I_p - AA^\dagger)BL^\dagger$, $G_{21} = (I_r - LL^\dagger)CA^\dagger$, $G_{22} = LL^\dagger$, $H_{11} = A^\dagger A - A^\dagger BL^\dagger C(I_q - A^\dagger A)$, $H_{12} = A^\dagger B(I_s - L^\dagger L)$, $H_{21} = L^\dagger C(I_q - A^\dagger A)$, and $H_{22} = L^\dagger L$.

First suppose (i), i.e., $E^\dagger = F$. Thus, $(EF)^* = EF$. For each $x \in \mathbb{C}^r$, then

$$(I_p - AA^\dagger)BL^\dagger x = G_{12}x = G^*_{21}x = (A^\dagger)^*C^*(I_r - LL^\dagger)x. \qquad (A.21)$$

The left-hand side of (A.21) belongs to $\mathcal{N}(A^\dagger)$ and consequently to $\mathcal{R}((A^\dagger)^*)^\perp$, and the right-hand side of (A.21) belongs to $\mathcal{R}((A^\dagger)^*)$. Hence $G_{12} = 0_{p \times r}$, $G_{21} = 0_{r \times p}$, and $G_{11} = AA^\dagger$. Because of $E^\dagger = F$, we have $E = EFE = GE$ and therefore $AA^\dagger B = B$, $LL^\dagger C = C$ and $LL^\dagger D = D$. Taking into account $(FE)^* = FE$ and $FEF = F$, we obtain analogously $B = BL^\dagger L$, $C = CA^\dagger A$ and $D = DL^\dagger L$. Thus, (ii) holds.

The implication "(ii) ⇒ (iii)" is trivial.

Now suppose that (iii) holds. Then the definition of the matrix L along with $LL^\dagger C = C$ and $BL^\dagger L = B$ imply $LL^\dagger D = D$ and $DL^\dagger L = D$. Thus, (ii) is fulfilled.

Finally, suppose that (ii) holds. Then we obtain $EF = G = \mathrm{diag}(AA^\dagger, LL^\dagger) \in \mathbb{C}_H^{(p+r) \times (p+r)}$ and

$$EFE = \begin{pmatrix} AA^\dagger A & AA^\dagger B \\ LL^\dagger C & LL^\dagger D \end{pmatrix} = E,$$

and similarly $(FE)^* = FE$ and $FEF = F$. Hence, $E^\dagger = F$. Consequently, the implication "(ii)⇒(i)" is verified.

From parts (a) and (b) of Lemma A.3 we see the equivalence of (ii) and (iv) as well as the equivalence of (iii) and (v). ☐

Corollary A.11. *Let $A \in \mathbb{C}^{p \times q}$, let $C \in \mathbb{C}^{r \times q}$, let $D \in \mathbb{C}^{r \times s}$, and let*

$$G := \begin{pmatrix} A & 0_{p \times s} \\ C & D \end{pmatrix} \qquad and \qquad H := \begin{pmatrix} A^\dagger & 0_{q \times r} \\ -D^\dagger CA^\dagger & D^\dagger \end{pmatrix}. \qquad (A.22)$$

(a) *The following statements are equivalent:*
   (i) *There are matrices $W \in \mathbb{C}^{q \times p}$, $Y \in \mathbb{C}^{s \times p}$, and $Z \in \mathbb{C}^{s \times r}$ such that*

$$G^\dagger = \begin{pmatrix} W & 0_{q \times r} \\ Y & Z \end{pmatrix}.$$

   (ii) $G^\dagger = H$.
   (iii) $\mathcal{R}(C) \subseteq \mathcal{R}(D)$ *and* $\mathcal{N}(A) \subseteq \mathcal{N}(C)$.
   (iv) $DD^\dagger C = C$ *and* $CA^\dagger A = C$.
(b) *If (i) is satisfied, then*

$$GG^\dagger = \mathrm{diag}\left(AA^\dagger, DD^\dagger\right) \qquad and \qquad G^\dagger G = \mathrm{diag}\left(A^\dagger A, D^\dagger D\right).$$

Proof. (a) "(i) $\Rightarrow$ (iii)". Using (i) and the equations $(GG^\dagger)^* = GG^\dagger$, $(G^\dagger G)^* = G^\dagger G$ and $GG^\dagger G = G$, we get $C = DZC$ and $C = CWA$. In particular, (iii) is true.

"(iii) $\Rightarrow$ (ii)". Use Proposition A.10.

"(ii) $\Rightarrow$ (i)". This implication is trivial.

"(iii) $\Leftrightarrow$ (iv)". Use parts (a) and (b) of Lemma A.3.

(b) Let (i) be satisfied. Then using (ii) and (iv), we get

$$GG^\dagger = \begin{pmatrix} AA^\dagger & 0_{p\times r} \\ CA^\dagger - DD^\dagger CA^\dagger & DD^\dagger \end{pmatrix} = \operatorname{diag}\left(AA^\dagger, DD^\dagger\right)$$

and, similarly, $G^\dagger G = \operatorname{diag}\left(A^\dagger A, D^\dagger D\right)$. $\qquad\square$

Corollary A.12. *Let* $A \in \mathbb{C}^{p\times q}$, *let* $B \in \mathbb{C}^{p\times s}$, *let* $D \in \mathbb{C}^{r\times s}$, *and let*

$$G := \begin{pmatrix} A & B \\ 0_{r\times q} & D \end{pmatrix} \qquad and \qquad H := \begin{pmatrix} A^\dagger & -A^\dagger BD^\dagger \\ 0_{s\times p} & D^\dagger \end{pmatrix}. \qquad (A.23)$$

(a) *The following statements are equivalent:*

(i) *There are matrices* $W \in \mathbb{C}^{q\times p}$, $X \in \mathbb{C}^{q\times r}$, *and* $Z \in \mathbb{C}^{s\times r}$ *such that*

$$G^\dagger = \begin{pmatrix} W & X \\ 0_{s\times p} & Z \end{pmatrix}.$$

(ii) $G^\dagger = H$.

(iii) $\mathcal{R}(B) \subseteq \mathcal{R}(A)$ *and* $\mathcal{N}(D) \subseteq \mathcal{N}(B)$.

(iv) $AA^\dagger B = B$ *and* $BD^\dagger D = B$.

(b) *If (i) is satisfied, then*

$$GG^\dagger = \operatorname{diag}\left(AA^\dagger, DD^\dagger\right) \qquad and \qquad G^\dagger G = \operatorname{diag}\left(A^\dagger A, D^\dagger D\right).$$

Corollary A.12 can be proved analogous to Corollary A.11.

An alternate approach to Corollary A.11 and Corollary A.12 can be found in Campbell/Meyer [4, Theorem 3.4.1].

Lemma A.13. *Let* $A \in \mathbb{C}^{p\times q}$, $C \in \mathbb{C}^{r\times q}$, *and* $D \in \mathbb{C}^{r\times s}$. *Let* G *be given by* (A.22).

(a) *The following statements are equivalent:*

(i) $\mathcal{R}(C) \subseteq \mathcal{R}(D)$.

(ii) $\mathcal{R}(G) = \mathcal{R}(\operatorname{diag}(A, D))$.

(b) *Let (i) be satisfied. Then* $\operatorname{rank} G = \operatorname{rank} A + \operatorname{rank} D$.

Proof. (a) "(i)$\Rightarrow$(ii)": Part (b) of Lemma A.3 yields

$$DD^\dagger C = C. \qquad (A.24)$$

Let $x \in \mathcal{R}(G)$. Then there are $y_1 \in \mathbb{C}^q$ and $y_2 \in \mathbb{C}^s$ such that $x = G(y_1^*, y_2^*)^*$. In view of (A.22) and (A.24), then

$$x = \begin{pmatrix} Ay_1 \\ Cy_1 + Dy_2 \end{pmatrix} = \begin{pmatrix} Ay_1 \\ DD^\dagger Cy_1 + Dy_2 \end{pmatrix} = \operatorname{diag}(A, D) \begin{pmatrix} y_1 \\ D^\dagger Cy_1 + y_2 \end{pmatrix}$$

follows. In particular, $x \in \mathcal{R}\left(\mathrm{diag}(A, D)\right)$. Hence,

$$\mathcal{R}\left(G\right) \subseteq \mathcal{R}\left(\mathrm{diag}(A, D)\right) \tag{A.25}$$

is proved. In order to check that

$$\mathcal{R}\left(\mathrm{diag}(A, D)\right) \subseteq \mathcal{R}\left(G\right), \tag{A.26}$$

we now consider an arbitrary $x \in \mathcal{R}\left(\mathrm{diag}(A, D)\right)$. Then there are $y_1 \in \mathbb{C}^q$ and $y_2 \in \mathbb{C}^s$ such that $x = \mathrm{diag}(A, D) \cdot (y_1^*, y_2^*)^*$. Using (A.24), this implies

$$x = \begin{pmatrix} Ay_1 \\ Dy_2 \end{pmatrix} = \begin{pmatrix} Ay_1 \\ Cy_1 + D(y_2 - D^\dagger Cy_1) \end{pmatrix} = \begin{pmatrix} A & 0_{p \times s} \\ C & D \end{pmatrix} \begin{pmatrix} y_1 \\ y_2 - D^\dagger Cy_1 \end{pmatrix}.$$

Hence, $x \in \mathcal{R}\left(G\right)$. Thus, (A.26) is proved. From (A.25) and (A.26) we get (ii).

"(ii)$\Rightarrow$(i)": Let $y \in \mathcal{R}(C)$. Then there is some $x \in \mathbb{C}^q$ such that $y = Cx$. Setting $z := Ax$, we obtain

$$\begin{pmatrix} z \\ y \end{pmatrix} = \begin{pmatrix} Ax \\ Cx \end{pmatrix} = \begin{pmatrix} A & 0_{p \times s} \\ C & D \end{pmatrix} \begin{pmatrix} x \\ 0_{s \times 1} \end{pmatrix} = G \begin{pmatrix} x \\ 0_{s \times 1} \end{pmatrix} \in \mathcal{R}\left(G\right).$$

Consequently, (ii) implies $\left(\begin{smallmatrix} z \\ y \end{smallmatrix}\right) \in \mathcal{R}(\mathrm{diag}(A, D))$ and therefore $y \in \mathcal{R}(D)$. Hence, $\mathcal{R}(C) \subseteq \mathcal{R}(D)$.

(b) This follows immediately from (a).   □

The following result can be proved in the same way as Lemma A.13.

Lemma A.14. *Let* $A \in \mathbb{C}^{p \times q}$, $B \in \mathbb{C}^{p \times s}$ *and* $D \in \mathbb{C}^{r \times s}$. *Let* G *be given by* (A.23).
  (a) *The following statements are all equivalent:*
    (i) $\mathcal{R}\left(B\right) \subseteq \mathcal{R}\left(A\right)$.
    (ii) $\mathcal{R}\left(G\right) = \mathcal{R}\left(\mathrm{diag}(A, D)\right)$.
  (b) *If condition* (i) *is met, then* $\mathrm{rank}\, G = \mathrm{rank}\, A + \mathrm{rank}\, D$.

We now characterize lower block triangular range-Hermitian matrices.

Proposition A.15. *Let* $A \in \mathbb{C}^{p \times p}$, $C \in \mathbb{C}^{q \times p}$, $D \in \mathbb{C}^{q \times q}$ *and* $G := \begin{pmatrix} A & 0_{p \times q} \\ C & D \end{pmatrix}$.
  (a) *The following statements are equivalent:*
    (i) $G \in \mathbb{C}_{\mathrm{EP}}^{(p+q) \times (p+q)}$.
    (ii) $A \in \mathbb{C}_{\mathrm{EP}}^{p \times p}$, $D \in \mathbb{C}_{\mathrm{EP}}^{q \times q}$, $\mathcal{R}\left(C\right) \subseteq \mathcal{R}\left(D\right)$ *and* $\mathcal{N}\left(A\right) \subseteq \mathcal{N}\left(C\right)$.
    (iii) $A \in \mathbb{C}_{\mathrm{EP}}^{p \times p}$, $D \in \mathbb{C}_{\mathrm{EP}}^{q \times q}$, $DD^\dagger C = C$ *and* $CA^\dagger A = C$.
  (b) *If* (ii) *is satisfied, then* $GG^\dagger = \mathrm{diag}\left(AA^\dagger, DD^\dagger\right)$, $G^\dagger G = \mathrm{diag}\left(A^\dagger A, D^\dagger D\right)$ *and*

$$G^\dagger = \begin{pmatrix} A^\dagger & 0_{p \times q} \\ -D^\dagger CA^\dagger & D^\dagger \end{pmatrix}.$$

Proof. (a) 1.) "(i) $\Rightarrow$ (ii)". In view of (i) the application of Proposition A.4 yields the existence of matrices $X, Y \in \mathbb{C}^{(p+q) \times (p+q)}$ such that

$$GX = G^* \tag{A.27}$$

and

$$G^*Y = G \tag{A.28}$$

are satisfied. We use the block partitions

$$X = \begin{pmatrix} P & Q \\ K & S \end{pmatrix} \qquad \text{and} \qquad Y = \begin{pmatrix} T & U \\ V & W \end{pmatrix}, \tag{A.29}$$

where $P \in \mathbb{C}^{p \times p}$ and $T \in \mathbb{C}^{p \times p}$. From (A.27)–(A.29) we obtain

$$\begin{pmatrix} A^* & C^* \\ 0_{q \times p} & D^* \end{pmatrix} = \begin{pmatrix} AP & AQ \\ CP + DR & CQ + DS \end{pmatrix} \tag{A.30}$$

and

$$\begin{pmatrix} A & 0_{p \times q} \\ C & D \end{pmatrix} = \begin{pmatrix} A^*T + C^*V & A^*U + C^*W \\ D^*V & D^*W \end{pmatrix}, \tag{A.31}$$

Comparing corresponding diagonal block on both sides of (A.30) and (A.31) we get

$$A^* = AP \qquad \text{and} \qquad D = D^*W, \tag{A.32}$$

respectively. In view of Proposition A.4 we see from (A.32) that

$$A \in \mathbb{C}_{\mathrm{EP}}^{p \times p} \qquad \text{and} \qquad D \in \mathbb{C}_{\mathrm{EP}}^{q \times q}, \tag{A.33}$$

respectively. Comparing the left lower $q \times p$-block on both sides of (A.31) we get $C = D^*V$. Thus, $\mathcal{R}(C) \subseteq \mathcal{R}(D^*)$. Otherwise, from (A.33), we have $\mathcal{R}(C) = \mathcal{R}(D^*)$. Hence,

$$\mathcal{R}(C) \subseteq \mathcal{R}(D). \tag{A.34}$$

Comparing the right upper $p \times q$-block on boths sides of (A.30), we obtain $C^* = AQ$. Thus, $\mathcal{R}(C^*) \subseteq \mathcal{R}(D)$. Otherwise, from (A.33), we get $\mathcal{R}(A) = \mathcal{R}(A^*)$. Hence, $\mathcal{R}(C^*) = \mathcal{R}(A^*)$. Forming orthogonal complements yields $\mathcal{N}(A) \subseteq \mathcal{N}(C)$. In view of this, (A.33) and (A.34), we see that (ii) holds.

   2.) "(ii) $\Rightarrow$ (i)". In view of (ii), we have $\mathcal{R}(C) \subseteq \mathcal{R}(D)$ and $\mathcal{N}(A) \subseteq \mathcal{N}(C)$. Thus, Corollary A.11 yields

$$GG^\dagger = \mathrm{diag}\left(AA^\dagger, DD^\dagger\right) \qquad \text{and} \qquad G^\dagger G = \mathrm{diag}\left(A^\dagger A, D^\dagger D\right). \tag{A.35}$$

In view of (ii) we have $A \in \mathbb{C}_{\mathrm{EP}}^{p \times p}$ and $D \in \mathbb{C}_{\mathrm{EP}}^{q \times q}$. Thus, Proposition A.4 yields $AA^\dagger = A^\dagger A$ and $DD^\dagger = D^\dagger D$. Combining this with (A.35) gives us

$$GG^\dagger = G^\dagger G. \tag{A.36}$$

Thus Proposition A.4 yields (i).

   3.) "(ii) $\Leftrightarrow$ (iii)". This follows from parts (a) and (b) of Lemma A.3.

   (b) The first two identites follow from (A.35) and (A.36). The formula for $G^\dagger$ is a consequence of Corollary A.11. $\square$

We now consider Proposition A.10 in the special case of a non-negative Hermitian block matrix, E.

Proposition A.16. *Let $E \in \mathbb{C}_{\geq}^{(p+r)\times(p+r)}$ and (A.19) be the block partition of E with $p \times p$-block A. If $F \in \mathbb{C}^{(p+r)\times(p+r)}$ is given by (A.20), then*

$$E^{\dagger} = F \qquad \text{if and only if} \qquad \mathcal{N}(L) \subseteq \mathcal{N}(D).$$

Proof. If $E^{\dagger} = F$, then $\mathcal{N}(L) \subseteq \mathcal{N}(D)$ follows immediately from Proposition A.10.

Conversely, suppose $\mathcal{N}(L) \subseteq \mathcal{N}(D)$. We apply a well-known block characterization of non-negative Hermitian block matrices to E (see, e.g., [7, Lemma 1.1.9]). Then we have $A \in \mathbb{C}_{\geq}^{p \times p}$, $D \in \mathbb{C}_{\geq}^{r \times r}$, $\mathcal{R}(B) \subseteq \mathcal{R}(A)$, $\mathcal{R}(C) \subseteq \mathcal{R}(D)$, $C = B^*$ and $L \in \mathbb{C}_{\geq}^{r \times r}$. Consequently, we see that $D - L = B^* A^{\dagger} B \in \mathbb{C}_{\geq}^{r \times r}$, and hence, $\mathcal{N}(D) \subseteq \mathcal{N}(L)$ and $\mathcal{R}(L) \subseteq \mathcal{R}(D)$. Thus, we have $\mathcal{N}(D) = \mathcal{N}(L)$, and we get

$$\mathcal{N}(L) = \mathcal{N}(D) = \mathcal{N}(D^*) = \mathcal{R}(D)^{\perp} \subseteq \mathcal{R}(C)^{\perp} = \mathcal{N}(C^*) = \mathcal{N}(B),$$
$$\mathcal{N}(A) = \mathcal{N}(A^*) = \mathcal{R}(A)^{\perp} \subseteq \mathcal{R}(B)^{\perp} = \mathcal{R}(C^*)^{\perp} = \mathcal{N}(C),$$
$$\mathcal{R}(D) = \mathcal{R}(D^*) = \mathcal{N}(D)^{\perp} = \mathcal{N}(L)^{\perp} = \mathcal{N}(L^*)^{\perp} = \mathcal{R}(L),$$

and hence $\mathcal{R}(C) \subseteq \mathcal{R}(D) = \mathcal{R}(L)$. From Proposition A.10, it then follows that $E^{\dagger} = F$. $\qquad\square$

It should be mentioned that in Groß [9] the block structure of the Moore-Penrose inverse of a partitioned non-negative Hermitian matrix is discussed. The special situation of Proposition A.16 can also be derived from [9, Theorem 1].

References

[1] A. Baricz, J. Vesti, and M. Vuorinen, *On Kaluza's sign criterion for reciprocal power series*, Arxiv Math. 1010, 5337v1 (26. October, 2010).

[2] A. Ben-Israel and T.N.E. Greville, *Generalized inverses – Theory and applications*, 2nd ed., CMS Books in Mathematics/Ouvrages de Mathématiques de la SMC, 15, Springer-Verlag, New York, 2003. MR1987382 (2004b:15008).

[3] E.H.M. Brietzke, *Monotonicity of coefficients of reciprocal power series*, Real Analysis Exchange **27** (2001/02), 41–48. MR1887680

[4] S.L. Campbell and C.D. Meyer Jr., *Generalized inverses of linear transformations*, Dover Publications Inc., New York, 1991. Corrected reprint of the 1979 original. MR1105324 (92a:15003).

[5] S. Cheng and Y. Tian, *Two sets of new characterizations for normal and EP matrices*, Linear Algebra Appl. 375 (2003), 181–195. MR2013464 (2004m:15006).

[6] J.B. Conway, *Functions of One Complex Variable*, Graduate Texts in Mathematics, vol. 11, Springer-Verlag, New York – Berlin – Heidelberg – Tokyo, 1978. MR0503901 (80c:30003).

[7] V.K. Dubovoj, B. Fritzsche, and B. Kirstein, *Matricial version of the classical Schur problem*, Teubner-Texte zur Mathematik [Teubner Texts in Mathematics], vol. 129, B.G. Teubner Verlagsgesellschaft mbH, Stuttgart, 1992. With German, French and Russian summaries. MR1152328 (93e:47021).

[8] B. Fritzsche, B. Kirstein, and U. Raabe, *On some interrelations between J-Potapov functions and J-Potapov sequences*, in: Characteristic Functions, Scattering Functions and Transfer Functions – The Moshe Livsic Memorial Volume (D. Alpay and V. Vinnikov, eds.), OT Series, vol. 197, Birkhäuser, Basel – Boston – Berlin, 2010, pp. 251–279. MR2647541.

[9] J. Groß, *The Moore-Penrose inverse of a partitioned nonnegative matrix*, Linear Algebra Appl. 321 (2000), 113–121. MR1799987 (2001j:15005).

[10] B.G. Hansen and F.W. Steutel, *On moment sequences and infinitely divisible sequences*, J. Math. Anal. Appl. **136** (1988), 304–313. MR0972601 (90a:60155)

[11] R.A. Horn, *On moment sequences and renewal sequences*, J. Math. Anal. Appl. **31** (1970), 130–135. MR0271658 (42 #6541)

[12] W. Jurkat, *Questions of signs in power series*, Proc. Amer. Math. Soc. **5** (1954), 964–970. MR0064890 (16,351f)

[13] Th. Kaluza, *Über die Koeffizienten reziproker Potenzreihen*, Math. Z. **28** (1928), 161–170 (German). MR1544949

[14] J. Lamperti, *On the coefficients of reciprocal power series*, Amer. Math. Monthly **65** (1958), 90–94. MR0097657 (20 #4125)

[15] C.D. Meyer Jr., *Some remarks on EP_r matrices, and the generalized inverses*, Linear Algebra Appl. 3 (1970), 275–278. MR0266935 (42 #1837).

[16] M.H. Pearl, *On generalized inverses of matrices*, Proc. Cambridge Phil. Soc. 51 (1950), 406–418. MR0197485 (33 #5650).

[17] C.R. Rao and S.K. Mitra, *Generalized inverse of matrices and its applications*, John Wiley & Sons, Inc., New York – London – Sydney, 1971. MR0338013 (49 #2780)

[18] H. Schwerdtfeger, *Introduction to Linear Algebra and the Theory of Matrices*, Noordhoff, Groningen, 1950. MR0038923 (12,470g).

[19] Y. Tian and H. Wang, *Characterizations of EP matrices and weighted EP matrices*, Linear Algebra Appl. 434 (2011), 1295–1318.

Bernd Fritzsche, Bernd Kirstein, Conrad Mädler and Tilo Schwarz
Mathematisches Institut
Universität Leipzig
Augustusplatz 10/11
D-04109 Leipzig, Germany
e-mail: fritzsche@math.uni-leipzig.de
        kirstein@math.uni-leipzig.de
        maedler@math.uni-leipzig.de
        tilo.schwarz@may-schwarz.de

Operator Theory:
Advances and Applications, Vol. 226, 57–115
© 2012 Springer Basel

On Reciprocal Sequences of Matricial Carathéodory Sequences and Associated Matrix Functions

Bernd Fritzsche, Bernd Kirstein, Andreas Lasarow and Armin Rahn

Abstract. This paper provides a detailed discussion of reciprocal sequences of finite and infinite matricial Carathéodory sequences, including an examination of the relationship a reciprocal sequence has to its original sequence. The properties of such reciprocal sequences are described. Of particular interest is the fact that the reciprocal sequence of a Carathéodory sequence is again a Carathéodory sequence. Later, these results are applied to matrix-valued functions. The natural focus is, in particular, on the matricial Carathéodory functions associated with a matricial Carathéodory sequence and its reciprocal sequence. These matrix functions are then shown to be Moore-Penrose inverses of each other. The implications of these results for non-negative Hermitian matrix measures on the unit circle are also discussed.

Mathematics Subject Classification (2010). Primary 47A56; Secondary 44A60.

Keywords. Reciprocal sequences, Moore-Penrose inverse, matrix-valued Carathéodory functions.

0. Introduction

Several classes of sequences of complex matrices were discussed in [22]. There, these classes were studied in the context of a generalization of the inversion of power series with matrix coefficients. In particular, each finite or infinite sequence in the set $\mathbb{C}^{p \times q}$ of all complex $p \times q$ matrices was assigned a reciprocal sequence. This was done in such a way (see Definition 1.1) that, for the special case $p = q$ and $\det s_0 \neq 0$, the assigned reciprocal sequence was none other than the usual reciprocal sequence of conventional power series inversion theory.

Our first main objective for this paper is to study the reciprocal sequences of finite and infinite $q \times q$ Carathéodory sequences. We will see that the reciprocal sequence of a Carathéodory sequence is itself a Carathéodory sequence. Furthermore, we will see that the properties of such reciprocal sequences can be fully expressed in terms of the original Carathéodory sequence. Later, starting with an infinite $q \times q$ Carathéodory sequence and its reciprocal $q \times q$ Carathéodory sequence, we consider their respective $q \times q$ Carathéodory functions.

Our second main goal is to show that these functions are Moore-Penrose inverses of one another on the complex open unit disk $\mathbb{D}$. This will, ultimately, yield a new proof for the holomorphicity of the Moore-Penrose inverse of a $q \times q$ Carathéodory function in $\mathbb{D}$. An earlier proof of this result in [5, Theorem 4.5] was based on an approach using the Cayley transform.

This paper is organized as follows. Section 1 offers a brief introduction to reciprocal sequences of finite and infinite sequences of matrices in $\mathbb{C}^{p \times q}$ (see Definition 1.1). We also examine the class of *first term dominant* matrix sequences (see Definition 1.4), which is particularly interesting and useful in the context of reciprocal sequences. These observations are continued from [22]. In this section we also look at *EP sequences*, earlier considered in [22, Section 7]. We build on and expand these results (see Proposition 1.11 and Theorem 1.12).

In Section 2 we present a summary of preparatory results on finite and infinite Toeplitz non-negative definite sequences in $\mathbb{C}^{q \times q}$. The emphasis, here, is on the inner structure of such sequences, which is described using matrix balls (see Proposition 2.9). We also introduce a number of important subclasses of Toeplitz non-negative definite sequences in $\mathbb{C}^{q \times q}$ (see Definition 2.12 and Definition 2.24). Relevant results are drawn from [11] and [8, Section 3.5]. Proposition 2.7 shows that Toeplitz non-negative definite sequences in $\mathbb{C}^{q \times q}$ are first term dominant. This proposition connects Section 2 to Section 1.

Section 3 centers on a discussion of matricial Carathéodory sequences, in light of their relationship to Toeplitz non-negative definite sequences. This relationship shows that a certain duality exists between the two types of sequences and makes it possible to define particular subclasses of matricial Carathéodory sequences using Section 2 (see Definition 3.15 and Definition 3.22).

In Section 4, we focus on reciprocal sequences of matricial $q \times q$ Carathéodory sequences. We show that $q \times q$ Carathéodory sequences are EP sequences and recognize that they are first term dominant (see Proposition 4.1). We then use this fact to apply the EP sequence results of Section 1 to $q \times q$ Carathéodory sequences. This leads directly to our first main result, namely, that the reciprocal sequence of a $q \times q$ Carathéodory sequence is itself a $q \times q$ Carathéodory sequence (see Theorem 4.4). This, in turn, brings us to the question of how a $q \times q$ Carathéodory sequence is related to its reciprocal sequence and to what extent one can be used to describe the other. In the second part of Section 4, we deal mainly with these questions. We will see that for the subclasses introduced in Section 3, both sequences either belong to the relevant subclass or neither of them do (see Theorem 4.8 and Theorem 4.10). Furthermore, we describe the matrix ball structure of reciprocal sequences to Carathéodory sequences in terms of the original Carathéodory sequence (see Lemma 4.9).

The focus of Section 5 is on a particular reciprocal sequence operation for matricial Toeplitz non-negative definite sequences. Because of how closely matricial Toeplitz non-negative definite sequences and matricial Carathéodory sequences are related, we are able to use Section 4 in constructing a new Toeplitz non-negative definite sequence to associate with a given initial Toeplitz non-negative definite sequence. Having established a procedure for constructing new sequences of this type, we proceed with an extensive examination of the relationship between an initial matricial Toeplitz non-negative definite sequence and the Toeplitz non-negative definite sequence generated from it.

Section 6 includes a short introduction to the theory for the matricial Carathéodory class $\mathcal{C}_q(\mathbb{D})$, where $\mathbb{D} := \{ w \in \mathbb{C} : |w| < 1 \}$ is the open unit disk in the complex plane $\mathbb{C}$. The $\mathcal{C}_q(\mathbb{D})$ class is made up of every $q \times q$ matrix function Ω that is holomorphic in $\mathbb{D}$ with non-negative real part at all $w \in \mathbb{D}$. Via Taylor coefficients, we see that there is a well-known one-to-one correspondence between the class $\mathcal{C}_q(\mathbb{D})$ and the set of all infinite $q \times q$ Carathéodory sequences (see Propositions 6.10 and 6.12). This suggests introducing a number of special subclasses of $\mathcal{C}_q(\mathbb{D})$, which we do with the help of Section 3.

Section 7 is dedicated to an analysis of the Moore-Penrose inverse Ω^+ of a function $\Omega \in \mathcal{C}_q(\mathbb{D})$. We show that $\Omega^+ \in \mathcal{C}_q(\mathbb{D})$ and also that the Taylor coefficient sequence for Ω^+ is the reciprocal sequence to Ω's Taylor coefficient sequence (see Theorem 7.3). We furthermore show that for the subclasses of $\mathcal{C}_q(\mathbb{D})$ defined in Section 6, both of the $\mathcal{C}_q(\mathbb{D})$ functions Ω and Ω^+ either belong to the relevant subclass or neither of them do (see Theorem 7.9 and Proposition 7.10).

We are also interested in applying these results to the theory of non-negative Hermitian $q \times q$ measures on the unit circle $\mathbb{T} := \{ z \in \mathbb{C} : |z| = 1 \}$. For this reason, we include a brief review of results in matricial harmonic analysis on $\mathbb{T}$. Because of the matricial version of a classical result by G. Herglotz (see Proposition 8.1), we know that (via Fourier coefficients) there is a one-to-one correspondence between the set of all infinite Toeplitz non-negative definite sequences in $\mathbb{C}^{q \times q}$ and the set of all non-negative Hermitian $q \times q$ measures on the Borel σ-algebra for the unit circle. This one-to-one correspondence makes it possible to generate a reciprocal non-negative Hermitian measure using an operation based on Section 5's method of generating a Toeplitz non-negative definite sequence "by reciprocation". This idea then informs how Section 8 develops. The aforementioned construction was already considered in [8, Section 3.6] for the special case of a non-negative Hermitian measure F with non-singular matrix $F(\mathbb{T})$. There, the focus was on the relationship between associated pairs of orthogonal systems of matrix polynomials.

Our closing Section 9 concentrates on a class of matrix functions that are holomorphic in the upper half-plane $\Pi_+ := \{ z \in \mathbb{C} : \operatorname{Im} z \in (0, +\infty) \}$. This matrix function class is particularly interesting in the context of the matrix version of the Hamburger Moment Problem. The class $\mathcal{R}_q(\Pi_+)$ discussed in Section 9 is comprised of all matrix functions $G : \Pi_+ \longrightarrow \mathbb{C}^{q \times q}$ that are holomorphic in Π_+ and having non-negative Hermitian imaginary part at all $z \in \Pi_+$. There is a one-to-one correspondence between $\mathcal{R}_q (\Pi_+)$ and $\mathcal{C}_q (\mathbb{D})$, which makes it possible to show that the Moore-Penrose inverse G^+ of a function $G \in \mathcal{R}_q (\Pi_+)$ is holomorphic and also that null space and range remain constant (see Theorem 9.4).

To conclude this introduction, we quickly review some notation. Throughout this paper, let p and q be positive integers. We use $\mathbb{N}$ and $\mathbb{N}_0$ to denote the sets of positive and non-negative integers, respectively. The set of all integers is $\mathbb{Z}$. For any $\alpha \in \mathbb{Z}$ and $\varkappa \in \mathbb{Z} \cup \{ +\infty \}$, we let $\mathbb{Z}_{\alpha, \varkappa} := \{ \ell \in \mathbb{Z} : \alpha \leq \ell \leq \varkappa \}$.

The set of all Hermitian matrices in $\mathbb{C}^{q \times q}$ will be denoted by $\mathbb{C}_{\mathrm{H}}^{q \times q}$, while $\mathbb{C}_{\geq}^{q \times q}$ and $\mathbb{C}_{>}^{q \times q}$ will stand for the sets of all non-negative and positive Hermitian matrices, respectively. The set of all contractive $p \times q$ matrices is defined as $\mathbb{K}_{p \times q} := \{ A \in \mathbb{C}^{p \times q} : \|A\|_{\mathrm{S}} \leq 1 \}$ and the set of all strictly contractive $p \times q$ matrices as $\mathbb{D}_{p \times q} := \{ A \in \mathbb{C}^{p \times q} : \|A\|_{\mathrm{S}} < 1 \}$, where $\|\cdot\|_{\mathrm{S}}$ is the operator norm in $\mathbb{C}^{p \times q}$. Suppose that $A \in \mathbb{C}^{p \times q}$, then $\mathcal{N}(A)$ is A's null space and $\mathcal{R}(A)$ is its range. The zero matrix in $\mathbb{C}^{p \times q}$ and the identity matrix in $\mathbb{C}^{q \times q}$ are denoted by $0_{p \times q}$ and I_q, respectively. Suppose $A \in \mathbb{C}^{p \times q}$, then its adjoint will be denoted by A^* and its Moore-Penrose inverse by A^+. For each $A \in \mathbb{C}^{q \times q}$, the matrices $\operatorname{Re} A := \frac{1}{2} (A + A^*)$ and $\operatorname{Im} A := \frac{1}{2i} (A - A^*)$ are, respectively, called the real and imaginary parts of A. The determinant of A will be denoted by $\det A$. For $n \in \mathbb{N}$ and any sequence $(A_j)_{j=1}^n$ of complex $p \times q$ matrices, we also define

$$\operatorname{row} (A_j)_{j=1}^n := (A_1, \quad A_2, \quad \cdots, \quad A_n) \quad \text{and} \quad \operatorname{col} (A_j)_{j=1}^n := \left[\operatorname{row} (A_j^{\mathrm{T}})_{j=1}^n \right]^{\mathrm{T}}.$$

For each $q \in \mathbb{N}$, we let

$$\mathcal{R}_{q, \geq} := \left\{ A \in \mathbb{C}^{q \times q} : \operatorname{Re} A \in \mathbb{C}_{\geq}^{q \times q} \right\} \quad \text{and} \quad \mathcal{I}_{q, \geq} := \left\{ A \in \mathbb{C}^{q \times q} : \operatorname{Im} A \in \mathbb{C}_{\geq}^{q \times q} \right\}.$$

We also define

$$\mathcal{R}_{q, >} := \left\{ A \in \mathbb{C}^{q \times q} : \operatorname{Re} A \in \mathbb{C}_{>}^{q \times q} \right\} \quad \text{and} \quad \mathcal{I}_{q, >} := \left\{ A \in \mathbb{C}^{q \times q} : \operatorname{Im} A \in \mathbb{C}_{>}^{q \times q} \right\}.$$

Clearly, the set $\mathbb{C}_{\geq}^{q \times q}$ of all non-negative Hermitian complex $q \times q$ matrices is a subset of $\mathcal{R}_{q, \geq}$ and the set $\mathbb{C}_{>}^{q \times q}$ of all positive Hermitian complex $q \times q$ matrices is a subset of $\mathcal{R}_{q, >}$. A complex $q \times q$ matrix A is called range-Hermitian (or an EP matrix) if $\mathcal{R}(A) = \mathcal{R}(A^*)$. The set of all range-Hermitian matrices in $\mathbb{C}^{q \times q}$ will be denoted by $\mathbb{C}_{\mathrm{EP}}^{q \times q}$.

1. Reciprocal sequences

The approach to constructing a special transformation for sequences of complex matrices considered here was introduced in [22]. The main theme of [22] was invertibility for sequences of complex matrices. In this section, we provide a quick review of relevant results from [22], because they will later be important.

Definition 1.1. Let $\varkappa \in \mathbb{N}_0 \cup \{+\infty\}$ and let $(s_j)_{j=0}^{\varkappa}$ be a sequence of complex $p \times q$ matrices. The sequence $(s_j^{\sharp})_{j=0}^{\varkappa}$ of $q \times p$ matrices defined by

$$
s_k^{\sharp} := \begin{cases} s_0^+, & \text{if } k = 0, \\ -s_0^+ \displaystyle\sum_{l=0}^{k-1} s_{k-l} s_l^{\sharp}, & \text{if } k \in \mathbb{Z}_{1, \varkappa}, \end{cases}
$$

is called the reciprocal sequence corresponding to $(s_j)_{j=0}^{\varkappa}$.

Remark 1.2. Let $\varkappa \in \mathbb{N}_0 \cup \{+\infty\}$ and let $(s_j)_{j=0}^{\varkappa}$ be a sequence of complex $p \times q$-matrices with reciprocal sequence $(s_j^{\sharp})_{j=0}^{\varkappa}$. Given an arbitrary $m \in \mathbb{Z}_{0, \varkappa}$, the reciprocal sequence corresponding to $(s_j)_{j=0}^{m}$ is then $(s_j^{\sharp})_{j=0}^{m}$.

Example 1.3. Let $\varkappa \in \mathbb{N}_0 \cup \{+\infty\}$. Suppose $(s_j)_{j=0}^{\varkappa}$ is a sequence in $\mathbb{C}^{p \times q}$ and that $u \in \mathbb{C}$. For $j \in \mathbb{Z}_{0, \varkappa}$, suppose that $s_{j, u} := u^j s_j$. By induction, we then obtain $(s_{j, u})^{\sharp} = u^j s_j^{\sharp}$, for $j \in \mathbb{Z}_{0, \varkappa}$.

We will see that reciprocal sequences are especially interesting when considered for the following class of matricial sequences.

Definition 1.4. Let $\varkappa \in \mathbb{N}_0 \cup \{+\infty\}$. A sequence $(s_j)_{j=0}^{\varkappa}$ in $\mathbb{C}^{p \times q}$ is called first term dominant if

$$
\mathcal{N}(s_0) \subseteq \bigcap_{j=0}^{\varkappa} \mathcal{N}(s_j) \qquad \text{and} \qquad \bigcup_{j=0}^{\varkappa} \mathcal{R}(s_j) \subseteq \mathcal{R}(s_0).
$$

The set of all such sequences $(s_j)_{j=0}^{\varkappa}$ in $\mathbb{C}^{p \times q}$ will be denoted by $\mathcal{D}_{p \times q, \varkappa}$.

Example 1.5. Let $\varkappa \in \mathbb{N}_0 \cup \{+\infty\}$. All sequences $(s_j)_{j=0}^{\varkappa}$ in $\mathbb{C}^{q \times q}$ with $\det s_0 \neq 0$ belong to $\mathcal{D}_{q \times q, \varkappa}$.

Remark 1.6. Let $\varkappa \in \mathbb{N}_0 \cup \{+\infty\}$ and $(s_j)_{j=0}^{\varkappa}$ be a sequence in $\mathbb{C}^{p \times q}$. Then $(s_j)_{j=0}^{\varkappa} \in \mathcal{D}_{p \times q, \varkappa}$ if and only if $(s_j)_{j=0}^{m} \in \mathcal{D}_{p \times q, m}$, for all $m \in \mathbb{Z}_{0, \varkappa}$.

Remark 1.7. Let $(s_j)_{j=0}^{\varkappa}$ be a sequence in $\mathbb{C}^{q \times q}$ and let $(\alpha_j)_{j=0}^{\varkappa}$ be a sequence in $\mathbb{C} \setminus \{0\}$. Then $(s_j)_{j=0}^{\varkappa} \in \mathcal{D}_{q \times q, \varkappa}$ if and only if $(\alpha_j s_j)_{j=0}^{\varkappa} \in \mathcal{D}_{q \times q, \varkappa}$.

Given a $\varkappa \in \mathbb{N}_0 \cup \{+\infty\}$ and a sequence $(s_j)_{j=0}^{\varkappa}$ in $\mathbb{C}^{p \times q}$, we define, for each $m \in \mathbb{Z}_{0, \varkappa}$, the lower triangular block Toeplitz matrices $S_m^{(s)}$ as

$$
S_m^{(s)} := \begin{pmatrix} s_0 & 0 & 0 & \cdots & 0 \\ s_1 & s_0 & 0 & \cdots & 0 \\ s_2 & s_1 & s_0 & \ddots & \vdots \\ \vdots & \vdots & \vdots & \ddots & 0 \\ s_m & s_{m-1} & s_{m-2} & \cdots & s_0 \end{pmatrix} \tag{1.1}
$$

and, whenever it is clear which sequence is meant, we will simply write S_m instead of $S_m^{(s)}$. For each $m \in \mathbb{Z}_{0, \varkappa}$, using $\left(s_j^{\sharp} \right)_{j=0}^{\varkappa}$ instead of $\left(s_j \right)_{j=0}^{\varkappa}$, we also set

$$S_m^{\sharp} := S_m^{\left(s^{\sharp} \right)}. \tag{1.2}$$

The following result (see [22, Proposition 4.20]) is particularly useful.

Proposition 1.8. *If* $\varkappa \in \mathbb{N}_0 \cup \{+\infty\}$ *and* $\left(s_j \right)_{j=0}^{\varkappa} \in \mathcal{D}_{p \times q, \varkappa}$, *then* $S_m^+ = S_m^{\sharp}$ *for each* $m \in \mathbb{Z}_{0, \varkappa}$.

If $r, s \in \mathbb{N}$, $A = (a_{jk})_{\substack{j=1,\ldots,p \\ k=1,\ldots,q}} \in \mathbb{C}^{p \times q}$ and $B \in \mathbb{C}^{r \times s}$, then the Kronecker product $A \otimes B$ of the matrices A and B is given by $A \otimes B := (a_{jk}B)_{\substack{j=1,\ldots,p \\ k=1,\ldots,q}}$.

It should be noted that, if $s_0 \in \mathbb{C}^{p \times q}$ and if $m \in \mathbb{N}$, then the complex $(m+1)p \times (m+1)q$ matrix $\operatorname{diag}(s_0, s_0, \ldots, s_0)$ can be expressed as $I_{m+1} \otimes s_0$.

Lemma 1.9. *If* $\varkappa \in \mathbb{N}_0 \cup \{+\infty\}$ *and* $\left(s_j \right)_{j=0}^{\varkappa} \in \mathcal{D}_{p \times q, \varkappa}$, *then, for each* $m \in \mathbb{Z}_{0, \varkappa}$,

$$S_m S_m^+ = I_{m+1} \otimes (s_0 s_0^+) \qquad \text{and} \qquad S_m^+ S_m = I_{m+1} \otimes (s_0^+ s_0).$$

If, furthermore, $p = q$ *and if* $s_0 \in \mathbb{C}_{\mathrm{EP}}^{q \times q}$, *then* $S_m S_m^+ = S_m^+ S_m$ *for every* $m \in \mathbb{Z}_{0, \varkappa}$.

Proof. Combine part (a) of [22, Theorem 4.21] with [22, Proposition 3.6]. $\qquad\square$

We now consider a special class of sequences of complex square matrices, namely the set of EP sequences (see [22, Section 7]). If $n \in \mathbb{N}_0$ and $\left(s_j \right)_{j=0}^{n}$ is a sequence in $\mathbb{C}^{q \times q}$, then $\left(s_j \right)_{j=0}^{n}$ is called an EP sequence when $S_n^{(s)} \in \mathbb{C}_{\mathrm{EP}}^{(n+1)q \times (n+1)q}$, where $S_n^{(s)}$ is defined by (1.1). If $n \in \mathbb{N}$ and $\left(s_j \right)_{j=0}^{n}$ is an EP sequence in $\mathbb{C}^{q \times q}$, then, for any $k \in \mathbb{Z}_{0, n-1}$, it follows that $\left(s_j \right)_{j=0}^{k}$ is also an EP sequence (see [22, Remark 7.2]). A sequence $\left(s_j \right)_{j=0}^{\infty}$ in $\mathbb{C}^{q \times q}$ is thus called an EP sequence when $\left(s_j \right)_{j=0}^{n}$ is an EP sequence for all $n \in \mathbb{N}_0$. If $\varkappa \in \mathbb{N}_0 \cup \{+\infty\}$, then $\mathcal{F}_{q, \varkappa}^{\mathrm{EP}}$ will stand for the set of all EP sequences $\left(s_j \right)_{j=0}^{\varkappa}$ in $\mathbb{C}^{q \times q}$. The following proposition (see [22, Proposition 7.4]) shows that EP sequences make up a subclass of first term dominant sequences (introduced in Definition 1.4).

Proposition 1.10. *Let* $\varkappa \in \mathbb{N}_0 \cup \{+\infty\}$ *and* $\left(s_j \right)_{j=0}^{\varkappa}$ *be a sequence in* $\mathbb{C}^{q \times q}$.

(a) *The following two conditions are equivalent to one another:*
   (i) $\left(s_j \right)_{j=0}^{\varkappa} \in \mathcal{F}_{q, \varkappa}^{\mathrm{EP}}$.
   (ii) $s_0 \in \mathbb{C}_{\mathrm{EP}}^{q \times q}$ *and* $\left(s_j \right)_{j=0}^{\varkappa} \in \mathcal{D}_{q \times q, \varkappa}$.

(b) *Suppose* (i) *is satisfied, then* $s_0 s_0^+ = s_0^+ s_0$ *and, for each* $n \in \mathbb{Z}_{0, \varkappa}$,

$$S_n S_n^+ = I_{n+1} \otimes \left(s_0 s_0^+ \right) \qquad \text{and} \qquad S_n^+ S_n = I_{n+1} \otimes \left(s_0^+ s_0 \right).$$

Proposition 1.11. *If* $\varkappa \in \mathbb{N}_0 \cup \{+\infty\}$ *and* $\left(s_j \right)_{j=0}^{\varkappa} \in \mathcal{F}_{q, \varkappa}^{\mathrm{EP}}$, *then* $\left(s_j \right)_{j=0}^{n} \in \mathcal{F}_{q, n}^{\mathrm{EP}}$ *and* $S_n^+ = S_n^{\sharp}$ *for each* $n \in \mathbb{Z}_{0, \varkappa}$.

Our next result will offer us further insight into the structure of EP sequences.

Theorem 1.12. *Let $q \geq 2$, $r \in \mathbb{Z}_{1, q-1}$ and $\varkappa \in \mathbb{N}_0 \cup \{+\infty\}$. Suppose, furthermore, that $(s_j)_{j=0}^{\varkappa} \in \mathcal{F}_{q, \varkappa}^{\mathrm{EP}}$ with $\operatorname{rank} s_0 = r$ (and therefore $\dim[\mathcal{N}(s_0)] = q - r$). Let $(u_s)_{s=1}^{r}$ be an orthonormal basis in $\mathcal{R}(s_0)$ and $(u_s)_{s=r+1}^{q}$ be an orthonormal basis in $\mathcal{N}(s_0)$. For each $\ell \in \mathbb{Z}_{1, q}$, let $U_\ell := (u_1, \quad u_2, \quad \ldots, \quad u_\ell)$ and, for each $j \in \mathbb{Z}_{0, \varkappa}$, let $\widetilde{s}_j := U_r^* s_j U_r$. Then:*

(a) $(u_s)_{s=1}^{q}$ *is an orthonormal basis in $\mathbb{C}^{q \times 1}$ and the matrix U_q is unitary.*

(b) $U_q^* s_j U_q = \operatorname{diag}\left(\widetilde{s}_j, 0_{(q-r) \times (q-r)}\right)$ *for each $j \in \mathbb{Z}_{0, \varkappa}$. The matrix $\widetilde{s}_0$ is, in particular, non-singular.*

Proof. By part (a) of Proposition 1.10, we get $s_0 \in \mathbb{C}_{\mathrm{EP}}^{q \times q}$. Thus, it follows by Proposition A.5 that $\mathcal{N}(s_0) = \mathcal{N}(s_0^*)$. Hence, we see that $\mathbb{C}^{q \times 1} = \mathcal{R}(s_0) \oplus \mathcal{N}(s_0)$, from which we obtain (a). Furthermore, by part (a) of Proposition 1.10, we have $(s_m)_{m=0}^{\varkappa} \in \mathcal{D}_{q \times q, \varkappa}$. Recalling Definition 1.4, we then obtain $\mathcal{N}(s_0) \subseteq \mathcal{N}(s_j)$ and, since $\mathcal{N}(s_0) = \mathcal{N}(s_0^*)$, it follows by [22, Proposition 5.1] that

$$\mathcal{N}(s_0) \subseteq \mathcal{N}(s_j^*). \tag{1.3}$$

For each $m \in \mathbb{Z}_{r+1, q}$, we have, by assumption, that $u_m \in \mathcal{N}(s_0)$, which, because of (1.3), implies $s_j u_m = 0_{q \times 1}$ and $s_j^* u_m = 0_{q \times 1}$ for each $j \in \mathbb{Z}_{0, \varkappa}$. Consequently,

$$U_q^* s_j U_q = \begin{pmatrix} (u_1, \cdots, u_r)^* s_j (u_1, \cdots, u_r), & (u_1, \cdots, u_r)^* s_j (u_{r+1}, \cdots, u_q) \\ (u_{r+1}, \cdots, u_q)^* s_j (u_1, \cdots, u_r), & (u_{r+1}, \cdots, u_q)^* s_j (u_{r+1}, \cdots, u_q) \end{pmatrix}$$

$$= \operatorname{diag}\left(\widetilde{s}_j, 0_{(q-r) \times (q-r)}\right)$$

for each $j \in \mathbb{Z}_{0, \varkappa}$. Since U_q is non-singular, it follows that

$$\operatorname{rank} \widetilde{s}_0 = \operatorname{rank}\left[\operatorname{diag}\left(\widetilde{s}_0, 0_{(q-r) \times (q-r)}\right)\right] = \operatorname{rank}\left[U_q^* s_0 U_q\right] = \operatorname{rank} s_0 = r.$$

Thus, the matrix $\widetilde{s}_0$ is non-singular. $\square$

2. Matricial Toeplitz non-negative definite sequences

Our focus in this section will be on a special class of sequences in $\mathbb{C}^{q \times q}$. Before we describe this class, we first introduce some notation. Given a $\varkappa \in \mathbb{N}_0$ and a sequence $(C_j)_{j=0}^{\varkappa}$ in $\mathbb{C}^{q \times q}$ we define, for each $n \in \mathbb{Z}_{0, \varkappa}$, the block Toeplitz matrix

$$\mathrm{T}_n^{(C)} := \begin{pmatrix} C_0 & C_1^* & \cdots & C_n^* \\ C_1 & C_0 & \ddots & \vdots \\ \vdots & \ddots & \ddots & C_1^* \\ C_n & \cdots & C_1 & C_0 \end{pmatrix}. \tag{2.1}$$

Whenever it is clear which sequence is meant, we will simply write T_n instead of $\mathrm{T}_n^{(C)}$. If $n \in \mathbb{N}_0$ and $(C_j)_{j=0}^{n}$ is a sequence in $\mathbb{C}^{q \times q}$, then $(C_j)_{j=0}^{n}$ is called a **Toeplitz non-negative definite sequence** (or **T-n.n.d. sequence**, for short) if T_n is non-negative Hermitian. We call $(C_j)_{j=0}^{n}$ a **Toeplitz positive definite sequence** (or **T-p.d. sequence**) if T_n is positive Hermitian. If $n \in \mathbb{N}$ and $(C_j)_{j=0}^{n}$ is a T-n.n.d.

sequence in $\mathbb{C}^{q \times q}$, then, for each $m \in \mathbb{Z}_{0,n-1}$, so is $(C_j)_{j=0}^m$. Similarly, if $(C_j)_{j=0}^n$ is T-p.d., the same holds for $(C_j)_{j=0}^m$. Thus, we say that a sequence $(C_j)_{j=0}^\infty$ is Toeplitz non-negative definite (T-n.n.d.) if, for each $n \in \mathbb{N}_0$, the sequence $(C_j)_{j=0}^n$ is Toeplitz non-negative definite and Toeplitz positive definite (T-p.d.) if, for each $n \in \mathbb{N}_0$, the sequence $(C_j)_{j=0}^n$ is Toeplitz positive definite. For $\varkappa \in \mathbb{N}_0 \cup \{+\infty\}$, the set of all T-n.n.d. sequences in $\mathbb{C}^{q \times q}$ will be denoted by $\mathcal{T}_{q,\varkappa}$ and the set of all T-p.d. sequences in $\mathbb{C}^{q \times q}$ by $\widetilde{\mathcal{T}}_{q,\varkappa}$.

Example 2.1. Suppose $K \in \mathbb{C}^{q \times q}$. For each $j \in \{0,1\}$, let $C_j := K^j$. A well-known characterization of non-negative Hermitian block matrices (see, for instance, [8, Lemma 1.1.9]) shows that $(C_j)_{j=0}^1 \in \mathcal{T}_{q,1}$ if and only if $K \in \mathbb{K}_{q \times q}$, and also that $(C_j)_{j=0}^1 \in \widetilde{\mathcal{T}}_{q,1}$ if and only if $K \in \mathbb{D}_{q \times q}$.

Example 2.2. Let $\varkappa \in \mathbb{N}_0 \cup \{+\infty\}$. If $A \in \mathbb{C}_{\geq}^{q \times q}$ and $C_j := A$ for each $j \in \mathbb{Z}_{0,\varkappa}$, then $(C_j)_{j=0}^\varkappa \in \mathcal{T}_{q,\varkappa}$, since $T_n = D^* A D \in \mathbb{C}_{\geq}^{(n+1)q \times (n+1)q}$ for all $n \in \mathbb{Z}_{0,\varkappa}$, where $D := (I_q, \ I_q, \ \cdots, \ I_q) \in \mathbb{C}^{q \times (n+1)q}$.

Remark 2.3. Let $\varkappa \in \mathbb{N}_0 \cup \{+\infty\}$ and $(C_j)_{j=0}^\varkappa \in \mathcal{T}_{q,\varkappa}$. Suppose, furthermore, that $A \in \mathbb{C}^{q \times p}$. Then $(A^* C_j A)_{j=0}^\varkappa \in \mathcal{T}_{p,\varkappa}$.

Remark 2.4. Let $\varkappa \in \mathbb{N}_0 \cup \{+\infty\}$ and $s \in \mathbb{N}$. For each $r \in \mathbb{Z}_{1,s}$, let $\alpha_r \in [0,+\infty)$ and $\left(C_j^{(r)}\right)_{j=0}^\varkappa \in \mathcal{T}_{q,\varkappa}$. Then $\left(\sum_{r=1}^s \alpha_r C_j^{(r)}\right)_{j=0}^\varkappa \in \mathcal{T}_{q,\varkappa}$. If, furthermore, there exists an $r_0 \in \mathbb{Z}_{1,s}$ such that $\alpha_{r_0} \in (0,+\infty)$ and $\left(C_j^{(r_0)}\right)_{j=0}^\varkappa \in \widetilde{\mathcal{T}}_{q,\varkappa}$, then

$$\left(\sum_{r=1}^s \alpha_r C_j^{(r)}\right)_{j=0}^\varkappa \in \widetilde{\mathcal{T}}_{q,\varkappa}.$$

Remark 2.5. Let $\varkappa \in \mathbb{N}_0 \cup \{+\infty\}$ and $s \in \mathbb{Z}_{2,+\infty}$. Suppose, furthermore, that $(q_j)_{j=1}^s$ is a sequence in $\mathbb{N}$ and that $q = \sum_{r=1}^s q_j$. For each $r \in \mathbb{Z}_{1,s}$, let $\left(C_j^{(r)}\right)_{j=0}^\varkappa$ be a sequence from $\mathbb{C}^{q_j \times q_j}$. Then:

(a) $\left(\text{diag}\left(C_j^{(1)}, C_j^{(2)}, \ldots, C_j^{(s)}\right)\right)_{j=0}^\varkappa \in \mathcal{T}_{q,\varkappa}$ if and only if $\left(C_j^{(r)}\right)_{j=0}^\varkappa \in \mathcal{T}_{q_r,\varkappa}$ for all $r \in \mathbb{Z}_{1,s}$.

(b) $\left(\text{diag}\left(C_j^{(1)}, C_j^{(2)}, \ldots, C_j^{(s)}\right)\right)_{j=0}^\varkappa \in \widetilde{\mathcal{T}}_{q,\varkappa}$ if and only if $\left(C_j^{(r)}\right)_{j=0}^\varkappa \in \widetilde{\mathcal{T}}_{q_r,\varkappa}$ for all $r \in \mathbb{Z}_{1,s}$.

Remark 2.6. Let $\varkappa \in \mathbb{N}_0 \cup \{+\infty\}$.

(a) Suppose $(b_j)_{j=0}^\varkappa \in \mathcal{T}_{1,\varkappa}$ and $(C_j)_{j=0}^\varkappa \in \mathcal{T}_{q,\varkappa}$, then $(b_j C_j)_{j=0}^\varkappa \in \mathcal{T}_{q,\varkappa}$.

(b) Suppose $(b_j)_{j=0}^\varkappa \in \widetilde{\mathcal{T}}_{1,\varkappa}$ and $(C_j)_{j=0}^\varkappa \in \widetilde{\mathcal{T}}_{q,\varkappa}$, then $(b_j C_j)_{j=0}^\varkappa \in \widetilde{\mathcal{T}}_{q,\varkappa}$.

With our next result, we establish a link to Section 1.

Proposition 2.7. *Let $\varkappa \in \mathbb{N}_0 \cup \{+\infty\}$. Then $\mathcal{T}_{q,\varkappa} \subseteq \mathcal{D}_{q \times q, \varkappa}$.*

Proof. The case in which $\varkappa = 0$ is trivial. We thus suppose that $\varkappa \geq 1$. Let $(C_j)_{j=0}^{\varkappa}$ be a T-n.n.d. sequence in $\mathbb{C}^{q \times q}$ and $\ell \in \mathbb{Z}_{1,\varkappa}$. We thus have $T_\ell \in \mathbb{C}_{\geq}^{(\ell+1)q \times (\ell+1)q}$. In particular, this implies that

$$\begin{pmatrix} C_0 & C_\ell^* \\ C_\ell & C_0 \end{pmatrix} \in \mathbb{C}_{\geq}^{2q \times 2q}.$$

It thus follows that $\mathcal{R}(C_\ell) \subseteq \mathcal{R}(C_0)$, $\mathcal{R}(C_\ell^*) \subseteq \mathcal{R}(C_0)$ and $C_0^* = C_0$ (see, e.g., [8, Lemma 1.1.9]). Hence, we obtain

$$\bigcup_{j=0}^{\varkappa} \mathcal{R}(C_j) \subseteq \mathcal{R}(C_0) \qquad \text{and} \qquad \bigcup_{j=0}^{\varkappa} \mathcal{R}(C_j^*) \subseteq \mathcal{R}(C_0^*).$$

Finally, applying [22, Proposition 5.1] yields $(C_j)_{j=0}^{\varkappa} \in \mathcal{D}_{q \times q, \varkappa}$. $\qquad \square$

A direct consequence of Proposition 2.7 is the following well-known fact:

Corollary 2.8. *If $\varkappa \in \mathbb{N}_0 \cup \{+\infty\}$ and $(C_j)_{j=0}^{\varkappa}$ is a T-n.n.d. sequence in $\mathbb{C}^{q \times q}$ with $C_0 = 0_{q \times q}$, then $C_j = 0_{q \times q}$ for all $j \in \mathbb{Z}_{0,\varkappa}$.*

We will now take a more detailed look at the structure of Toeplitz non-negative definite sequences in $\mathbb{C}^{q \times q}$. We draw from [11] and [8, Section 3.4], where the structure of Toeplitz non-negative definite sequences is discussed in detail. This structure is described using special matrices. For each $\varkappa \in \mathbb{N}_0 \cup \{+\infty\}$ and any sequence $(C_j)_{j=0}^{\varkappa}$ in $\mathbb{C}^{q \times q}$, we set

$$M_1 := 0_{q \times q}, \qquad L_1 := C_0 \qquad \text{and} \qquad R_1 := C_0. \qquad (2.2)$$

If $\varkappa \geq 1$, then, for each $n \in \mathbb{Z}_{1,\varkappa}$, we furthermore define

$$Z_n := \text{row} \, (C_{n+1-j})_{j=1}^{n} \qquad \text{and} \qquad Y_n := \text{col} \, (C_j)_{j=1}^{n} \qquad (2.3)$$

as well as

$$L_{n+1} := C_0 - Z_n T_{n-1}^+ Z_n^*, \qquad R_{n+1} := C_0 - Y_n^* T_{n-1}^+ Y_n \qquad (2.4)$$

(where we use the block Toeplitz matrix in (2.1)) and

$$M_{n+1} := Z_n T_{n-1}^+ Y_n. \qquad (2.5)$$

The following proposition describes the inherent structure of a Toeplitz non-negative definite sequence in $\mathbb{C}^{q \times q}$.

Proposition 2.9. *Let $\varkappa \in \mathbb{N}_0 \cup \{+\infty\}$ and let $(C_j)_{j=0}^{\varkappa}$ be a T-n.n.d. sequence in $\mathbb{C}^{q \times q}$. Then:*

(a) *$(L_{j+1})_{j=0}^{\varkappa}$ and $(R_{j+1})_{j=0}^{\varkappa}$ are monotonically decreasing sequences, where*

$$\text{rank} \, L_{j+1} = \text{rank} \, R_{j+1} \qquad \text{and} \qquad \det L_{j+1} = \det R_{j+1},$$

for each $j \in \mathbb{Z}_{0,\varkappa}$. If $\varkappa = +\infty$, then the sequences $(L_{j+1})_{j=0}^{\infty}$ and $(R_{j+1})_{j=0}^{\infty}$ converge to non-negative Hermitian matrices L and R, respectively. When this is the case, $\det L = \det R$.

(b) *Suppose that $\varkappa \geq 1$. For each $n \in \mathbb{Z}_{0,\varkappa-1}$, the matrix*

$$K_{n+1} := \left(\sqrt{L_{n+1}} \right)^{+} (C_{n+1} - M_{n+1}) \left(\sqrt{R_{n+1}} \right)^{+}$$

is contractive and

$$C_{n+1} = M_{n+1} + \sqrt{L_{n+1}} K_{n+1} \sqrt{R_{n+1}}.$$

If $n \geq 1$, then

$$L_{n+2} = \sqrt{L_{n+1}} \left(I_q - K_{n+1} K_{n+1}^{*} \right) \sqrt{L_{n+1}},$$

$$R_{n+2} = \sqrt{R_{n+1}} \left(I_q - K_{n+1}^{*} K_{n+1} \right) \sqrt{R_{n+1}}$$

and all of the following conditions are equivalent:
   (i) $L_{n+1} = L_{n+2}$.
   (ii) $R_{n+1} = R_{n+2}$.
   (iii) $C_{n+1} = M_{n+2}$.
   (iv) $K_{n+1} = 0_{q \times q}$.
(c) *If $(C_j)_{j=0}^{\varkappa}$ is T-p.d. then, for each $n \in \mathbb{Z}_{0,\varkappa-1}$, the matrices L_{n+1} and R_{n+1} are positive Hermitian and K_{n+1} is strictly contractive.*

Proof. See [8, Remark 3.4.1, Theorem 3.4.1 and Remark 3.4.3]. □

   If $\varkappa \in \mathbb{N} \cup \{+\infty\}$ and $(C_j)_{j=0}^{\varkappa}$ is a T-n.n.d. sequence in $\mathbb{C}^{q \times q}$, then the sequence $(K_j)_{j=1}^{\varkappa}$ defined in Proposition 2.9 is called the Schur parameter sequence of $(C_j)_{j=0}^{\varkappa}$. Note that Schur parameter sequences appear, for instance, for Toeplitz positive definite sequences in [7, Definition 2.3], where a sequence of this type is referred to as the *sequence of canonical moments*.

Definition 2.10. A sequence $(C_j)_{j=0}^{\infty} \in \mathcal{T}_{q,\infty}$ is called totally Toeplitz positive definite if the matrix L from part (a) of Proposition 2.9 satisfies $\det L \neq 0$. The set of all totally Toeplitz positive definite sequences in $\mathcal{T}_{q,\infty}$ will be denoted by $\mathcal{T}_{q,\infty}^{t}$.

   Proposition 2.9 shows that

$$\mathcal{T}_{q,\infty}^{t} \subseteq \tilde{\mathcal{T}}_{q,\infty}. \tag{2.6}$$

   We next concentrate on the extension problem for finite Toeplitz non-negative definite matricial sequences. If $M, A, B \in \mathbb{C}^{q \times q}$, then the set

$$\mathcal{K}(M; A, B) := \{ M + AKB : K \in \mathbb{K}_{q \times q} \}$$

is called the (closed) matrix ball with center M, left semi-radius A and right semi-radius B and the set

$$\overset{\circ}{\mathcal{K}}(M; A, B) := \{ M + AKB : K \in \mathbb{D}_{q \times q} \}$$

is called the open matrix ball with center M, left semi-radius A and right semi-radius B. The following theorem (see [8, Theorem 3.4.1]) gives us a complete answer to the extension problem for finite Toeplitz non-negative definite sequences.

Theorem 2.11. *Let $n \in \mathbb{N}_0$ and $(C_j)_{j=0}^{n}$ be a sequence in $\mathbb{C}^{q \times q}$. Then:*

(a) *The set*

$$\mathcal{C}_{\geq}\left[\,(\,C_j\,)_{j=0}^n\,\right] := \left\{ C_{n+1} \in \mathbb{C}^{q \times q} \;:\; (\,C_j\,)_{j=0}^{n+1} \in \mathcal{T}_{q,\,n+1} \right\}$$

is non-empty if and only if $(\,C_j\,)_{j=0}^n \in \mathcal{T}_{q,\,n}$.

(b) *Suppose* $(\,C_j\,)_{j=0}^n \in \mathcal{T}_{q,\,n}$. *Then* $L_{n+1},\,R_{n+1} \in \mathbb{C}_{\geq}^{q \times q}$ *and*

$$\mathcal{C}_{\geq}\left[\,(\,C_j\,)_{j=0}^n\,\right] = \mathfrak{K}\left(M_{n+1};\; \sqrt{L_{n+1}},\; \sqrt{R_{n+1}} \right).$$

In particular, $M_{n+1} \in \mathcal{C}_{\geq}\left[\,(\,C_j\,)_{j=0}^n\,\right]$.

(c) *The set*

$$\mathcal{C}_{>}\left[\,(\,C_j\,)_{j=0}^n\,\right] := \left\{ C_{n+1} \in \mathbb{C}^{q \times q} \;:\; (\,C_j\,)_{j=0}^{n+1} \in \tilde{\mathcal{T}}_{q,\,n+1} \right\}$$

is non-empty if and only if $(\,C_j\,)_{j=0}^n \in \tilde{\mathcal{T}}_{q,\,n}$.

(d) *Suppose* $(\,C_j\,)_{j=0}^n \in \tilde{\mathcal{T}}_{q,\,n}$. *Then* $L_{n+1},\,R_{n+1} \in \mathbb{C}_{>}^{q \times q}$ *and*

$$\mathcal{C}_{>}\left[\,(\,C_j\,)_{j=0}^n\,\right] = \overset{\circ}{\mathfrak{K}}\left(M_{n+1};\; \sqrt{L_{n+1}},\; \sqrt{R_{n+1}} \right).$$

In particular, $M_{n+1} \in \mathcal{C}_{>}\left[\,(\,C_j\,)_{j=0}^n\,\right]$.

The following definition is motivated by the role which the matrix M_k (introduced in (2.2) and (2.5)) plays in Theorem 2.11; more precisely, the fact that M_k appears as the center of the matrix balls in parts (b) and (d).

Definition 2.12. Let $\varkappa \in \mathbb{N} \cup \{+\infty\}$ and $(\,C_j\,)_{j=0}^{\varkappa}$ be a sequence in $\mathbb{C}^{q \times q}$.

(a) Suppose $k \in \mathbb{Z}_{1,\,\varkappa}$. We say that $(\,C_j\,)_{j=0}^{\varkappa}$ is a central sequence of order k (or an order k central sequence), if $C_k = M_k$ for all $j \in \mathbb{Z}_{k,\,\varkappa}$, where M_k is given by (2.2) and (2.5).

(b) Let $k \in \mathbb{Z}_{1,\,\varkappa}$. We call $(\,C_j\,)_{j=0}^{\varkappa}$ a central sequence of minimal order k (or a minimal order k central sequence) if it has both of the following two properties:

   (i) $(\,C_j\,)_{j=0}^{\varkappa}$ is order k central.

   (ii) If $\varkappa \geq 2$ and $k \in \mathbb{Z}_{2,\,\varkappa}$, then $(\,C_j\,)_{j=0}^{\varkappa}$ is not order $k - 1$ central.

(c) The sequence $(\,C_j\,)_{j=0}^{\varkappa}$ is simply called a central sequence, if there exists a $k \in \mathbb{Z}_{1,\,\varkappa}$ such that $(\,C_j\,)_{j=0}^{\varkappa}$ is order k central.

Remark 2.13. Let $\varkappa \in \mathbb{N} \cup \{+\infty\}$ and $k \in \mathbb{Z}_{1,\,\varkappa}$. Furthermore, suppose that $(\,C_j\,)_{j=0}^{\varkappa}$ is an order k central sequence in $\mathbb{C}^{q \times q}$. For each $\ell \in \mathbb{Z}_{k,\,\varkappa}$, the sequence $(\,C_j\,)_{j=0}^{\varkappa}$ is then also order ℓ central.

It is easy to characterize order 1 central sequences in $\mathbb{C}^{q \times q}$.

Remark 2.14. Let $\varkappa \in \mathbb{N} \cup \{+\infty\}$ and $(\,C_j\,)_{j=0}^{\varkappa}$ be a sequence in $\mathbb{C}^{q \times q}$.

(a) Inductively, we see from (2.4) and (2.5) that $(\,C_j\,)_{j=0}^{\varkappa}$ is order 1 central when $C_j = 0_{q \times q}$, for all $j \in \mathbb{Z}_{1,\,\varkappa}$.

(b) From (a) we see that $(C_j)_{j=0}^{\varkappa}$ is T-n.n.d. and order 1 central if and only if $C_0 \in \mathbb{C}_{\geq}^{q \times q}$ and $C_j = 0_{q \times q}$, for all $j \in \mathbb{Z}_{1,\varkappa}$.

(c) From (a) we furthermore see that $(C_j)_{j=0}^{\varkappa}$ is T-p.d. and order 1 central when $C_0 \in \mathbb{C}_{>}^{q \times q}$ and $C_j = 0_{q \times q}$, for all $j \in \mathbb{Z}_{1,\varkappa}$.

Our next steps will be towards characterizing central Toeplitz non-negative definite sequences.

Proposition 2.15. *Let $\varkappa \in \mathbb{N} \cup \{+\infty\}$ and $(C_j)_{j=0}^{\varkappa}$ be a T-n.n.d. sequence in $\mathbb{C}^{q \times q}$ with Schur parameter sequence $(K_j)_{j=1}^{\varkappa}$. Let, furthermore, $k \in \mathbb{Z}_{1,\varkappa}$. All of the following conditions are equivalent:*

(i) $(C_j)_{j=0}^{\varkappa}$ *is order k central.*
(ii) $L_{j+1} = L_k$ *for all $j \in \mathbb{Z}_{k,\varkappa}$.*
(iii) $R_{j+1} = R_k$ *for all $j \in \mathbb{Z}_{k,\varkappa}$.*
(iv) $K_j = 0_{q \times q}$ *for all $j \in \mathbb{Z}_{k,\varkappa}$.*

Proof. The equivalence of (i)–(iv) follows directly from Proposition 2.9. $\quad\square$

Next, we arrive at recursion formulas for the elements of a central Toeplitz non-negative definite sequence. The corresponding result, the following Theorem 2.16, is an immediate consequence of [14, Proposition 1 and Remark 1]. A result equivalent to Theorem 2.16 is found in [15, Theorem 32] as well as in [8, Theorem 3.4.3]. For the special case of a Toeplitz positive definite sequence, an equivalent result to Theorem 2.16 was already included in [12, Theorem 20] with two different proofs. The first of these proofs is based on applying an extension problem result (for the Wiener algebra $W(\mathbb{T})$) by Dym/Gohberg [9, Theorem 6.1]. The second of these proofs uses results for orthogonal matrix polynomials on the unit circle by Delsarte/Genin/Kamp [28]. For the special case of a Toeplitz positive definite sequence, a further equivalent result to Theorem 2.16 can be found in Ellis/Gohberg [10, Theorem 2.2].

Theorem 2.16. *Let $\varkappa \in \mathbb{Z}_{2,\infty} \cup \{+\infty\}$, $k \in \mathbb{Z}_{1,\varkappa-1}$ and $(C_j)_{j=0}^{\varkappa}$ be an order k central T-n.n.d. sequence in $\mathbb{C}^{q \times q}$. Let $s \in \mathbb{Z}_{k+1,\varkappa}$. Suppose that*

$$Z_{s,k} := \text{row } (C_{s-j})_{j=1}^{k} \qquad \text{and} \qquad Y_{s,k} := \text{col } (C_{s-1-k+j})_{j=1}^{k}$$

and that Z_k and Y_k are given by (2.3). Then

$$C_s = Z_{s,k} T_{k-1}^{+} Y_k \qquad \text{and} \qquad C_s = Z_k T_{k-1}^{+} Y_{s,k}.$$

Corollary 2.17. *Let $\varkappa \in \mathbb{Z}_{2,\varkappa} \cup \{+\infty\}$ and $(C_j)_{j=0}^{\varkappa}$ be an order 2 central T-n.n.d. sequence in $\mathbb{C}^{q \times q}$. Suppose that $s \in \mathbb{Z}_{2,\varkappa}$. Then:*

(a) $C_s = C_{s-1} C_0^{+} C_1$ *and* $C_s = C_1 C_0^{+} C_{s-1}$.
(b) $C_s = C_1 \left(C_0^{+} C_1 \right)^{s-1}$ *and* $C_s = \left(C_1 C_0^{+} \right)^{s-1} C_1$.

Proof. Part (a) follows directly from Theorem 2.16, while (b) follows from (a). $\quad\square$

Given an $n \in \mathbb{N}_0$ and a sequence $(C_j)_{j=0}^n$ in $\mathbb{C}^{q \times q}$, we next consider the unique extension of $(C_j)_{j=0}^n$ to an order $n+1$ central sequence.

Remark 2.18. Let $n \in \mathbb{N}_0$ and $(C_j)_{j=0}^n$ be a sequence in $\mathbb{C}^{q \times q}$. There exists a unique order $n+1$ central sequence $(\widetilde{C}_j)_{j=0}^\infty$ in $\mathbb{C}^{q \times q}$ such that $\widetilde{C}_j = C_j$ for each $j \in \mathbb{Z}_{0,n}$. This sequence is called the central sequence corresponding to $(C_j)_{j=0}^n$.

We have now arrived at an idea central to our topic. Specifically, we now consider central sequences for the case that our initial sequence is Toeplitz non-negative or positive definite.

Lemma 2.19. *Let $k \in \mathbb{N}$ and $(C_j)_{j=0}^{k-1}$ be a sequence in $\mathbb{C}^{q \times q}$. Suppose that $(\widetilde{C}_j)_{j=0}^\infty$ is the central sequence corresponding to $(C_j)_{j=0}^{k-1}$. Then:*

(a) *$(\widetilde{C}_j)_{j=0}^\infty \in \mathcal{T}_{q,\infty}$ if and only if $(C_j)_{j=0}^{k-1} \in \mathcal{T}_{q,k-1}$.*
(b) *All of the following conditions are equivalent:*
   (i) *$(\widetilde{C}_j)_{j=0}^\infty \in \mathcal{T}_{q,\infty}^{\mathrm{t}}$.*
   (ii) *$(\widetilde{C}_j)_{j=0}^\infty \in \widetilde{\mathcal{T}}_{q,\infty}$.*
   (iii) *$(C_j)_{j=0}^{k-1} \in \widetilde{\mathcal{T}}_{q,k-1}$.*

Proof. Part (a) follows immediately from parts (b) and (d) of Theorem 2.11.
(b) "(iii)$\Longrightarrow$(i)". Because of (iii), part (c) of Proposition 2.9 implies $L_k \in \mathbb{C}_>^{q \times q}$. Suppose that $(\widetilde{L}_{s+1})_{s=0}^\infty$ is the sequence constructed from $(\widetilde{C}_j)_{j=0}^\infty$ via (2.2) - (2.4) and that $\widetilde{L} := \lim_{s \to \infty} \widetilde{L}_{s+1}$. Since Proposition 2.15 yields $\widetilde{L}_s = L_k$ for each $s \in \mathbb{Z}_{k,\infty}$, we get that $\widetilde{L} = L_k$. Because the matrix L_k is positive Hermitian, it follows that $\det \widetilde{L} \neq 0$ and we have (i).
"(i)$\Longrightarrow$(ii)" follows directly from (2.6).
"(ii)$\Longrightarrow$(iii)" follows from $(\widetilde{C}_j)_{j=0}^{k-1} = (C_j)_{j=0}^{k-1}$ and the definition of $\widetilde{\mathcal{T}}_{q,\infty}$. □

If $k \in \mathbb{N}$ and $(C_j)_{j=0}^{k-1} \in \mathcal{T}_{q,k-1}$, then Theorem 2.16 gives us a method for recursively constructing the central sequence corresponding to $(C_j)_{j=0}^{k-1}$. In anticipation of later applications, we formulate this result for the special case $k = 1$.

Corollary 2.20. *Suppose that $(C_j)_{j=0}^1$ is a T-n.n.d. sequence in $\mathbb{C}^{q \times q}$ and that $(\widetilde{C}_j)_{j=0}^\infty$ is the central sequence corresponding to $(C_j)_{j=0}^1$. Then*

$$\widetilde{C}_s = C_1 \left(C_0^+ C_1 \right)^{s-1} \qquad \text{and} \qquad \widetilde{C}_s = \left(C_1 C_0^+ \right)^{s-1} C_1$$

for each $s \in \mathbb{Z}_{2,+\infty}$.

Proof. We need only combine Corollary 2.17 with Remark 2.18. □

Example 2.21. Let $K \in \mathbb{K}_{q \times q}$. For each $j \in \mathbb{N}_0$, let $C_j := K^j$. Using Example 2.1, Corollary 2.20 and Lemma 2.19, it is easily verified that:

(a) $(C_j)_{j=0}^{\infty}$ is an order 2 central T-n.n.d. sequence in $\mathbb{C}^{q \times q}$ and if (and only if) $K \neq 0_{q \times q}$, then $(C_j)_{j=0}^{\infty}$ is minimal order 2 central.

(b) $L_1 = I_q$, $R_1 = I_q$ and, for each $k \in \mathbb{N}$, furthermore

$$L_{k+1} = I_q - KK^* \qquad \text{and} \qquad R_{k+1} = I_q - K^*K.$$

(c) $(C_j)_{j=0}^{\infty} \in \mathcal{T}_{q,\infty}^{t}$ if and only if $K \in \mathbb{D}_{q \times q}$.

Example 2.22. Let $K \in \mathbb{K}_{q \times q} \cap \mathbb{C}_{\geq}^{q \times q}$ and $r \in \mathbb{N}$. For each $j \in \mathbb{N}_0$, suppose $C_j := K^{r+j}$. Then $(C_j)_{j=0}^{\infty}$ is an order 2 central T-n.n.d. sequence in $\mathbb{C}^{q \times q}$. This can be recognized as follows. Since $K \in \mathbb{C}_{\geq}^{q \times q}$, part (b) of Lemma A.4 implies $KK^+ = K^+K$. Thus, by induction, we obtain

$$(K^r)^+ K^{r+1} = K. \tag{2.7}$$

For each $j \in \mathbb{N}_0$, we have $C_j = (\sqrt{K^r})^* K^j \sqrt{K^r}$. Combining this with the fact that Example 2.21 yields $(K^j)_{j=0}^{\infty} \in \mathcal{T}_{q,\infty}$, we see by Remark 2.3 that $(C_j)_{j=0}^{\infty} \in \mathcal{T}_{q,\infty}$. Therefore, $(C_j)_{j=0}^{1} \in \mathcal{T}_{q,1}$. Thus, (2.7) implies

$$C_1 (C_0^+ C_1)^{s-1} = K^{r+1} \left[(K^r)^+ K^{r+1} \right]^{s-1} = K^{r+1} K^{s-1} = K^{r+s} = C_s$$

for each $s \in \mathbb{Z}_{2,\infty}$. Thus, Corollary 2.20 shows that $(C_j)_{j=0}^{\infty}$ is the central sequence corresponding to $(C_j)_{j=0}^{1}$. Hence, $(C_j)_{j=0}^{\infty}$ is order 2 central.

Remark 2.23. Let $\varkappa \in \mathbb{N}_0 \cup \{+\infty\}$.

(a) If $(C_j)_{j=0}^{\varkappa} \in \mathcal{T}_{q,\varkappa}$ and $w \in \mathbb{D} \cup \mathbb{T}$, then from part (a) of Remark 2.6 and Example 2.21, it follows that $(w^j C_j)_{j=0}^{\varkappa} \in \mathcal{T}_{q,\varkappa}$.

(b) If $(C_j)_{j=0}^{\varkappa} \in \widetilde{\mathcal{T}}_{q,\varkappa}$ and $w \in \mathbb{D}$, then from part (b) of Remark 2.6 and Example 2.21, it follows that $(w^j C_j)_{j=0}^{\varkappa} \in \widetilde{\mathcal{T}}_{q,\varkappa}$.

We now introduce another class of matricial sequences which will be particularly interesting when we again look at Toeplitz non-negative definite sequences.

Definition 2.24. Let $\varkappa \in \mathbb{N} \cup \{+\infty\}$ and $(C_j)_{j=0}^{\varkappa}$ be a sequence in $\mathbb{C}^{q \times q}$.

(a) Suppose $k \in \mathbb{Z}_{1,\varkappa}$. We will say that $(C_j)_{j=0}^{\varkappa}$ is a canonical sequence of order k (or an order k canonical sequence), if $\operatorname{rank} T_{k-1} = \operatorname{rank} T_k$.

(b) Suppose $k \in \mathbb{Z}_{1,\varkappa}$. We call $(C_j)_{j=0}^{\varkappa}$ a canonical sequence of minimal order k (or a minimal order k canonical sequence) if it has the following two properties:
   (i) $(C_j)_{j=0}^{\varkappa}$ is order k canonical.
   (ii) If $\varkappa \geq 2$ and $k \in \mathbb{Z}_{2,\varkappa}$, then for each $\ell \in \mathbb{Z}_{1,k}$, the sequence $(C_j)_{j=0}^{\varkappa}$ is not order ℓ canonical.

(c) $(C_j)_{j=0}^{\varkappa}$ is simply called a canonical sequence, if there exists a $k \in \mathbb{Z}_{1,\varkappa}$ such that $(C_j)_{j=0}^{\varkappa}$ is order k canonical.

We next arrive at a characterization of canonical Toeplitz non-negative definite sequences.

Lemma 2.25. *If $\varkappa \in \mathbb{N}_0 \cup \{+\infty\}$ and $(C_j)_{j=0}^{\varkappa}$ is a T-n.n.d. sequence in $\mathbb{C}^{q \times q}$, then $\operatorname{rank} T_n = \operatorname{rank} T_{n-1} + \operatorname{rank} L_{n+1}$ for all $n \in \mathbb{Z}_{1, \varkappa}$.*

Proof. Apply [8, Lemma 1.1.7]. $\quad\square$

Proposition 2.26. *Let $\varkappa \in \mathbb{N} \cup \{+\infty\}$ and $(C_j)_{j=0}^{\varkappa}$ be a T-n.n.d. sequence in $\mathbb{C}^{q \times q}$. Let, furthermore, $k \in \mathbb{Z}_{1, \varkappa}$. All of the following conditions are equivalent:*

(i) $(C_j)_{j=0}^{\varkappa}$ *is order k canonical.*
(ii) $L_{k+1} = 0_{q \times q}$.
(iii) $R_{k+1} = 0_{q \times q}$.
(iv) $L_{j+1} = 0_{q \times q}$ *for all $j \in \mathbb{Z}_{k, \varkappa}$.*
(v) $R_{j+1} = 0_{q \times q}$ *for all $j \in \mathbb{Z}_{k, \varkappa}$.*

Proof. Combining Definition 2.24 and Lemma 2.25 with part (b) of Proposition 2.9 yields the proof. $\quad\square$

Corollary 2.27. *Let $\varkappa \in \mathbb{N} \cup \{+\infty\}$ and $k \in \mathbb{Z}_{1, \varkappa}$. Furthermore, suppose that $(C_j)_{j=0}^{\varkappa}$ is an order k canonical T-n.n.d. sequence in $\mathbb{C}^{q \times q}$. Then:*

(a) *For each $\ell \in \mathbb{Z}_{k, \varkappa}$, the sequence $(C_j)_{j=0}^{\varkappa}$ is order ℓ canonical.*
(b) *If $\varkappa \geq 2$ and $k \in \mathbb{Z}_{1, \varkappa-1}$, then, for each $\ell \in \mathbb{Z}_{k+1, \varkappa}$, the sequence $(C_j)_{j=0}^{\varkappa}$ is order ℓ central.*

Proof. Use Propositions 2.26 and 2.15. $\quad\square$

Example 2.28. Let $\varkappa \in \mathbb{N} \cup \{+\infty\}$ and $A \in \mathbb{C}_{\mathrm{H}}^{q \times q}$. For each $j \in \mathbb{Z}_{0, \varkappa}$ let $C_j := A$. Then, for each $n \in \mathbb{Z}_{0, \varkappa}$, it follows that $\operatorname{rank} T_n = \operatorname{rank} A$. Therefore, $(C_j)_{j=0}^{\varkappa}$ is order 1 canonical. In particular, if $A \in \mathbb{C}_{\geq}^{q \times q}$, then it follows from Example 2.2 that $(C_j)_{j=0}^{\varkappa}$ is an order 1 canonical T-n.n.d. sequence.

Example 2.29. Let $K \in \mathbb{K}_{q \times q}$. For each $j \in \mathbb{N}_0$, let $C_j := K^j$. Then $(C_j)_{j=0}^{\infty}$ is canonical if and only if K is unitary. When this is the case, $(C_j)_{j=0}^{\infty}$ is order 1 canonical. This can be recognized as follows: From part (a) of Example 2.21, we see that $(C_j)_{j=0}^{\infty} \in \mathcal{T}_{1, \infty}$. If K is not unitary, then part (b) of Example 2.21 shows us that $L_{k+1} \neq 0_{q \times q}$, for each $k \in \mathbb{N}_0$. By Proposition 2.26, it thus follows that $(C_j)_{j=0}^{\infty}$ is not canonical. If K is unitary, then part (b) of Example 2.21 shows us that $L_{k+1} = 0_{q \times q}$, for each $k \in \mathbb{N}$. By Proposition 2.26, we thus see that $(C_j)_{j=0}^{\infty}$ is order 1 canonical.

The following lemma demonstrates an important approach to constructing canonical $\mathcal{T}_{q, \infty}$ sequences.

Lemma 2.30. *Suppose* $r \in \mathbb{N}$, $(A_s)_{s=1}^r$ *is a sequence in* $\mathbb{C}_{\ge}^{q \times q}$ *and that* $(z_s)_{s=1}^r$ *is a sequence of pairwise different points in* $\mathbb{T}$. *For each* $j \in \mathbb{N}_0$, *let*

$$C_j := \sum_{s=1}^r z_s^{-j} A_s.$$

Then $(C_j)_{j=0}^\infty \in \mathcal{T}_{q, \infty}$ *and* $\operatorname{rank} T_n = \sum_{s=1}^r \operatorname{rank} A_s$ *for* $n \in \mathbb{Z}_{r-1, \infty}$. *The sequence* $(C_j)_{j=0}^\infty$ *is, furthermore, order* r *canonical and* $(C_j)_{j=0}^{r-1} \in \tilde{\mathcal{T}}_{q, r-1}$ *if and only if* $(A_s)_{s=1}^r$ *is a sequence in* $\mathbb{C}_{>}^{q \times q}$.

Proof. Let $n \in \mathbb{N}_0$. Considering the Vandermonde matrix

$$V_{q, n}((z_s)_{s=1}^r) := \begin{pmatrix} z_1^0 I_q & z_1^1 I_q & \cdots & z_1^n I_q \\ z_2^0 I_q & z_2^1 I_q & \cdots & z_2^n I_q \\ \vdots & \vdots & & \vdots \\ z_r^0 I_q & z_r^1 I_q & \cdots & z_r^n I_q \end{pmatrix},$$

we obtain

$$T_n = [V_{q, n}((z_s)_{s=1}^r)]^* [\operatorname{diag}(A_1, A_2, \ldots, A_r)] [V_{q, n}((z_s)_{s=1}^r)]. \quad (2.8)$$

Since $(A_s)_{s=1}^r$ is a sequence in $\mathbb{C}_{\ge}^{q \times q}$, this implies $T_n \in \mathbb{C}_{\ge}^{(n+1)q \times (n+1)q}$. Hence, $(C_j)_{j=0}^\infty \in \mathcal{T}_{q, \infty}$. Since the elements of $(z_s)_{s=1}^r$ are pairwise different, it follows that

$$\operatorname{rank}[V_{q, n}((z_s)_{s=1}^r)] = r \cdot q, \quad (2.9)$$

for all $n \in \mathbb{Z}_{r-1, \infty}$ and thus, from (2.8) that

$$\operatorname{rank} T_n = \operatorname{rank}[\operatorname{diag}(A_1, A_2, \ldots, A_r)] = \sum_{s=1}^r \operatorname{rank} A_s.$$

We thus also see that $(C_j)_{j=0}^\infty$ is order r canonical. For $n = r - 1$ it follows from (2.9) that $\det [V_{q, r-1}((z_s)_{s=1}^r)] \ne 0$. Therefore, from (2.8), we see that $(C_j)_{j=0}^{r-1} \in \tilde{\mathcal{T}}_{q, r-1}$ if and only if $(A_s)_{s=1}^r$ is a sequence in $\mathbb{C}_{>}^{q \times q}$. $\square$

The next result shows (see [20, Corollary 1.14]) that every canonical sequence $(C_j)_{j=0}^n \in \mathcal{T}_{q, \infty}$ is of the form described in Lemma 2.30.

Theorem 2.31. *Suppose that* $(C_j)_{j=0}^\infty$ *is a sequence in* $\mathbb{C}^{q \times q}$. *Then both of the following two conditions are equivalent:*

(i) $(C_j)_{j=0}^\infty$ *is a canonical T-n.n.d. sequence.*

(ii) *There are some* $r \in \mathbb{N}$, *pairwise different points* $z_1, z_2, \ldots, z_r \in \mathbb{T}$ *and matrices* $A_1, A_2, \ldots, A_r \in \mathbb{C}_{\ge}^{q \times q}$ *such that*

$$C_j = \sum_{m=1}^r z_m^{-j} A_m \qquad j \in \mathbb{N}_0.$$

The following proposition can be seen as an addendum to Theorem 2.11.

Proposition 2.32. *Let* $n \in \mathbb{N}_0$ *and* $(C_j)_{j=0}^n \in \widetilde{\mathcal{T}}_{q,n}$. *Suppose, furthermore, that* $C_{n+1} \in \mathbb{C}^{q \times q}$. *Then both of the following two conditions are equivalent:*

(i) *The sequence* $(C_j)_{j=0}^{n+1}$ *is an order* $n+1$ *canonical T-n.n.d. sequence.*

(ii) *There exists a unitary matrix* $U_{n+1} \in \mathbb{C}^{q \times q}$ *such that*

$$C_{n+1} = M_{n+1} + \sqrt{L_{n+1}} U_{n+1} \sqrt{R_{n+1}}.$$

Proof. Combine part (b) of Proposition 2.9, part (b) of Theorem 2.11 and Proposition 2.26. $\square$

For a detailed discussion of canonical Toeplitz non-negative definite sequences in $\mathbb{C}^{q \times q}$, we refer the reader to [20, Section 1], where, for arbitrary $n \in \mathbb{N}_0$ and $(C_j)_{j=0}^n \in \mathcal{T}_{q,n}$, the set of all $C_{n+1} \in \mathcal{C}_{\geq}\left[(C_j)_{j=0}^n\right]$ for which $(C_j)_{j=0}^{n+1}$ is order n canonical is described. This set is never empty.

The canonical extension problem for sequences in $\mathcal{T}_{q,n}$ is a special case of the problem of determining all rank-preserving extensions of such sequences. This more general problem was dealt with in [11, Theorem 3]. For discussions of this topic in the scalar case, we refer the reader to the monograph Iohvidov [24] as well as the article Akimoto/Ito [1].

3. Matricial Carathéodory sequences

In this section, we present some basic facts on matricial Carathéodory sequences and their relationship to Toeplitz non-negative definite sequences of matrices. If $n \in \mathbb{N}_0$, then a sequence $(s_j)_{j=0}^n$ in $\mathbb{C}^{q \times q}$ is called a $q \times q$ **Carathéodory sequence** (or simply a $q \times q$ **C-sequence**) if the real part $\mathrm{Re}\, S_n$ of the matrix S_n given by (1.1) and $S_n := S_n^{(s)}$ is non-negative Hermitian, i.e., if $S_n \in \mathcal{R}_{(n+1)q, \geq}$, and a **strict** $q \times q$ **Carathéodory sequence** (or **strict** $q \times q$ **C-sequence**) if $\mathrm{Re}\, S_n$ is positive Hermitian, i.e., if $S_n \in \mathcal{R}_{(n+1)q, >}$.

Letting $n \in \mathbb{N}$, $m \in \mathbb{Z}_{0, n-1}$ and $(s_j)_{j=0}^n$ be a $q \times q$ C-sequence, we see from (1.1) that S_m is the upper left $(m+1)q \times (m+1)q$ block of S_n. Hence, $(s_j)_{j=0}^m$ is also a $q \times q$ C-sequence. Similarly, if $(s_j)_{j=0}^n$ is a strict C-sequence, then $(s_j)_{j=0}^m$ is also a strict C-sequence. For this reason, we call a sequence $(s_j)_{j=0}^\infty$ in $\mathbb{C}^{q \times q}$ a $q \times q$ **Carathéodory sequence** if, for each $n \in \mathbb{N}_0$, the sequence $(s_j)_{j=0}^n$ is a $q \times q$ Carathéodory sequence. A Carathéodory sequence $(s_j)_{j=0}^\infty$ is called **strict** if, for each $n \in \mathbb{N}_0$, the sequence $(s_j)_{j=0}^n$ is a strict $q \times q$ Carathéodory sequence. For each $\varkappa \in \mathbb{N}_0 \cup \{+\infty\}$, the set of all $q \times q$ Carathéodory sequences will be denoted by $\mathcal{C}_{q, \varkappa}$ and the set of all strict $q \times q$ Carathéodory sequences by $\widetilde{\mathcal{C}}_{q, \varkappa}$.

Remark 3.1. Let $\varkappa \in \mathbb{N}_0 \cup \{+\infty\}$. If $(s_j)_{j=0}^\varkappa$ is a $q \times q$ C-sequence, then $s_0 \in \mathcal{R}_{q, \geq}$. If $(s_j)_{j=0}^\varkappa$ is a strict $q \times q$ C-sequence, then $s_0 \in \mathcal{R}_{q, >}$.

Remark 3.2. Let $\varkappa \in \mathbb{N}_0 \cup \{+\infty\}$ and let $(s_j)_{j=0}^{\varkappa}$ be a sequence in $\mathbb{C}^{q \times q}$ with $s_j = 0_{q \times q}$, for each $j \in \mathbb{Z}_{1, \varkappa}$. Then $(s_j)_{j=0}^{\varkappa}$ is a $q \times q$ C-sequence if and only if $s_0 \in \mathcal{R}_{q, \geq}$. Moreover, $(s_j)_{j=0}^{\varkappa}$ is a strict $q \times q$ C-sequence if and only if $s_0 \in \mathcal{R}_{q, >}$.

Remark 3.3. Let $\varkappa \in \mathbb{N}_0 \cup \{+\infty\}$. If $(s_j)_{j=0}^{\varkappa}$ is a $q \times q$ C-sequence and $A \in \mathbb{C}^{q \times p}$, then $(A^* s_j A)_{j=0}^{\varkappa}$ is a $p \times p$ C-sequence.

Remark 3.4. Let $\varkappa \in \mathbb{N}_0 \cup \{+\infty\}$ and let $m \in \mathbb{N}$. For each $r \in \mathbb{Z}_{1, m}$, suppose that $\alpha_r \in [0, +\infty)$ and $\left(s_j^{(r)} \right)_{j=0}^{\varkappa} \in \mathcal{C}_{q, \varkappa}$. Then $\left(\sum_{r=1}^{m} \alpha_r s_j^{(r)} \right)_{j=0}^{\varkappa} \in \mathcal{C}_{q, \varkappa}$. If, furthermore, there exists an $r_0 \in \mathbb{Z}_{1, m}$ such that $\alpha_{r_0} \in (0, +\infty)$ and $\left(s_j^{(r_0)} \right)_{j=0}^{\varkappa} \in \widetilde{\mathcal{C}}_{q, \varkappa}$, then $\left(\sum_{r=1}^{m} \alpha_r s_j^{(r)} \right)_{j=0}^{\varkappa} \in \widetilde{\mathcal{C}}_{q, \varkappa}$.

Remark 3.5. Let $\varkappa \in \mathbb{N}_0 \cup \{+\infty\}$ and $r \in \mathbb{N}$. Suppose that $(q_j)_{j=1}^{r}$ is a sequence in $\mathbb{N}$ such that $q = \sum_{j=1}^{r} q_j$. For each $m \in \mathbb{Z}_{1, r}$, let $(s_j^{(m)})_{j=0}^{\varkappa}$ be a sequence in $\mathbb{C}^{q_m \times q_m}$. Then $(d_j)_{j=0}^{\varkappa} := \left(\mathrm{diag}\left(s_j^{(1)}, s_j^{(2)}, \ldots, s_j^{(r)} \right) \right)_{j=0}^{\varkappa}$ is a $q \times q$ C-sequence if and only if $(s_j^{(m)})_{j=0}^{\varkappa}$ is a $q_m \times q_m$ C-sequence for all $m \in \mathbb{Z}_{1, r}$. The sequence $(d_j)_{j=0}^{\varkappa}$ is a strict $q \times q$ C-sequence if and only if $(s_j^{(m)})_{j=0}^{\varkappa}$ is a strict $q_m \times q_m$ C-sequence for all $m \in \mathbb{Z}_{1, r}$.

Matricial Carathéodory sequences are closely related to Toeplitz non-negative definite sequences of matrices.

Remark 3.6. Let $\varkappa \in \mathbb{N}_0 \cup \{+\infty\}$. Let $(s_j)_{j=0}^{\varkappa}$ be a sequence in $\mathbb{C}^{q \times q}$ and let $(C_j)_{j=0}^{\varkappa}$ be defined by

$$C_\ell := \begin{cases} \mathrm{Re}\, s_0, & \text{if } \ell = 0, \\ \frac{1}{2} s_\ell, & \text{if } \ell \in \mathbb{Z}_{1, \varkappa}. \end{cases} \tag{3.1}$$

Then $\mathrm{Re}\, S_k = T_k$ for each $k \in \mathbb{Z}_{0, \varkappa}$. Thus, $(s_j)_{j=0}^{\varkappa}$ is a $q \times q$ C-sequence if and only if $(C_j)_{j=0}^{\varkappa}$ is Toeplitz non-negative definite. Furthermore, $(s_j)_{j=0}^{\varkappa}$ is a strict $q \times q$ C-sequence if and only if $(C_j)_{j=0}^{\varkappa}$ is Toeplitz positive definite.

Remark 3.7. Let $\varkappa \in \mathbb{N}_0 \cup \{+\infty\}$ and let $(C_j)_{j=0}^{\varkappa}$ be a sequence in $\mathbb{C}^{q \times q}$. Suppose that $s_0 := C_0$ and that $s_j := 2C_j$, for each $j \in \mathbb{Z}_{1, \varkappa}$. For each $k \in \mathbb{Z}_{0, \varkappa}$, it then follows that $T_k = \mathrm{Re}\, S_k$. Thus, $(C_j)_{j=0}^{\varkappa}$ is a T-n.n.d. sequence if and only if $(s_j)_{j=0}^{\varkappa}$ is a $q \times q$ Carathéodory sequence and $s_0 = s_0^*$.

Remark 3.8. Let $\varkappa \in \mathbb{N}_0 \cup \{+\infty\}$ and $(s_j)_{j=0}^{\varkappa}$ be a sequence in $\mathbb{C}^{q \times q}$ and let $(\widehat{s}_j)_{j=0}^{\varkappa}$ be defined by $\widehat{s}_0 := \mathrm{Re}\, s_0$ and $\widehat{s}_\ell := s_\ell$ for each $\ell \in \mathbb{Z}_{1, \varkappa}$. Then Remark 3.6

shows that $(s_j)_{j=0}^\varkappa$ is a $q \times q$ C-sequence if and only if $(\widehat{s}_j)_{j=0}^\varkappa$ is a $q \times q$ C-sequence. $(s_j)_{j=0}^\varkappa$ is a strict $q \times q$ C-sequence if and only if $(\widehat{s}_j)_{j=0}^\varkappa$ is a strict $q \times q$ C-sequence.

Remark 3.9. Let $\varkappa \in \mathbb{N}_0 \cup \{+\infty\}$.

(a) If $(r_j)_{j=0}^\varkappa \in \mathcal{C}_{1,\varkappa}$ and $(s_j)_{j=0}^\varkappa \in \mathcal{C}_{q,\varkappa}$, then, by part (a) of Remark 2.6 and Remark 3.6, it follows that $(r_j s_j)_{j=0}^\varkappa \in \mathcal{C}_{q,\varkappa}$.

(b) If $(r_j)_{j=0}^\varkappa \in \widetilde{\mathcal{C}}_{1,\varkappa}$ and $(s_j)_{j=0}^\varkappa \in \widetilde{\mathcal{C}}_{q,\varkappa}$, then, by part (b) of Remark 2.6 and Remark 3.6, it follows that $(r_j s_j)_{j=0}^\varkappa \in \widetilde{\mathcal{C}}_{q,\varkappa}$.

Remark 3.10. Let $\varkappa \in \mathbb{N}_0 \cup \{+\infty\}$.

(a) Suppose $u \in \mathbb{D} \cup \mathbb{T}$ and $(s_j)_{j=0}^\varkappa \in \mathcal{C}_{q,\varkappa}$. By part (a) of Remark 2.23 and part (a) of Remark 3.9 it follows that $(u^j s_j)_{j=0}^\varkappa \in \mathcal{C}_{q,\varkappa}$.

(b) Suppose $u \in \mathbb{D}$ and $(s_j)_{j=0}^\varkappa \in \widetilde{\mathcal{C}}_{q,\varkappa}$. By part (b) of Remark 2.23 and part (b) of Remark 3.9 it follows that $(u^j s_j)_{j=0}^\varkappa \in \widetilde{\mathcal{C}}_{q,\varkappa}$.

Lemma 3.11. *Let* $\varkappa \in \mathbb{N}_0 \cup \{+\infty\}$. *Then* $\mathcal{T}_{q,\varkappa} \subseteq \mathcal{C}_{q,\varkappa}$. *Furthermore, a sequence* $(C_j)_{j=0}^\varkappa \in \mathcal{T}_{q,\varkappa}$ *belongs to* $\widetilde{\mathcal{C}}_{q,\varkappa}$ *if and only if* $C_0 \in \mathbb{C}_>^{q \times q}$.

Proof. Let $(C_j)_{j=0}^\varkappa \in \mathcal{T}_{q,\varkappa}$. Thus, $C_0 \in \mathbb{C}_\geq^{q \times q}$ and, in particular, it follows that $C_0 \in \mathcal{R}_{q,\geq}$. Hence, by Remark 3.2, it follows that the sequence $(r_j)_{j=0}^\varkappa$ given by $r_j := \delta_{j,0} C_0$, where $\delta_{j,k}$ is the Kronecker delta, belongs to $\mathcal{C}_{q,\varkappa}$. If we define the sequence $(s_j)_{j=0}^\varkappa$ as we did in Remark 3.7 using $(C_j)_{j=0}^\varkappa$, then by the same remark $(s_j)_{j=0}^\varkappa$ belongs to $\mathcal{C}_{q,\varkappa}$. Since $C_j = \frac{1}{2}(r_j + s_j)$ for each $j \in \mathbb{Z}_{0,\varkappa}$, it follows from Remark 3.4 that $(C_j)_{j=0}^\varkappa \in \mathcal{C}_{q,\varkappa}$. If $C_0 \in \mathbb{C}_>^{q \times q}$, then $C_0 \in \mathcal{R}_{q,>}$. Since $C_0 \in \mathbb{C}_>^{q \times q}$, it therefore follows by Remark 3.2 that $(r_j)_{j=0}^\varkappa \in \widetilde{\mathcal{C}}_{q,\varkappa}$ and thus by Remark 3.4 that $(C_j)_{j=0}^\varkappa \in \widetilde{\mathcal{C}}_{q,\varkappa}$. Conversely, if we suppose that $(C_j)_{j=0}^\varkappa \in \widetilde{\mathcal{C}}_{q,\varkappa}$, then it immediately follows that $C_0 = \operatorname{Re} C_0 \in \mathbb{C}_>^{q \times q}$. $\square$

Example 3.12. Let $K \in \mathbb{K}_{q \times q}$ and let $C_j := K^j$, for each $j \in \mathbb{N}_0$. We then have $C_0 = I_q \in \mathbb{C}_>^{q \times q}$ and see from Example 2.21 that $(C_j)_{j=0}^\infty \in \mathcal{T}_{q,\infty}$. Thus, applying Lemma 3.11, we see that $(C_j)_{j=0}^\infty \in \widetilde{\mathcal{C}}_{q,\infty}$.

Example 3.13. Let $K \in \mathbb{K}_{q \times q} \cap \mathbb{C}_\geq^{q \times q}$, $r \in \mathbb{N}$ and $C_j := K^{r+j}$, for each $j \in \mathbb{N}_0$. Then $(C_j)_{j=0}^\infty \in \mathcal{C}_{q,\infty}$. Furthermore, $(C_j)_{j=0}^\infty \in \widetilde{\mathcal{C}}_{q,\infty}$ if and only if $K \in \mathbb{C}_>^{q \times q}$. This follows by combining Example 2.22 and Lemma 3.11, while observing that $K^r \in \mathbb{C}_>^{q \times q}$ if and only if $K \in \mathbb{C}_>^{q \times q}$.

Motivated by Remark 3.6, we now go about implementing special modifications of Definitions 2.10, 2.12 and 2.24.

Definition 3.14. A sequence $(s_j)_{j=0}^\infty$ in $\mathbb{C}^{q \times q}$ is called a **totally strict** $q \times q$ Carathéodory sequence if the sequence $(C_j)_{j=0}^\infty$ defined by (3.1) with $\varkappa = +\infty$ is a totally

Toeplitz positive definite sequence in $\mathbb{C}^{q \times q}$. The set of all totally strict $q \times q$ Carathéodory sequences will be denoted by $\mathcal{C}^t_{q, \infty}$.

Formula (2.6) and Remark 3.6 imply that

$$\mathcal{C}^t_{q, \infty} \subseteq \widetilde{\mathcal{C}}_{q, \infty}. \tag{3.2}$$

Definition 3.15. Let $\varkappa \in \mathbb{N} \cup \{+\infty\}$ and $(s_j)_{j=0}^{\varkappa}$ be a sequence in $\mathbb{C}^{q \times q}$. Furthermore, let the sequence $(C_j)_{j=0}^{\varkappa}$ be defined by (3.1).

(a) Suppose $k \in \mathbb{Z}_{1, \varkappa}$. We say that $(s_j)_{j=0}^{\varkappa}$ is a $q \times q$ Carathéodory-central sequence of order k (or a $q \times q$ order k C-central sequence), if $(C_j)_{j=0}^{\varkappa}$ is order k central.

(b) Suppose $k \in \mathbb{Z}_{1, \varkappa}$. We say that $(s_j)_{j=0}^{\varkappa}$ is a $q \times q$ Carathéodory-central sequence of minimal order k (or a $q \times q$ minimal order k C-central sequence) if $(C_j)_{j=0}^{\varkappa}$ is central of minimal order k.

(c) The sequence $(s_j)_{j=0}^{\varkappa}$ is simply called a $q \times q$ Carathéodory-central sequence (or a $q \times q$ C-central sequence), if there exists a $k \in \mathbb{Z}_{1, \varkappa}$ such that $(s_j)_{j=0}^{\varkappa}$ is order k central.

Remark 3.16. Let $\varkappa \in \mathbb{N} \cup \{+\infty\}$ and $k \in \mathbb{Z}_{1, \varkappa}$. Furthermore, suppose that $(s_j)_{j=0}^{\varkappa}$ is a $q \times q$ order k C-central sequence. For any $\ell \in \mathbb{Z}_{k, \varkappa}$, Remark 2.13 then shows that $(s_j)_{j=0}^{\varkappa}$ is also ℓ order C-central.

The following is the analogous result to Remark 2.18 for C-centrality.

Remark 3.17. Let $n \in \mathbb{N}_0$ and $(s_j)_{j=0}^n$ be a sequence in $\mathbb{C}^{q \times q}$.

(a) There exists a unique order $n + 1$ C-central sequence $(\widetilde{s}_j)_{j=0}^{\infty}$ in $\mathbb{C}^{q \times q}$ such that $\widetilde{s}_j = s_j$, for each $j \in \mathbb{Z}_{0, n}$. This sequence $(\widetilde{s}_j)_{j=0}^{\infty}$ is called the C-central sequence corresponding to $(s_j)_{j=0}^n$.

(b) Let $(C_j)_{j=0}^n$ be defined by (3.1) with $\varkappa = n$. Then the central sequence $(\widetilde{C}_j)_{j=0}^{\infty}$ corresponding to $(C_j)_{j=0}^n$ is given by $\widetilde{C}_0 = \mathrm{Re}\, \widetilde{s}_0$ and $\widetilde{C}_\ell = \frac{1}{2}\widetilde{s}_\ell$ for each $\ell \in \mathbb{N}$.

Remark 3.18. Let $k \in \mathbb{N}$ and $(s_j)_{j=0}^{\infty}$ be an order k C-central sequence in $\mathbb{C}^{q \times q}$. By part (a) of Remark 3.17 it then follows that $(s_j)_{j=0}^{\infty}$ is the C-central sequence corresponding to $(s_j)_{j=0}^{k-1}$.

We now consider C-central sequences in more detail.

Lemma 3.19. *Let $k \in \mathbb{N}$ and $(s_j)_{j=0}^{k-1}$ be a sequence in $\mathbb{C}^{q \times q}$. Suppose that $(\widetilde{s}_j)_{j=0}^{\infty}$ is the C-central sequence corresponding to $(s_j)_{j=0}^{k-1}$. Then:*

(a) $(\widetilde{s}_j)_{j=0}^{\infty} \in \mathcal{C}_{q, \infty}$ *if and only if* $(s_j)_{j=0}^{k-1} \in \mathcal{C}_{q, k-1}$.

(b) *The following three conditions are all equivalent:*

(i) $(s_j)_{j=0}^{k-1} \in \tilde{\mathcal{C}}_{q,\,k-1}$.
(ii) $(\tilde{s}_j)_{j=0}^{k-1} \in \mathcal{C}_{q,\,k-1}^{\mathsf{t}}$.
(iii) $(\tilde{s}_j)_{j=0}^{\infty} \in \tilde{\mathcal{C}}_{q,\,\infty}$.

Proof. Combine Remark 3.6, Remark 3.17 and Lemma 2.19. $\qquad\square$

Example 3.20. Let $K \in \mathbb{K}_{q \times q}$ and the sequence $(C_j)_{j=0}^{\infty}$ be given by $s_0 := I_q$ and $s_j := 2K^j$ for each $j \in \mathbb{N}$. From Example 2.21 and Remark 3.6 we see that:

(a) $(s_j)_{j=0}^{\infty}$ is an order 2 C-central $q \times q$ C-sequence and if (and only if) $K \neq 0_{q \times q}$, then $(s_j)_{j=0}^{\infty}$ is minimal order 2 C-central.
(b) $(s_j)_{j=0}^{\infty} \in \mathcal{C}_{q,\,\infty}^{\mathsf{t}}$ if and only if $K \in \mathbb{D}_{q \times q}$.

Example 3.21. Let $K \in \mathbb{K}_{q \times q} \cap \mathbb{C}_{\geq}^{q \times q}$, $r \in \mathbb{N}$ and $(C_j)_{j=0}^{\infty}$ be given by $s_0 := K^r$ and $s_j := 2K^{r+j}$ for each $j \in \mathbb{N}$. From Example 2.22 and Remark 3.6 we see that $(C_j)_{j=0}^{\infty}$ is an order 2 C-central $q \times q$ Carathéodory sequence.

Recalling Remark 3.6, we now see how Definition 2.24 carries over to $q \times q$ Carathéodory sequences.

Definition 3.22. Let $\varkappa \in \mathbb{N} \cup \{+\infty\}$ and $(s_j)_{j=0}^{\varkappa}$ be a sequence in $\mathbb{C}^{q \times q}$. Furthermore, let the sequence $(C_j)_{j=0}^{\varkappa}$ be defined by (3.1).

(a) Suppose $k \in \mathbb{Z}_{1,\,\varkappa}$. We say that $(s_j)_{j=0}^{\varkappa}$ is a $q \times q$ Carathéodory-canonical sequence of order k (or $q \times q$ order k C-canonical sequence), if $(C_j)_{j=0}^{\varkappa}$ is order k canonical.
(b) Suppose $k \in \mathbb{Z}_{1,\,\varkappa}$. We say that $(s_j)_{j=0}^{\varkappa}$ is $q \times q$ Carathéodory-canonical of minimal order k (or $q \times q$ minimal order k C-canonical) if $(C_j)_{j=0}^{\varkappa}$ is minimal order k canonical.
(c) $(s_j)_{j=0}^{\varkappa}$ is simply called a $q \times q$ Carathéodory-canonical sequence (or a $q \times q$ C-canonical sequence), if there exists a $k \in \mathbb{Z}_{1,\,\varkappa}$ such that $(s_j)_{j=0}^{\varkappa}$ is order k canonical.

Combining Definition 3.22 and Corollary 2.27 (while recalling Remark 3.6 and Definition 3.15), we obtain the following remark.

Remark 3.23. Let $\varkappa \in \mathbb{N} \cup \{+\infty\}$ and $k \in \mathbb{Z}_{1,\,\varkappa}$. Suppose that $(s_j)_{j=0}^{\varkappa}$ is an order k C-canonical $q \times q$ Carathéodory sequence. Then:

(a) For each $\ell \in \mathbb{Z}_{k,\,\varkappa}$, the sequence $(s_j)_{j=0}^{\varkappa}$ is order ℓ C-canonical.
(b) Let $\varkappa \geq 2$ and $k \in \mathbb{Z}_{1,\,\varkappa-1}$. For any $\ell \in \mathbb{Z}_{k+1,\,\varkappa}$, it then follows that $(s_j)_{j=0}^{\varkappa}$ is order ℓ C-central.

Example 3.24. Let $\varkappa \in \mathbb{N} \cup \{+\infty\}$ and $A \in \mathbb{C}_{\geq}^{q \times q}$. Suppose, furthermore, that $s_0 := A$ and $s_j := 2A$ for each $j \in \mathbb{Z}_{1,\,\varkappa}$. Then Example 2.28 and Remark 3.6 show that $(s_j)_{j=0}^{\varkappa}$ is an order 1 C-central $q \times q$ Carathéodory sequence.

Example 3.25. Let $K \in \mathbb{K}_{q \times q}$. Suppose, furthermore, that $s_0 := I_q$ and $s_j := 2K^j$ for $j \in \mathbb{N}$. Then part (a) of Example 3.20 shows that $(s_j)_{j=0}^\infty$ is a $q \times q$ Carathéodory sequence. Example 2.29 furthermore shows that $(s_j)_{j=0}^\infty$ is C-canonical if and only if K is unitary. When this is the case, $(s_j)_{j=0}^\infty$ is order 1 C-canonical.

4. Reciprocal sequences of matricial Carathéodory sequences

We will, in this section, discuss reciprocal sequences of matricial Carathéodory sequences. Our first task will be to show that the reciprocal sequence of a Carathéodory sequence is itself a Carathéodory sequence. We will draw heavily on results established in Section 1. We next show that every Carathéodory sequence belongs to the class of matrices introduced in Definition 1.4, thus establishing the aforementioned connection to Section 1.

Proposition 4.1. *Suppose* $\varkappa \in \mathbb{N}_0 \cup \{+\infty\}$.

(a) $\mathcal{C}_{q, \varkappa} \subseteq \mathcal{F}_{q, \varkappa}^{\mathrm{EP}}$.

(b) $\mathcal{C}_{q, \varkappa} \subseteq \mathcal{D}_{q \times q, \varkappa}$.

Proof. Suppose that $(s_j)_{j=0}^\varkappa$ is in $\mathcal{C}_{q, \varkappa}$. It then follows, for each $n \in \mathbb{N}_0$, that $S_n \in \mathcal{R}_{(n+1)q, \geq}$. Thus, by Lemma A.8, we have $S_n \in \mathbb{C}_{\mathrm{EP}}^{(n+1)q \times (n+1)q}$, for each $n \in \mathbb{N}_0$, and therefore $(s_j)_{j=0}^\varkappa \in \mathcal{F}_{q, \varkappa}^{\mathrm{EP}}$. The proof of part (a) is complete. Part (b) follows by part (a) of Proposition 1.10 from the part already proved. $\square$

Corollary 4.2. *Suppose* $\varkappa \in \mathbb{N}_0 \cup \{+\infty\}$. *Then* $\mathcal{T}_{q, \varkappa} \subseteq \mathcal{F}_{q, \varkappa}^{\mathrm{EP}}$.

Proof. Use part (a) of Proposition 4.1 and Lemma 3.11. $\square$

From Corollory 4.2 and part (a) of Proposition 1.10 we furthermore obtain $\mathcal{T}_{q, \varkappa} \subseteq \mathcal{D}_{q \times q, \varkappa}$, as was earlier shown in Proposition 2.7. We next consider the reciprocal sequence to the reciprocal sequence of a Carathéodory sequence $(s_j)_{j=0}^\varkappa$, i.e., its second reciprocal sequence $\left((s_j^\sharp)^\sharp \right)_{j=0}^\varkappa$. This sequence must coincide with the original Carathéodory sequence.

Corollary 4.3. *If* $\varkappa \in \mathbb{N}_0 \cup \{+\infty\}$ *and* $(s_j)_{j=0}^\varkappa$ *is a* $q \times q$ *Carathéodory sequence, then* $\left((s_j^\sharp)^\sharp \right)_{j=0}^\varkappa = (s_j)_{j=0}^\varkappa$.

Proof. Combine Proposition 4.1 with [22, Proposition 5.13, Remark 4.7]. $\square$

We now come to the first main result of this section.

Theorem 4.4. *Let* $\varkappa \in \mathbb{N}_0 \cup \{+\infty\}$ *and* $(s_j)_{j=0}^\varkappa$ *be a* $q \times q$ *Carathéodory sequence. The reciprocal sequence* $(s_j^\sharp)_{j=0}^\varkappa$ *is then also a* $q \times q$ *Carathéodory sequence. If* $(s_j)_{j=0}^\varkappa$ *is a strict* $q \times q$ *Carathéodory sequence, then* $(s_j^\sharp)_{j=0}^\varkappa$ *is also a strict* $q \times q$ *Carathéodory sequence.*

Proof. By Proposition 4.1, it follows that $(s_j)_{j=0}^{\varkappa} \in \mathcal{F}_{q,\varkappa}^{EP}$. Let $n \in \mathbb{Z}_{0,\varkappa}$. Propositions A.6 and 1.11 now yield

$$\operatorname{Re} S_n^{\sharp} = (S_n^{\sharp})^* (\operatorname{Re} S_n) S_n^{\sharp}. \tag{4.1}$$

Thus, since $\operatorname{Re} S_n \in \mathbb{C}_{\geq}^{(n+1)q \times (n+1)q}$, we see from (4.1) that $\operatorname{Re} S_n^{\sharp} \in \mathbb{C}_{\geq}^{(n+1)q \times (n+1)q}$. Therefore, $S_n^{\sharp} \in \mathcal{R}_{(n+1)q, \geq}$. Thus, $(s_j^{\sharp})_{j=0}^{\varkappa}$ is a $q \times q$ C-sequence.
Suppose $(s_j)_{j=0}^{\varkappa}$ is a strict $q \times q$ C-sequence. By Remark 3.1, it follows that $s_0 \in \mathcal{R}_{q, >}$. Thus, by part (a) of Lemma A.12, we have $\det s_0 \neq 0$. Using Definition 1.1, we see that $s_0^{\sharp} = s_0^{-1}$. Thus, $\det S_n^{\sharp} = (\det s_0)^{-(n+1)} \neq 0$. Because $\operatorname{Re} S_n \in \mathbb{C}_{>}^{(n+1)q \times (n+1)q}$, we see from (4.1) that $\operatorname{Re} S_n^{\sharp} \in \mathbb{C}_{>}^{(n+1)q \times (n+1)q}$. Hence, $S_n^{\sharp} \in \mathcal{R}_{(n+1)q, >}$ and $(s_j^{\sharp})_{j=0}^{\varkappa}$ is a strict $q \times q$ C-sequence. □

Corollary 4.5. *Let $\varkappa \in \mathbb{N}_0 \cup \{+\infty\}$ and let $(s_j)_{j=0}^{\varkappa}$ be a sequence in $\mathbb{C}^{q \times q}$. The following two conditions are equivalent to one another:*

(i) *$(s_j)_{j=0}^{\varkappa}$ is a $q \times q$ Carathéodory sequence.*

(ii) *$(s_j)_{j=0}^{\varkappa} \in \mathcal{D}_{q \times q, \varkappa}$ and its reciprocal sequence $(s_j^{\sharp})_{j=0}^{\varkappa}$ is a $q \times q$ Carathéodory sequence.*

Proof. "(i)$\Longrightarrow$(ii)". Because of (i), Proposition 4.1 implies $(s_j)_{j=0}^{\varkappa} \in \mathcal{D}_{q \times q, \varkappa}$, while it follows by Theorem 4.4 that $(s_j^{\sharp})_{j=0}^{\varkappa}$ is a $q \times q$ Carathéodory sequence. "(ii)$\Longrightarrow$(i)". Because of (ii), it follows by Theorem 4.4 and Corollary 4.3 that $(s_j)_{j=0}^{\varkappa}$ is a $q \times q$ Carathéodory sequence. □

Together, Proposition 4.1 and Theorem 1.12 will give us a better picture of the structure of a $q \times q$ Carathéodory sequence.

Theorem 4.6. *Let $q \geq 2$, $r \in \mathbb{Z}_{1, q-1}$ and $\varkappa \in \mathbb{N}_0 \cup \{+\infty\}$. Furthermore, let $(s_j)_{j=0}^{\varkappa}$ be a $q \times q$ Carathéodory sequence with rank $s_0 = r$. Suppose $(\tilde{s}_j)_{j=0}^{\varkappa}$ is defined as in Theorem 1.12. Then, parts (a) and (b) of Theorem 1.12 both hold true and $(\tilde{s}_j)_{j=0}^{\varkappa}$ is an $r \times r$ Carathéodory sequence.*

Proof. By Proposition 4.1, we have $(s_j)_{j=0}^{\varkappa} \in \mathcal{F}_{q,\varkappa}^{EP}$. Parts (a) and (b) of Theorem 1.12 therefore both hold true. From Remark 3.3 it, furthermore, follows that $(\tilde{s}_j)_{j=0}^{\varkappa}$ is an $r \times r$ Carathéodory sequence. □

We now study the relationship a Carathéodory sequence has to its the reciprocal sequence.

Lemma 4.7. *Let $\varkappa \in \mathbb{N} \cup \{+\infty\}$ and let $(s_j)_{j=0}^{\varkappa}$ be a $q \times q$ Carathéodory sequence. Furthermore, let $(s_j^{\sharp})_{j=0}^{\varkappa}$ be the reciprocal sequence to $(s_j)_{j=0}^{\varkappa}$. Suppose that*

$$C_j := \begin{cases} \operatorname{Re} s_0 & \text{if } j = 0, \\ \frac{1}{2}s_j & \text{if } j \in \mathbb{Z}_{1, \varkappa} \end{cases} \quad \text{and} \quad C_j^{[\sharp]} := \begin{cases} \operatorname{Re} s_0^{\sharp} & \text{if } j = 0, \\ \frac{1}{2}s_j^{\sharp} & \text{if } j \in \mathbb{Z}_{1, \varkappa} \end{cases} \tag{4.2}$$

for each $j \in \mathbb{Z}_{0, \varkappa}$. Then:

(a) $(C_j)_{j=0}^{\varkappa}$ and $\left(C_j^{[\sharp]}\right)_{j=0}^{\varkappa}$ are both T-n.n.d. sequences in $\mathbb{C}^{q\times q}$.

(b) Let $k \in \mathbb{Z}_{0,\varkappa}$. Recalling (2.1), let
$$T_k := T_k^{(C)} \qquad and \qquad T_k^{[\sharp]} := T_k^{(C^{[\sharp]})}.$$
Then $\operatorname{rank} T_k = \operatorname{rank} T_k^{[\sharp]}$,
$$T_k = \operatorname{Re} S_k, \qquad T_k^{[\sharp]} = \operatorname{Re} S_k^{\sharp}, \tag{4.3}$$
$$T_k^{[\sharp]} S_k = \left(S_k^+\right)^* T_k, \qquad S_k T_k^{[\sharp]} = T_k \left(S_k^+\right)^*, \tag{4.4}$$
$$S_k^* T_k^{[\sharp]} = T_k S_k^+ \qquad and \qquad T_k^{[\sharp]} S_k^* = S_k^+ T_k. \tag{4.5}$$

(c) For each $k \in \mathbb{Z}_{1,\varkappa}$, the matrices Z_k and Y_k defined by (2.3) satisfy
$$Z_k T_{k-1}^+ T_{k-1} = Z_k \qquad and \qquad T_{k-1} T_{k-1}^+ Y_k = Y_k. \tag{4.6}$$

(d) For each $k \in \mathbb{Z}_{1,\varkappa}$, the matrices $Z_k^{[\sharp]}$ and $Y_k^{[\sharp]}$ defined by
$$Z_k^{[\sharp]} := \operatorname{row}\left(C_{k+1-j}^{[\sharp]}\right)_{j=1}^{k} \qquad and \qquad Y_k^{[\sharp]} := \operatorname{col}\left(C_j^{[\sharp]}\right)_{j=1}^{k}$$
satisfy
$$Z_k^{[\sharp]} = -s_0^+ Z_k S_{k-1}^+, \qquad Y_k^{[\sharp]} = -S_{k-1}^+ Y_k s_0^+, \tag{4.7}$$
$$Z_k^{[\sharp]} = -s_0^+ Z_k T_{k-1}^+ S_{k-1}^* T_{k-1}^{[\sharp]} \quad and \quad Y_k^{[\sharp]} = -T_{k-1}^{[\sharp]} S_{k-1}^* T_{k-1}^+ Y_k s_0^+. \tag{4.8}$$

Proof. Part (a) follows from Remark 3.6 using Theorem 4.4. The equations in (4.3) follow directly from the definitions of the relevant matrices. From Proposition 4.1, we see that $(s_j)_{j=0}^{\varkappa} \in \mathcal{F}_{q,\varkappa}^{EP}$. Combining this with (4.3), we need only apply Proposition 1.11 along with parts (g), (i), (j), (h) and (k) of Corollary A.7 to obtain the remaining equations of part (b). Let $k \in \mathbb{Z}_{1,\varkappa}$. The matrix T_k is non-negative Hermitian and admits the block-partitions
$$T_k = \begin{pmatrix} T_{k-1} & Z_k^* \\ Z_k & C_0 \end{pmatrix} \qquad and \qquad T_k = \begin{pmatrix} C_0 & Y_k^* \\ Y_k & T_{k-1} \end{pmatrix}. \tag{4.9}$$
Using well-known properties of non-negative Hermitian block matrices (see, for instance, [8, Lemma 1.1.9]), we obtain both equations in (4.6). Because of Proposition 4.1, the equations in (4.7) follow directly from [22, Corollary 4.23]. Using (b) and (c), we then see that
$$-s_0^+ Z_k T_{k-1}^+ S_{k-1}^* T_{k-1}^{[\sharp]} = -s_0^+ Z_k T_{k-1}^+ T_{k-1} S_{k-1}^+ = -s_0^+ Z_k S_{k-1}^+ = Z_k^{[\sharp]},$$
and similarly,
$$-T_{k-1}^{[\sharp]} S_{k-1}^* T_{k-1}^+ Y_k s_0^+ = -S_{k-1}^+ T_{k-1} T_{k-1}^+ Y_k s_0^+ = -S_{k-1}^+ Y_k s_0^+ = Y_k^{[\sharp]}.$$
Thus, the proof is complete. $\square$

Using part (b) of Lemma 4.7, we will see that the $q\times q$ Carathéodory sequence properties introduced in Definition 3.22 carry over to reciprocal sequences.

Theorem 4.8. *Let* $\varkappa \in \mathbb{N} \cup \{+\infty\}$ *and let* $(s_j)_{j=0}^{\varkappa}$ *be a* $q \times q$ *Carathéodory sequence. Furthermore, let* $\left(s_j^{\sharp}\right)_{j=0}^{\varkappa}$ *be the reciprocal sequence to* $(s_j)_{j=0}^{\varkappa}$. *Then:*

(a) $(s_j)_{j=0}^{\varkappa}$ *is C-canonical if and only if* $\left(s_j^{\sharp}\right)_{j=0}^{\varkappa}$ *is C-canonical.*

(b) *Let* $k \in \mathbb{Z}_{1,\varkappa}$. *Then:*

  (b1) $(s_j)_{j=0}^{\varkappa}$ *is order* k *C-canonical if and only if* $\left(s_j^{\sharp}\right)_{j=0}^{\varkappa}$ *is order* k *C-canonical.*

  (b2) $(s_j)_{j=0}^{\varkappa}$ *is minimal order* k *C-canonical if and only if* $\left(s_j^{\sharp}\right)_{j=0}^{\varkappa}$ *is minimal order* k *C-canonical.*

Proof. Part (b) follows directly from Definition 3.22 and part (b) of Lemma 4.7, while part (a) follows immediately from (b). $\qquad\square$

We more closely analysed the structure of Toeplitz non-negative definite sequences in Proposition 2.9. In particular, we reviewed how this structure could be described by certain matrix balls. We now suppose that we have a $q \times q$ Carathéodory sequence $(s_j)_{j=0}^{\varkappa}$. In Lemma 4.7, we introduced two Toeplitz non-negative sequences $(C_j)_{j=0}^{\varkappa}$ and $\left(C_j^{[\sharp]}\right)_{j=0}^{\varkappa}$, which we defined with the help of $(s_j)_{j=0}^{\varkappa}$ and $\left(s_j^{\sharp}\right)_{j=0}^{\varkappa}$. Our next objective is to express the matrices which describe the inner structure of $\left(s_j^{[\sharp]}\right)_{j=0}^{\varkappa}$ in terms of the sequences $(s_j)_{j=0}^{\varkappa}$ and $(C_j)_{j=0}^{\varkappa}$. For this reason, we recall that for each $A \in \mathbb{C}^{q \times q}$, well-known results on left and right polar decompositions of square matrices say that there exist unitary $q \times q$ matrices U and V, such that

$$A = \sqrt{AA^*}\,U \qquad \text{and} \qquad A = V\sqrt{A^*A}.$$

Lemma 4.9. *Let* $\varkappa \in \mathbb{N} \cup \{+\infty\}$. *Suppose* $(s_j)_{j=0}^{\varkappa}$ *is a* $q \times q$ *Carathéodory sequence with reciprocal sequence* $\left(s_j^{\sharp}\right)_{j=0}^{\varkappa}$. *Suppose, furthermore, that* $(C_j)_{j=0}^{\varkappa}$ *and* $\left(C_j^{[\sharp]}\right)_{j=0}^{\varkappa}$ *are the matricial T-n.n.d. sequences introduced in Lemma 4.7. Then:*

(a) *Let* $L_1^{[\sharp]} := C_0^{[\sharp]}$, $R_1^{[\sharp]} := C_0^{[\sharp]}$ *and, for each* $k \in \mathbb{Z}_{1,\varkappa}$, *let*

$$L_{k+1}^{[\sharp]} := C_0^{[\sharp]} - Z_{k-1}^{[\sharp]}\left[T_{k-1}^{[\sharp]}\right]^{+}\left(Z_k^{[\sharp]}{}_1\right)^{*}$$

and

$$R_{k+1}^{[\sharp]} := C_0^{[\sharp]} - \left(Y_{k-1}^{[\sharp]}\right)^{*}\left[T_{k-1}^{[\sharp]}\right]^{+}Y_{k-1}^{[\sharp]}.$$

Then the matrices L_{k+1}, R_{k+1}, $L_{k+1}^{[\sharp]}$ *and* $R_{k+1}^{[\sharp]}$ *are all non-negative Hermitian and satisfy*

$$L_{k+1}^{[\sharp]} = s_0^{+} L_{k+1}\left(s_0^{+}\right)^{*}, \qquad R_{k+1}^{[\sharp]} = \left(s_0^{+}\right)^{*} R_{k+1} s_0^{+}, \tag{4.10}$$

$$L_{k+1} = s_0 L_{k+1}^{[\sharp]} s_0^{*}, \qquad R_{k+1} = s_0^{*} R_{k+1}^{[\sharp]} s_0 \tag{4.11}$$

and, in particular,

$$\operatorname{rank} L_{k+1}^{[\sharp]} = \operatorname{rank} L_{k+1} = \operatorname{rank} R_{k+1} = \operatorname{rank} R_{k+1}^{[\sharp]}. \tag{4.12}$$

If U_{k+1} and V_{k+1} are unitary $q \times q$ matrices such that U_{k+1} produces a left polar decomposition of $s_0^+ \sqrt{L_{k+1}}$ and V_{k+1} a right polar decomposition of $\sqrt{R_{k+1}} s_0^+$, then

$$\sqrt{L_{k+1}^{[\sharp]}} U_{k+1} = s_0^+ \sqrt{L_{k+1}} \qquad \text{and} \qquad V_{k+1} \sqrt{R_{k+1}^{[\sharp]}} = \sqrt{R_{k+1}} s_0^+. \qquad (4.13)$$

(b) Let $M_1^{[\sharp]} := 0_{q \times q}$ and, for each $m \in \mathbb{Z}_{1,\varkappa}$, let

$$M_{m+1}^{[\sharp]} := Z_m^{[\sharp]} \left(T_{m-1}^{[\sharp]} \right)^+ Y_m^{[\sharp]}.$$

Suppose that $k \in \mathbb{Z}_{1,\varkappa}$. Then:

$$M_{k+1}^{[\sharp]} = s_0^+ Z_k S_{k-1}^+ S_{k-1}^* T_{k-1}^+ Y_k s_0^+, \qquad (4.14)$$

$$M_{k+1}^{[\sharp]} = s_0^+ Z_k T_{k-1}^+ S_{k-1}^* S_{k-1}^+ Y_k s_0^+, \qquad (4.15)$$

$$M_{k+1}^{[\sharp]} = s_0^+ \left(2 Z_k S_{k-1}^+ Y_k - M_{k+1} \right) s_0^+, \qquad (4.16)$$

$$C_k^{[\sharp]} - M_k^{[\sharp]} = -s_0^+ \left(C_k - M_k \right) s_0^+ \qquad (4.17)$$

and

$$C_k - M_k = -s_0 \left(C_k^{[\sharp]} - M_k^{[\sharp]} \right) s_0. \qquad (4.18)$$

(c) Let $(K_j)_{j=1}^{\varkappa}$ be the Schur parameter sequence for $(C_j)_{j=0}^{\varkappa}$ and $(K_j^{[\sharp]})_{j=1}^{\varkappa}$ the Schur parameter sequence for $(C_j^{[\sharp]})_{j=0}^{\varkappa}$. Suppose $s \in \mathbb{Z}_{1,\varkappa}$ and that U_s and V_s are unitary $q \times q$ matrices such that U_s produces a left polar decomposition of $s_0^+ \sqrt{L_s}$ and V_s a right polar decomposition of $\sqrt{R_s} s_0^+$. Then:

$$\left(\sqrt{L_s^{[\sharp]}} \right)^+ \sqrt{L_s^{[\sharp]}} U_s K_s = U_s K_s, \qquad (4.19)$$

$$K_s V_s \left(\sqrt{R_s^{[\sharp]}} \right)^+ \sqrt{R_s^{[\sharp]}} = K_s V_s \qquad (4.20)$$

and

$$K_s^{[\sharp]} = -U_s K_s V_s. \qquad (4.21)$$

Proof. (a) By Remark 3.1, we see that $s_0 \in \mathcal{R}_{q,\geq}$. By part (c) of Lemma A.8 and Proposition A.6, it then follows that

$$\text{Re} \left(s_0^+ \right) = s_0^+ \left(\text{Re} \, s_0 \right) \left(s_0^+ \right)^* \qquad \text{and} \qquad \text{Re} \left(s_0^+ \right) = \left(s_0^+ \right)^* \left(\text{Re} \, s_0 \right) s_0^+.$$

By Definition 1.1 and (4.2), we obtain

$$C_0^{[\sharp]} = \text{Re} \, s_0^{\sharp} = \text{Re} \left(s_0^+ \right) = s_0^+ \left(\text{Re} \, s_0 \right) \left(s_0^+ \right)^* = s_0^+ C_0 \left(s_0^+ \right)^* \qquad (4.22)$$

and, similarly,

$$C_0^{[\sharp]} = \left(s_0^+ \right)^* C_0 s_0^+. \qquad (4.23)$$

Recalling (2.2), we see that this implies

$$L_1^{[\#]} = s_0^+ L_1 \left(s_0^+ \right)^* \qquad \text{and} \qquad R_1^{[\#]} = \left(s_0^+ \right)^* R_1 s_0^+ . \tag{4.24}$$

Suppose now that $k \in \mathbb{Z}_{1,\varkappa}$. Part (d) of Lemma 4.7 yields that the equations in (4.8) hold true. From part (a) of Lemma 4.7, it follows that T_{k-1} and $T_{k-1}^{[\#]}$ are both Hermitian matrices. Thus, from equations (4.8) we obtain

$$\left(Z_k^{[\#]} \right)^* = -T_{k-1}^{[\#]} S_{k-1} T_{k-1}^+ Z_k^* \left(s_0^+ \right)^* \tag{4.25}$$

and

$$\left(Y_k^{[\#]} \right)^* = - \left(s_0^+ \right)^* Y_k^* T_{k-1}^+ S_{k-1} T_{k-1}^{[\#]} . \tag{4.26}$$

Part (b) of Lemma 4.7 give us

$$S_{k-1}^* T_{k-1}^{[\#]} = T_{k-1} S_{k-1}^+ \qquad \text{and} \qquad T_{k-1}^{[\#]} S_{k-1}^* = S_{k-1}^+ T_{k-1} . \tag{4.27}$$

By Proposition 4.1, we have $(s_j)_{j=0}^{\varkappa} \in \mathcal{F}_{q,\varkappa}^{\mathrm{EP}}$. Parts (b) and (d) of Corollary A.7 along with Proposition 1.11 give us

$$(\operatorname{Re} S_{k-1}) S_{k-1}^+ S_{k-1} = \operatorname{Re} S_{k-1} \qquad \text{and} \qquad S_{k-1} S_{k-1}^+ (\operatorname{Re} S_{k-1}) = \operatorname{Re} S_{k-1} .$$

Thus, it follows by Remark 1.2 and part (b) of Lemma 4.7 that

$$T_{k-1} S_{k-1}^+ S_{k-1} = T_{k-1} \qquad \text{and} \qquad S_{k-1} S_{k-1}^+ T_{k-1} = T_{k-1} . \tag{4.28}$$

Using (4.8), (4.25), (4.27) and (4.28), we see that

$$Z_k^{[\#]} \left(T_{k-1}^{[\#]} \right)^+ \left(Z_k^{[\#]} \right)^*$$

$$= \left[-s_0^+ Z_k T_{k-1}^+ S_{k-1}^* T_{k-1}^{[\#]} \right] \left(T_{k-1}^{[\#]} \right)^+ \left[-T_{k-1}^{[\#]} S_{k-1} T_{k-1}^+ Z_k^* \left(s_0^+ \right)^* \right]$$

$$= s_0^+ Z_k T_{k-1}^+ S_{k-1}^* T_{k-1}^{[\#]} S_{k-1} T_{k-1}^+ Z_k^* \left(s_0^+ \right)^*$$

$$= s_0^+ Z_k T_{k-1}^+ T_{k-1} S_{k-1}^+ S_{k-1} T_{k-1}^+ Z_k^* \left(s_0^+ \right)^*$$

$$= s_0^+ Z_k T_{k-1}^+ T_{k-1} T_{k-1}^+ Z_k^* \left(s_0^+ \right)^*$$

$$= s_0^+ Z_k T_{k-1}^+ Z_k^* \left(s_0^+ \right)^* \tag{4.29}$$

and, similarly, using (4.26), (4.8), (4.27) and (4.28) also that

$$\left(Y_k^{[\#]} \right)^* \left(T_{k-1}^{[\#]} \right)^+ Y_k^{[\#]} = \left(s_0^+ \right)^* Y_k^* T_{k-1}^+ Y_k s_0^+ . \tag{4.30}$$

From (4.22), (4.29) and (2.4) we see that

$$L_{k+1}^{[\#]} = C_0^{[\#]} - Z_k^{[\#]} \left(T_{k-1}^{[\#]} \right)^+ \left(Z_k^{[\#]} \right)^*$$

$$= s_0^+ C_0 \left(s_0^+ \right)^* - s_0^+ Z_k T_{k-1}^+ Z_k^* \left(s_0^+ \right)^*$$

$$= s_0^+ L_{k+1} \left(s_0^+ \right)^* . \tag{4.31}$$

Similarly, from (4.23), (4.30) and (2.4) we also see that

$$R_{k+1}^{[\#]} = \left(s_0^+ \right)^* R_{k+1} s_0^+ . \tag{4.32}$$

The sequences $(L_{j+1})_{j=0}^{\varkappa}$, $(R_{j+1})_{j=0}^{\varkappa}$, $(L_{j+1}^{[\#]})_{j=0}^{\varkappa}$ and $(R_{j+1}^{[\#]})_{j=0}^{\varkappa}$ are monotonically decreasing sequences of non-negative Hermitian matrices, by part (a) of Lemma 4.7 and [8, Remark 3.4.3]. Therefore,

$$\mathcal{R}(L_{k+1}) \subseteq \mathcal{R}(L_1) = \mathcal{R}(C_0) = \mathcal{R}(\operatorname{Re} s_0)$$

and, similarly, $\mathcal{R}(R_{k+1}) \subseteq \mathcal{R}(\operatorname{Re} s_0)$. Because of $s_0 \in \mathcal{R}_{q,\geq}$ and part (b) of Lemma A.8, we have $\mathcal{R}(\operatorname{Re} s_0) \subseteq \mathcal{R}(s_0)$. Thus,

$$\mathcal{R}(L_{k+1}) \subseteq \mathcal{R}(s_0) \qquad \text{and} \qquad \mathcal{R}(R_{k+1}) \subseteq \mathcal{R}(s_0). \qquad (4.33)$$

Consequently, part (a) of Lemma A.2 implies

$$s_0 s_0^+ L_{k+1} = L_{k+1} \qquad \text{and} \qquad s_0 s_0^+ R_{k+1} = R_{k+1}. \qquad (4.34)$$

Since L_{k+1} and R_{k+1} are non-negative Hermitian, we have, in particular,

$$L_{k+1}^* = L_{k+1} \qquad \text{and} \qquad R_{k+1}^* = R_{k+1}. \qquad (4.35)$$

By considering orthogonal complements, we see from (4.33) and (4.35) that

$$\mathcal{N}(s_0^*) \subseteq \mathcal{N}(L_{k+1}) \qquad \text{and} \qquad \mathcal{N}(s_0^*) \subseteq \mathcal{N}(R_{k+1}). \qquad (4.36)$$

Because of $s_0 \in \mathcal{R}_{q,\geq}$, part (c) of Lemma A.8 yields $s_0 \in \mathbb{C}_{\mathrm{EP}}^{q\times q}$. It thus follows by Proposition A.5 that $\mathcal{N}(s_0) = \mathcal{N}(s_0^*)$. Along with (4.36), this implies $\mathcal{N}(s_0) \subseteq \mathcal{N}(L_{k+1})$ and $\mathcal{N}(s_0) \subseteq \mathcal{N}(R_{k+1})$. Hence, part (b) of Lemma A.2 implies

$$L_{k+1} s_0^+ s_0 = L_{k+1} \qquad \text{and} \qquad R_{k+1} s_0^+ s_0 = R_{k+1}. \qquad (4.37)$$

Using (4.10), (4.34) and (4.35), it follows that

$$s_0 L_{k+1}^{[\#]} s_0^* = s_0 s_0^+ L_{k+1}\left(s_0^+\right)^* s_0^* = L_{k+1}\left(s_0^+\right)^* s_0^* = L_{k+1}^*\left(s_0^+\right)^* s_0^*$$
$$= \left(s_0 s_0^+ L_{k+1}\right)^* = L_{k+1}^* = L_{k+1}.$$

Similarly, (4.10), (4.37) and (4.35) imply $s_0^* R_{k+1}^{[\#]} s_0 = R_{k+1}$. From (4.10) and (4.11), we see that $\operatorname{rank} L_{k+1} = \operatorname{rank} L_{k+1}^{[\#]}$ and $\operatorname{rank} R_{k+1} = \operatorname{rank} R_{k+1}^{[\#]}$. Because part (a) of Lemma 4.7 and part (a) of Proposition 2.9 yield $\operatorname{rank} L_{k+1} = \operatorname{rank} R_{k+1}$, we get (4.12). From (4.10), we see that

$$\left(s_0^+ \sqrt{L_{k+1}}\right)\left(s_0^+ \sqrt{L_{k+1}}\right)^* = s_0^+ L_{k+1}\left(s_0^+\right)^* = L_{k+1}^{[\#]}$$

and

$$\left(\sqrt{R_{k+1}}\, s_0^+\right)^*\left(\sqrt{R_{k+1}}\, s_0^+\right) = \left(s_0^+\right)^* R_{k+1} s_0^+ = R_{k+1}^{[\#]}.$$

By our choice of U_{k+1} and V_{k+1}, it thus follows that

$$s_0^+ \sqrt{L_{k+1}} = \sqrt{\left(s_0^+ \sqrt{L_{k+1}}\right)\left(s_0^+ \sqrt{L_{k+1}}\right)^*}\, U_{k+1} = \sqrt{L_{k+1}^{[\#]}}\, U_{k+1}$$

and

$$\sqrt{R_{k+1}}\, s_0^+ = V_{k+1} \sqrt{\left(\sqrt{R_{k+1}}\, s_0^+\right)^*\left(\sqrt{R_{k+1}}\, s_0^+\right)} = V_{k+1} \sqrt{R_{k+1}^{[\#]}}.$$

This completes the proof of (a).

(b) Recalling part (d) of Lemma 4.7, we see that

$$M_{k+1}^{[\sharp]} = Z_k^{[\sharp]} \left(T_{k-1}^{[\sharp]} \right)^+ Y_k^{[\sharp]}$$

$$= \left(-s_0^+ Z_k T_{k-1}^+ S_{k-1}^* T_{k-1}^{[\sharp]} \right) \left(T_{k-1}^{[\sharp]} \right)^+ \left(-T_{k-1}^{[\sharp]} S_{k-1}^* T_{k-1}^+ Y_k s_0^+ \right)$$

$$= s_0^+ Z_k T_{k-1}^+ S_{k-1}^* T_{k-1}^{[\sharp]} S_{k-1}^* T_{k-1}^+ Y_k s_0^+ \tag{4.38}$$

for $k \in \mathbb{Z}_{1,\varkappa}$. By part (c) of Lemma 4.7, it follows that (4.6) holds true. From (4.38), the first equation in (4.27) and (4.6), we obtain

$$M_{k+1}^{[\sharp]} = s_0^+ Z_k T_{k-1}^+ S_{k-1}^* T_{k-1}^{[\sharp]} S_{k-1}^* T_{k-1}^+ Y_k s_0^+$$

$$= s_0^+ Z_k T_{k-1}^+ T_{k-1} S_{k-1}^+ S_{k-1}^* T_{k-1}^+ Y_k s_0^+$$

$$= s_0^+ Z_k S_{k-1}^+ S_{k-1}^* T_{k-1}^+ Y_k s_0^+.$$

Thus, (4.14) is proved. From (4.38), the second equation in (4.27) and the second equation in (4.6), we conclude that

$$M_{k+1}^{[\sharp]} = s_0^+ Z_k T_{k-1}^+ S_{k-1}^* S_{k-1}^+ Y_k s_0^+.$$

By part (b) of Lemma 4.7, it follows that

$$T_{k-1} = \operatorname{Re} S_{k-1} \tag{4.39}$$

and hence

$$2T_{k-1} = S_{k-1} + S_{k-1}^*. \tag{4.40}$$

Using the second equation in (4.6) and (4.40), we now see that

$$2Y_k = 2T_{k-1} T_{k-1}^+ Y_k = S_{k-1} T_{k-1}^+ Y_k + S_{k-1}^* T_{k-1}^+ Y_k,$$

i.e.,

$$S_{k-1}^* T_{k-1}^+ Y_k = 2Y_k - S_{k-1} T_{k-1}^+ Y_k. \tag{4.41}$$

From (4.14) and (4.41), we obtain

$$M_{k+1}^{[\sharp]} = s_0^+ Z_k S_{k-1}^+ \left(2Y_k - S_{k-1} T_{k-1}^+ Y_k \right) s_0^+$$

$$= s_0^+ \left(2Z_k S_{k-1}^+ Y_k - Z_k S_{k-1}^+ S_{k-1} T_{k-1}^+ Y_k \right) s_0^+. \tag{4.42}$$

Since $(s_j)_{j=0}^{\varkappa}$ is a $q \times q$ Carathéodory sequence, we have $S_{k-1} \in \mathcal{R}_{kq, \geq}$. Thus, it follows by part (d) of Lemma A.8 that

$$S_{k-1} S_{k-1}^+ = S_{k-1}^+ S_{k-1} \tag{4.43}$$

and from part (b) of Lemma A.8 that $\mathcal{R}(\operatorname{Re} S_{k-1}) \subseteq \mathcal{R}(S_{k-1})$. Therefore, (4.39) implies $\mathcal{R}(T_{k-1}) \subseteq \mathcal{R}(S_{k-1})$. Because $T_{k-1}^* = T_{k-1}$, it follows from Lemma A.3 that $\mathcal{R}(T_{k-1}) = \mathcal{R}(T_{k-1}^+)$. Hence, $\mathcal{R}(T_{k-1}^+) \subseteq \mathcal{R}(S_{k-1})$. Thus, by part (a) of Lemma A.2, we have

$$S_{k-1} S_{k-1}^+ T_{k-1}^+ = T_{k-1}^+.$$

This, along with (4.43) and (2.5), implies that

$$Z_k S_{k-1}^+ S_{k-1} T_{k-1}^+ Y_k = Z_k S_{k-1} S_{k-1}^+ T_{k-1}^+ Y_k = M_{k+1}.$$

Thus, using (4.42), we obtain (4.16).

In order to prove (4.17), we first consider the case $\varkappa = 1$. From Definition 1.1, we see that $s_1^\sharp = -s_0^+ s_1 s_0^+$. Recalling that $M_1 = 0_{q \times q}$, $M_1^{[\sharp]} = 0_{q \times q}$ and $\frac{1}{2} s_1 = C_1$ (which follows from (4.2)) we now see that

$$C_1^{[\sharp]} - M_1^{[\sharp]} = \frac{1}{2} s_1^\sharp = -s_0^+ \left(\frac{1}{2} s_1 \right) s_0^+ = -s_0^+ C_1 s_0^+ = -s_0^+ \left(C_1 - M_1 \right) s_0^+.$$

Thus, (4.17) is proved for $\varkappa = 1$. Suppose now that $\varkappa \geq 2$ and $k \in \mathbb{Z}_{2, \varkappa}$. Since (4.2) shows that $s_j = 2C_j$ for each $j \in \mathbb{Z}_{1, \varkappa}$, it follows by [22, Corollary 4.24] that

$$C_k^{[\sharp]} = \frac{1}{2} s_0^\sharp = s_0^+ \left(-C_k + 2 Z_{k-1} S_{k-2}^+ Y_{k-1} \right) s_0^+.$$

Equation (4.16) yields

$$M_k^{[\sharp]} = s_0^+ \left(2 Z_{k-1} S_{k-2}^+ Y_{k-1} - M_k \right) s_0^+.$$

Therefore, we also obtain (4.17) for $\varkappa \geq 2$ and $k \in \mathbb{Z}_{2, \varkappa}$.

Now we again consider an arbitrary $\varkappa \in \mathbb{N} \cup \{ +\infty \}$ and an arbitrary $m \in \mathbb{Z}_{1, \varkappa}$. We know from Proposition 4.1 that $(s_j)_{j=0}^{\varkappa} \in \mathcal{D}_{q \times q, \varkappa}$. Thus,

$$\mathcal{N} (s_0) \subseteq \mathcal{N} (s_m) \qquad \text{and} \qquad \mathcal{R} (s_m) \subseteq \mathcal{R} (s_0). \qquad (4.44)$$

Since (4.2) shows that $C_m = \frac{1}{2} s_m$, we have $\mathcal{N} (s_m) = \mathcal{N} (C_m)$ and $\mathcal{R} (s_m) = \mathcal{R} (C_m)$. Therefore, (4.44) yields $\mathcal{N} (s_0) \subseteq \mathcal{N} (C_m)$ and $\mathcal{R} (C_m) \subseteq \mathcal{R} (s_0)$. By Lemma A.2, it thus follows that

$$C_m s_0^+ s_0 = C_m \qquad \text{and} \qquad s_0 s_0^+ C_m = C_m. \qquad (4.45)$$

From (4.17), $M_1 = 0_{q \times q}$ and (4.45), we see that

$$-s_0 \left(C_1^{[\sharp]} - M_1^{[\sharp]} \right) s_0 = -s_0 \left[-s_0^+ \left(C_1 - M_1 \right) s_0^+ \right] s_0$$
$$= s_0 s_0^+ C_1 s_0^+ s_0 = C_1 = C_1 - M_1.$$

We have thus shown (4.18) for $\varkappa = 1$. Suppose now that $\varkappa \geq 2$ and that $k \in \mathbb{Z}_{2, \varkappa}$. It follows from (4.45) that

$$s_0 s_0^+ Z_{k-1} = Z_{k-1} \qquad \text{and} \qquad Y_{k-1} s_0^+ s_0 = Y_{k-1}.$$

Because of (2.5), we then get

$$s_0 s_0^+ M_k s_0^+ s_0 = s_0 s_0^+ Z_{k-1} T_{k-2}^+ Y_{k-1} s_0^+ s_0 = Z_{k-1} T_{k-2}^+ Y_{k-1} = M_k.$$

Thus, using (4.17) and (4.45), we now obtain

$$-s_0 \left(C_k^\sharp - M_k^\sharp \right) s_0 = -s_0 \left[-s_0^+ \left(C_k - M_k \right) s_0^+ \right] s_0$$
$$= s_0 s_0^+ C_k s_0^+ s_0 - s_0 s_0^+ M_k s_0^+ s_0 = C_k - M_k.$$

The proof of (4.18) is thus complete.

(c) From part (a) of Lemma A.4 and (4.33), we see that

$$\mathcal{R}\left(\sqrt{L_s}\right) = \mathcal{R}(L_s) \subseteq \mathcal{R}(s_0).$$

Part (a) of Lemma A.2, therefore implies that

$$s_0 s_0^+ \sqrt{L_s} = \sqrt{L_s}.$$

Thus, it now follows from (4.13) that

$$s_0 \sqrt{L_s^{[\sharp]}} U_s = s_0 s_0^+ \sqrt{L_s} = \sqrt{L_s}.$$

Considering adjoints of these matrices, we obtain

$$\sqrt{L_s} = U_s^* \sqrt{L_s^{[\sharp]}} s_0^*. \tag{4.46}$$

By definition of the Schur parameters for $(C_j)_{j=0}^{\varkappa}$ and part (b) of Lemma A.4, we have

$$\sqrt{L_s}\left(\sqrt{L_s}\right)^+ K_s = \sqrt{L_s}\left(\sqrt{L_s}\right)^+ \left(\sqrt{L_s}\right)^+ (C_s - M_s)\left(\sqrt{R_s}\right)^+$$

$$= \left(\sqrt{L_s}\right)^+ \sqrt{L_s}\left(\sqrt{L_s}\right)^+ (C_s - M_s)\left(\sqrt{R_s}\right)^+$$

$$= K_s.$$

From (4.46), we thus obtain

$$K_s = U_s^* \sqrt{L_s^{[\sharp]}} s_0^* \left(\sqrt{L_s}\right)^+ K_s. \tag{4.47}$$

Equation (4.47), the unitarity of the matrix U_s, part (c) of Lemma A.4 and (4.47) again imply that

$$\left(\sqrt{L_s^{[\sharp]}}\right)^+ \sqrt{L_s^{[\sharp]}} U_s K_s = \left(\sqrt{L_s^{[\sharp]}}\right)^+ \sqrt{L_s^{[\sharp]}} U_s U_s^* \sqrt{L_s^{[\sharp]}} s_0^* \left(\sqrt{L_s}\right)^+ K_s$$

$$= \left(\sqrt{L_s^{[\sharp]}}\right)^+ L_s^{[\sharp]} s_0^* \left(\sqrt{L_s}\right)^+ K_s$$

$$= \sqrt{L_s^{[\sharp]}} s_0^* \left(\sqrt{L_s}\right)^+ K_s$$

$$= U_s K_s.$$

This completes the proof of (4.19). Similarly, we obtain (4.20). By part (a) of Lemma 4.7 and Proposition 2.9, we have

$$C_s - M_s = \sqrt{L_s} K_s \sqrt{R_s}.$$

Thus, using the definition of the Schur parameters for $(C_j)_{j=0}^{\varkappa}$ and $(C_j^{[\sharp]})_{j=0}^{\varkappa}$ as well as formulas (4.17), (4.13), (4.19) and (4.20), we obtain

$$
\begin{aligned}
K_s^{[\sharp]} &= \left(\sqrt{L_s^{[\sharp]}}\right)^+ \left(C_s^{[\sharp]} - M_s^{[\sharp]}\right) \left(\sqrt{R_s^{[\sharp]}}\right)^+ \\
&= \left(\sqrt{L_s^{[\sharp]}}\right)^+ \left[-s_0^+ \left(C_s - M_s\right) s_0^+\right] \left(\sqrt{R_s^{[\sharp]}}\right)^+ \\
&= -\left(\sqrt{L_s^{[\sharp]}}\right)^+ s_0^+ \sqrt{L_s} K_s \sqrt{R_s} s_0^+ \left(\sqrt{R_s^{[\sharp]}}\right)^+ \\
&= -\left(\sqrt{L_s^{[\sharp]}}\right)^+ \sqrt{L_s^{[\sharp]}} U_s K_s V_s \sqrt{R_s^{[\sharp]}} \left(\sqrt{R_s^{[\sharp]}}\right)^+ \\
&= -U_s K_s V_s \sqrt{R_s^{[\sharp]}} \left(\sqrt{R_s^{[\sharp]}}\right)^+ = -U_s K_s V_s.
\end{aligned}
$$

Thus, we have shown (4.21). This completes the proof. $\square$

Lemma 4.9 leads us to our next result. We will see that the reciprocal sequences of Carathéodory sequences inherit the C-centrality properties of their generating sequences.

Theorem 4.10. *Let $\varkappa \in \mathbb{N} \cup \{+\infty\}$ and $(s_j)_{j=0}^{\varkappa}$ be a $q \times q$ Carathéodory sequence. Furthermore, suppose $(s_j^{\sharp})_{j=0}^{\varkappa}$ is the reciprocal sequence to $(s_j)_{j=0}^{\varkappa}$. Then:*

(a) *$(s_j)_{j=0}^{\varkappa}$ is C-central if and only if $(s_j^{\sharp})_{j=0}^{\varkappa}$ is C-central.*

(b) *Let $k \in \mathbb{Z}_{1,\varkappa}$. Then:*

    (b1) *$(s_j)_{j=0}^{\varkappa}$ is order k C-central if and only if $(s_j^{\sharp})_{j=0}^{\varkappa}$ is order k C-central.*

    (b2) *$(s_j)_{j=0}^{\varkappa}$ is minimal order k C-central if and only if $(s_j^{\sharp})_{j=0}^{\varkappa}$ is minimal order k C-central.*

Proof. (b) follows directly from Definition 3.15 and part (b) of Lemma 4.9 while (a) follows immediately from (b). $\square$

Our next result shows that generating reciprocal sequences, as an operation, is compatible with the operation of generating the central $q \times q$ Carathéodory sequence of a finite $q \times q$ Carathéodory sequence.

Theorem 4.11. *Let $n \in \mathbb{N}_0$, let $(s_j)_{j=0}^{n}$ be a $q \times q$ Carathéodory sequence and let $(s_j^{\sharp})_{j=0}^{n}$ be the reciprocal sequence to $(s_j)_{j=0}^{n}$. Recalling Theorem 4.4, let $(\tilde{s}_j)_{j=0}^{\infty}$ and $(\tilde{s}_{j,\sharp})_{j=0}^{\infty}$ be the C-central sequences for $(s_j)_{j=0}^{n}$ and $(s_j^{\sharp})_{j=0}^{n}$, respectively. Then $(\tilde{s}_{j,\sharp})_{j=0}^{\infty}$ is the reciprocal sequence to $(\tilde{s}_j)_{j=0}^{\infty}$.*

Proof. By Remark 3.17, we have $\tilde{s}_j = s_j$ for each $j \in \mathbb{Z}_{0,n}$ and the sequence $(\tilde{s}_j)_{j=0}^{\infty}$ is order $n+1$ C-central. Suppose that $(\tilde{s}_j^{\sharp})_{j=0}^{\infty}$ is the reciprocal sequence to $(\tilde{s}_j)_{j=0}^{\infty}$. Because of Remark 1.2, we have $(\tilde{s}_j^{\sharp})_{j=0}^{n} = (s_j^{\sharp})_{j=0}^{n}$ and part (a) of

Lemma 3.19, Theorem 4.4 and part (a) of Theorem 4.10 we see that $\left(\widetilde{s}_j^{\sharp}\right)_{j=0}^{\infty}$ is a $q \times q$ order $n+1$ C-central Carathéodory sequence. It thus follows by Remark 3.17 that $\left(\widetilde{s}_j^{\sharp}\right)_{j=0}^{\infty}$ is the C-central $q \times q$ sequence for $\left(s_j^{\sharp}\right)_{j=0}^{n}$. We therefore have $\left(\widetilde{s}_j^{\sharp}\right)_{j=0}^{\infty} = \left(\widetilde{s}_{j,\sharp}\right)_{j=0}^{n}$. $\qquad\square$

We next consider reciprocal sequences of totally strict Carathéodory sequences.

Lemma 4.12. *Let* $(s_j)_{j=0}^{\infty} \in \mathcal{C}_{q,\infty}$ *and suppose that* $\left(s_j^{\sharp}\right)_{j=0}^{\infty}$ *is the reciprocal sequence to* $(s_j)_{j=0}^{\infty}$. *Then the sequences* $(L_{k+1})_{k=0}^{\infty}$, $(R_{k+1})_{k=0}^{\infty}$, $\left(L_{k+1}^{[\sharp]}\right)_{k=0}^{\infty}$ *and* $\left(R_{k+1}^{[\sharp]}\right)_{k=0}^{\infty}$ *are monotonically decreasing and convergent. Their limits*

$$L := \lim_{k \to \infty} L_{k+1}, \qquad\qquad R := \lim_{k \to \infty} R_{k+1},$$

$$L^{[\sharp]} := \lim_{k \to \infty} L_{k+1}^{[\sharp]} \qquad and \qquad R^{[\sharp]} := \lim_{k \to \infty} R_{k+1}^{[\sharp]}$$

are, furthermore, all non-negative Hermitian. Moreover,

$$L^{[\sharp]} = s_0^{+} L \left(s_0^{+}\right)^{*}, \qquad\qquad R^{[\sharp]} = \left(s_0^{+}\right)^{*} R s_0^{+},$$

$$L = s_0 L^{[\sharp]} s_0^{*}, \qquad and \qquad R = s_0^{*} R^{[\sharp]} s_0.$$

In particular, $\operatorname{rank} L = \operatorname{rank} L^{[\sharp]}$ *and* $\operatorname{rank} R = \operatorname{rank} R^{[\sharp]}$.

Proof. Use Proposition 2.9, Theorem 4.4 and Lemma 4.9. $\qquad\square$

Theorem 4.13. *Let* $(s_j)_{j=0}^{\infty} \in \mathcal{C}_{q,\infty}$ *and suppose that* $\left(s_j^{\sharp}\right)_{j=0}^{\infty}$ *is the reciprocal sequence to* $(s_j)_{j=0}^{\infty}$. *Then* $(s_j)_{j=0}^{\infty} \in \mathcal{C}_{q,\infty}^{t}$ *if and only if* $\left(s_j^{\sharp}\right)_{j=0}^{\infty} \in \mathcal{C}_{q,\infty}^{t}$.

Proof. Use Lemma 4.12. $\qquad\square$

5. Matricial Toeplitz non-negative definite sequences generated by reciprocation

In this section, we will soon arrive at a particular operation, which, given a Toeplitz non-negative definite sequence $(C_j)_{j=0}^{\varkappa}$ in $\mathbb{C}^{q \times q}$, will make it possible to construct a new Toeplitz non-negative definite sequence $\left(C_j^{\square}\right)_{j=0}^{\varkappa}$ from $(C_j)_{j=0}^{\varkappa}$. We next provide a detailed look into this construction.

Proposition 5.1. *Let* $\varkappa \in \mathbb{N}_0 \cup \{+\infty\}$ *and* $(C_j)_{j=0}^{\varkappa}$ *be a T-n.n.d. sequence in* $\mathbb{C}^{q \times q}$. *Suppose that*

$$s_j := \begin{cases} C_0 & \text{if } j = 0, \\ 2C_j & \text{if } j \in \mathbb{Z}_{1,\varkappa}. \end{cases} \tag{5.1}$$

Then $(s_j)_{j=0}^{\varkappa}$ is a $q \times q$ Carathéodory sequence with $s_0 = s_0^*$ and this sequence's reciprocal sequence $(s_j^\sharp)_{j=0}^{\varkappa}$ is also a $q \times q$ Carathéodory sequence with $s_0^\sharp = (s_0^\sharp)^*$. Furthermore, if $(C_j^\square)_{j=0}^{\varkappa}$ is defined as

$$C_j^\square := \begin{cases} s_0^\sharp & \text{if } j = 0, \\ \frac{1}{2} s_j^\sharp & \text{if } j \in \mathbb{Z}_{1,\varkappa}, \end{cases} \tag{5.2}$$

then $(C_j^\square)_{j=0}^{\varkappa}$ is a $q \times q$ T-n.n.d. sequence.

Proof. By Remark 3.7, it follows that $(s_j)_{j=0}^{\varkappa} \in \mathcal{C}_{q,\varkappa}$ and $s_0 = s_0^*$. Theorem 4.4 thus implies $(s_j^\sharp)_{j=0}^{\varkappa} \in \mathcal{C}_{q,\varkappa}$, where $s_0 = s_0^*$ leads to $s_0^\sharp = (s_0^\sharp)^*$ in view of Definition 1.1 (note also Lemma A.3). It therefore follows by Remark 3.7 that $(C_j^\square)_{j=0}^{\varkappa}$ is a $q \times q$ T-n.n.d. sequence. □

Proposition 5.1 leads us to the following definition.

Definition 5.2. Let $\varkappa \in \mathbb{N}_0 \cup \{+\infty\}$ and let $(C_j)_{j=0}^{\varkappa}$ be a T-n.n.d. sequence in $\mathbb{C}^{q \times q}$. We will call the sequence $(C_j^\square)_{j=0}^{\varkappa}$ defined in (formula (5.2) of) Proposition 5.1, the T-n.n.d. sequence in $\mathbb{C}^{q \times q}$ generated from $(C_j)_{j=0}^{\varkappa}$ by reciprocation.

A T-n.n.d. sequence generated by reciprocation is not the same as a reciprocal sequence, as the following example demonstrates.

Example 5.3. Let $K \in \mathbb{K}_{q \times q}$. Suppose, for $j \in \mathbb{N}_0$, that $C_j := K^j$. Example 2.21 yields $(C_j)_{j=0}^{\infty} \in \mathcal{T}_{q,\infty}$. On the one hand, it follows by induction that the reciprocal sequence $(C_j^\sharp)_{j=0}^{\infty}$ to $(C_j)_{j=0}^{\infty}$ is

$$C_j^\sharp = \begin{cases} I_q, & \text{if } j = 0, \\ -K, & \text{if } j = 1, \\ 0_{q \times q}, & \text{if } j \in \mathbb{Z}_{2,+\infty}. \end{cases}$$

On the other hand, it also follows by induction that the sequence $(C_j^\square)_{j=0}^{\infty}$ generated from $(C_j)_{j=0}^{\infty}$ by reciprocation is given by

$$C_j^\square = (-K)^j, \qquad j \in \mathbb{N}_0.$$

Proposition 5.4. Let $\varkappa \in \mathbb{N}_0 \cup \{+\infty\}$ and let $(C_j)_{j=0}^{\varkappa}$ be a T-n.n.d. sequence in $\mathbb{C}^{q \times q}$. Suppose that $(C_j^\square)_{j=0}^{\varkappa}$ is the T-n.n.d. sequence in $\mathbb{C}^{q \times q}$ generated from $(C_j)_{j=0}^{\varkappa}$ by reciprocation. If $((C_j^\square)^\square)_{j=0}^{\varkappa}$ is the T-n.n.d. sequence in $\mathbb{C}^{q \times q}$ generated from $(C_j^\square)_{j=0}^{\varkappa}$ by reciprocation, then

$$\left((C_j^\square)^\square\right)_{j=0}^{\varkappa} = (C_j)_{j=0}^{\varkappa}.$$

Proof. For each $j \in \mathbb{Z}_{0,\varkappa}$, let

$$s_j := \begin{cases} C_0 & \text{if } j = 0, \\ 2C_j & \text{if } j \in \mathbb{Z}_{1,\varkappa}. \end{cases} \tag{5.3}$$

Because of (5.3) and (5.2), for each $j \in \mathbb{Z}_{0,\varkappa}$, we have,

$$s_j^\sharp = \begin{cases} C_0^\square & \text{if } j = 0, \\ 2C_j^\square & \text{if } j \in \mathbb{Z}_{1,\varkappa}. \end{cases}$$

For each $j \in \mathbb{Z}_{0,\varkappa}$, we thus see, by the definition of $\left((C_j^\square)^\square \right)_{j=0}^\varkappa$, that

$$(C_j^\square)^\square = \begin{cases} (s_0^\sharp)^\sharp & \text{if } j = 0, \\ \frac{1}{2}(s_j^\sharp)^\sharp & \text{if } j \in \mathbb{Z}_{1,\varkappa}. \end{cases} \tag{5.4}$$

From (5.1), it follows by Remark 3.7 that $(s_j)_{j=0}^\varkappa$ is a $q \times q$ Carathéodory sequence. Therefore, by Corollary 4.3, we have

$$\left((s_j^\sharp)^\sharp \right)_{j=0}^\varkappa = (s_j)_{j=0}^\varkappa. \tag{5.5}$$

Finally, from (5.4), (5.5) and (5.3), we obtain

$$\left((C_j^\square)^\square \right)_{j=0}^\varkappa = (C_j)_{j=0}^\varkappa,$$

which completes the proof. □

We now look at how Definition 5.2 relates to the previous section.

Lemma 5.5. *Let $\varkappa \in \mathbb{N}_0 \cup \{+\infty\}$ and let $(C_j)_{j=0}^\varkappa$ be a T-n.n.d. sequence in $\mathbb{C}^{q\times q}$. Suppose that $(C_j^\square)_{j=0}^\varkappa$ is the T-n.n.d. sequence in $\mathbb{C}^{q\times q}$ generated from $(C_j)_{j=0}^\varkappa$ by reciprocation. For $j \in \mathbb{Z}_{0,\varkappa}$, let the matrix s_j be given by (5.3). Then $(s_j)_{j=0}^\varkappa$ is a $q \times q$ Carathéodory sequence and*

$$C_j = \begin{cases} \operatorname{Re} s_0 & \text{if } j = 0, \\ \frac{1}{2}s_j & \text{if } j \in \mathbb{Z}_{1,\varkappa} \end{cases} \qquad \text{and} \qquad C_j^\square = \begin{cases} \operatorname{Re} s_0^\sharp & \text{if } j = 0, \\ \frac{1}{2}s_j^\sharp & \text{if } j \in \mathbb{Z}_{1,\varkappa}. \end{cases}$$

Proof. By Proposition 5.1, $(s_j)_{j=0}^\varkappa$ is a $q \times q$ Carathéodory sequence. By definition of $(C_j)_{j=0}^\varkappa$, we have $C_0^* = C_0$. Thus, $s_0^* = s_0$. Recalling Definition 1.1, we therefore have $(s_0^\sharp)^* = (s_0^*)^* = (s_0^*)^+ = s_0^+ = s_0^\sharp$. We thus see that $\operatorname{Re} s_0 = s_0$ and $\operatorname{Re} s_0^\sharp = s_0^\sharp$. Along with Definition 5.2, this completes the proof. □

Using Lemma 5.5, we can now translate Theorems 4.8, 4.10, 4.11 and 4.13 so that they apply to Definition 5.2.

Theorem 5.6. *Let $\varkappa \in \mathbb{N} \cup \{+\infty\}$ and let $(C_j)_{j=0}^\varkappa$ be a T-n.n.d. sequence in $\mathbb{C}^{q\times q}$. Let, furthermore, $(C_j^\square)_{j=0}^\varkappa$ be the T-n.n.d. sequence in $\mathbb{C}^{q\times q}$ generated from $(C_j)_{j=0}^\varkappa$ by reciprocation. Then:*

(a) $(C_j)_{j=0}^{\varkappa}$ is canonical if and only if $\left(C_j^{\square}\right)_{j=0}^{\varkappa}$ is canonical.

(b) Let $k \in \mathbb{Z}_{1,\varkappa}$. Then:

    (b1) $(C_j)_{j=0}^{\varkappa}$ is order k canonical if and only if $\left(C_j^{\square}\right)_{j=0}^{\varkappa}$ is order k canonical.

    (b2) $(C_j)_{j=0}^{\varkappa}$ is minimal order k canonical if and only if $\left(C_j^{\square}\right)_{j=0}^{\varkappa}$ is minimal order k canonical.

Proof. Combine Lemma 5.5 with Theorem 4.8, recalling Definition 3.22. $\square$

Theorem 5.7. *Let* $\varkappa \in \mathbb{N} \cup \{+\infty\}$ *and let* $(C_j)_{j=0}^{\varkappa}$ *be a* T-n.n.d. *sequence in* $\mathbb{C}^{q \times q}$. *Let, furthermore,* $\left(C_j^{\square}\right)_{j=0}^{\varkappa}$ *be the* T-n.n.d. *sequence in* $\mathbb{C}^{q \times q}$ *generated from* $(C_j)_{j=0}^{\varkappa}$ *by reciprocation. Then:*

(a) $(C_j)_{j=0}^{\varkappa}$ *is central if and only if* $\left(C_j^{\square}\right)_{j=0}^{\varkappa}$ *is central.*

(b) *Let* $k \in \mathbb{Z}_{1,\varkappa}$. *Then:*

    (b1) $(C_j)_{j=0}^{\varkappa}$ *is order k central if and only if* $\left(C_j^{\square}\right)_{j=0}^{\varkappa}$ *is order k central.*

    (b2) $(C_j)_{j=0}^{\varkappa}$ *is minimal order k central if and only if* $\left(C_j^{\square}\right)_{j=0}^{\varkappa}$ *is minimal order k central.*

Proof. Combine Lemma 5.5 with Theorem 4.10, recalling Definition 3.15. $\square$

Theorem 5.8. *Let* $n \in \mathbb{N}_0$ *and* $(C_j)_{j=0}^{n}$ *be a* T-n.n.d. *sequence in* $\mathbb{C}^{q \times q}$. *Suppose that* $\left(C_j^{\square}\right)_{j=0}^{n}$ *is the* T-n.n.d. *sequence in* $\mathbb{C}^{q \times q}$ *generated from* $(C_j)_{j=0}^{n}$ *by reciprocation. Let, furthermore,* $(D_j)_{j=0}^{\infty}$ *and* $(E_j)_{j=0}^{\infty}$ *be the central* T-n.n.d. *sequences for* $(C_j)_{j=0}^{n}$ *and* $\left(C_j^{\square}\right)_{j=0}^{n}$, *respectively. Suppose, also, that* $\left(D_j^{\square}\right)_{j=0}^{\infty}$ *is the* T-n.n.d. *sequence in* $\mathbb{C}^{q \times q}$ *generated from* $(D_j)_{j=0}^{\infty}$ *by reciprocation. Then*

$$\left(D_j^{\square}\right)_{j=0}^{\infty} = (E_j)_{j=0}^{\infty}.$$

Proof. Combine Lemma 5.5 with Theorem 4.11. $\square$

Theorem 5.9. *Let* $(C_j)_{j=0}^{\infty}$ *be a* T-n.n.d. *sequence in* $\mathbb{C}^{q \times q}$. *Suppose, furthermore, that* $\left(C_j^{\square}\right)_{j=0}^{\infty}$ *is the* T-n.n.d. *sequence in* $\mathbb{C}^{q \times q}$ *generated from* $(C_j)_{j=0}^{\infty}$ *by reciprocation. Then* $(C_j)_{j=0}^{\infty} \in \mathcal{C}_{q,\infty}^{\mathrm{t}}$ *if and only if* $\left(C_j^{\square}\right)_{j=0}^{\infty} \in \mathcal{C}_{q,\infty}^{\mathrm{t}}$.

Proof. Combine Lemma 5.5 and Theorem 4.13, recalling Definition 3.14. $\square$

6. Matricial Carathéodory functions

The set of $q \times q$ Carathéodory sequences is closely related to the following class of holomorphic $q \times q$ matrix functions. A function $\Omega : \mathbb{D} \longrightarrow \mathbb{C}^{q \times q}$ is called a $q \times q$ Carathéodory function in $\mathbb{D}$ if it is holomorphic in $\mathbb{D}$ and if $\Omega(w) \in \mathcal{R}_{q,\geq}$ for all $w \in \mathbb{D}$. The set of all $q \times q$ Carathéodory functions in $\mathbb{D}$ will be denoted by $\mathcal{C}_q(\mathbb{D})$.

Remark 6.1. If $\Omega \in \mathcal{C}_q(\mathbb{D})$ and $A \in \mathbb{C}^{q \times p}$, then $A^* \Omega A \in \mathcal{C}_p(\mathbb{D})$.

Remark 6.2. If $n \in \mathbb{N}$, $(\alpha_j)_{j=1}^n$ is a sequence in $[0, +\infty)$ and $(\Omega_j)_{j=1}^n$ is a sequence in $\mathcal{C}_q(\mathbb{D})$, then $\sum_{j=1}^n \alpha_j \Omega_j \in \mathcal{C}_q(\mathbb{D})$.

The class $\mathcal{C}_q(\mathbb{D})$ is closely related to the *Schur* class. A matrix function $S : \mathbb{D} \longrightarrow \mathbb{C}^{p \times q}$ is called a $p \times q$ Schur function in $\mathbb{D}$ if it is holomorphic in $\mathbb{D}$ and if $S(w)$ is contractive for all $w \in \mathbb{D}$. The set of all $p \times q$ Schur functions in $\mathbb{D}$ will be denoted by $\mathcal{S}_{p \times q}(\mathbb{D})$. We furthermore define

$$\mathcal{S}'_{p \times q}(\mathbb{D}) := \{f \in \mathcal{S}_{p \times q}(\mathbb{D}) : f(w) \in \mathbb{D}_{p \times q} \text{ for all } w \in \mathbb{D}\},$$

as well as $\mathcal{S}(\mathbb{D}) := \mathcal{S}_{1 \times 1}(\mathbb{D})$ and $\mathcal{S}'(\mathbb{D}) := \mathcal{S}'_{1 \times 1}(\mathbb{D})$. We next review a well-known relationship that exists between the classes $\mathcal{C}_q(\mathbb{D})$ and $\mathcal{S}_{q \times q}(\mathbb{D})$ (see [8, Proposition 2.1.3]).

Lemma 6.3. *Let* $\Omega \in \mathcal{C}_q(\mathbb{D})$. *Then* $\det [I_q + \Omega(w)] \neq 0$ *for each* $w \in \mathbb{D}$, *and the function*

$$S := (I_q - \Omega)(I_q + \Omega)^{-1} \tag{6.1}$$

belongs to $\mathcal{S}_{q \times q}(\mathbb{D})$. *Furthermore*, $\det [I_q + S(w)] \neq 0$ *for each* $w \in \mathbb{D}$, *and*

$$\Omega = (I_q - S)(I_q + S)^{-1}. \tag{6.2}$$

For each $\Omega \in \mathcal{C}_q(\mathbb{D})$, the matrix-valued function in (6.1) is called the Cayley transform of Ω. Lemma A.13 immediately leads us to another well-known relationship for the sets $\mathcal{S}_{q \times q}(\mathbb{D})$ and $\mathcal{C}_q(\mathbb{D})$.

Lemma 6.4. *Suppose that* $S \in \mathcal{S}_{q \times q}(\mathbb{D})$, *then* $I_q + S \in \mathcal{C}_q(\mathbb{D})$.

Example 6.5. Suppose that $K \in \mathbb{K}_{q \times q}$ and, furthermore, that $\Omega : \mathbb{D} \longrightarrow \mathbb{C}^{q \times q}$ is defined as $w \longmapsto I_q + wK$. Then $\Omega \in \mathcal{C}_q(\mathbb{D})$.

One of our goals will later be to generalize the following well-known result for the class $\mathcal{C}_q(\mathbb{D})$. Because it relates to our later approach, we include a proof.

Proposition 6.6. *Let* $\Omega \in \mathcal{C}_q(\mathbb{D})$ *be such that* $\det \Omega$ *does not vanish identically in* $\mathbb{D}$. *Then* $\det [\Omega(w)] \neq 0$ *for any* $w \in \mathbb{D}$, *and* $\Omega^{-1} \in \mathcal{C}_q(\mathbb{D})$.

Proof. Let S be the Cayley transform of Ω. By Lemma 6.3, it follows that

$$S \in \mathcal{S}_{q \times q}(\mathbb{D}) \tag{6.3}$$

and we obtain (6.2). Since $\det \Omega$ is not the zero function in $\mathbb{D}$, we see from (6.2) that $\det (I_q - S)$ is not the zero function in $\mathbb{D}$. Combining this with (6.3), we see from [8, Lemma 2.1.7] that $\det (I_q - S)$ does not vanish anywhere in $\mathbb{D}$. Thus, it follows from (6.2) that Ω does not vanish anywhere in $\mathbb{D}$. Therefore (note also the formulas after Remark A.1), we see that $\Omega^{-1} \in \mathcal{C}_q(\mathbb{D})$. $\qquad\square$

Remark 6.7. Let $\Omega \in \mathcal{C}_q(\mathbb{D})$ and $f \in \mathcal{S}'(\mathbb{D})$. The properties of holomorphic functions yield $\Omega \circ f \in \mathcal{C}_q(\mathbb{D})$.

Example 6.8. Let $\Omega \in \mathcal{C}_q(\mathbb{D})$ and $u \in \mathbb{D} \cup \mathbb{T}$. Suppose that $\Omega_u : \mathbb{D} \longrightarrow \mathbb{C}^{q \times q}$ is defined as $\Omega_u : w \longmapsto \Omega(uw)$. From Remark 6.7, we see that $\Omega_u \in \mathcal{C}_q(\mathbb{D})$.

Example 6.9. Let $\Omega \in \mathcal{C}_q(\mathbb{D})$ and $m \in \mathbb{Z}_{2,\infty}$. Suppose that $\Omega_{[m]} : \mathbb{D} \longrightarrow \mathbb{C}^{q \times q}$ is defined as $\Omega_{[m]} : w \longmapsto \Omega(w^m)$. From Remark 6.7, we see that $\Omega_{[m]} \in \mathcal{C}_q(\mathbb{D})$.

For more clarity later on, we next describe the well-known relationship that exists between $q \times q$ Carathéodory functions and $q \times q$ Carathéodory sequences (see, for instance, [11]).

Proposition 6.10. *Let* $\Omega \in \mathcal{C}_q(\mathbb{D})$ *and, for each* $w \in \mathbb{D}$, *let*

$$\Omega(w) = \sum_{j=0}^{\infty} s_j w^j \tag{6.4}$$

be the Taylor series representation of Ω. *Then* $(s_j)_{j=0}^{\infty}$ *is a* $q \times q$ *Carathéodory sequence.*

Example 6.11. Let $K \in \mathbb{K}_{q \times q}$. The sequence $(s_j)_{j=0}^{\infty}$, given by $s_j := \delta_{j,0} I_q + \delta_{j,1} K$, for each $j \in \mathbb{N}_0$, is then a $q \times q$ Carathéodory sequence. Indeed, if $\Omega : \mathbb{D} \longrightarrow \mathbb{C}^{q \times q}$ is defined as $\Omega(w) := I_q + wK$, then Example 6.5 yields $\Omega \in \mathcal{C}_q(\mathbb{D})$ and $(s_j)_{j=0}^{\infty}$ is the Taylor coefficient sequence for Ω. Proposition 6.10 thus shows that $(s_j)_{j=0}^{\infty} \in \mathcal{C}_{q,\infty}$.

Proposition 6.12. *Let* $(s_j)_{j=0}^{\infty}$ *be a* $q \times q$ *Carathéodory sequence. Then:*

(a) *The sequence* $\left(\sum_{j=0}^{n} s_j w^j \right)_{n=0}^{\infty}$ *converges for any* $w \in \mathbb{D}$.

(b) *If* $\Omega : \mathbb{D} \longrightarrow \mathbb{C}^{q \times q}$ *is defined as* $\Omega(w) := \lim_{n \to \infty} \sum_{j=0}^{n} s_j w^j$, *then* $\Omega \in \mathcal{C}_q(\mathbb{D})$.

If $(s_j)_{j=0}^{\infty}$ is a $q \times q$ Carathéodory sequence, then the function $\Omega \in \mathcal{C}_q(\mathbb{D})$ defined in part (b) of Proposition 6.12 will be referred to as the function in $\mathcal{C}_q(\mathbb{D})$ corresponding to $(s_j)_{j=0}^{\infty}$.

Example 6.13. Let $(s_j)_{j=0}^{\infty} \in \mathcal{C}_{q,\infty}$ and $u \in \mathbb{D} \cup \mathbb{T}$. Suppose, for each $j \in \mathbb{N}_0$, that $s_{j,u} := u^j s_j$. Part (a) of Remark 3.10 then yields $(s_{j,u})_{j=0}^{\infty} \in \mathcal{C}_{q,\infty}$. If Ω is the function in $\mathcal{C}_q(\mathbb{D})$ corresponding to $(s_j)_{j=0}^{\infty}$, then we see, for the function Ω_u defined in Example 6.8, that

$$\Omega_u(w) = \sum_{j=0}^{\infty} s_{j,u} w^j.$$

Thus, Ω_u is the function in $\mathcal{C}_q(\mathbb{D})$ corresponding to $(s_{j,u})_{j=0}^{\infty}$.

Example 6.14. Let $(s_j)_{j=0}^{\infty} \in \mathcal{C}_{q,\infty}$ and Ω be the function in $\mathcal{C}_q(\mathbb{D})$ corresponding to $(s_j)_{j=0}^{\infty}$. Suppose, furthermore, that $m \in \mathbb{Z}_{2,+\infty}$ and that

$$t_j := \begin{cases} s_{\frac{j}{m}}, & \text{if there exists a } k \in \mathbb{N}_0 \text{ such that } j = k \cdot m, \\ 0_{q \times q}, & \text{if } j \in \mathbb{N}_0 \setminus \{ k \cdot m : k \in \mathbb{N}_0 \}. \end{cases}$$

Let $\Omega_{[m]} \in \mathcal{C}_q(\mathbb{D})$ be the function defined in Example 6.9. For each $w \in \mathbb{D}$, then

$$\Omega_{[m]}(w) = \sum_{j=0}^{\infty} t_j w^j.$$

Thus, $(t_j)_{j=0}^{\infty} \in \mathcal{C}_{q,\infty}$ and $\Omega_{[m]}$ is the function in $\mathcal{C}_q(\mathbb{D})$ corresponding to $(t_j)_{j=0}^{\infty}$.

Propositions 6.10 and 6.12 suggest introducing a number of special subclasses for $\mathcal{C}_q(\mathbb{D})$ via Taylor coefficient sequences.

Definition 6.15. Let $\varkappa \in \mathbb{N}_0 \cup \{+\infty\}$. Suppose $\Omega \in \mathcal{C}_q(\mathbb{D})$ with Taylor series representation (6.4) for each $w \in \mathbb{D}$. If $(s_j)_{j=0}^{\varkappa} \in \tilde{\mathcal{C}}_{q,\varkappa}$, then Ω is called a **strict** $q \times q$ **Carathéodory function of order** $\varkappa$ (or strict order $\varkappa$ Carathéodory function). The set of all strict Carathéodory functions of order $\varkappa$ will be denoted by $\mathcal{C}_q^{(\varkappa)}(\mathbb{D})$.

Definition 6.16. Suppose $\Omega \in \mathcal{C}_q(\mathbb{D})$ with Taylor series representation (6.4), for each $w \in \mathbb{D}$. If $(s_j)_{j=0}^{\infty} \in \mathcal{C}_{q,\infty}^{t}$, then Ω is called a **totally strict** $q \times q$ **Carathéodory function in** $\mathbb{D}$. The set of all totally strict $q \times q$ Carathéodory functions in $\mathbb{D}$ will be denoted by $\mathcal{C}_q^{t}(\mathbb{D})$.

From the above two Definitions 6.15 and 6.16, it follows, via (3.2), that

$$\mathcal{C}_q^{t}(\mathbb{D}) \subseteq \mathcal{C}_q^{(\infty)}(\mathbb{D}) = \bigcap_{\varkappa=0}^{\infty} \mathcal{C}_q^{(\varkappa)}(\mathbb{D}).$$

Definition 6.17. Let $\Omega \in \mathcal{C}_q(\mathbb{D})$ with Taylor series representation (6.4) for $w \in \mathbb{D}$.
 (a) Let $k \in \mathbb{N}$. If the sequence $(s_j)_{j=0}^{\infty}$ is C-central of order k, then the function Ω is called **central of order** k (or order k central).
 (b) Let $k \in \mathbb{N}$. If the sequence $(s_j)_{j=0}^{\infty}$ is C-central of minimal order k, then the function Ω is called **central of minimal order** k (or minimal order k central).
 (c) If there exists a $k \in \mathbb{N}$ such that Ω is order k central, then Ω is simply called a **central function**.

Remark 6.18. Let $k \in \mathbb{N}$ and $\ell \in \mathbb{Z}_{k+1,\infty}$. Suppose that $\Omega \in \mathcal{C}_q(\mathbb{D})$ is central of order k. Remark 3.16 implies that Ω is also central of order ℓ.

Recalling Proposition 6.12 and Lemma 3.19, we are led to the following.

Definition 6.19. Let $n \in \mathbb{N}$ and $(s_j)_{j=0}^{n}$ be a $q \times q$ Carathéodory sequence. Furthermore, let $(\tilde{s}_j)_{j=0}^{\infty}$ be the C-central sequence corresponding to $(s_j)_{j=0}^{n}$. The function $\Omega_{(s_j)_{j=0}^{n}}$ in $\mathcal{C}_q(\mathbb{D})$ corresponding to $(\tilde{s}_j)_{j=0}^{\infty}$ is then called the **central** $q \times q$ **Carathéodory function for** $(s_j)_{j=0}^{n}$.

Remark 6.20. Let $k \in \mathbb{N}$. Suppose that $\Omega \in \mathcal{C}_q(\mathbb{D})$ is order k central with Taylor series representation (6.4) for each $w \in \mathbb{D}$. Because of Definition 6.19 and Remark 3.18, it follows that $(s_j)_{j=0}^{k-1} \in \mathcal{C}_{q,k-1}$ and Ω is the central $q \times q$ Carathéodory function $\Omega_{(s_j)_{j=0}^{k-1}}$ for $(s_j)_{j=0}^{k-1}$.

The function defined in Definition 6.19 is discussed extensively in [16]. In particular, it was there shown that $\Omega_{(s_j)_{j=0}^n}$ can always be interpreted as the restriction of a rational $q \times q$ matrix function to $\mathbb{D}$. Using the $q \times q$ Carathéodory sequence $(s_j)_{j=0}^n$, particular quadruples of $q \times q$ matrix polynomials were constructed. These quadruples were then used to obtain left and right quotient representations of $\Omega_{(s_j)_{j=0}^n}$. These representations were, in turn, used in [18] to parametrize the solution set of the Carathéodory problem associated with $(s_j)_{j=0}^n$.

Proposition 6.21. *Let $n \in \mathbb{N}_0$ and $(s_j)_{j=0}^n \in \mathcal{C}_{q,n}$. Then all of the following conditions are equivalent:*

(i) $(s_j)_{j=0}^n \in \widetilde{\mathcal{C}}_{q,n}$.

(ii) $\Omega_{(s_j)_{j=0}^n} \in \mathcal{C}_q^{\mathrm{t}}(\mathbb{D})$.

(iii) $\Omega_{(s_j)_{j=0}^n} \in \mathcal{C}_q^{(\infty)}(\mathbb{D})$.

Proof. Applying part (b) of Lemma 3.19 yields the proof.   □

Definition 6.22. Let $\Omega \in \mathcal{C}_q(\mathbb{D})$ with Taylor series representation (6.4) for $w \in \mathbb{D}$.

(a) Let $k \in \mathbb{N}$. If the sequence $(s_j)_{j=0}^\infty$ is order k C-canonical, then the function Ω is also called canonical of order k (or order k canonical).

(b) Let $k \in \mathbb{N}$. If the sequence $(s_j)_{j=0}^\infty$ is minimal order k C-canonical, then the function Ω is also called canonical of minimal order k (or minimal order k canonical).

(c) If there exists a $k \in \mathbb{N}$ such that Ω is order k canonical, then Ω is simply called a canonical function.

Remark 6.23. Let $k \in \mathbb{N}$, $\ell \in \mathbb{Z}_{k+1,\infty}$ and $\Omega \in \mathcal{C}_q(\mathbb{D})$ be order k canonical.

(a) Part (a) of Corollary 2.27 implies that Ω is also order ℓ canonical.

(b) It follows, by part (b) of Corollary 2.27, that Ω is also central of order ℓ.

For the $q \times q$ Carathéodory sequences in Examples 3.20 and 3.21, we now determine the corresponding functions in $\mathcal{C}_q(\mathbb{D})$.

Example 6.24. Let $K \in \mathbb{K}_{q \times q}$. Then:

(a) It follows for each $w \in \mathbb{D}$ that $-wK \in \mathbb{D}_{q \times q}$ and that $\det(I_q - wK) \neq 0$.

(b) The function $\Omega_K : \mathbb{D} \longrightarrow \mathbb{C}^{q \times q}$, defined via

$$w \longmapsto (I_q + wK)(I_q - wK)^{-1} \tag{6.5}$$

belongs to $\mathcal{C}_q(\mathbb{D})$.

(c) Since $wK \in \mathbb{D}_{q \times q}$, the Neumann series representation yields

$$\Omega_K(w) = (I_q + wK)\left[\sum_{j=0}^{\infty}(wK)^j\right] = I_q + 2\sum_{j=1}^{\infty} K^j w^j. \qquad (6.6)$$

Example 3.20 shows that the function Ω_K is order 2 central and, if (and only if) $K \neq 0_{q \times q}$, then Ω_K is minimal order 2 central. Furthermore, because of Remark 6.20, Example 3.20 and Proposition 6.21, the following three conditions are all equivalent:

(i) $K \in \mathbb{D}_{q \times q}$.

(ii) $\Omega_K \in \mathcal{C}_q^{t}(\mathbb{D})$.

(iii) $\Omega_K \in \mathcal{C}_q^{(\infty)}(\mathbb{D})$.

(d) For each $w \in \mathbb{D}$, we have $\det[\Omega_K(w)] \neq 0$, $-K \in \mathbb{K}_{q \times q}$ and $\Omega_K^{-1} = \Omega_{-K}$.

(e) Example 3.25 shows that the function Ω_K is canonical if and only if K is unitary. When this is the case, Ω_K is order 1 canonical.

Example 6.25. Let $K \in \mathbb{K}_{q \times q} \cap \mathbb{C}_{\geq}^{q \times q}$ and $r \in \mathbb{N}$. Then:

(a) Recalling part (a) of Example 6.24, let $\Omega_{K,r} : \mathbb{D} \longrightarrow \mathbb{C}^{q \times q}$ be defined as

$$w \longmapsto \left(\sqrt{K}\right)^r (I_q + wK)(I_q - wK)^{-1}\left(\sqrt{K}\right)^r. \qquad (6.7)$$

(b) Then

$$\Omega_{K,r} = \left[\left(\sqrt{K}\right)^r\right]^* \Omega_K\left(\sqrt{K}\right)^r, \qquad (6.8)$$

where Ω_K is the $q \times q$ Carathéodory function given in Example 6.24. It follows from Example 6.24 and (6.8) that, for each $w \in \mathbb{D}$, the function $\Omega_{K,r}$ belongs to $\mathcal{C}_q(\mathbb{D})$ and admits the Taylor series representation

$$\Omega_{K,r}(w) = K^r + 2\sum_{j=0}^{\infty} K^{r+j} w^j.$$

(c) Example 6.24 shows that the function $\Omega_{K,r}$ is order 2 central.

For the $q \times q$ Carathéodory sequences in Examples 3.12 and 3.13, we now determine the corresponding functions in $\mathcal{C}_q(\mathbb{D})$.

Example 6.26. Let $K \in \mathbb{K}_{q \times q}$ and $(C_j)_{j=0}^{\infty}$ be the $q \times q$ C-sequence defined in Example 3.12, i.e., $C_j := K^j$ for each $j \in \mathbb{N}_0$. Suppose that H_K is the function in $\mathcal{C}_q(\mathbb{D})$ corresponding to $(C_j)_{j=0}^{\infty}$. Then, for each $w \in \mathbb{D}$, we see from the Neumann series representation that $\det(I_q - wK) \neq 0$ and

$$H_K(w) = (I_q - wK)^{-1}.$$

Example 6.27. Let $K \in \mathbb{K}_{q \times q} \cap \mathbb{C}_{\geq}^{q \times q}$ and $r \in \mathbb{N}$. Let $(C_j)_{j=0}^{\infty}$ be the $q \times q$ C-sequence defined in Example 3.13, i.e., $C_j := K^{r+j}$ for each $j \in \mathbb{N}_0$. Suppose that $H_{K,r}$ is the function in $\mathcal{C}_q(\mathbb{D})$ corresponding to $(C_j)_{j=0}^{\infty}$. It follows that $C_j = \left(\sqrt{K}\right)^r K^j \left(\sqrt{K}\right)^r$ for any $j \in \mathbb{N}_0$. Therefore, it follows from Example 6.26, for each $w \in \mathbb{D}$, that $\det(I_q - wK) \neq 0$ and

$$H_{K,r}(w) = \left(\sqrt{K}\right)^r (I_q - wK)^{-1}\left(\sqrt{K}\right)^r.$$

7. Moore-Penrose Inverses of Matricial Carathéodory Functions

Our next steps will be towards showing that the Moore-Penrose inverse Ω^+ of a function $\Omega \in \mathcal{C}_q(\mathbb{D})$ is itself a member of $\mathcal{C}_q(\mathbb{D})$. We will, furthermore, show that the sequence of Ω^+'s Taylor coefficients is the reciprocal sequence to Ω's Taylor coefficient sequence. Before we show that the Moore-Penrose inverse of a $q \times q$ Carathéodory function is holomorphic, we first formulate and prove a result which will help us reach our aforementioned goal and which is also of interest on its own.

Lemma 7.1. *Let* $(s_j)_{j=0}^{\infty}$ *be a sequence in* $\mathbb{C}^{q \times q}$ *and* $(s_j^{\sharp})_{j=0}^{\infty}$ *be its reciprocal sequence. Suppose that the following conditions are met:*

(I) $(s_j)_{j=0}^{\infty} \in \mathcal{D}_{q \times q, \infty}$.

(II) *There exists a (in* $\mathbb{D}$*) holomorphic matrix function* $\Omega : \mathbb{D} \longrightarrow \mathbb{C}^{q \times q}$ *with Taylor expansion (6.4) for each* $w \in \mathbb{D}$.

(III) *There exists a (in* $\mathbb{D}$*) holomorphic matrix function* $\Omega^{\sharp} : \mathbb{D} \longrightarrow \mathbb{C}^{q \times q}$, *which can be expressed, for each* $w \in \mathbb{D}$, *as*

$$\Omega^{\sharp}(w) = \sum_{j=0}^{\infty} s_j^{\sharp} w^j. \tag{7.1}$$

Then $\Omega^+ = \Omega^{\sharp}$ *and, for each* $w \in \mathbb{D}$

$$(\Omega\Omega^+)(w) = (\Omega\Omega^+)(0), \qquad (\Omega^+\Omega)(w) = (\Omega^+\Omega)(0), \tag{7.2}$$

$$\mathcal{R}(\Omega(w)) = \mathcal{R}(\Omega(0)) \qquad and \qquad \mathcal{N}(\Omega(w)) = \mathcal{N}(\Omega(0)). \tag{7.3}$$

Proof. Because of (I), it follows by Proposition 1.8 that $S_n^+ = S_n^{\sharp}$, for each $n \in \mathbb{N}_0$, where S_n and $S_n^{\sharp}$ are given by (1.1) and (1.2). Thus, for each $n \in \mathbb{N}_0$, we obtain

$$S_n S_n^{\sharp} S_n = S_n \qquad and \qquad S_n^{\sharp} S_n S_n^{\sharp} = S_n^{\sharp}. \tag{7.4}$$

Also, recalling (I) and using Lemma 1.9, for each $n \in \mathbb{N}_0$, we get

$$S_n S_n^{\sharp} = I_{n+1} \otimes (s_0 s_0^+) \qquad and \qquad S_n^{\sharp} S_n = I_{n+1} \otimes (s_0^+ s_0). \tag{7.5}$$

Let $w \in \mathbb{D}$. Because of (7.4), (7.5), (II) and (III), it therefore follows by [8, Lemmas 1.1.19 and 1.1.21] that

$$\Omega(w)\Omega^{\sharp}(w)\Omega(w) = \Omega(w) \qquad and \qquad \Omega^{\sharp}(w)\Omega(w)\Omega^{\sharp}(w) = \Omega^{\sharp}(w) \tag{7.6}$$

and also that

$$\Omega(w)\Omega^{\sharp}(w) = s_0 s_0^+ \qquad and \qquad \Omega^{\sharp}(w)\Omega(w) = s_0^+ s_0. \tag{7.7}$$

In particular, (7.7) implies

$$(\Omega(w)\Omega^{\sharp}(w))^* = \Omega(w)\Omega^{\sharp}(w) \qquad and \qquad (\Omega^{\sharp}(w)\Omega(w))^* = \Omega^{\sharp}(w)\Omega(w). \tag{7.8}$$

From (7.6) and (7.8), we obtain $\Omega^+ = \Omega^{\sharp}$. Furthermore, [22, Proposition 8.4] yields (7.2) and (7.3). $\qquad \square$

We continue with a first observation on the ranges and null spaces of a function $\Omega \in \mathcal{C}_q(\mathbb{D})$.

Lemma 7.2. *Let* $\Omega \in \mathcal{C}_q(\mathbb{D})$ *and* $w \in \mathbb{D}$. *Then*

$$\mathcal{R}\left([\Omega(w)]^*\right) = \mathcal{R}\left(\Omega(w)\right) = \mathcal{R}\left([\Omega(w)]^+\right)$$

and

$$\mathcal{N}\left([\Omega(w)]^+\right) = \mathcal{N}\left(\Omega(w)\right) = \mathcal{N}\left([\Omega(w)]^*\right).$$

Proof. Since $\Omega \in \mathcal{C}_q(\mathbb{D})$, we see that $\Omega(w) \in \mathcal{R}_{q,\geq}$. By Lemma A.8, it therefore follows that $\Omega(w) \in \mathbb{C}_{\mathrm{EP}}^{q\times q}$. The lemma thus follows from Proposition A.5. $\qquad\square$

Our next theorem is also the first main result of this section. It includes the earlier-mentioned and quite versatile generalization of Proposition 6.6. The proof of the theorem, moreover, demonstrates an alternative approach to Proposition 6.6.

Theorem 7.3. *Let* $\Omega \in \mathcal{C}_q(\mathbb{D})$ *with Taylor series representation* (6.4) *for each* $w \in \mathbb{D}$. *Furthermore, let* $\left(s_j^\sharp\right)_{j=0}^{\infty}$ *be the reciprocal sequence to* $\left(s_j\right)_{j=0}^{\infty}$. *Then* Ω^+ *belongs to* $\mathcal{C}_q(\mathbb{D})$ *and* Ω^+ *admits the Taylor series representation*

$$\Omega^+(w) = \sum_{j=0}^{\infty} s_j^\sharp w^j. \tag{7.9}$$

Furthermore (7.2), (7.3),

$$\mathcal{R}\left(\Omega^+(w)\right) = \mathcal{R}\left(\Omega(0)\right) \quad and \quad \mathcal{N}\left(\Omega^+(w)\right) = \mathcal{N}\left(\Omega(0)\right) \tag{7.10}$$

all hold, for each $w \in \mathbb{D}$.

Proof. By assumption, we have (6.4) for each $w \in \mathbb{D}$. Since $\Omega \in \mathcal{C}_q(\mathbb{D})$, we see by Proposition 6.10 that $\left(s_j\right)_{j=0}^{\infty}$ is a $q \times q$ Carathéodory sequence. Thus, it follows by Theorem 4.4 that $\left(s_j^\sharp\right)_{j=0}^{\infty}$ is also a $q \times q$ Carathéodory sequence. From Proposition 6.12, we now see that there is a function $\Omega^\sharp \in \mathcal{C}_q(\mathbb{D})$ which admits the representation (7.1) for each $w \in \mathbb{D}$. By Proposition 4.1, it follows that $\left(s_j\right)_{j=0}^{\infty} \in \mathcal{D}_{q\times q,\infty}$, since $\left(s_j\right)_{j=0}^{\infty}$ is a $q \times q$ Carathéodory sequence. Thus, using (7.9), (7.10) and Lemma 7.1, we obtain $\Omega^+ = \Omega^\sharp$, as well as (7.2) and (7.3). Equations (7.10) follow by Lemma 7.2. $\qquad\square$

In [5, Theorem 4.5] it was already shown, using an alternate approach, that if $\Omega \in \mathcal{C}_q(\mathbb{D})$, then $\Omega^+ \in \mathcal{C}_q(\mathbb{D})$. Likewise, parts of Theorem 7.3 were also proved as part of [5, Theorem 4.5]. The approach used for [5, Theorem 4.5] was largely based on specific properties of the Schur class $\mathcal{S}_{q\times q}(\mathbb{D})$ and how these properties carry over to the class $\mathcal{C}_q(\mathbb{D})$ via Cayley transform.

Corollary 7.4. *Let* $\left(s_j\right)_{j=0}^{\infty}$ *be a* $q \times q$ *Carathéodory sequence and* Ω *be the function in* $\mathcal{C}_q(\mathbb{D})$ *associated with* $\left(s_j\right)_{j=0}^{\infty}$. *Furthermore, let* $\left(s_j^\sharp\right)_{j=0}^{\infty}$ *be the reciprocal sequence to* $\left(s_j\right)_{j=0}^{\infty}$. *Then,* $\left(s_j^\sharp\right)_{j=0}^{\infty}$ *is a* $q \times q$ *Carathéodory sequence and, if* $\Omega^\sharp$ *is the function in* $\mathcal{C}_q(\mathbb{D})$ *associated with* $\left(s_j^\sharp\right)_{j=0}^{\infty}$, *then* $\Omega^\sharp = \Omega^+$.

Proof. Proposition 6.12 shows that $\Omega \in \mathcal{C}_q(\mathbb{D})$ and that (6.4) holds true for each $w \in \mathbb{D}$. Thus, the corollary follows directly from Theorem 7.3. □

We next consider several applications of Theorem 7.3.

Corollary 7.5. *Suppose* $n \in \mathbb{N}$, $(\alpha_j)_{j=1}^n$ *is a sequence in* $[0, +\infty)$ *and* $(\Omega_j)_{j=1}^n$ *is a sequence in* $\mathcal{C}_q(\mathbb{D})$. *Then*

$$\left(\sum_{j=1}^n \alpha_j \Omega_j^+ \right)^+ \in \mathcal{C}_q(\mathbb{D}).$$

Proof. By Theorem 7.3, it follows that $(\Omega_j^+)_{j=1}^n$ is a sequence in $\mathcal{C}_q(\mathbb{D})$. Because of Remark 6.2, we therefore have

$$\sum_{j=1}^n \alpha_j \Omega_j^+ \in \mathcal{C}_q(\mathbb{D}).$$

Hence, applying Theorem 7.3 completes the proof. □

Corollary 7.6. *Suppose* $S \in \mathcal{S}_{q \times q}(\mathbb{D})$. *Then* $(I_q + S)^+ \in \mathcal{C}_q(\mathbb{D})$ *and the equations*

$$[I_q + S(w)][I_q + S(w)]^+ = [I_q + S(0)][I_q + S(0)]^+$$

and

$$[I_q + S(w)]^+[I_q + S(w)] = [I_q + S(0)]^+[I_q + S(0)]$$

hold true for each $w \in \mathbb{D}$. *Furthermore*,

$$\mathcal{R}(I_q + S(w)) = \mathcal{R}(I_q + S(0)), \qquad \mathcal{N}(I_q + S(w)) = \mathcal{N}(I_q + S(0)),$$

$$\mathcal{R}\left([I_q + S(w)]^+\right) = \mathcal{R}(I_q + S(0)) \quad \text{and} \quad \mathcal{N}\left([I_q + S(w)]^+\right) = \mathcal{N}(I_q + S(0))$$

hold true for each $w \in \mathbb{D}$.

Proof. By Lemma 6.4, we see that $I_q + S$ belongs to $\mathcal{C}_q(\mathbb{D})$. Thus, applying Theorem 7.3 completes the proof. □

Our next result tells us more about the structure of Carathéodory functions.

Theorem 7.7. *Let* $q \in \mathbb{Z}_{2,+\infty}$ *and* $r \in \mathbb{Z}_{1,q-1}$. *Suppose that* $\Omega \in \mathcal{C}_q(\mathbb{D})$ *with* rank $[\Omega(0)] = r$. *Furthermore, let* $(u_s)_{s=1}^r$ *be an orthonormal basis in* $\mathcal{R}(\Omega(0))$ *and let* $(u_s)_{s=r+1}^q$ *be an orthonormal basis in* $\mathcal{N}(\Omega(0))$. *For each* $\ell \in \mathbb{Z}_{1,q}$, *let* $U_\ell := (u_1, u_2, \ldots, u_\ell)$. *Finally, let* $\widetilde{\Omega} := U_r^* \Omega U_r$. *Then:*

   (a) $(u_s)_{s=1}^q$ *is an orthonormal basis in* $\mathbb{C}^{q \times 1}$ *and the matrix* U_q *is unitary.*
   (b) *The function* $\widetilde{\Omega}$ *belongs to* $\mathcal{C}_r(\mathbb{D})$ *and* $\widetilde{\Omega}(w)$ *is non-singular for all* $w \in \mathbb{D}$. *Furthermore,*

$$\Omega = U_q^* \left[\mathrm{diag}\left(\widetilde{\Omega}, 0_{(q-r) \times (q-r)} \right) \right] U_q$$

and

$$\Omega^+ = U_q^* \left[\mathrm{diag}\left(\widetilde{\Omega}^{-1}, 0_{(q-r) \times (q-r)} \right) \right] U_q.$$

Proof. Let $(s_j)_{j=0}^{\infty}$ be the Taylor coefficient sequence for Ω. Then, for each $w \in \mathbb{D}$, let (6.4) be the Taylor series representation of Ω. Then $\Omega(0) = s_0$, and, in particular, rank $s_0 = r$. Since $\Omega \in \mathcal{C}_q(\mathbb{D})$, it follows by Proposition 6.10 that $(s_j)_{j=0}^{\infty} \in \mathcal{C}_{q,\infty}$. Thus, Proposition 4.1 and part (a) of Theorem 1.12 yield part (a). By Remark 6.1, it follows that $\widetilde{\Omega} \in \mathcal{C}_r(\mathbb{D})$. For each $j \in \mathbb{N}_0$, we set

$$\widetilde{s}_j := U_r^* s_j U_r. \tag{7.11}$$

Since rank $s_0 = r$, we then see from Theorem 4.6 that $\det \widetilde{s}_0 \neq 0$. Let $w \in \mathbb{D}$. Using (6.4) and (7.11), we then obtain

$$\widetilde{\Omega}(w) = \sum_{j=0}^{\infty} U_r^* s_j U_r w^j = \sum_{j=0}^{\infty} \widetilde{s}_j w^j. \tag{7.12}$$

In particular, $\widetilde{\Omega}(0) = \widetilde{s}_0$. Thus, $\det \widetilde{s}_0 \neq 0$ implies $\det \widetilde{\Omega}(0) \neq 0$. Since $\widetilde{\Omega}$ belongs to $\mathcal{C}_r(\mathbb{D})$ and from Proposition 6.6 we now obtain $\det \widetilde{\Omega}(w) \neq 0$ for any $w \in \mathbb{D}$. Because of (7.11) and Theorem 4.6, it follows that

$$s_j = U_q^* \left[\operatorname{diag} \left(\widetilde{s}_j, 0_{(q-r) \times (q-r)} \right) \right] U_q \tag{7.13}$$

for each $j \in \mathbb{N}_0$. Using (6.4), (7.13) and (7.12), we obtain

$$\Omega(w) = \sum_{j=0}^{\infty} s_j w^j = \sum_{j=0}^{\infty} U_q^* \left[\operatorname{diag} \left(\widetilde{s}_j, 0_{(q-r) \times (q-r)} \right) \right] U_q w^j$$

$$= U_q^* \left(\sum_{j=0}^{\infty} \left[\operatorname{diag} \left(\widetilde{s}_j, 0_{(q-r) \times (q-r)} \right) \right] w^j \right) U_q$$

$$= U_q^* \left(\operatorname{diag} \left(\sum_{j=0}^{\infty} \widetilde{s}_j w^j, 0_{(q-r) \times (q-r)} \right) \right) U_q$$

$$= U_q^* \left[\operatorname{diag} \left(\widetilde{\Omega}(w), 0_{(q-r) \times (q-r)} \right) \right] U_q$$

for each $w \in \mathbb{D}$. The remainder of the theorem is a direct consequence of a well-known property of the Moore-Penrose inverse (see, e.g., [8, Lemma 1.1.3]). □

A result similar to Theorem 7.7 was obtained in [5, Theorem 3.5]. The approach used in [5] was based on the Cayley transform and is thus entirely different from the approach used in the above proof of Theorem 4.4, which relies mainly on the structural properties of $q \times q$ Carathéodory sequences described in Theorem 4.6. It should be noted that the approach used in [5, Theorem 4.5] (to show that the Moore-Penrose inverse of a function $\Omega \in \mathcal{C}_q(\mathbb{D})$ is holomorphic) relies on [5, Theorem 3.5]. A closer look at this approach reveals that this can also be shown using the above Theorem 7.7.

Let $\Omega \in \mathcal{C}_q(\mathbb{D})$ and, for each $w \in \mathbb{D}$, let (6.4) be the Taylor series representation of Ω. Theorem 7.3 implies that $\Omega^+ \in \mathcal{C}_q(\mathbb{D})$ and that Ω^+ has Taylor series representation (7.9) for each $w \in \mathbb{D}$. We next explore the relationship between Ω and Ω^+. Since $(s_j)_{j=0}^{\infty}$ and $(s_j^{\sharp})_{j=0}^{\infty}$ are, by Proposition 6.10, Carathéodory sequences, we can use the results of Section 3. Our next task will be to find sub-classes $\mathcal{C}_q(\mathbb{D})$ which are invariant with respect to the Moore-Penrose inverse, i.e., if a function Ω belongs to our subclass, so should its Moore-Penrose inverse Ω^+.

Proposition 7.8. *Let* $\Omega \in \mathcal{C}_q(\mathbb{D})$. *Then:*

(a) *Let* $\varkappa \in \mathbb{N}_0 \cup \{+\infty\}$. *Then* $\Omega \in \mathcal{C}_q^{(\varkappa)}(\mathbb{D})$ *if and only if* $\Omega^+ \in \mathcal{C}_q^{(\varkappa)}(\mathbb{D})$.

(b) $\Omega \in \mathcal{C}_q^{\mathrm{t}}(\mathbb{D})$ *if and only if* $\Omega^+ \in \mathcal{C}_q^{\mathrm{t}}(\mathbb{D})$.

Proof. Suppose that Ω has the Taylor series representation (6.4) for each $w \in \mathbb{D}$. By Proposition 6.10, it follows that $(s_j)_{j=0}^{\infty}$ is a $q \times q$ C-sequence. Theorem 7.3 thus implies that $\Omega^+ \in \mathcal{C}_q(\mathbb{D})$ and also that Ω^+'s Taylor coefficient sequence is $(s_j^{\sharp})_{j=0}^{\infty}$. Recalling Definition 6.15 and Definition 6.16, we see that (a) and (b) follow from Lemma 4.7 and Theorem 4.13, respectively. $\qquad\square$

We will next see that for any $\Omega \in \mathcal{C}_q(\mathbb{D})$, the Moore-Penrose inverse Ω^+ will have the same centrality properties.

Theorem 7.9. *Let* $\Omega \in \mathcal{C}_q(\mathbb{D})$.

(a) Ω *is a central function if and only if* Ω^+ *is central.*

(b) *Let* $k \in \mathbb{N}$.

   (b1) *The function* Ω *is order k central if and only if* Ω^+ *is order k central.*

   (b2) *The function* Ω *is minimal order k central if and only if* Ω^+ *is minimal order k central.*

Proof. We first proceed as in the proof of Proposition 7.8. Recalling Definition 6.17, we then see that parts (a), (b1) and (b2) follow, respectively, from parts (a), (b1) and (b2) of Theorem 4.10. $\qquad\square$

Proposition 7.10. *Let* $n \in \mathbb{N}_0$ *and* $(s_j)_{j=0}^{n}$ *be a $q \times q$ Carathéodory sequence. Furthermore, let* $\Omega_{(s_j)_{j=0}^n}$ *be the central $q \times q$ Carathéodory function for* $(s_j)_{j=0}^{n}$. *Then,* $(s_j^{\sharp})_{j=0}^{n}$ *is a $q \times q$ Carathéodory sequence and, if* $\Omega_{(s_j^{\sharp})_{j=0}^n}$ *is the central $q \times q$ Carathéodory function for* $(s_j^{\sharp})_{j=0}^{n}$, *then*

$$\Omega_{\left(s_j^{\sharp}\right)_{j=0}^{n}} = \left[\Omega_{(s_j)_{j=0}^{n}}\right]^{+}. \tag{7.14}$$

Proof. By Theorem 4.4, it follows that $(s_j^{\sharp})_{j=0}^{n}$ is a $q \times q$ C-sequence. Suppose that $(t_j)_{j=0}^{\infty}$ and $(t_{j,\sharp})_{j=0}^{\infty}$ are the central $q \times q$ C-sequences associated with $(s_j)_{j=0}^{n}$ and $(s_j^{\sharp})_{j=0}^{n}$, respectively. Because of Definition 6.19, we see that $(t_j)_{j=0}^{\infty}$ and $(t_{j,\sharp})_{j=0}^{\infty}$ are the Taylor coefficient sequences for $\Omega_{(s_j)_{j=0}^n}$ and $\Omega_{(s_j^{\sharp})_{j=0}^n}$, respectively. By Theorem 4.11, it follows that $(t_j^{\sharp})_{j=0}^{\infty} = (t_{j,\sharp})_{j=0}^{\infty}$, where $(t_j^{\sharp})_{j=0}^{\infty}$ is the reciprocal sequence to $(t_j)_{j=0}^{\infty}$. Therefore, (7.14) follows by Theorem 7.3. $\qquad\square$

Proposition 7.11. *Let* $\Omega \in \mathcal{C}_q(\mathbb{D})$.

(a) Ω *is a canonical function if and only if* Ω^+ *is canonical.*

(b) *Let* $k \in \mathbb{N}$.

    (b1) *The function* Ω *is minimal order* k *canonical if and only if* Ω^+ *is minimal order* k *canonical.*

    (b2) *The function* Ω *is order* k *canonical if and only if* Ω^+ *is order* k *canonical.*

Proof. We first proceed as in the proof of Proposition 7.8. Recalling Definition 6.22, we then see that parts (a), (b1) and (b2) follow, respectively, from parts (a), (b1) and (b2) of Theorem 4.8. $\square$

8. An approach to constructing the reciprocal of a non-negative Hermitian $q \times q$ measure

Using the results of previous sections, we now consider applications in non-negative Hermitian $q \times q$ measures on the unit circle $\mathbb{T}$. In order to establish a connection to previous sections, we review the relationships that exist between non-negative Hermitian $q \times q$ measures on $\mathbb{T}$, Toeplitz non-negative definite sequences in $\mathbb{C}^{q \times q}$ and $q \times q$ Carathéodory functions.

The σ-algebra of all Borel subsets of the unit circle $\mathbb{T}$ will be denoted by $\mathfrak{B}_{\mathbb{T}}$ and the set of all non-negative Hermitian $q \times q$ measures on $(\mathbb{T}, \mathfrak{B}_{\mathbb{T}})$ by $\mathcal{M}_{\geq}^q(\mathbb{T}, \mathfrak{B}_{\mathbb{T}})$. If $F \in \mathcal{M}_{\geq}^q(\mathbb{T}, \mathfrak{B}_{\mathbb{T}})$, then, for each $j \in \mathbb{Z}$, the matrix

$$C_j^{(F)} := \int_{\mathbb{T}} z^{-j} F(dz)$$

is called the jth Fourier coefficient of F. The Fourier coefficients allow us to construct a bijection between the set $\mathcal{M}_{\geq}^q(\mathbb{T}, \mathfrak{B}_{\mathbb{T}})$ and the set $\mathcal{T}_{q, \infty}$ of all infinite Toeplitz non-negative definite sequences in $\mathbb{C}^{q \times q}$. This is expressed in the following well-known matricial version of a classical result by Herglotz (see, for instance, [8, Theorem 2.2.1]).

Proposition 8.1. *A sequence* $(C_j)_{j=0}^{\infty}$ *in* $\mathbb{C}^{q \times q}$ *is Toeplitz non-negative definite if and only if there exists an* $F \in \mathcal{M}_{\geq}^q(\mathbb{T}, \mathfrak{B}_{\mathbb{T}})$ *such that*

$$\left(C_j^{(F)} \right)_{j=0}^{\infty} = (C_j)_{j=0}^{\infty}. \tag{8.1}$$

If such an $F \in \mathcal{M}_{\geq}^q(\mathbb{T}, \mathfrak{B}_{\mathbb{T}})$ *exists, then it is unique.*

If $(C_j)_{j=0}^{\infty}$ is a T-n.n.d. sequence in $\mathbb{C}^{q \times q}$, then, by Proposition 8.1, the unique $F \in \mathcal{M}_{\geq}^q(\mathbb{T}, \mathfrak{B}_{\mathbb{T}})$ such that (8.1) is met is called the spectral measure for $(C_j)_{j=0}^{\infty}$. Proposition 8.1 suggests paying particular attention to the subclass

$$\left\{ F \in \mathcal{M}_{\geq}^q(\mathbb{T}, \mathfrak{B}_{\mathbb{T}}) : \left(C_j^{(F)} \right)_{j=0}^{\infty} \in \tilde{\mathcal{T}}_{q, \infty} \right\}. \tag{8.2}$$

This subclass is particularly interesting for its role in methods of constructing orthogonal matrix polynomials (see Delsarte/Genin/Kamp [6], [8, Section 3.6], [11]). Let $F \in \mathcal{M}_{\geq}^q(\mathbb{T}, \mathfrak{B}_{\mathbb{T}})$. Because of Proposition 8.1, Definition 2.12 can be carried over to non-negative Hermitian measures.

Definition 8.2. Let $F \in \mathcal{M}_{\geq}^q(\mathbb{T}, \mathfrak{B}_{\mathbb{T}})$ with Fourier coefficient sequence $\left(C_j^{(F)} \right)_{j=0}^{\infty}$.

(a) Let $k \in \mathbb{N}$. We say that the measure F is central of order k (or order k central) if $\left(C_j^{(F)} \right)_{j=0}^{\infty}$ is a central sequence of order k.

(b) Let $k \in \mathbb{N}$. We say that the measure F is central of minimal order k (or minimal order k central) if the sequence $\left(C_j^{(F)} \right)_{j=0}^{\infty}$ is central of minimal order k.

(c) We say that F is a central measure if there exists a $k \in \mathbb{N}$ such that F is central of minimal order k.

Remark 8.3. Suppose that $k \in \mathbb{N}$ and that $F \in \mathcal{M}_{\geq}^q(\mathbb{T}, \mathfrak{B}_{\mathbb{T}})$ is order k central. For each $\ell \in \mathbb{Z}_{k+1, \infty}$, Remark 2.13 shows that F is also order ℓ central.

A discussion of central measures belonging to the set in (8.2) can be found in [8, Section 3.6]. There it was shown that such measures are, in particular, absolutely continuous with respect to Lebesgue–Borel measure on the unit circle. The density functions for these measures were, furthermore, determined and expressed in terms of special matrix polynomials which play an important role in the theory of orthogonal matrix polynomials on the unit circle.

A standard result for stationary sequences in Hilbert modules says that every $F \in \mathcal{M}_{\geq}^q(\mathbb{T}, \mathfrak{B}_{\mathbb{T}})$ can be interpreted as a non-stochastic spectral measure of a stationary sequence in a Hilbert module. In [12, Theorem 9], it was shown that F is order $k \in \mathbb{N}$ central if and only if the stationary sequence associated with F is autoregressive of order k. Central sequences and Proposition 8.1 next lead us to consider central measures.

Remark 8.4. Let $n \in \mathbb{N}_0$. Suppose $\left(C_j \right)_{j=0}^{n}$ is a T-n.n.d. sequence in $\mathbb{C}^{q \times q}$ and that $\left(\tilde{C}_j \right)_{j=0}^{\infty}$ is the central sequence corresponding to $\left(C_j \right)_{j=0}^{n}$. Part (a) of Lemma 2.19 then implies that $\left(\tilde{C}_j \right)_{j=0}^{\infty} \in \mathcal{T}_{q, \infty}$. The spectral measure $F_{(C_j)_{j=0}^n} \in \mathcal{M}_{\geq}^q(\mathbb{T}, \mathfrak{B}_{\mathbb{T}})$ for $\left(\tilde{C}_j \right)_{j=0}^{\infty}$ is then called the central measure for $\left(C_j \right)_{j=0}^{n}$.

Remark 8.5. Let $k \in \mathbb{N}$ and $F \in \mathcal{M}_{\geq}^q(\mathbb{T}, \mathfrak{B}_{\mathbb{T}})$ be order k central. It then, clearly, follows that $\left(C_j^{(F)} \right)_{j=0}^{k-1} \in \mathcal{T}_{q, k-1}$ and $F = F_{\left(C_j^{(F)} \right)_{j=0}^{k-1}}$.

Proposition 8.6. *Let $n \in \mathbb{N}_0$ and $\left(C_j \right)_{j=0}^{n} \in \mathcal{T}_{q, n}$. Then $F_{(C_j)_{j=0}^n}$ belongs to the set in (8.2) if and only if $\left(C_j \right)_{j=0}^{n} \in \tilde{\mathcal{T}}_{q, n}$.*

Proof. The theorem follows from Remark 8.4 and Lemma 2.19.   □

Definition 8.7. Let $F \in \mathcal{M}^q_{\geq}(\mathbb{T}, \mathfrak{B}_{\mathbb{T}})$ with Fourier coefficient sequence $(C_j^{(F)})_{j=0}^{\infty}$.

(a) Let $k \in \mathbb{N}$. We say that F is canonical of order k (or order k canonical) if $(C_j^{(F)})_{j=0}^{\infty}$ is order k canonical.

(b) Let $k \in \mathbb{N}$. We say that F is canonical of minimal order k (or minimal order k canonical) if $(C_j^{(F)})_{j=0}^{\infty}$ is minimal order k canonical.

(c) We say that F is canonical if there is a $k \in \mathbb{N}$ such that F is order k canonical.

Remark 8.8. Let $k \in \mathbb{N}$ and $F \in \mathcal{M}^q_{\geq}(\mathbb{T}, \mathfrak{B}_{\mathbb{T}})$ be order k canonical. Corollary 2.27 then implies:

(a) For each $\ell \in \mathbb{Z}_{k+1, \infty}$, the measure F is order ℓ canonical.

(b) For each $\ell \in \mathbb{Z}_{k+1, \infty}$, the measure F is order ℓ central.

For a given $z \in \mathbb{T}$, we will denote the Dirac measure on $(\mathbb{T}, \mathfrak{B}_{\mathbb{T}})$ with unit mass in z by ε_z. Combining Theorem 2.31 and Proposition 8.1 leads us to the following characterization of canonical measures in $\mathcal{M}^q_{\geq}(\mathbb{T}, \mathfrak{B}_{\mathbb{T}})$.

Theorem 8.9. *Let $F \in \mathcal{M}^q_{\geq}(\mathbb{T}, \mathfrak{B}_{\mathbb{T}})$. All of the following conditions are equivalent:*

(i) *F is a canonical measure.*

(ii) *There exists an $r \in \mathbb{N}$, a sequence $(A_s)_{s=1}^r$ in $\mathbb{C}^{q \times q}_{\geq}$ and a sequence $(z_s)_{s=1}^r$ of pairwise different points in $\mathbb{T}$ such that*

$$F = \sum_{s=1}^r \varepsilon_{z_s} A_s.$$

For a detailed discussion of canonical measures in $\mathcal{M}^q_{\geq}(\mathbb{T}, \mathfrak{B}_{\mathbb{T}})$, we refer the reader to [19], [20] and [21]. It is shown, in these articles, that, given a finite Toeplitz positive definite sequence $(C_j)_{j=0}^n$ in $\mathbb{C}^{q \times q}$ and a $u \in \mathbb{T}$, there exists a unique solution F_u of the trigonometric matricial moment problem corresponding to $(C_j)_{j=0}^n$ which has maximal mass with respect to the singleton $\{u\}$. More precisely, this solution is such that, for any F from the solution set, the inequality $F(\{u\}) \leq F_u(\{u\})$ holds. This F_u is canonical and can be constructed explicity from the sequence $(C_j)_{j=0}^n$.

The class $\mathcal{C}_q(\mathbb{D})$ will often feature prominently. We therefore provide a quick survey of results from [8, Section 2.2] on the relationship between $\mathcal{C}_q(\mathbb{D})$ and $\mathcal{M}^q_{\geq}(\mathbb{T}, \mathfrak{B}_{\mathbb{T}})$.

Proposition 8.10. *Let $F \in \mathcal{M}^q_{\geq}(\mathbb{T}, \mathfrak{B}_{\mathbb{T}})$. The function $\Omega_F : \mathbb{D} \longrightarrow \mathbb{C}^{q \times q}$ given by*

$$\Omega_F(w) := \int_{\mathbb{T}} \frac{z + w}{z - w} F(dz)$$

belongs to $\mathcal{C}_q(\mathbb{D})$ and, for any $w \in \mathbb{D}$, furthermore

$$\Omega_F(w) = C_0^{(F)} + 2 \sum_{j=1}^{\infty} C_j^{(F)} w^j. \tag{8.3}$$

If $F \in \mathcal{M}^q_{\geq}(\mathbb{T}, \mathfrak{B}_{\mathbb{T}})$ and Ω_F is defined as in Proposition 8.10, then Ω_F is called the Riesz-Herglotz transform of F. The close relationship that exists between

$\mathcal{M}^q_\geq(\mathbb{T}, \mathfrak{B}_\mathbb{T})$ and $\mathcal{C}_q(\mathbb{D})$ is evidenced by the following matricial generalization of a classical theorem by F. Riesz and G. Herglotz.

Theorem 8.11.
(a) *If $F \in \mathcal{M}^q_\geq(\mathbb{T}, \mathfrak{B}_\mathbb{T})$ and $H \in \mathbb{C}^{q \times q}_\mathrm{H}$, then $\Omega := \Omega_F + iH$ belongs to $\mathcal{C}_q(\mathbb{D})$ and $\mathrm{Im}\,[\Omega(0)] = H$.*
(b) *If $\Omega \in \mathcal{C}_q(\mathbb{D})$, then there exists a unique $F \in \mathcal{M}^q_\geq(\mathbb{T}, \mathfrak{B}_\mathbb{T})$ such that*
$$\Omega = \Omega_F + i\mathrm{Im}\,[\Omega(0)]. \tag{8.4}$$

If $\Omega \in \mathcal{C}_q(\mathbb{D})$, then the unique $F \in \mathcal{M}^q_\geq(\mathbb{T}, \mathfrak{B}_\mathbb{T})$ such that condition (8.4) is met is called the **Riesz-Herglotz measure** of Ω.

Suppose $F \in \mathcal{M}^q_\geq(\mathbb{T}, \mathfrak{B}_\mathbb{T})$. Let $\left(\left(C_j^{(F)} \right)^\square \right)_{j=0}^\infty$ be the Toeplitz non-negative definite sequence in $\mathbb{C}^{q \times q}$ generated from $\left(C_j^{(F)} \right)_{j=0}^\infty$ by reciprocation (see Definition 5.2 and Proposition 8.1). We are next most interested in describing the properties of the spectral measure $F^\square$ belonging to the sequence $\left(\left(C_j^{(F)} \right)^\square \right)_{j=0}^\infty$. It should, at this juncture, be mentioned that [8, Section 3.6] includes a discussion of this subject matter for the case in which F belongs to the set in (8.2). We now come to the main result of this section.

Theorem 8.12. *Let $F \in \mathcal{M}^q_\geq(\mathbb{T}, \mathfrak{B}_\mathbb{T})$ and Ω_F be the Riesz-Herglotz transform of F. Suppose that $\left(\left(C_j^{(F)} \right)^\square \right)_{j=0}^\infty$ is the T-n.n.d. sequence in $\mathbb{C}^{q \times q}$ generated from $\left(C_j^{(F)} \right)_{j=0}^\infty$ by reciprocation. Let $F^\square$ be the spectral measure for $\left(\left(C_j^{(F)} \right)^\square \right)_{j=0}^\infty$. Then $F^\square \in \mathcal{M}^q_\geq(\mathbb{T}, \mathfrak{B}_\mathbb{T})$ and $\Omega_{F^\square} = \Omega_F^+$.*

Proof. Let $w \in \mathbb{D}$. By Proposition 8.10, we have (8.3) and
$$\Omega_{F^\square}(w) = C_0^{(F^\square)} + 2 \sum_{j=1}^\infty C_j^{(F^\square)} w^j. \tag{8.5}$$

Since $F^\square$ is the spectral measure for $\left(\left(C_j^{(F)} \right)^\square \right)_{j=0}^\infty$, we have $C_j^{(F^\square)} = \left(C_j^{(F)} \right)^\square$ for each $j \in \mathbb{N}_0$. From (8.5) it therefore follows that
$$\Omega_{F^\square}(w) = \left(C_0^{(F)} \right)^\square + 2 \sum_{j=1}^\infty \left[\left(C_j^{(F)} \right)^\square w^j \right]. \tag{8.6}$$

We now define $(s_j)_{j=0}^\infty$ as $s_0 := C_0^{(F)}$ and, for each $j \in \mathbb{N}$, as $s_j := 2C_j^{(F)}$. From (8.3) we then see that
$$\Omega_F(w) = \sum_{j=0}^\infty s_j w^j. \tag{8.7}$$

Part (a) of Theorem 8.11 yields $\Omega_F \in \mathcal{C}_q(\mathbb{D})$. Because of (8.7) and Theorem 7.3, we have
$$\Omega_F^+(w) = \sum_{j=0}^\infty s_j^\sharp w^j. \tag{8.8}$$

By definition of $\left(\left(C_j^{(F)} \right)^{\square} \right)_{j=0}^{\infty}$, we have

$$s_0^{\sharp} := \left(C_0^{(F)} \right)^{\square} \quad \text{and} \quad s_j^{\sharp} := 2 \left(C_j^{(F)} \right)^{\square}$$

for each $j \in \mathbb{N}$ (see Proposition 5.1). Because of (8.6) and (8.8), this implies $\Omega_{F\square} = \Omega_F^+$, which completes the proof. $\qquad\square$

Theorem 8.12 shows us that Ω_F leads to a generalization of the classical concept of a reciprocal measure $F \in \mathcal{M}_{\geq}^q(\mathbb{T}, \mathfrak{B}_{\mathbb{T}})$ with $\det [F(\mathbb{T})] \neq 0$ (see, for instance, [8, Definition 3.6.10]). Thus, we arrive at the following definition.

Definition 8.13. Let $F \in \mathcal{M}_{\geq}^q(\mathbb{T}, \mathfrak{B}_{\mathbb{T}})$. The measure $F^{\square} \in \mathcal{M}_{\geq}^q(\mathbb{T}, \mathfrak{B}_{\mathbb{T}})$ defined in Theorem 8.12 is called the reciprocal measure to F.

Theorem 8.14. *Suppose that $F \in \mathcal{M}_{\geq}^q(\mathbb{T}, \mathfrak{B}_{\mathbb{T}})$ and let $F^{\square}$ be the reciprocal measure to F. Then:*

(a) *The measure F is central if and only if $F^{\square}$ is central.*
(b) *Let $k \in \mathbb{N}$.*
  (b1) *F is order k central if and only if $F^{\square}$ is order k central.*
  (b2) *F is minimal order k central if and only if $F^{\square}$ is minimal order k central.*

Proof. Recalling Definition 8.13 and Definition 8.2, we see that all parts of the theorem follow from Theorem 5.7. $\qquad\square$

Proposition 8.15. *Let $n \in \mathbb{N}_0$. Furthermore, let $(C_j)_{j=0}^{n}$ be a T-n.n.d. sequence in $\mathbb{C}^{q \times q}$ and $\left(C_j^{\square} \right)_{j=0}^{n}$ be the T-n.n.d. sequence in $\mathbb{C}^{q \times q}$ generated from $(C_j)_{j=0}^{n}$ by reciprocation. Let $F_{(C_j)_{j=0}^{n}}$ and $F_{\left(C_j^{\square} \right)_{j=0}^{n}}$ be the central measures in $\mathcal{M}_{\geq}^q(\mathbb{T}, \mathfrak{B}_{\mathbb{T}})$ for $(C_j)_{j=0}^{n}$ and $\left(C_j^{\square} \right)_{j=0}^{n}$, respectively. If $\left[F_{(C_j)_{j=0}^{n}} \right]^{\square}$ is the reciprocal measure to $F_{(C_j)_{j=0}^{n}}$, then*

$$\left[F_{(C_j)_{j=0}^{n}} \right]^{\square} = F_{\left(C_j^{\square} \right)_{j=0}^{n}}.$$

Proof. From Remark 8.4, we see that applying Theorem 5.8 and Theorem 8.12 completes the proof. $\qquad\square$

Proposition 8.16. *Suppose that $F \in \mathcal{M}_{\geq}^q(\mathbb{T}, \mathfrak{B}_{\mathbb{T}})$ and that $F^{\square}$ is the reciprocal measure to F. Then:*

(a) *The measure F is canonical if and only if $F^{\square}$ is canonical.*
(b) *Let $k \in \mathbb{N}$.*
  (b1) *F is order k canonical if and only if $F^{\square}$ is order k canonical.*
  (b2) *F is minimal order k canonical if and only if $F^{\square}$ is minimal order k canonical.*

Proof. Definition 8.7, Proposition 8.1 and Theorem 5.6 yield the proof. $\qquad\square$

9. Matricial R-functions in the open upper half-plane

In this section, we turn our attention to another class of matrix functions. More specifically, we are interested in a special subclass of $q \times q$ matrix functions that are holomorphic in $\Pi_+ := \{ z \in \mathbb{C} \; : \; \operatorname{Im} z \in (0, +\infty) \}$. This class of matrix functions is particularly interesting in the context of the matricial version of the Hamburger Moment Problem and is closely related to $\mathcal{C}_q(\mathbb{D})$, as we will later see.

A function $G \; : \; \Pi_+ \longrightarrow \mathbb{C}^{q \times q}$ is called a $q \times q$ R-function, if G is holomorphic in Π_+ and if the imaginary part $\operatorname{Im} G(z)$ of $G(z)$ is non-negative Hermitian, i.e., if $G(z) \in \mathcal{I}_{q, \geq}$ for all $z \in \Pi_+$. The set of all $q \times q$ R-functions will be denoted by $\mathcal{R}_q(\Pi_+)$. The name R-*function* is here adopted from M.G. Krein. In the literature, it is not uncommon to see functions of this type referred to instead as *Nevanlinna, Pick* or *Herglotz functions*. We now proceed with some preliminary observations on the ranges and null spaces of functions in $\mathcal{R}_q(\Pi_+)$.

Lemma 9.1. *Let* $G \in \mathcal{R}_q(\Pi_+)$. *For each* $z \in \Pi_+$,

$$\mathcal{R}(G(z)) = \mathcal{R}([G(z)]^*), \qquad \mathcal{R}(G(z)) = \mathcal{R}([G(z)]^+),$$
$$\mathcal{N}(G(z)) = \mathcal{N}([G(z)]^*) \quad and \quad \mathcal{N}(G(z)) = \mathcal{N}([G(z)]^+).$$

Proof. By assumption, $G(z) \in \mathcal{I}_{q, \geq}$. For each $z \in \Pi_+$, Lemma A.10 thus implies $G(z) \in \mathbb{C}^{q \times q}_{\mathrm{EP}}$. Applying Proposition A.5 thus completes the proof. $\square$

We now take a closer look at how the classes $\mathcal{R}_q(\Pi_+)$ and $\mathcal{C}_q(\mathbb{D})$ are related.

Remark 9.2. The function $\gamma : \mathbb{D} \longrightarrow \mathbb{C}$, defined as

$$\gamma(w) := i\frac{1 - w}{1 + w}$$

is a bijection between $\mathbb{D}$ and Π_+. The inverse function $\delta : \Pi_+ \longrightarrow \mathbb{C}$ to γ is given by

$$\delta(v) = \frac{1 + iv}{1 - iv}.$$

Lemma 9.3. *Let the mappings* γ *and* δ *be defined as in Remark* 9.2.
  (a) *If* Ω *belongs to* $\mathcal{C}_q(\mathbb{D})$ *and* $G := i(\Omega \circ \delta)$, *then* G *belongs to* $\mathcal{R}_q(\Pi_+)$ *and* $\Omega = (-i)(G \circ \gamma)$.
  (b) *If* $G \in \mathcal{R}_q(\Pi_+)$ *and* $\Omega := (-i)(G \circ \gamma)$, *then* Ω *belongs to* $\mathcal{C}_q(\mathbb{D})$ *and* $G = i(\Omega \circ \delta)$.

Proof. (a) Recalling Remark 9.2, we see that G is holomorphic in Π_+. Let $v \in \Pi_+$. For any $A \in \mathbb{C}^{q \times q}$, we have $\operatorname{Im}(iA) = \operatorname{Re} A$. Thus, it follows that

$$\operatorname{Im}[G(v)] = \operatorname{Im}[i(\Omega \circ \delta)(v)] = \operatorname{Im}[i\Omega(\delta(v))] = \operatorname{Re}[\Omega(\delta(v))].$$

Thus, $G(v) \in \mathcal{I}_{q, \geq}$, since $\Omega(\delta(v)) \in \mathcal{R}_{q, \geq}$. From Remark 9.2 and the definition of G, we thus see that $\Omega = (-i)(G \circ \gamma)$, which completes the proof of (a).
(b) Using Remark 9.2 and the fact that $\operatorname{Re}(-iA) = \operatorname{Im} A$ for any $A \in \mathbb{C}^{q \times q}$, we obtain (b) in much the same way as we did (a) in the first part of the proof. $\square$

Our next theorem is the $\mathcal{R}_q\left(\Pi_+\right)$ counterpart to Theorem 7.3.

Theorem 9.4. *Let $G \in \mathcal{R}_q\left(\Pi_+\right)$. Then:*

(a) $-G^+ \in \mathcal{R}_q\left(\Pi_+\right)$.

(b) $\left(GG^+\right)(v) = \left(GG^+\right)(i)$ *and* $\left(G^+G\right)(v) = \left(G^+G\right)(i)$, *for* $v \in \Pi_+$.

(c) $\mathcal{R}\left(G(v)\right) = \mathcal{R}\left(G(i)\right)$ *and* $\mathcal{N}\left(G(v)\right) = \mathcal{N}\left(G(i)\right)$, *for* $v \in \Pi_+$.

(d) $\mathcal{R}\left(G^+(v)\right) = \mathcal{R}\left(G(i)\right)$ *and* $\mathcal{N}\left(G^+(v)\right) = \mathcal{N}\left(G(i)\right)$, *for* $v \in \Pi_+$.

Proof. (a) Let $\Omega := \left(-i\right)\left(G \circ \gamma\right)$. By part (b) of Lemma 9.3, it thus follows that $\Omega \in \mathcal{C}_q\left(\mathbb{D}\right)$ and that $G = i\left(\Omega \circ \delta\right)$ and, consequently, that

$$G^+ = \left(-i\right)\left(\Omega \circ \delta\right)^+ = \left(-i\right)\left(\Omega^+ \circ \delta\right). \tag{9.1}$$

By Theorem 7.3, it follows that $\Omega^+ \in \mathcal{C}_q\left(\mathbb{D}\right)$. Therefore, by part (a) of Lemma 9.3, we have

$$i\left(\Omega^+ \circ \delta\right) \in \mathcal{R}_q\left(\Pi_+\right). \tag{9.2}$$

Finally, from (9.1) and (9.2) we obtain $-G^+ \in \mathcal{R}_q\left(\Pi_+\right)$, which completes the proof of (a).

(b)–(c) By assumption, both G and G^+ are holomorphic in the non-empty open and connected set Π_+. Parts (b) and (c) therefore follow by [22, Proposition 8.4].

(d) Because of (c), part (d) follows immediately by Lemma 9.1. $\qquad\square$

For a comprehensive survey on the class $\mathcal{R}_q\left(\Pi_+\right)$, we refer the reader to the paper Gesztesy/Tsekanovskii [23].

Appendix A. Some facts from matrix theory

Remark A.1. Let $A \in \mathbb{C}^{q \times q}$. Since $\mathrm{Re}\left(A^*\right) = \mathrm{Re}\,A$, it follows that $A \in \mathcal{R}_{q,\geq}$ if and only if $A^* \in \mathcal{R}_{q,\geq}$. Similarly, since $\mathrm{Im}\left(A^*\right) = -\mathrm{Im}\,A$, it follows that $A \in \mathcal{I}_{q,\geq}$ if and only if $-A^* \in \mathcal{I}_{q,\geq}$.

If $A \in \mathbb{C}^{q \times q}$ is a non-singular matrix, then it is well known that

$$\mathrm{Re}\left(A^{-1}\right) = A^{-1}\left(\mathrm{Re}\,A\right)\left(A^{-1}\right)^*, \qquad \mathrm{Re}\left(A^{-1}\right) = \left(A^{-1}\right)^*\left(\mathrm{Re}\,A\right)A^{-1},$$

$$\mathrm{Im}\left(A^{-1}\right) = -A^{-1}\left(\mathrm{Im}\,A\right)\left(A^{-1}\right)^* \quad \text{and} \quad \mathrm{Im}\left(A^{-1}\right) = -\left(A^{-1}\right)^*\left(\mathrm{Im}\,A\right)A^{-1},$$

so that $A \in \mathcal{R}_{q,\geq}$ implies $A^{-1} \in \mathcal{R}_{q,\geq}$ and $A \in \mathcal{I}_{q,\geq}$ implies $-A^{-1} \in \mathcal{I}_{q,\geq}$. In order to generalize these results for arbitrary complex $q \times q$ matrices, we give a very brief overview of standard results for Moore-Penrose inverses. We will often use the properties of the Moore-Penrose inverse described in these lemmas. For a detailed discussion of the Moore-Penrose inverse and its properties, we refer the reader to [2] and [8, Section 1.1].

Lemma A.2. *Suppose $A \in \mathbb{C}^{p \times q}$.*

(a) *Let $B \in \mathbb{C}^{p \times r}$. The following are all equivalent:*

(i) $\mathcal{R}\left(B\right) \subseteq \mathcal{R}\left(A\right)$.

(ii) $AA^+B = B$.

(iii) *There exists an $X \in \mathbb{C}^{q \times r}$ such that $AX = B$.*

(b) *Let $C \in \mathbb{C}^{r \times q}$. The following are all equivalent:*
   (iv) $\mathcal{N}(A) \subseteq \mathcal{N}(C)$.
   (v) $CA^+A = C$.
   (vi) *There exists a $Y \in \mathbb{C}^{r \times p}$ such that $YA = C$.*

Lemma A.3. *Suppose $A \in \mathbb{C}_{\mathrm{H}}^{q \times q}$, i.e., that A is a Hermitian matrix. Then the Moore-Penrose inverse A^+ of A is also Hermitian and the following equations hold true:*

$$\mathcal{R}(A) = \mathcal{R}(A^+), \qquad \mathcal{N}(A) = \mathcal{N}(A^+) \qquad and \qquad AA^+ = A^+A.$$

Lemma A.4. *Suppose $A \in \mathbb{C}_{\geq}^{q \times q}$. Then $A^+ \in \mathbb{C}_{\geq}^{q \times q}$ and:*

(a) $\mathcal{R}(A) = \mathcal{R}(\sqrt{A})$ *and* $\mathcal{N}(A) = \mathcal{N}(\sqrt{A})$.
(b) $AA^+ = A^+A = \sqrt{A}(\sqrt{A})^+ = (\sqrt{A})^+\sqrt{A}$.
(c) $\sqrt{A} = A(\sqrt{A})^+ = (\sqrt{A})^+A$ *and* $(\sqrt{A})^+ = A^+\sqrt{A} = \sqrt{A}A^+$.

We recall that a complex $q \times q$ matrix A is an EP matrix if $\mathcal{R}(A) = \mathcal{R}(A^*)$. We also recall that $\mathbb{C}_{\mathrm{EP}}^{q \times q}$ denotes the set of all EP matrices in $\mathbb{C}^{q \times q}$. Schwerdt-feger [28] first introduced the class $\mathbb{C}_{\mathrm{EP}}^{q \times q}$ (EP stands for "Equal Projectors"). This class of matrices is important in the theory of generalized inverses (see, e.g., the monographs Campbell/Meyer [3, Chapter 4, Section 3] and Ben-Israel/Greville [2, Chapter 4, Section 4], and the papers Pearl [27] and Meyer [26]). We use the following characterizations of the class $\mathbb{C}_{\mathrm{EP}}^{q \times q}$, which can be found in [4] and [29].

Proposition A.5. *If $A \in \mathbb{C}^{q \times q}$, then all of the following conditions are equivalent:*

(i) $A \in \mathbb{C}_{\mathrm{EP}}^{q \times q}$.   (viii) $A^+ \in \mathbb{C}_{\mathrm{EP}}^{q \times q}$.
(ii) $\mathcal{R}(A) \subseteq \mathcal{R}(A^*)$.   (ix) $\mathcal{R}(A) = \mathcal{R}(A^+)$.
(iii) $\mathcal{R}(A^*) \subseteq \mathcal{R}(A)$.   (x) $\mathcal{N}(A) = \mathcal{N}(A^+)$.
(iv) $\mathcal{N}(A) \subseteq \mathcal{N}(A^*)$.   (xi) $A^+A^2 = A$.
(v) $\mathcal{N}(A^*) \subseteq \mathcal{N}(A)$.   (xii) $A^2A^+ = A$.
(vi) $\mathcal{N}(A^*) = \mathcal{N}(A)$.   (xiii) $A(A^+)^2 = A^+$.
(vii) $AA^+ = A^+A$.   (xiv) $(A^+)^2A = A^+$.

We also require a few additional ways of characterizing the class $\mathbb{C}_{\mathrm{EP}}^{q \times q}$. We will need matricial real and imaginary parts for these characterizations.

Proposition A.6. *Let $A \in \mathbb{C}^{q \times q}$. The following conditions are all equivalent:*

(i) $A \in \mathbb{C}_{\mathrm{EP}}^{q \times q}$.
(ii) $\alpha A^+ + \beta(A^+)^* = A^+(\beta A + \alpha A^*)(A^+)^*$ *for all $\alpha, \beta \in \mathbb{C}$.*
(iii) $\mathrm{Re}(A^+) = A^+(\mathrm{Re}\,A)(A^+)^*$ *and* $\mathrm{Im}(A^+) = -A^+(\mathrm{Im}\,A)(A^+)^*$.
(iv) $\alpha A^+ + \beta(A^+)^* = (A^+)^*(\beta A + \alpha A^*)A^+$ *for all $\alpha, \beta \in \mathbb{C}$.*
(v) $\mathrm{Re}(A^+) = (A^+)^*(\mathrm{Re}\,A)A^+$ *and* $\mathrm{Im}(A^+) = -(A^+)^*(\mathrm{Im}\,A)A^+$.
(vi) $\alpha A + \beta A^* = A(\beta A^+ + \alpha(A^+)^*)A^*$ *for all $\alpha, \beta \in \mathbb{C}$.*
(vii) $\mathrm{Re}\,A = A \cdot \mathrm{Re}(A^+) \cdot A^*$ *and* $\mathrm{Im}\,A = -A \cdot \mathrm{Im}(A^+) \cdot A^*$.
(viii) $\alpha A + \beta A^* = A^*(\beta A^+ + \alpha(A^+)^*)A$ *for all $\alpha, \beta \in \mathbb{C}$.*
(ix) $\mathrm{Re}\,A = A^* \cdot \mathrm{Re}(A^+) \cdot A$ *and* $\mathrm{Im}\,A = -A^* \cdot \mathrm{Im}(A^+) \cdot A$.

Proof. "(i)$\Longrightarrow$(ii)". It follows from (i), by Proposition A.5, that $AA^+ = A^+A$. It now follows that

$$A^+A\left(A^+\right)^* = \left(AA^+\right)^*\left(A^+\right)^* = \left(A^+AA^+\right)^* = \left(A^+\right)^*$$

and

$$A^+A^*\left(A^+\right)^* = A^+\left(A^+A\right)^* = A^+AA^+ = A^+.$$

For any α, $\beta \in \mathbb{C}$, we therefore have

$$\alpha A^+ + \beta\left(A^+\right)^* = \alpha A^+A^*\left(A^+\right)^* + \beta A^+A\left(A^+\right)^* = A^+\left(\beta A + \alpha A^*\right)\left(A^+\right)^*.$$

"(ii)$\Longrightarrow$(iii)". For $\alpha = \frac{1}{2} = \beta$, part (ii) yields

$$\text{Re}\left(A^+\right) = \frac{1}{2}A^+ + \frac{1}{2}\left(A^+\right)^* = A^+ \cdot \frac{1}{2}\left(A + A^*\right)\left(A^+\right)^* = A^+\left(\text{Re}\,A\right)\left(A^+\right)^*.$$

Similarly, for $\alpha = \frac{1}{2i}$ and $\beta = -\frac{1}{2i}$, part (ii) also yields the second equation in (iii). "(iii)$\Longrightarrow$(i)". Since $A = \text{Re}\,A + i\text{Im}\,A$, $\text{Re}\left(A^*\right) = \text{Re}\,A$ and $\text{Im}\left(A^*\right) = -\text{Im}\,A$, it follows that $A^* = \text{Re}\,A - i\text{Im}\,A$. Using $A = \text{Re}\,A + i\text{Im}\,A$ and (iii), we thus obtain

$$\begin{aligned}
A^+ &= \text{Re}\left(A^+\right) + i\text{Im}\left(A^+\right) \\
&= A^+\left(\text{Re}\,A\right)\left(A^+\right)^* - iA^+\left(\text{Im}\,A\right)\left(A^+\right)^* \\
&= A^+\left(\text{Re}\,A - i\text{Im}\,A\right)\left(A^+\right)^* \\
&= A^+A^*\left(A^+\right)^* = A^+\left(A^+A\right)^* = \left(A^+\right)^2 A.
\end{aligned}$$

Therefore, by Proposition A.5 we obtain (i). Conditions (i), (ii) and (iii) are therefore all equivalent. Similarly, the same holds for conditions (i), (iv) and (v). We next consider the following condition:

(x) $A^+ \in \mathbb{C}_{\text{EP}}^{q\times q}$.

Using the identity $\left(A^+\right)^+ = A$, we see from the first part of the proof that (vi)–(x) are all equivalent. By Proposition A.5, it follows that (i) and (x) are equivalent to one another. Thus, (i)–(ix) are all equivalent. $\square$

Corollary A.7. *Suppose* $A \in \mathbb{C}_{\text{EP}}^{q\times q}$, *then:*

(a) $\mathcal{R}\left(\text{Re}\,A\right) \subseteq \mathcal{R}\left(A\right)$.

(b) $AA^+ \cdot \text{Re}\,A = \text{Re}\,A$.

(c) $\mathcal{N}\left(A\right) \subseteq \mathcal{N}\left(\text{Re}\,A\right)$.

(d) $\text{Re}\,A \cdot A^+A = \text{Re}\,A$.

(e) $A^*\left(A^+\right)^* \text{Re}\,A = \text{Re}\,A$.

(f) $\text{Re}\,A \cdot \left(A^+\right)^* A^* = \text{Re}\,A$.

(g) $\text{rank}\left(\text{Re}\,A\right) = \text{rank}\left[\text{Re}\left(A^+\right)\right]$.

(h) $A^* \cdot \text{Re}\left(A^+\right) = \text{Re}\,A \cdot A^+$.

(i) $\text{Re}\left(A^+\right) \cdot A = \left(A^+\right)^* \cdot \text{Re}\,A$.

(j) $A \cdot \text{Re}\left(A^+\right) = \text{Re}\,A \cdot \left(A^+\right)^*$.

(k) $\text{Re}\left(A^+\right) \cdot A^* = A^+ \cdot \text{Re}\,A$.

(l) $\mathcal{R}\left(\text{Im}\,A\right) \subseteq \mathcal{R}\left(A\right)$.

(m) $AA^+ \cdot \text{Im}\,A = \text{Im}\,A$.

(n) $\mathcal{N}\left(A\right) \subseteq \mathcal{N}\left(\text{Im}\,A\right)$.

(o) $\text{Im}\,A \cdot A^+A = \text{Im}\,A$.

(p) $A^*\left(A^+\right)^* \text{Im}\,A = \text{Im}\,A$.

(q) $\text{Im}\,A \cdot \left(A^+\right)^* A^* = \text{Im}\,A$.

(r) $\text{rank}\left(\text{Im}\,A\right) = \text{rank}\left[\text{Im}\left(A^+\right)\right]$.

Proof. By Proposition A.6, we have

$$\operatorname{Re} A = A \left[\operatorname{Re} \left(A^+ \right) \right] A^* \qquad \text{and} \qquad \operatorname{Re} A = A^* \left[\operatorname{Re} \left(A^+ \right) \right] A. \qquad \text{(A.1)}$$

Because of (A.1), parts (a), (b) and (c) thus follow by Lemma A.2. Recalling that $(\operatorname{Re} A)^* = \operatorname{Re} A$ and using parts (d) and (b), we see that

$$A^* \left(A^+ \right)^* \operatorname{Re} A = A^* \left(A^+ \right)^* \left(\operatorname{Re} A \right)^* = \left(\operatorname{Re} A \cdot A^+ A \right)^* = \left(\operatorname{Re} A \right)^* = \operatorname{Re} A$$

and, similarly, that $\operatorname{Re} A \cdot \left(A^+ \right)^* A^* = \operatorname{Re} A$. Thus, the proof of (e) and (f) is complete. Proposition A.6 gives us $\operatorname{Re} \left(A^+ \right) = A^+ \left[\operatorname{Re} A \right] \left(A^+ \right)^*$. From (A.1), we thus obtain (g).

(h) Using (A.1), the identity $\left(A^+ \right)^+ = A$ and (e), we see that

$$A^* \left(\operatorname{Re} A^+ \right) = A^* \left(A^+ \right)^* \cdot \operatorname{Re} A \cdot A^+ = \operatorname{Re} A \cdot A^+.$$

Similarly, (A.1) and (d) imply (i), whereas (A.1) and (b) show that (j) is true, whereas (A.1) and (f) yield (k). The proofs of (l)–(r) are similar. $\qquad \square$

One of our next goals is to show that the sets $\mathcal{R}_{q,\geq}$ and $\mathcal{I}_{q,\geq}$ are subsets of $\mathbb{C}_{\mathrm{EP}}^{q\times q}$. This will serve as part of our motivation for the next few results. Once we have proved that these inclusions are true, we will be able to apply all of our results for $\mathbb{C}_{\mathrm{EP}}^{q\times q}$ to the classes $\mathcal{R}_{q,\geq}$ and $\mathcal{I}_{q,\geq}$.

Lemma A.8. *Let $A \in \mathcal{R}_{q,\geq}$. Then:*

(a) $\mathcal{N}(A) \subseteq \mathcal{N}(\operatorname{Re} A)$.
(b) $\mathcal{R}(\operatorname{Re} A) \subseteq \mathcal{R}(A)$.
(c) $A \in \mathbb{C}_{\mathrm{EP}}^{q\times q}$.
(d) $AA^+ = A^+ A$.

Proof. (a) Suppose $x \in \mathcal{N}(A)$. Then $Ax = 0_{q\times 1}$. Since $\operatorname{Re} A \in \mathbb{C}_{\geq}^{q\times q}$, this implies

$$\left(\sqrt{\operatorname{Re} A} x \right)^* \sqrt{\operatorname{Re} A} x = x^* \left(\operatorname{Re} A \right) x = \frac{1}{2} \left(x^* A x + \left(A x \right)^* x \right) = 0.$$

By part (a) of Lemma A.4 we then get $x \in \mathcal{N}(\operatorname{Re} A)$.
(b) Since $A \in \mathcal{R}_{q,\geq}$, it follows by Remark A.1 that $A^* \in \mathcal{R}_{q,\geq}$. Thus, (a) implies $\mathcal{N}(A^*) \subseteq \mathcal{N}(\operatorname{Re}(A^*))$. Considering orthogonal complements now yields $\mathcal{R}\left(\left[\operatorname{Re}(A^*) \right]^* \right) \subseteq \mathcal{R}(A)$. Because of $\operatorname{Re}(A^*) = \operatorname{Re} A$ and $\operatorname{Re}(A^*) \in \mathbb{C}_{\mathrm{H}}^{q\times q}$, we have $\left[\operatorname{Re}(A^*) \right]^* = \operatorname{Re}(A^*) = \operatorname{Re} A$. Thus, we finally obtain $\mathcal{R}(\operatorname{Re} A) \subseteq \mathcal{R}(A)$.
(c) Let $x \in \mathcal{N}(A)$. Then $Ax = 0_{q\times 1}$ and, by (a) also $(\operatorname{Re} A)x = 0_{q\times 1}$. Thus,

$$A^* x = Ax + A^* x = 2 \left(\operatorname{Re} A \right) x = 0_{q\times 1}.$$

Hence, $\mathcal{N}(A) \subseteq \mathcal{N}(A^*)$. Consequently, Proposition A.5 implies $A \in \mathbb{C}_{\mathrm{EP}}^{q\times q}$.
(d) Because of (c), part (d) follows from Proposition A.5. $\qquad \square$

Proposition A.9. *Let $A \in \mathbb{C}^{q\times q}$. Then $A \in \mathcal{R}_{q,\geq}$ if and only if $A^+ \in \mathcal{R}_{q,\geq}$.*

Proof. Use part (c) of Lemma A.8 and Proposition A.6. $\qquad \square$

The next two results are the $\mathcal{I}_{q,\geq}$ counterparts to Lemma A.8 and Proposition A.9 and their proofs are thus quite similar to those for the corresponding $\mathcal{R}_{q,\geq}$ results.

Lemma A.10. *If $A \in \mathcal{I}_{q,\geq}$, then*

(a) $\mathcal{N}(A) \subseteq \mathcal{N}(\operatorname{Im} A)$.
(b) $\mathcal{R}(\operatorname{Im} A) \subseteq \mathcal{R}(A)$.
(c) $A \in \mathbb{C}_{\mathrm{EP}}^{q \times q}$.
(d) $AA^{+} = A^{+}A$.

Proposition A.11. *Let $A \in \mathbb{C}^{q \times q}$. Then $A \in \mathcal{I}_{q,\geq}$ if and only if $-A^{+} \in \mathcal{I}_{q,\geq}$.*

Lemma A.12. (a) *Let $A \in \mathcal{R}_{q,>}$, then $\det A \neq 0$ and $A^{-1} \in \mathcal{R}_{q,>}$.*
(b) *Let $A \in \mathcal{I}_{q,>}$, then $\det A \neq 0$ and $-A^{-1} \in \mathcal{I}_{q,>}$.*

Proof. Since $A \in \mathcal{R}_{q,>}$, it follows that $\mathcal{N}(\operatorname{Re} A) = \{0_{q \times 1}\}$. By part (a) of Lemma A.8, we thus obtain $\mathcal{N}(A) = \{0_{q \times 1}\}$. Therefore, $\det A \neq 0$. Finally, part (c) of Lemma A.8 and Proposition A.6 yield $A^{-1} \in \mathcal{R}_{q,>}$. The proof of (b) is similar to the proof of (a), but using parts (a) and (c) of Lemma A.10. □

We next establish a few additional connections between $\mathbb{K}_{q \times q}$ and $\mathcal{R}_{q,\geq}$ as well as between $\mathbb{D}_{q \times q}$ and $\mathcal{R}_{q,>}$.

Lemma A.13. *Suppose $A \in \mathbb{K}_{q \times q}$. Then*

$$\operatorname{Re} A \in \mathbb{K}_{q \times q}, \qquad I_q + A \in \mathcal{R}_{q,\geq} \qquad and \qquad I_q + \operatorname{Re} A \in \mathcal{R}_{q,\geq}$$

Proof. We see that $\|\operatorname{Re} A\|_{\mathrm{S}} \leq \|A\|_{\mathrm{S}} \leq 1$. Hence, $\operatorname{Re} A \in \mathbb{K}_{q \times q}$. We have

$$\operatorname{Re}(I_q + A) = I_q + \operatorname{Re} A = [1 - \|\operatorname{Re} A\|_{\mathrm{S}}] \cdot I_q + [\operatorname{Re} A + \|\operatorname{Re} A\|_{\mathrm{S}} \cdot I_q]. \quad (\text{A.2})$$

From $\|\operatorname{Re} A\|_{\mathrm{S}} \leq 1$, we obtain

$$[1 - \|\operatorname{Re} A\|_{\mathrm{S}}] \cdot I_q \in \mathbb{C}_{\geq}^{q \times q}. \quad (\text{A.3})$$

We recall that $\operatorname{Re} A \in \mathbb{C}_{\mathrm{H}}^{q \times q}$. Thus, using the Bunjakowski-Cauchy-Schwarz Inequality, we get

$$\operatorname{Re} A + \|\operatorname{Re} A\|_{\mathrm{S}} \cdot I_q \in \mathbb{C}_{\geq}^{q \times q}. \quad (\text{A.4})$$

From (A.2) - (A.4) it follows that $\operatorname{Re}(I_q + A) \in \mathbb{C}_{\geq}^{q \times q}$. Therefore, $I_q + A \in \mathcal{R}_{q,\geq}$. Since $\operatorname{Re} A \in \mathbb{K}_{q \times q}$, it follows that $I_q + \operatorname{Re} A \in \mathcal{R}_{q,\geq}$. □

Similar to Lemma A.13, we can show that the following lemma is true.

Lemma A.14. *Suppose $A \in \mathbb{D}_{q \times q}$. Then*

$$\operatorname{Re} A \in \mathbb{D}_{q \times q}, \qquad I_q + A \in \mathcal{R}_{q,>} \qquad and \qquad I_q + \operatorname{Re} A \in \mathcal{R}_{q,>}.$$

Acknowledgement

The third author's work on the present paper was supported by the German Research Foundation (Deutsche Forschungsgemeinschaft) on badge LA 1386/3–1.

References

[1] A. Arimoto and T. Ito, *Singularly positive definite sequences and parametrization of extreme points*, Linear Algebra Appl. **239** (1996), 127–149. MR1384918

[2] A. Ben-Israel and T.N.E. Greville, *Generalized inverses – Theory and applications*, CMS Books in Mathematics/Ouvrages de Mathématiques de la SMC, vol. 15, Springer-Verlag, New York, 2003. MR1987382

[3] S.L. Campbell and C.D. Meyer Jr., *Generalized inverses of linear transformations*, Dover Publications Inc., New York, 1991. MR1105324

[4] S. Cheng and Y. Tian, *Two sets of new characterizations for normal and EP matrices*, Linear Algebra Appl. **375** (2003), 181–195. MR2013464

[5] A.E. Choque Rivero, A. Lasarow, and A. Rahn, *On ranges and Moore-Penrose inverses related to matrix Carathéodory and Schur functions* **5** (2011), 145–160.

[6] P. Delsarte, Y. Genin, and Y. Kamp, *Orthogonal polynomial matrices on the unit circle*, IEEE Trans. Circuits and Systems **25** (1978), 145–160. MR0481886

[7] H. Dette and J. Wagener, *Matrix measures on the unit circle, moment spaces, orthogonal polynomials and the Geronimus relations*, Linear Algebra Appl. **432** (2010), 1609–1026. MR2592906

[8] V.K. Dubovoj, B. Fritzsche, and B. Kirstein, *Matricial Version of the Classical Schur Problem*, Teubner-Texte zur Mathematik, vol. 129, Teubner, Leipzig, 1992. MR1152328

[9] H. Dym and I. Gohberg, *Extensions of matrix-valued functions with rational polynomial inverses*, Integral Equations and Operator Theory **2** (1979), 503–528. MR0555776

[10] R. Ellis and I. Gohberg, *Extensions of matrix-valued inner products on the modules and the inversion formula for block Toeplitz matrices*, Operator Theory and Analysis – The M.A. Kaashoek Anniversary Volume (H. Bart, I. Gohberg, and A.M. Ran, eds.), Operator Theory: Advances and Applications, vol. 122, Birkhäuser Verlag, Basel – Boston – Berlin, 2001, pp. 192–227. MR1846058

[11] B. Fritzsche and B. Kirstein, *An extension problem for non-negative Hermitian block Toeplitz matrices*, Math. Nachr. **130** (1987), 121–135. MR0885621

[12] B. Fritzsche and B. Kirstein, *An extension problem for non-negative Hermitian block Toeplitz matrices. II*, Math. Nachr. **131** (1987), 287–297. MR0908816

[13] B. Fritzsche and B. Kirstein, *An extension problem for non-negative Hermitian block Toeplitz matrices. III*, Math. Nachr. **135** (1988), 319–341. MR0944234

[14] B. Fritzsche and B. Kirstein, *On the structure of maximum entropy extension*, Optimization **20** (1989), 177–191. MR0981961

[15] B. Fritzsche and B. Kirstein, *An extension problem for non-negative Hermitian block Toeplitz matrices. V*, Math. Nachr. **144** (1989), 283–308. MR1037173

[16] B. Fritzsche and B. Kirstein, *Representation of central matrix-valued Carathéodory functions in both non-degenerate and degenerate cases*, Integral Equations and Operator Theory **50** (2004), 333–361. MR2104258

[17] B. Fritzsche, B. Kirstein, and A. Lasarow, *On rank invariance of Schwarz-Pick-Potapov block matrices of matrix-valued Carathéodory functions*, Toeplitz Matrices and Singular Integral Equations (A. Böttcher, I. Gohberg, and P. Junghanns, eds.), Operator Theory: Advances and Applications, vol. 135, Birkhäuser Verlag, Basel, 2002, pp. 161–181. MR1935763

[18] B. Fritzsche, B. Kirstein, and A. Lasarow, *The matricial Carathéodory problem in both nondegenerate and degenerate cases*, Interpolation, Schur Functions and Moment Problems (D. Alpay and I. Gohberg, eds.), Operator Theory: Advances and Applications, vol. 165, Birkhäuser Verlag, Basel – Boston – Berlin, 2006, pp. 251–290. MR2222523

[19] B. Fritzsche, B. Kirstein, and A. Lasarow, *On a class of extremal solutions of the nondegenerate matricial Carathéodory problem*, Analysis **27** (2007), 109–164. MR2350712

[20] B. Fritzsche, B. Kirstein, and A. Lasarow, *On canonical solutions of a moment problem for rational matrix-valued functions*, Proceedings IWOTA (2010), to appear.

[21] B. Fritzsche, B. Kirstein, and A. Lasarow, *On maximal weight solutions of a moment problem for rational matrix-valued functions*, This Volume.

[22] B. Fritzsche, B. Kirstein, C. Mädler, and T. Schwarz, *On the concept of invertibility for sequences of complex $p \times q$ matrices and its applications to holomorphic $p \times q$-matrix-valued functions*, This Volume.

[23] F. Gesztesy and E. Tsekanovskii, *On matrix-valued Herglotz functions*, Math. Nachr. **218** (2000), 61–138. MR1784638 (2001j:47018)

[24] I.S. Iohvidov, *Hankel and Toeplitz Matrices and Forms*, Algebraic Theory, Birkhäuser, Boston – Berlin – Stuttgart, 1982. MR0677503

[25] I.V. Kovalishina, *Analytic theory of a class of interpolation problems*, Izv. Akad. Nauk SSSR, Ser. Mat. **47** (1983), 455–497 (Russian). MR0703593

[26] C.D. Meyer Jr., *Some remarks on EP_r matrices, and the generalized inverses*, Linear Algebra Appl. **3** (1970), 275–278. MR0266935

[27] M.H. Pearl, *On generalized inverses of matrices*, Proc. Cambridge Phil. Soc. **51** (1950), 406–418. MR0197485

[28] H. Schwerdtfeger, *Introduction to Linear Algebra and the Theory of Matrices*, Noordhoff, Groningen, 1950. MR0038923

[29] Y. Tian and H. Wang, *Characterizations of EP matrices and weighted EP matrices*, Linear Algebra Appl. **434** (2011), 1295–1318. MR2763588

Bernd Fritzsche, Bernd Kirstein, Andreas Lasarow and Armin Rahn
Mathematisches Institut
Universität Leipzig
Postfach: 10 09 20
D-04009 Leipzig, Germany
e-mail: fritzsche@math.uni-leipzig.de
        kirstein@math.uni-leipzig.de
        lasarow@math.uni-leipzig.de
        rahn@math.uni-leipzig.de

Operator Theory:
Advances and Applications, Vol. 226, 117–192
© 2012 Springer Basel

On a Schur-type Algorithm for Sequences of Complex $p \times q$-matrices and its Interrelations with the Canonical Hankel Parametrization

Bernd Fritzsche, Bernd Kirstein, Conrad Mädler and Tilo Schwarz

Abstract. Building on work started in [12], we further examine the structure of the set $\mathcal{H}^{\geq}_{q,2n}$ of all Hankel non-negative definite sequences $(s_j)^{2n}_{j=0}$ of complex $q \times q$-matrices. We furthermore examine the important subclasses $\mathcal{H}^{\geq,e}_{q,2n}$ and $\mathcal{H}^{>}_{q,2n}$, consisting of all Hankel non-negative definite and Hankel positive definite extendable sequences, respectively. These sequence-classes appear naturally when discussing matrix versions of the truncated Hamburger moment problem.

In [12] and [15] a canonical Hankel parametrization $[(C_k)^n_{k=1}, (D_k)^n_{k=0}]$, consisting of two sequences of complex matrices, was associated with every sequence $(s_j)^{2n}_{j=0}$ of complex $p \times q$-matrices. There is a bijective correspondence between the sequence and its canonical Hankel parametrization.

Chen and Hu [9] constructed a Schur-type algorithm for a special class of holomorphic matrix-valued functions in the upper half-plane so that matrix versions of the truncated Hamburger moment problem might be dealt with in the degenerate and non-degenerate cases, simultaneously. A closer analysis of their algorithm showed that it implicitly contains an interesting procedure for sequences belonging to $\mathcal{H}^{\geq,e}_{q,2n}$. This procedure serves as the focus of our work here, although we have chosen a slightly different and more general setting.

Our approach is based on a suitable extension of the concept of reciprocal sequences, which are used in power series inversions. We will show that, given n as a positive integer, this concept rests on a particular method for producing sequences belonging to $\mathcal{H}^{\geq}_{q,2(n-1)}$, starting from a sequence $(s_j)^{2n}_{j=0} \in \mathcal{H}^{\geq}_{q,2n}$. Using this, we develop a Schur-type algorithm for finite sequences of complex $p \times q$-matrices. We show that the Schur-type algorithm preserves specific subclasses of $\mathcal{H}^{\geq}_{q,2n}$, for example: $\mathcal{H}^{\geq,e}_{q,2n}$ and $\mathcal{H}^{>}_{q,2n}$. One of our main results (see Theorem 9.15) expresses that, given a sequence $(s_j)^{2n}_{j=0} \in \mathcal{H}^{\geq,e}_{q,2n}$, the Schur-type algorithm produces, exactly, its canonical Hankel parametrization. This leads us to a deeper understanding of the canonical Hankel parametrization.

Mathematics Subject Classification (2010). Primary 44A60, 47A57.

Keywords. Non-negative Hermitian measures, truncated matricial Hamburger moment problem, non-negative Hermitian block Hankel matrices, Hankel non-negative definite sequences, Hankel non-negative definite extendable sequences, Hankel positive definite sequences, canonical Hankel parametrization, reciprocal sequences, Schur-type algorithm.

1. Introduction

Our ultimate goal in this paper is to construct and study a Schur-type algorithm for finite and infinite sequences of complex $p \times q$-matrices. The impetus and idea for a construction of this type came from the paper by Chen/Hu [9], which deals with matrix versions of the classical Hamburger Moment Problem. In [9], Chen and Hu construct a Schur-type algorithm for special classes of holomorphic matrix-functions which made an approach for solving the above-mentioned matricial moment problems possible. A closer look at this algorithm reveals that it is based on a particular transformation for finite sequences of quadratic matrices. These sequences of quadratic matrices are obtained via infinite series expansions of holomorphic matrix-functions, namely, the matrix-functions which occur as part of the Schur-type algorithm described in [9]. Such sequences can, by way of Stieltjes transformation, be interpreted as moments of particular matricial measures.

We quickly realized that these results could, quite naturally, be extended to sequences of rectangular matrices. We were furthermore, early on, able to recognize that a particular transformation for finite sequences of complex $p \times q$-matrices would be central to the construction of the desired Schur-type algorithm for matrix-sequences. The reciprocal sequences (of matrices in the set $\mathbb{C}^{q \times p}$ of all complex $q \times p$-matrices associated with a sequence of matrices in $\mathbb{C}^{p \times q}$), discussed at length in the paper [16], play this central role in our construction.

When $p = q$ and F is a $q \times q$-matrix-function, holomorphic at the origin and with non-vanishing determinant, the relationship between the Taylor-coefficient sequences for the series expansions of F and F^{-1} is determined, in a domain around the origin, by the relationship between a sequence of matrices and its reciprocal sequence.

One application, which at the same time serves as further motivation, for the development of our Schur-type algorithm is in matrix versions of particular classical moment problems. We first present a quick summary of the notation used in this paper as well as some preliminary remarks regarding matricial moment problems.

Let $\mathbb{C}$, $\mathbb{R}$, $\mathbb{N}_0$ and $\mathbb{N}$ be the set of all complex numbers, the set of all real numbers, the set of all non-negative integers, and the set of all positive integers, respectively. For every $\alpha \in \mathbb{R}$ and $\beta \in \mathbb{R} \cup \{+\infty\}$ let $\mathbb{Z}_{\alpha,\beta}$ be the set of all integers k for which $\alpha \leq k \leq \beta$. Throughout this paper, let p and q be positive integers. If $\mathfrak{B}_{\mathbb{R}}$ denotes the σ-algebra of all Borel subsets of $\mathbb{R}$, then each countably additive mapping whose domain is $\mathfrak{B}_{\mathbb{R}}$ and whose values belong to the set $\mathbb{C}_{\geq}^{q \times q}$ of all non-negative Hermitian complex $q \times q$-matrices is called a non-negative Hermitian

$q \times q$-Borel-measure on $\mathbb{R}$. We will, at certain points, be using basic facts from the integration theory for non-negative Hermitian measures (see Kats [18] and Rosenberg [28–30]).

The sequences of matrices we work with in this paper are closely related to the following matrix versions of truncated moment problems.

P($\mathbb{R}$, $2n$, $\leq$): Let $n \in \mathbb{N}_0$ and let $(s_j)_{j=0}^{2n}$ be a sequence of complex $q \times q$-matrices. Describe the set $\mathcal{M}_{\geq}^q[\mathbb{R}; (s_j)_{j=0}^{2n}, \leq]$ of all non-negative Hermitian $q \times q$-Borel-measures σ on $\mathbb{R}$ such that, for each $j \in \mathbb{Z}_{0,2n}$, the integral

$$s_j^{[\sigma]} := \int_{\mathbb{R}} t^j \sigma(\mathrm{d}t) \tag{1.1}$$

exists, $s_k^{[\sigma]} = s_k$ is satisfied for each $k \in \mathbb{Z}_{0,2n-1}$, and the matrix $s_{2n} - s_{2n}^{[\sigma]}$ is non-negative Hermitian.

P($\mathbb{R}$, m, $=$): Let $m \in \mathbb{N}_0$ and let $(s_j)_{j=0}^{m}$ be a sequence of complex $q \times q$-matrices. Describe the set $\mathcal{M}_{\geq}^q[\mathbb{R}; (s_j)_{j=0}^{m}, =]$ of all non-negative Hermitian $q \times q$-Borel-measures σ on $\mathbb{R}$ such that for each $j \in \mathbb{Z}_{0,m}$ the integral (1.1) exists and $s_j^{[\sigma]} = s_j$ holds.

The study of the moment problems P($\mathbb{R}$, $2n$, $\leq$) and P($\mathbb{R}$, m, $=$) was a central theme in recent research (see, e.g., Bolotnikov [7], Chen/Hu [9]; and Dyukarev/Fritzsche/Kirstein/Mädler/Thiele [12]).

In [9] the authors presented a common method of solving simultaneously the non-degenerate and degenerate versions of problems P($\mathbb{R}$, $2n$, $\leq$) and P($\mathbb{R}$, m, $=$) based on the use of a matrix version of the classical Schur-type algorithm which already occurred in a fundamental memoir of Nevanlinna [25, Section 1]. In the non-degenerate case of the truncated matrix moment problems P($\mathbb{R}$, $2n$, $\leq$) and P($\mathbb{R}$, m, $=$), another Schur-type algorithm was worked out by I.V. Kovalishina [24, pp. 479–486]. Her approach is based on V.P. Potapov's method of fundamental matrix inequalities. V.P. Potapov interpreted the Schur algorithm from the view of J-theory, namely as some multiplicative decomposition of a full rank J-elementary factor in full rank J-elementary factors of the simplest kind. I.V. Kovalishina [24, pp. 483–486] followed this strategy. According to recent indefinite generalizations of Nevanlinna's algorithm, we refer to the papers Alpay/Dijksma/Langer [1] and Alpay/Dijksma/Langer/Shondin [2].

It should be mentioned that, in contrast with [12], V.P. Potapov's method of Fundamental Matrix Inequality (FMI) will not be applied directly in this paper. For the use of this method in the context of moment problems and related areas, we refer the reader to Kovalishina [24], Katsnel'son [19–23], Dyukarev/Katsnel'son [13, 14], and Dyukarev [11].

Central to this paper is the development of a Schur-type algorithm for sequences of complex matrices using new perspectives gained from the papers [12] and [15]. An important part of the strategy chosen in [12] and [15] is based on a thorough study of several classes of finite sequences $(s_j)_{j=0}^{2n}$ of complex $q \times q$-matrices which are closely related to the solvability of problems P($\mathbb{R}$, $2n$,

$\leq$) and P($\mathbb{R}$, m, $=$), respectively. Since these classes of finite sequences of complex $q \times q$-matrices contain much information about the problems P($\mathbb{R}$, $2n$, $\leq$) and P($\mathbb{R}$, m, $=$) we were inspired to rewrite the Schur-type algorithm created in [9] in sequence form to get new insights about the structure of the sequence classes under consideration. A similar approach was already used in the papers Bogner/Fritzsche/Kirstein [5,6] against the background of a matrix version of the classical Schur problem; namely the Schur-Potapov algorithm for matricial Schur functions in the unit disk was rewritten there in terms of matricial Schur sequences. It should be mentioned that, opposite to the papers [5,6] which are connected with the non-degenerate situation of the matricial Schur problem, the considerations in this paper handle the general cases of the problems P($\mathbb{R}$, $2n$, $\leq$) and P($\mathbb{R}$, m, $=$).

In order to describe the content of this paper in more detail we are going to give now exact definitions of some classes of finite sequences of complex $q \times q$-matrices which are the central objects in this paper. Moreover, we will review essential facts about them which will be used in the sequel. This material is taken from [12].

Let $n \in \mathbb{N}_0$ and let $(s_j)_{j=0}^{2n}$ be a sequence of complex $q \times q$-matrices. Then the sequence $(s_j)_{j=0}^{2n}$ is called *Hankel non-negative definite* if the block Hankel matrix

$$H_n := [s_{j+k}]_{j,k=0}^{n} \tag{1.2}$$

is non-negative Hermitian (or *Hankel positive definite* if H_n is positive Hermitian). We will use the notation $\mathcal{H}_{q,2n}^{\geq}$ and $\mathcal{H}_{q,2n}^{>}$ for the set of all Hankel non-negative definite and Hankel positive definite sequences $(s_j)_{j=0}^{2n}$ of complex $q \times q$-matrices, respectively. The importance of the set $\mathcal{H}_{q,2n}^{\geq}$ is based on the fact that problem P($\mathbb{R}$, $2n$, $\leq$) has a solution if and only if the sequence $(s_j)_{j=0}^{2n}$ belongs to $\mathcal{H}_{q,2n}^{\geq}$ (see, e.g., [12, Theorem 4.16]). If $n \in \mathbb{N}_0$ and if $(s_j)_{j=0}^{2n} \in \mathcal{H}_{q,2n}^{\geq}$ (respectively, $\mathcal{H}_{q,2n}^{>}$), then $(s_j)_{j=0}^{2m} \in \mathcal{H}_{q,2m}^{\geq}$ (respectively, $\mathcal{H}_{q,2m}^{>}$) for each $m \in \mathbb{Z}_{0,n}$. We will use $\mathcal{H}_{q,\infty}^{\geq}$ (respectively, $\mathcal{H}_{q,\infty}^{>}$) to denote the set of all sequences $(s_j)_{j=0}^{\infty}$ of complex $q \times q$-matrices such that $(s_j)_{j=0}^{2m} \in \mathcal{H}_{q,2m}^{\geq}$ (respectively, $(s_j)_{j=0}^{2m} \in \mathcal{H}_{q,2m}^{>}$) for each $m \in \mathbb{N}_0$.

Moreover, the Hankel non-negative definite extendability of Hankel non-negative definite sequences of complex $q \times q$-matrices is important for the study of problems P($\mathbb{R}$, $2n$, $\leq$) and P($\mathbb{R}$, m, $=$). For this reason, for each $n \in \mathbb{N}_0$, we will use the notation $\mathcal{H}_{q,2n}^{\geq,\mathrm{e}}$ for the set of all sequences $(s_j)_{j=0}^{2n}$ of complex $q \times q$-matrices for which there are complex $q \times q$-matrices s_{2n+1} and s_{2n+2} such that $(s_j)_{j=0}^{2(n+1)}$ belongs to $\mathcal{H}_{q,2(n+1)}^{\geq}$. Furthermore, if $n \in \mathbb{N}_0$, then by $\mathcal{H}_{q,2n+1}^{\geq,\mathrm{e}}$ we will designate the set of all sequences $(s_j)_{j=0}^{2n+1}$ of complex $q \times q$-matrices for which there is a complex $q \times q$ matrix s_{2n+2} such that $(s_j)_{j=0}^{2(n+1)}$ belongs to $\mathcal{H}_{q,2(n+1)}^{\geq}$. Observe that if a non-negative integer m and a sequence $(s_j)_{j=0}^{m}$ of complex $q \times q$-matrices are given, then problem P($\mathbb{R}$, m, $=$) has a solution if and only if $(s_j)_{j=0}^{m}$ belongs to $\mathcal{H}_{q,m}^{\geq,\mathrm{e}}$ (see [12, Theorem 4.17]). Obviously, $\mathcal{H}_{q,2k}^{\geq,\mathrm{e}} \subseteq \mathcal{H}_{q,2k}^{\geq}$ for each $k \in \mathbb{N}_0$ and

$\mathcal{H}_{q,0}^{\geq,e} = \mathcal{H}_{q,0}^{\geq}$. If $k \in \mathbb{N}$, then the example that $s_j := 0_{q \times q}$ for each $j \in \mathbb{Z}_{0,2k-1}$ and that $s_{2k} := I_q$ shows that the sets $\mathcal{H}_{q,2k}^{\geq,e}$ and $\mathcal{H}_{q,2k}^{\geq}$ do not coincide. Obviously, if $n \in \mathbb{N}_0$ and if $(s_j)_{j=0}^n \in \mathcal{H}_{q,n}^{\geq,e}$, then $(s_j)_{j=0}^m \in \mathcal{H}_{q,m}^{\geq,e}$ for each $m \in \mathbb{Z}_{0,n}$. In view of [12, Remark 2.8], for each $n \in \mathbb{N}_0$, the inclusion $\mathcal{H}_{q,2n}^{>} \subseteq \mathcal{H}_{q,2n}^{\geq,e}$ holds true.

An important topic of [12] was the study of the intrinsic structure of sequences belonging to one of the sets $\mathcal{H}_{q,2n}^{\geq}$, $\mathcal{H}_{q,2n}^{>}$, and $\mathcal{H}_{q,2n}^{\geq,e}$. For this reason, there was introduced in [12] a canonical Hankel parametrization for sequences $(s_j)_{j=0}^{2n}$ of complex $q \times q$-matrices. This concept of canonical Hankel parametrization was extended in [15] to sequences of complex $p \times q$-matrices. The canonical Hankel parametrization was used in [12, Proposition 2.30] to characterize the membership of a sequence $(s_j)_{j=0}^{2n}$ of complex $q \times q$-matrices to one of the classes $\mathcal{H}_{q,2n}^{\geq}$, $\mathcal{H}_{q,2n}^{>}$ and $\mathcal{H}_{q,2n}^{\geq,e}$.

In [12, Chapter 3] monic right (or, alternatively, left) orthogonal systems of $q \times q$-matrix polynomials were associated with a sequence $(s_j)_{j=0}^{2n} \in \mathcal{H}_{q,2n}^{\geq}$. In [15, Section 5] monic right (or left) orthogonal systems of the desired form were constructed recursively, using the canonical Hankel parametrization of $(s_j)_{j=0}^{2n}$.

In this paper, we focus on the more important properties of the canonical Hankel parametrization for sequences in $\mathcal{H}_{q,2n}^{\geq}$, as well as for sequences belonging to particular subclasses of $\mathcal{H}_{q,2n}^{\geq}$. More precisely, we will show that the canonical Hankel parametrization is closely related to the Schur-type algorithm (for sequences of complex matrices) that we later discuss as part of this paper. Our construction of this Schur-type algorithm is based on the concept of reciprocal sequences as developed in [16]. As a byproduct, we derive several identities which describe the interplay between various types of dual block Hankel matrices generated by a sequence from $\mathbb{C}^{p \times q}$ on one side and the corresponding reciprocal sequence on the other (see Section 6).

This paper is organized as follows: In Section 2, we summarize some essential facts on the canonical Hankel parametrization of finite or infinite sequences of complex $p \times q$-matrices.

In Section 3, we continue the study of the class of Hankel non negative definite sequences of complex $q \times q$-matrices and some of its important subclasses, which was started in [12] and [15].

There is a well-known method (taken from the theory of power series inversions) for constructing a reciprocal sequence $(A_j^\sharp)_{j=0}^n$ of complex $q \times q$-matrices satisfying $\det A_0 \neq 0$. Using Moore-Penrose inverses of complex $p \times q$-matrices, we extended this concept to rectangular matrices in [16]. There, we also introduced reciprocal sequences for sequences of complex $p \times q$-matrices (see also Definition 4.1). Reciprocal sequences are also central to this paper. In Section 4 we summarize some basic facts on reciprocal sequences as well as on the, for our purposes, particularly interesting class $\mathcal{D}_{p \times q, \kappa}$ of first term dominant sequences of complex $p \times q$-matrices (see Definition 4.5). These sections draw heavily from [16].

Section 5 contains some identities for block Hankel matrices which are associated with Cauchy products of sequences of complex matrices.

Section 6 plays a key role for further considerations. The main theme of this section is the study of interplay between various types of dual block Hankel matrices which are generated in an equal way by a sequence of complex $p \times q$-matrices and the corresponding reciprocal sequence of complex $q \times p$-matrices, respectively (see Theorem 6.1, Theorem 6.8, Theorem 6.9 and Theorem 6.13).

The investigations of Section 7 are inspired by Theorem 6.13 and its corollaries, which suggest a more careful study of the shortened negative reciprocal sequence. It is shown that, for a given sequence belonging to one of the classes $\mathcal{H}^{\geq}_{q,2n}$ and $\mathcal{H}^{\geq,e}_{q,2n}$, the shortened negative reciprocal sequence belongs to the classes $\mathcal{H}^{\geq}_{q,2(n-1)}$ and $\mathcal{H}^{\geq,e}_{q,2(n-1)}$, respectively (see Proposition 7.6 and Proposition 7.7).

In Section 8, the shortened negative reciprocal sequence will be replaced by a slightly modified sequence. In this way, we consider a transformation which associates with a sequence $(s_j)_{j=0}^{\kappa}$ from $\mathbb{C}^{p \times q}$ a particular sequence $(s_j^{(1)})_{j=0}^{\kappa-2}$ from $\mathbb{C}^{p \times q}$, which is called the first Schur transform of $(s_j)_{j=0}^{\kappa}$. We show that if a sequence $(s_j)_{j=0}^{2n}$ belongs to one of the classes $\mathcal{H}^{\geq}_{q,2n}$ and $\mathcal{H}^{\geq,e}_{q,2n}$, then its first Schur transform $(s_j^{(1)})_{j=0}^{2n-2}$ belongs to the classes $\mathcal{H}^{\geq}_{q,2(n-1)}$ and $\mathcal{H}^{\geq,e}_{q,2(n-1)}$, respectively (see Proposition 8.12 and Proposition 8.13).

The iteration of the first Schur transform introduced in Section 8 leads us to a particular Schur-type algorithm for finite or infinite sequences from $\mathbb{C}^{p \times q}$. The main theme of Section 9 is the investigation of this Schur-type algorithm. We show that this algorithm preserves the membership of a sequence to the class of Hankel non-negative definite sequences and its prominent subclasses (see Proposition 9.3, Proposition 9.4 and Proposition 9.5). Central results of the paper are Theorem 9.14 and Theorem 9.15 which contain explicit descriptions of the canonical Hankel parametrization of a sequence $(s_j)_{j=0}^{2n}$ belonging to one of the classes $\mathcal{H}^{\geq}_{q,2n}$ and $\mathcal{H}^{\geq,e}_{q,2n}$ in terms of the sequence of its Schur transforms. A further group of main results (see Theorem 9.18, Theorem 9.19 and Theorem 9.20) is focussed on the determination of the canonical Hankel parametrization of the kth Schur transform of a sequence belonging to one of the classes $\mathcal{H}^{\geq}_{q,2n}$, $\mathcal{H}^{\geq,e}_{q,2n}$ and $\mathcal{H}^{\geq}_{q,\infty}$. The last mentioned results indicate that, under the view of the considered Schur-type algorithm, the canonical Hankel parametrization can be interpreted as Schur parametrization.

In Section 10, we study several aspects of inversion of our Schur-type algorithm for sequences from $\mathbb{C}^{p \times q}$. This can be roughly described as follows: Let $\kappa \in \mathbb{N}_0 \cup \{+\infty\}$ and let $(t_j)_{j=0}^{\kappa}$ be a sequence from $\mathbb{C}^{p \times q}$. Furthermore, let $A, B \in \mathbb{C}^{p \times q}$. Then we will construct a sequence $(t_j^{(-1,A,B)})_{j=0}^{\kappa+2}$ from $\mathbb{C}^{p \times q}$ such that its first Schur transform coincides in generic cases with the original sequence $(t_j)_{j=0}^{\kappa}$. Particular attention will be paid to the case that the original sequence belongs to the class $\mathcal{H}^{\geq}_{q,2n}$ or one of its interesting subclasses.

2. The canonical Hankel parametrization of sequences of $p \times q$-Matrices

In this section, we present a summary of basic facts related to the canonical Hankel parametrization for sequences of complex $p \times q$-matrices. This concept was introduced in [12] for the case $p = q$ and then extended to the general case in [15]. The contents of this section draw heavily on [15].

If $\mathcal{X}$ is a non-empty set, then let $\mathcal{X}^{p \times q}$ be the set of $p \times q$-matrices with elements in $\mathcal{X}$. We will write $0_{p \times q}$ for the zero matrix in $\mathbb{C}^{p \times q}$ and I_q for the unit matrix in $\mathbb{C}^{q \times q}$. The sets of all Hermitian and positive Hermitian complex $q \times q$-matrices will be denoted by $\mathbb{C}_{\mathrm{H}}^{q \times q}$ and $\mathbb{C}_{>}^{q \times q}$, respectively. If $A \in \mathbb{C}^{p \times q}$, then $A^{\dagger}$ is the Moore-Penrose inverse of A. If $n \in \mathbb{N}$ and $(v_j)_{j=1}^{n}$ is a sequence of complex $p \times q$-matrices, then let

$$\mathrm{row}\,(v_j)_{j=1}^{n} := [v_1, v_2, \ldots, v_n] \qquad \text{and} \qquad \mathrm{col}\,(v_j)_{j=1}^{n} := \begin{bmatrix} v_1 \\ v_2 \\ \vdots \\ v_n \end{bmatrix}.$$

If $n \in \mathbb{N}$ and if $(p_j)_{j=1}^{n}$ and $(q_j)_{j=1}^{n}$ are sequences of positive integers such that $A \in \mathbb{C}^{p_j \times q_j}$ for each $j \in \mathbb{Z}_{1,n}$, then let

$$\mathrm{diag}\,(A_j)_{j=1}^{n} := \begin{bmatrix} A_1 & 0 & \cdots & 0 \\ 0 & A_2 & \cdots & 0 \\ \vdots & \vdots & \ddots & \vdots \\ 0 & 0 & \cdots & A_n \end{bmatrix}.$$

We will see that given a sequence $(s_j)_{j=0}^{2n}$ of complex $p \times q$-matrices, then an essential role is played by the Schur complements

$$L_0 := s_0 \qquad \text{and} \qquad L_k := s_{2k} - z_{k,2k-1} H_{k-1}^{\dagger} y_{k,2k-1}, \qquad k \in \mathbb{Z}_{1,n}, \qquad (2.1)$$

where the block Hankel matrix H_{k-1} is defined via (1.2) and where

$$z_{l,m} := \mathrm{row}\,(s_j)_{j=l}^{m} \qquad \text{and} \qquad y_{l,m} := \mathrm{col}\,(s_j)_{j=l}^{m} \qquad (2.2)$$

for all integers l and m with $0 \le l \le m \le 2n$.

To explain the canonical Hankel parametrization of a sequence of complex $p \times q$-matrices we need some further matrices. If $n \in \mathbb{N}$ and a sequence $(s_j)_{j=0}^{2n-1}$ of complex $p \times q$ matrices is given, then, for each $k \in \mathbb{Z}_{1,n}$, let the block Hankel matrix K_{k-1} be defined by

$$K_{k-1} := [s_{l+m+1}]_{l,m=0}^{k-1}$$

and let

$$\Sigma_k := z_{k,2k-1} H_{k-1}^{\dagger} K_{k-1} H_{k-1}^{\dagger} y_{k,2k-1}. \qquad (2.3)$$

If $n \in \mathbb{N}$ and if $(s_j)_{j=0}^{2n}$ is a sequence of complex $p \times q$ matrices, then, for each $k \in \mathbb{Z}_{1,n}$, let

$$M_k := z_{k,2k-1} H_{k-1}^\dagger y_{k+1,2k}, \quad N_k := z_{k+1,2k} H_{k-1}^\dagger y_{k,2k-1}, \quad P_k := L_k L_k^\dagger, \quad (2.4)$$

and let

$$\Lambda_k := M_k + N_k - \Sigma_k. \qquad (2.5)$$

Moreover, let

$$M_0 := 0_{p \times q}, \quad N_0 := 0_{p \times q}, \quad P_0 := L_0 L_0^\dagger, \quad \Sigma_0 := 0_{p \times q}, \quad \text{and} \quad \Lambda_0 := 0_{p \times q}. \quad (2.6)$$

Remark 2.1. Observe that if $n \in \mathbb{N}_0$ and if $(s_j)_{j=0}^{2n}$ is a sequence of Hermitian complex $q \times q$-matrices, then for each $k \in \mathbb{Z}_{0,n}$ we have

$$M_k^* = N_k, \qquad \Sigma_k^* = \Sigma_k, \qquad \text{and} \qquad \Lambda_k^* = \Lambda_k. \qquad (2.7)$$

In particular, (2.7) holds true if $(s_j)_{j=0}^{2n}$ belongs to $\mathcal{H}_{q,2n}^{\geq}$.

Remark 2.2. Let $\kappa \in \mathbb{N} \cup \{+\infty\}$ and let $(s_j)_{j=0}^{2\kappa}$ be a sequence of complex $p \times q$ matrices. Then one can easily see that there are unique sequences $(C_k)_{k=1}^\kappa$ and $(D_k)_{k=0}^\kappa$ of complex $p \times q$ matrices such that $s_0 = D_0$ and such that for each $k \in \mathbb{Z}_{1,\kappa}$ we have $s_{2k-1} = \Lambda_{k-1} + C_k$ and $s_{2k} = z_{k,2k-1} H_{k-1}^\dagger y_{k,2k-1} + D_k$. In particular, we see that $D_0 = L_0$ and, for each $k \in \mathbb{Z}_{1,\kappa}$, moreover $C_k = s_{2k-1} - \Lambda_{k-1}$ and $D_k = L_k$.

Remark 2.2 leads us to the following notion which will play a central role in the rest of the paper.

Definition 2.3. Let $\kappa \in \mathbb{N} \cup \{+\infty\}$ and let $(s_j)_{j=0}^{2\kappa}$ be a sequence of complex $p \times q$-matrices. Then the pair $[(C_k)_{k=1}^\kappa, (D_k)_{k=0}^\kappa]$ introduced in Remark 2.2 is called the *canonical Hankel parametrization* of $(s_j)_{j=0}^{2\kappa}$.

It should be mentioned that in [12] the canonical Hankel parametrization of a sequence $(s_j)_{j=0}^{2n}$ from $\mathbb{C}^{q \times q}$ was originally introduced in a slightly different way. From Remark 2.1 it is clear that both notions coincide in the case of a sequence $(s_j)_{j=0}^{2n}$ from $\mathbb{C}_H^{q \times q}$. For this reason all results about the canonical Hankel parametrization which were obtained in [12] for sequences from $\mathbb{C}_H^{q \times q}$ are also correct with respect to the canonical Hankel parametrization introduced in Definition 2.3.

Remark 2.4. Let $\kappa \in \mathbb{N} \cup \{+\infty\}$ and let $(C_k)_{k=1}^\kappa$ and $(D_k)_{k=0}^\kappa$ be sequences of complex $p \times q$ matrices. Then one can easily see that there is a unique sequence $(s_j)_{j=0}^{2\kappa}$ of complex $p \times q$ matrices such that $[(C_k)_{k=1}^\kappa, (D_k)_{k=0}^\kappa]$ is the canonical Hankel parametrization of $(s_j)_{j=0}^{2\kappa}$, namely the sequence given by $s_0 := D_0$ and, for each $k \in \mathbb{Z}_{1,\kappa}$, by $s_{2k-1} := \Lambda_{k-1} + C_k$ and $s_{2k} := z_{k,2k-1} H_{k-1}^\dagger y_{k,2k-1} + D_k$.

Remark 2.5. Let $\kappa \in \mathbb{N} \cup \{+\infty\}$, let $(s_j)_{j=0}^{2\kappa}$ be a sequence of complex $p \times q$-matrices, and let $[(C_k)_{k=1}^\kappa, (D_k)_{k=0}^\kappa]$ be the canonical Hankel parametrization of $(s_j)_{j=0}^{2\kappa}$. Then one can easily see that, for each $n \in \mathbb{Z}_{1,\kappa}$, the pair $[(C_k)_{k=1}^n, (D_k)_{k=0}^n]$ is the canonical Hankel parametrization of $(s_j)_{j=0}^{2n}$.

Remark 2.6. Let $\kappa \in \mathbb{N} \cup \{+\infty\}$, let $(s_j)_{j=0}^{2\kappa}$ be a sequence of complex $p \times q$-matrices, and let $[(C_k)_{k=1}^{\kappa}, (D_k)_{k=0}^{\kappa}]$ be the canonical Hankel parametrization of $(s_j)_{j=0}^{2\kappa}$. Let $m \in \mathbb{Z}_{p,\infty}$ and let $r \in \mathbb{Z}_{q,\infty}$. Lengthy, but straightforward calculations then show that, for each complex $m \times p$ matrix U with $U^*U = I_p$ and each complex $q \times r$ matrix V with $VV^* = I_q$, the pair $[(UC_kV)_{k=1}^{\kappa}, (UD_kV)_{k=0}^{\kappa}]$ is the canonical Hankel parametrization of the sequence $(Us_jV)_{j=0}^{2\kappa}$.

Remark 2.7. Let $\kappa \in \mathbb{N} \cup \{+\infty\}$, let $(s_j)_{j=0}^{2\kappa}$ be a sequence of complex $p \times q$-matrices, and let $[(C_k)_{k=1}^{\kappa}, (D_k)_{k=0}^{\kappa}]$ be the canonical Hankel parametrization of $(s_j)_{j=0}^{2\kappa}$. Then straightforward calculations show that $[(C_k^*)_{k=1}^{\kappa}, (D_k^*)_{k=0}^{\kappa}]$ is the canonical Hankel parametrization of $(s_j^*)_{j=0}^{2\kappa}$.

Remark 2.8. Let $\kappa \in \mathbb{N} \cup \{+\infty\}$, let $(s_j)_{j=0}^{2\kappa}$ be a sequence of complex $p \times q$-matrices, and let $[(C_k)_{k=1}^{\kappa}, (D_k)_{k=0}^{\kappa}]$ be the canonical Hankel parametrization of $(s_j)_{j=0}^{2\kappa}$. Straightforward calculations then show that $[(C_k^{\mathrm{T}})_{k=1}^{\kappa}, (D_k^{\mathrm{T}})_{k=0}^{\kappa}]$ is the canonical Hankel parametrization of $(s_j^{\mathrm{T}})_{j=0}^{2\kappa}$.

Remark 2.9. Let $\kappa \in \mathbb{N} \cup \{+\infty\}$ and let $n \in \mathbb{N}$. For each $m \in \mathbb{Z}_{1,n}$, let $p_m \in \mathbb{N}$, let $q_m \in \mathbb{N}$, let $(s_j^{(m)})_{j=0}^{2\kappa}$ be a sequence of complex $p_m \times q_m$ matrices, and let $[(C_k^{(m)})_{k=1}^{\kappa}, (D_k^{(m)})_{k=0}^{\kappa}]$ be the canonical Hankel parametrization of $(s_j^{(m)})_{j=0}^{2\kappa}$. By lengthy, but straightforward calculations one can check then that

$$[(\mathrm{diag}\,(C_k^{(m)})_{m=1}^n)_{k=1}^{\kappa}, (\mathrm{diag}\,(D_k^{(m)})_{m=1}^n)_{k=0}^{\kappa}]$$

is the canonical Hankel parametrization of $(\mathrm{diag}\,(s_j^{(m)})_{m=1}^n)_{j=0}^{2\kappa}$.

3. Some observations on finite and infinite sequences of matrices with particular Hankel-properties

A detailed examination of Hankel non-negative definite and Hankel non-negative definite extendable sequences of matrices in $\mathbb{C}^{q \times q}$ can be found in the papers [12] and [15]. In this section we provide a small addendum to these results and establish some facts that will later prove useful.

As usual, if $\mathcal{U}$ is a linear subspace of $\mathbb{C}^q$, then we will write $\mathcal{U}^{\perp}$ for the orthogonal complement of $\mathcal{U}$ in $\mathbb{C}^q$ with respect to the usual Euclidean inner product. Furthermore, for each $A \in \mathbb{C}^{p \times q}$, let $\mathcal{R}(A)$ and $\mathcal{N}(A)$ be the *range of* A and the *null space of* A, respectively. Clearly, for each $A \in \mathbb{C}_{\mathrm{H}}^{q \times q}$, we have $[\mathcal{R}(A)]^{\perp} = \mathcal{N}(A)$.

In order to realize a simultaneous exposition of the cases of a finite or infinite sequence of complex $q \times q$-matrices (for technical reasons) we set $\mathcal{H}_{q,\infty}^{\geq,\mathrm{e}} := \mathcal{H}_{q,\infty}^{\geq}$.

Lemma 3.1. *Let* $\kappa \in \mathbb{N}_0 \cup \{+\infty\}$ *and let* $(s_j)_{j=0}^{\kappa} \in \mathcal{H}_{q,\kappa}^{\geq,\mathrm{e}}$. *Then*

(a) $s_{2k} \in \mathbb{C}_{\geq}^{q \times q}$ *for each* $k \in \mathbb{Z}_{0,\frac{\kappa}{2}}$.
(b) $s_{\ell}^* = s_{\ell}$ *for each* $\ell \in \mathbb{Z}_{0,\kappa}$.

(c) $\displaystyle\bigcup_{j=2k}^{\kappa} \mathcal{R}(s_j) \subseteq \mathcal{R}(s_{2k})$ for each $k \in \mathbb{Z}_{0,\frac{\kappa}{2}}$.

(d) $\mathcal{N}(s_{2k}) \subseteq \displaystyle\bigcap_{j=2k}^{\kappa} \mathcal{N}(s_j)$ for each $k \in \mathbb{Z}_{0,\frac{\kappa}{2}}$.

(e) $(\mathcal{R}(s_{2k}))_{k=0}^{\kappa}$ is an antitone sequence.

(f) $(\mathcal{N}(s_{2k}))_{k=0}^{\kappa}$ is an isotone sequence.

Proof. If $\kappa \in \mathbb{N}_0$, then we know from [12, Lemma 3.34] that there exists a sequence $(s_j)_{j=\kappa+1}^{\infty}$ of complex $p \times q$-matrices such that $(s_j)_{j=0}^{\infty} \in \mathcal{H}_{q,\kappa}^{\geq}$. Then, for each non-negative integer ℓ, we have $H_\ell \in \mathbb{C}_{\geq}^{(\ell+1)q \times (\ell+1)q}$. Hence, we obtain the assertions of (a) and (b). Furthermore, it follows for $\ell \in \mathbb{N}_0$ that

$$\begin{bmatrix} s_0 & s_\ell \\ s_\ell & s_{2\ell} \end{bmatrix} \in \mathbb{C}_{\geq}^{2q \times 2q}. \tag{3.1}$$

Using a well-known result on non-negative Hermitian block matrices (see, e.g., [10, Lemma 1.1.9]), we conclude from (3.1) that $\mathcal{R}(s_\ell) \subseteq \mathcal{R}(s_0)$. Thus,

$$\bigcup_{j=0}^{\infty} \mathcal{R}(s_j) \subseteq \mathcal{R}(s_0). \tag{3.2}$$

For each $k \in \mathbb{N}_0$, the sequence $(s_j)_{j=2k}^{\infty}$ belongs to $\mathcal{H}_{q,\infty}^{\geq}$. Thus, (3.2) implies (c).

We obtain (d) from (c) by using (b) and $\mathcal{N}(s_l)^{\perp} = \mathcal{R}(s_l)$ for each $\ell \in \mathbb{N}_0$. (e) and (f) follow immediately from (c) and (d), respectively. □

Lemma 3.2. *Let* $\kappa \in \mathbb{N} \cup \{+\infty\}$ *and* $(s_j)_{j=0}^{2\kappa} \in \mathcal{H}_{q,2\kappa}^{\geq}$. *Then:*

(a) $s_{2k} \in \mathbb{C}_{\geq}^{q \times q}$ for each $k \in \mathbb{Z}_{0,\kappa}$.

(b) $s_\ell^* = s_\ell$ for each $\ell \in \mathbb{Z}_{0,2\kappa}$.

(c) $\displaystyle\bigcup_{j=0}^{2\kappa-1} \mathcal{R}(s_j) \subseteq \mathcal{R}(s_0)$ *and* $\mathcal{N}(s_0) \subseteq \displaystyle\bigcap_{j=0}^{2\kappa-1} \mathcal{N}(s_j)$.

Proof. Because of $\mathcal{H}_{q,\infty}^{\geq} = \mathcal{H}_{q,\infty}^{\geq,e}$, the case $\kappa = +\infty$ is already considered in Lemma 3.1. Let us now suppose $\kappa \in \mathbb{N}$. Then the matrix H_κ is non-negative Hermitian. Consequently, part (a) and part (b) hold true. Since $(s_j)_{j=0}^{2\kappa}$ belongs to $\mathcal{H}_{q,2\kappa}^{\geq}$, the sequence $(s_j)_{j=0}^{2\kappa-1}$ is a member of $\mathcal{H}_{q,2\kappa-1}^{\geq}$. Thus, part (c) follows from Lemma 3.1. □

Based on the matrices defined via (2.1) we now introduce an important subclass of $\mathcal{H}_{q,2n}^{\geq}$. If $n \in \mathbb{N}_0$ and if $(s_j)_{j=0}^{2n} \in \mathcal{H}_{q,2n}^{\geq}$, then $(s_j)_{j=0}^{2n}$ is called *completely degenerate* if $L_n = 0$. For each $n \in \mathbb{N}_0$, the set $\mathcal{H}_{q,2n}^{\geq,cd}$ of all completely degenerate sequences belonging to $\mathcal{H}_{q,2n}^{\geq}$ is a subset of $\mathcal{H}_{q,2n}^{\geq,e}$ (see [12, Corollary 2.14]). If $m \in \mathbb{N}_0$ and $(s_j)_{j=0}^{2m} \in \mathcal{H}_{q,2m}^{\geq,e}$ are given, then from [12, Proposition 2.13] one can easily see that $(s_j)_{j=0}^{2m}$ belongs to $\mathcal{H}_{q,2m}^{\geq,cd}$ if and only if there is some $n \in \mathbb{Z}_{0,m}$

such that $L_n = 0_{q \times q}$. A Hankel non-negative definite sequence $(s_j)_{j=0}^\infty$ of complex $q \times q$-matrices is said to be *completely degenerate* if there is some non-negative integer n such that $(s_j)_{j=0}^{2n}$ is a completely degenerate Hankel non-negative definite sequence. By $\mathcal{H}_{q,\infty}^{\geq,\mathrm{cd}}$ we denote the set of all completely degenerate Hankel non-negative definite sequences $(s_j)_{j=0}^\infty$ of complex $q \times q$-matrices. A Hankel non-negative definite sequence $(s_j)_{j=0}^\infty$ of complex $q \times q$-matrices is called *completely degenerate of order n* if the sequence $(s_j)_{j=0}^{2n}$ is completely degenerate. By $\mathcal{H}_{q,\infty}^{\geq,\mathrm{cd},n}$ we denote the set of all Hankel non-negative definite sequences $(s_j)_{j=0}^\infty$ from $\mathbb{C}^{q \times q}$ which are completely degenerate of order n. If $n \in \mathbb{N}_0$ and $(s_j)_{j=0}^\infty \in \mathcal{H}_{q,\infty}^{\geq,\mathrm{cd},n}$, then $(s_j)_{j=0}^{2m} \in \mathcal{H}_{q,2m}^{\geq,\mathrm{cd}}$ for each $m \in \mathbb{Z}_{n,\infty}$.

In the rest of this section, we discuss some later used properties of the class of Hankel non-negative definite sequences and their distinguished subclasses. This material is new and complements Section 2 in [12]. We start with several relatively simple, but useful observations.

Let $n \in \mathbb{N}_0$, let $(s_j)_{j=0}^{2n}$ be a sequence of complex $q \times q$-matrices, let $p \in \mathbb{N}$, and let $A \in \mathbb{C}^{p \times q}$. We now summarize some relations between the block Hankel matrices $H_{n,A} := [As_{j+k}A^*]_{j,k=0}^n$ and H_n defined in (1.2). First we observe that

$$H_{n,A} := [\mathrm{diag}(A, \ldots, A)] \cdot H_n \cdot [\mathrm{diag}(A, \ldots, A)]^* .$$

Remark 3.3. Let $\kappa \in \mathbb{N}_0 \cup \{+\infty\}$, let $(s_j)_{j=0}^{2\kappa} \in \mathcal{H}_{q,2\kappa}^\geq$, let $p \in \mathbb{N}$, and let $A \in \mathbb{C}^{p \times q}$. Then $(As_j A^*)_{j=0}^{2\kappa}$ belongs to $\mathcal{H}_{p,2\kappa}^\geq$.

Remark 3.4. Let $\kappa \in \mathbb{N}_0 \cup \{+\infty\}$ and let $(s_j)_{j=0}^\kappa \in \mathcal{H}_{q,\kappa}^{\geq,\mathrm{e}}$. For each $A \in \mathbb{C}^{p \times q}$, then $(As_j A^*)_{j=0}^\kappa$ belongs to $\mathcal{H}_{p,\kappa}^{\geq,\mathrm{e}}$.

Remark 3.5. Let $\kappa \in \mathbb{N}_0 \cup \{+\infty\}$ and let $(s_j)_{j=0}^{2\kappa} \in \mathcal{H}_{q,2\kappa}^>$. Further, let $A \in \mathbb{C}^{q \times q}$. If A is non-singular, then $(As_j A^*)_{j=0}^{2\kappa}$ belongs to $\mathcal{H}_{q,2\kappa}^>$. If A is singular, then $(As_j A^*)_{j=0}^{2\kappa}$ belongs to $\mathcal{H}_{q,2\kappa}^\geq \setminus \mathcal{H}_{q,2\kappa}^>$.

Remark 3.6. Let $\kappa \in \mathbb{N}_0 \cup \{+\infty\}$, let $(s_j)_{j=0}^\kappa \in \mathcal{H}_{p,\kappa}^{\geq,\mathrm{e}}$, and let $(t_j)_{j=0}^\kappa \in \mathcal{H}_{q,\kappa}^{\geq,\mathrm{e}}$. Then $(\mathrm{diag}(s_j, t_j))_{j=0}^\kappa \in \mathcal{H}_{p+q,\kappa}^{\geq,\mathrm{e}}$.

Remark 3.7. Let $n \in \mathbb{N}_0$ and let $(s_j)_{j=0}^{2n}$ and $(t_j)_{j=0}^{2n}$ be sequences from $\mathbb{C}^{p \times p}$ and $\mathbb{C}^{q \times q}$, respectively. Then:

(a) $(s_j)_{j=0}^{2n} \in \mathcal{H}_{p,2n}^\geq$ and $(t_j)_{j=0}^{2n} \in \mathcal{H}_{q,2n}^\geq$ if and only if $(\mathrm{diag}(s_j, t_j))_{j=0}^{2n} \in \mathcal{H}_{p+q,2n}^\geq$.

(b) $(s_j)_{j=0}^{2n} \in \mathcal{H}_{p,2n}^>$ and $(t_j)_{j=0}^{2n} \in \mathcal{H}_{q,2n}^>$ if and only if $(\mathrm{diag}(s_j, t_j))_{j=0}^{2n} \in \mathcal{H}_{p+q,2n}^>$.

(c) $(s_j)_{j=0}^{2n} \in \mathcal{H}_{p,2n}^{\geq,\mathrm{cd}}$ and $(t_j)_{j=0}^{2n} \in \mathcal{H}_{q,2n}^{\geq,\mathrm{cd}}$ if and only if $(\mathrm{diag}(s_j, t_j))_{j=0}^{2n} \in \mathcal{H}_{p+q,2n}^{\geq,\mathrm{cd}}$.

We now turn our attention to generalized Schur products of sequences with particular Hankel properties.

Lemma 3.8. *Let* $\kappa \in \mathbb{N}_0 \cup \{+\infty\}$ *and let* $(s_j)_{j=0}^{2\kappa}$ *and* $(t_j)_{j=0}^{2\kappa}$ *be sequences from* $\mathbb{C}$ *and* $\mathbb{C}^{q \times q}$, *respectively.*

(a) *If* $(s_j)_{j=0}^{2\kappa} \in \mathcal{H}_{1,2\kappa}^{\geq}$ *and* $(t_j)_{j=0}^{2\kappa} \in \mathcal{H}_{q,2\kappa}^{\geq}$, *then* $(s_j t_j)_{j=0}^{2\kappa} \in \mathcal{H}_{q,2\kappa}^{\geq}$.
(b) *If* $(s_j)_{j=0}^{2\kappa} \in \mathcal{H}_{1,2\kappa}^{>}$ *and* $(t_j)_{j=0}^{2\kappa} \in \mathcal{H}_{q,2\kappa}^{>}$, *then* $(s_j t_j)_{j=0}^{2\kappa} \in \mathcal{H}_{q,2\kappa}^{>}$.

Lemma 3.9. *Let* $m \in \mathbb{N}_0$ *and let* $(s_j)_{j=0}^{m} \in \mathcal{H}_{1,m}^{\geq,e}$ *and* $(t_j)_{j=0}^{m} \in \mathcal{H}_{q,m}^{\geq,e}$. *Then* $(s_j t_j)_{j=0}^{m} \in \mathcal{H}_{q,m}^{\geq,e}$.

Since $(\frac{1}{j+1})_{j=0}^{\infty}$ is the sequence of moments for the continuous uniform distribution on $[0, 1]$, it follows that $(\frac{1}{j+1})_{j=0}^{\infty} \in \mathcal{H}_{1,\infty}^{>}$. Thus, Lemmas 3.8 and 3.9 lead to the following examples.

Example. Let $\kappa \in \mathbb{N}_0 \cup \{+\infty\}$. If $(t_j)_{j=0}^{2\kappa}$ is a sequence from $\mathcal{H}_{q,2\kappa}^{\geq}$, then $(\frac{1}{j+1} t_j)_{j=0}^{\infty}$ is in $\mathcal{H}_{q,2\kappa}^{\geq}$. If $(t_j)_{j=0}^{2\kappa}$ is a sequence from $\mathcal{H}_{q,2\kappa}^{>}$, then $(\frac{1}{j+1} t_j)_{j=0}^{\infty}$ is in $\mathcal{H}_{q,2\kappa}^{>}$.

Example. Let $m \in \mathbb{N}_0$ and $(t_j)_{j=0}^{m} \in \mathcal{H}_{q,m}^{\geq,e}$, then $(\frac{1}{j+1} t_j)_{j=0}^{m} \in \mathcal{H}_{q,m}^{\geq,e}$.

To derive a result similar to Remarks 3.3, 3.4 and 3.5 for the class $\mathcal{H}_{q,2n}^{\geq,cd}$, we need an inequality for Schur complements (in the sense of the Löwner semi-ordering in the set $\mathbb{C}_{\mathrm{H}}^{q \times q}$ of all Hermitian complex $q \times q$-matrices).

Lemma 3.10. *Let* $p, r, s \in \mathbb{N}$, *let* $A \in \mathbb{C}_{\geq}^{p \times p}$, *let* $B \in \mathbb{C}^{p \times q}$ *be such that* $\mathcal{R}(B) \subseteq \mathcal{R}(A)$ *holds, let* $D \in \mathbb{C}^{q \times q}$, *let* $X \in \mathbb{C}^{p \times r}$, *and let* $Y \in \mathbb{C}^{q \times s}$. *Then*

$$Y^* DY - (X^* BY)^* (X^* AX)^\dagger X^* BY \geq Y^* (D - B^* A^\dagger B) Y.$$

Proof. Because of $A^* = A$ and $\mathcal{R}(B) \subseteq \mathcal{R}(A)$ we have $\mathcal{N}(A) = \mathcal{R}(A)^\perp \subseteq \mathcal{R}(B)^\perp = \mathcal{N}(B^*)$. Taking additionally into account $A \in \mathbb{C}_{\geq}^{q \times q}$, $\mathcal{R}(B) \subseteq \mathcal{R}(A)$, part (b) and part (c) of Lemma A.1, and $A^\dagger = A^\dagger A (A^\dagger)^* \in \mathbb{C}_{\geq}^{p \times p}$, we obtain

$$\begin{bmatrix} X^* AX & X^* B \\ (X^* B)^* & B^* A^\dagger B \end{bmatrix} = (AX, B)^* A^\dagger (AX, B) \in \mathbb{C}_{\geq}^{(r+q) \times (r+q)}.$$

This relation yields $B^* A^\dagger B - (X^* B)^* (X^* AX)^\dagger X^* B \in \mathbb{C}_{\geq}^{r \times r}$ (see, e.g., [10, Lemma 1.1.9 (a)]). Consequently,

$$
\begin{aligned}
& Y^* DY - (X^* BY)^* (X^* AX)^\dagger X^* BY - Y^* (D - B^* A^\dagger B) Y \\
&= Y^* DY - Y^* (X^* B)^* (X^* AX)^\dagger X^* BY - Y^* (D - B^* A^\dagger B) Y \\
&= Y^* \left[B^* A^\dagger B - (X^* B)^* (X^* AX)^\dagger X^* B \right] Y \in \mathbb{C}_{\geq}^{s \times s},
\end{aligned}
$$

which implies the asserted inequality. $\qquad \square$

Proposition 3.11. *Let* $n \in \mathbb{N}_0$, *let* $(s_j)_{j=0}^{2n} \in \mathcal{H}_{q,2n}^{\geq}$, *let* $p \in \mathbb{N}$, *and let* $C \in \mathbb{C}^{p \times q}$. *Let* $L_{n,C}$ *be the matrix given by (2.1) with* $(s_j)_{j=0}^{2n}$ *replaced by* $(C s_j C^*)_{j=0}^{2n}$. *Then*

$L_{n,C} \geq CL_nC^*$. *Furthermore, if $n = 0$ or if $n > 0$ and*

$$\mathcal{N}(C) \subseteq \bigcap_{j=0}^{2n-1} \mathcal{N}(s_j), \tag{3.3}$$

then $L_{n,C} = CL_nC^*$.

Proof. In view of (2.1), the case $n = 0$ is trivial. Now we consider the case $n > 0$. Because of $(s_j)_{j=0}^{2n} \in \mathcal{H}_{q,2n}^{\geq}$ we have $z_{n,2n-1} = y_{n,2n-1}^*$ and, additionally using [12, Remark 2.1 (a)], moreover $H_{n-1} \in \mathbb{C}_{\geq}^{nq \times nq}$ and $\mathcal{R}(y_{n,2n-1}) \subseteq \mathcal{R}(H_{n-1})$. Thus,

$$L_{n,C} \geq CL_nC^* \tag{3.4}$$

follows in view of (2.1) from Lemma 3.10 with $A := H_{n-1}$, $B := y_{n,2n-1}$, $D := s_{2n}$, $X := \mathrm{diag}(C^*, C^*, \ldots, C^*) \in \mathbb{C}^{nq \times np}$, and $Y := C^*$. Now suppose (3.3). Then in view of part (b) of Lemma A.1 for each $j \in \mathbb{Z}_{0,2n-1}$, we have then $s_jC^{\dagger}C = s_j$ and because of part (b) of Lemma 3.2 moreover $s_j^* = s_j$. Consequently, for each $j \in \mathbb{Z}_{0,2n-1}$, we get

$$C^{\dagger}Cs_j(C^{\dagger}C)^* = (C^{\dagger}C)^*s_jC^{\dagger}C = (C^{\dagger}C)^*s_j = (C^{\dagger}C)^*s_j^* = (s_jC^{\dagger}C)^*$$
$$= s_j^* = s_j. \tag{3.5}$$

Using part (a) of Lemma A.1, we obtain

$$CC^{\dagger}Cs_{2n}(C^{\dagger}C)^*C^* = Cs_{2n}C^*(C^{\dagger})^*C^* = Cs_{2n}C^*(C^*)^{\dagger}C^* = Cs_{2n}C^*. \tag{3.6}$$

Let $L_{n,C^{\dagger}C}$ be the matrix given by (2.1) with $(s_j)_{j=0}^{2n}$ replaced by the sequence $(C^{\dagger}Cs_j(C^{\dagger}C)^*)_{j=0}^{2n}$. In view of (2.1), (3.6) and (3.5), we have then

$$CL_{n,C^{\dagger}C}C^* = CL_nC^*. \tag{3.7}$$

Applying the obtained inequality (3.4) to the sequence $(Cs_jC^*)_{j=0}^{2n}$ and the matrix $C^{\dagger}$, we get

$$L_{n,C^{\dagger}C} \geq C^{\dagger}L_{n,C}(C^{\dagger})^*. \tag{3.8}$$

From (3.7), (3.8), (3.4) and part (a) of Lemma A.1 we conclude that

$$CL_nC^* = CL_{n,C^{\dagger}C}C^* \geq CC^{\dagger}L_{n,C}(C^{\dagger})^*C^* \geq CC^{\dagger}CL_nC^*(C^{\dagger})^*C^*$$
$$= CC^{\dagger}CL_nC^*(C^*)^{\dagger}C^* = CL_nC^*.$$

Consequently, $CC^{\dagger}L_{n,C}(C^{\dagger})^*C^* = CL_nC^*$ holds. Taking into account (2.1) and part (a) of Lemma A.1, this implies

$$L_{n,C} = CC^{\dagger}L_{n,C}(C^*)^{\dagger}C^* = CC^{\dagger}L_{n,C}(C^{\dagger})^*C^* = CL_nC^*. \qquad \square$$

Proposition 3.12. *Let $n \in \mathbb{N}_0$, let $(s_j)_{j=0}^{2n} \in \mathcal{H}_{q,2n}^{\geq,\mathrm{cd}}$, let $p \in \mathbb{N}$, and let C be a complex $p \times q$ matrix such that in the case $n > 0$ the relation (3.3) holds. Then $(Cs_jC^*)_{j=0}^{2n}$ belongs to $\mathcal{H}_{p,2n}^{\geq,\mathrm{cd}}$.*

Proof. Because of $(s_j)_{j=0}^{2n} \in \mathcal{H}_{q,2n}^{\geq,\mathrm{cd}}$, we have $(s_j)_{j=0}^{2n} \in \mathcal{H}_{q,2n}^{\geq}$. Consequently, Remark 3.3 provides us with $(Cs_jC^*)_{j=0}^{2n} \in \mathcal{H}_{p,2n}^{\geq}$. Let $L_{n,C}$ be the matrix given by

(2.1) with $(s_j)_{j=0}^{2n}$ replaced by $(Cs_jC^*)_{j=0}^{2n}$. In view of $(s_j)_{j=0}^{2n} \in \mathcal{H}_{q,2n}^{\geq}$ and (3.3), Proposition 3.11 yields $L_{n,C} = CL_nC^*$. Since $(s_j)_{j=0}^{2n} \in \mathcal{H}_{q,2n}^{\geq,\mathrm{cd}}$ holds, we have $L_n = 0_{q\times q}$ and hence $L_{n,C} = 0_{p\times p}$. In view of $(Cs_jC^*)_{j=0}^{2n} \in \mathcal{H}_{p,2n}^{\geq}$, this completes the proof. $\qquad\square$

The following example shows that the condition (3.3) in Proposition 3.12 cannot be omitted.

Example 3.13. Let $s_0 := I_2$, $s_1 := \left[\begin{smallmatrix}1&1\\1&0\end{smallmatrix}\right]$, $s_2 := \left[\begin{smallmatrix}2&1\\1&1\end{smallmatrix}\right]$ and $C := \left[\begin{smallmatrix}1&0\\0&0\end{smallmatrix}\right]$. Then

$$H_1 = \begin{bmatrix} 1 & 0 & 1 & 1 \\ 0 & 1 & 1 & 0 \\ 1 & 1 & 2 & 1 \\ 1 & 0 & 1 & 1 \end{bmatrix} = \begin{bmatrix} 1 & 0 & 1 & 1 \\ 0 & 1 & 1 & 0 \end{bmatrix}^* \begin{bmatrix} 1 & 0 & 1 & 1 \\ 0 & 1 & 1 & 0 \end{bmatrix} \in \mathbb{C}_{\geq}^{4\times 4}$$

and

$$L_1 = \begin{bmatrix} 2 & 1 \\ 1 & 1 \end{bmatrix} - \begin{bmatrix} 1 & 1 \\ 1 & 0 \end{bmatrix} I_2^\dagger \begin{bmatrix} 1 & 1 \\ 1 & 0 \end{bmatrix} = 0_{2\times 2},$$

which implies $(s_j)_{j=0}^{2} \in \mathcal{H}_{2,2}^{\geq,\mathrm{cd}}$. However, $Cs_0C^* = \left[\begin{smallmatrix}1&0\\0&0\end{smallmatrix}\right]$, $Cs_1C^* = \left[\begin{smallmatrix}1&0\\0&0\end{smallmatrix}\right]$ and $Cs_2C^* = \left[\begin{smallmatrix}2&0\\0&0\end{smallmatrix}\right]$, and hence

$$H_{1,C} = \begin{bmatrix} 1 & 0 & 1 & 0 \\ 0 & 0 & 0 & 0 \\ 1 & 0 & 2 & 0 \\ 0 & 0 & 0 & 0 \end{bmatrix} = \begin{bmatrix} 1 & 0 & 1 & 0 \\ 0 & 0 & 1 & 0 \end{bmatrix}^* \begin{bmatrix} 1 & 0 & 1 & 0 \\ 0 & 0 & 1 & 0 \end{bmatrix} \in \mathbb{C}_{\geq}^{4\times 4}$$

and

$$L_{1,C} = \begin{bmatrix} 2 & 0 \\ 0 & 0 \end{bmatrix} - \begin{bmatrix} 1 & 0 \\ 0 & 0 \end{bmatrix} \begin{bmatrix} 1 & 0 \\ 0 & 0 \end{bmatrix}^\dagger \begin{bmatrix} 1 & 0 \\ 0 & 0 \end{bmatrix} = \begin{bmatrix} 1 & 0 \\ 0 & 0 \end{bmatrix} \neq 0_{2\times 2},$$

which implies $(Cs_jC^*)_{j=0}^{2} \in \mathcal{H}_{2,2}^{\geq} \setminus \mathcal{H}_{2,2}^{\geq,\mathrm{cd}}$.

Remark 3.14. Let $n \in \mathbb{N}_0$, let $(s_j)_{j=0}^{\infty} \in \mathcal{H}_{q,\infty}^{\geq}$ be completely degenerate of order n, let $p \in \mathbb{N}$, and let C be a complex $p \times q$ matrix such that in the case $n > 0$ the relation (3.3) holds. In view of Remark 3.3 and Proposition 3.12, then the sequence $(Cs_jC^*)_{j=0}^{\infty}$ belongs to $\mathcal{H}_{p,\infty}^{\geq}$ and it is completely degenerate of order n.

The following two results are taken from Chen/Hu [9, Lemma 2.3, Corollary 2.5]. These results, which can be proved using Lemma 3.1, play an important role in the conception used in [9], namely in certain proofs they allow a reduction to the case of sequences $(s_j)_{j=0}^{2n}$ with positive Hermitian matrix s_0.

Lemma 3.15. *Let* $n \in \mathbb{N}$, *let* $(s_j)_{j=0}^{2n} \in \mathcal{H}_{q,2n}^{\geq}$, *and let* $r := \mathrm{rank}\, s_0$.

(a) *If* $r = 0$, *then* $s_j = 0_{q\times q}$ *for every* $j \in \mathbb{Z}_{0,2n-1}$ *and* $s_{2n} \in \mathbb{C}_{\geq}^{q\times q}$.

(b) *If* $0 < r < q$, *then there is a unitary* $q \times q$ *matrix* U, *a sequence* $(\tilde{s}_j)_{j=0}^{2n} \in \mathcal{H}_{r,2n}^{\geq}$ *with* $\tilde{s}_0 \in \mathbb{C}_{>}^{r\times r}$, *and a matrix* $D \in \mathbb{C}_{\geq}^{(q-r)\times(q-r)}$ *such that* $U^*s_jU = \mathrm{diag}(\tilde{s}_j, 0_{(q-r)\times(q-r)})$ *for every* $j \in \mathbb{Z}_{0,2n-1}$ *and* $U^*s_{2n}U = \mathrm{diag}(\tilde{s}_{2n}, D)$.

Lemma 3.16. *Let* $n \in \mathbb{N}$, *let* $(s_j)_{j=0}^{2n} \in \mathcal{H}_{q,2n}^{\geq,e}$, *and let* $r := \operatorname{rank} s_0$.

(a) *If* $r = 0$, *then* $s_j = 0_{q \times q}$ *for each* $j \in \mathbb{Z}_{0,2n}$.

(b) *If* $r = q$, *then* s_0 *is positive Hermitian.*

(c) *Let* $0 < r < q$. *Then:*

    (c1) *There are a unitary complex* $q \times q$ *matrix* U *and an* $\tilde{s}_0 \in \mathbb{C}_{\geq}^{r \times r}$ *such that*

$$U^* s_0 U = \operatorname{diag}(\tilde{s}_0, 0_{(q-r) \times (q-r)}). \qquad (3.9)$$

    (c2) *Let* $U \in \mathbb{C}^{q \times q}$ *and* $\tilde{s}_0 \in \mathbb{C}^{r \times r}$ *be arbitrary matrices such that* (3.9) *holds. Then there is a unique sequence* $(\tilde{s}_j)_{j=1}^{2n}$ *of complex* $r \times r$ *matrices such that* $U^* s_j U = \operatorname{diag}(\tilde{s}_j, 0_{(q-r) \times (q-r)})$ *for each* $j \in \mathbb{Z}_{0,2n}$. *The sequence* $(\tilde{s}_j)_{j=0}^{2n}$ *belongs necessarily to* $\mathcal{H}_{r,2n}^{\geq,e}$. *If the matrix* U *is non-singular, then the matrix* $\tilde{s}_0$ *is positive Hermitian.*

4. Reciprocal sequences

The concept used in this section of constructing a special transformation for finite and infinite sequences of complex $p \times q$-matrices is presented in [16]. The cited paper deals with the question of invertibility as it applies to finite and infinite sequences of complex $p \times q$-matrices.

    Our next definition will prove to be of particular importance throughout this paper.

Definition 4.1. Let $\kappa \in \mathbb{N}_0 \cup \{+\infty\}$ and let $(s_j)_{j=0}^{\kappa}$ be a sequence of complex $p \times q$-matrices. The sequence $(s_j^\sharp)_{j=0}^{\kappa}$ defined by

$$s_k^\sharp := \begin{cases} s_0^\dagger, & \text{if } k = 0 \\ -s_0^\dagger \sum_{l=0}^{k-1} s_{k-l} s_l^\sharp, & \text{if } k \in \mathbb{Z}_{1,\kappa} \end{cases} \qquad (4.1)$$

is called the *reciprocal sequence corresponding to* $(s_j)_{j=0}^{\kappa}$.

Remark 4.2. Let $\kappa \in \mathbb{N}_0 \cup \{+\infty\}$ and let $(s_j)_{j=0}^{\kappa}$ be a sequence of complex $p \times q$-matrices with reciprocal sequence $(s_j^\sharp)_{j=0}^{\kappa}$. For each $m \in \mathbb{Z}_{0,\kappa}$ then $(s_j^\sharp)_{j=0}^{m}$ is the reciprocal sequence corresponding to $(s_j)_{j=0}^{m}$.

Example 4.3. Let $\kappa \in \mathbb{N} \cup \{+\infty\}$ and let $(s_j)_{j=0}^{\kappa}$ be a sequence in $\mathbb{C}^{p \times q}$ such that $s_j = 0_{p \times q}$ for each $j \in \mathbb{Z}_{1,\kappa}$. Then $s_j^\sharp = 0_{q \times p}$ for each $j \in \mathbb{Z}_{1,\kappa}$.

    The following result (see [16, Proposition 5.11]) gives a characterization of the coincidence of the reciprocal sequences.

Lemma 4.4. *Let* $\kappa \in \mathbb{N}_0 \cup \{+\infty\}$. *If* $(s_j)_{j=0}^{\kappa}$ *and* $(t_j)_{j=0}^{\kappa}$ *are sequences in* $\mathbb{C}^{p \times q}$, *then the following two statements are equivalent to one another.*

(i) $s_0 s_0^\dagger s_j s_0^\dagger s_0 = t_0 t_0^\dagger s_j t_0^\dagger t_0$ *for every* $j \in \mathbb{Z}_{0,\kappa}$.

(ii) $s_j^\sharp = t_j^\sharp$ *for every* $j \in \mathbb{Z}_{0,\kappa}$.

We will see that reciprocal sequences are especially interesting when considered for the class of finite or infinite sequences of complex $p \times q$-matrices defined in Definition 4.5.

Definition 4.5. Let $\kappa \in \mathbb{N}_0 \cup \{+\infty\}$ and $(s_j)_{j=0}^{\kappa}$ be a sequence in in $\mathbb{C}^{p \times q}$. We then say that $(s_j)_{j=0}^{\kappa}$ is *dominated by its first term* (or, simply, that it is *first term dominant*) when

$$\mathcal{N}(s_0) \subseteq \bigcap_{j=0}^{\kappa} \mathcal{N}(s_j) \qquad \text{and} \qquad \bigcup_{j=0}^{\kappa} \mathcal{R}(s_j) \subseteq \mathcal{R}(s_0). \qquad (4.2)$$

The set of all first term dominant sequences $(s_j)_{j=0}^{\kappa}$ in $\mathbb{C}^{p \times q}$ will be denoted by $\mathcal{D}_{p \times q, \kappa}$.

Example 4.6. Let $\kappa \in \mathbb{N}_0 \cup \{+\infty\}$. All sequences $(s_j)_{j=0}^{\kappa}$ of complex $q \times q$-matrices such that $\det s_0 \neq 0$ belong to the class $\mathcal{D}_{q \times q, \kappa}$.

Remark 4.7. Let $\kappa \in \mathbb{N}_0 \cup \{+\infty\}$ and $(s_j)_{j=0}^{\kappa}$ be a sequence of complex $p \times q$-matrices. Then

$$(s_j)_{j=0}^{\kappa} \in \mathcal{D}_{p \times q, \kappa} \qquad \text{if and only if} \qquad (s_j)_{j=0}^{m} \in \mathcal{D}_{p \times q, m}, \text{ for all } m \in \mathbb{Z}_{0, \kappa}.$$

The following result which is taken from [16, Proposition 5.1] contains several characterizations of the membership of a sequence of complex $p \times q$-matrices to the class $\mathcal{D}_{p \times q, \kappa}$.

Proposition 4.8. *If $\kappa \in \mathbb{N}_0 \cup \{+\infty\}$ and $(s_j)_{j=0}^{\kappa}$ is a sequence in $\mathbb{C}^{p \times q}$, then the following statements are all equivalent:*

(i) $(s_j)_{j=0}^{\kappa} \in \mathcal{D}_{p \times q, \kappa}$.
(ii) $\mathcal{N}(s_0) \subseteq \bigcap_{j=0}^{\kappa} \mathcal{N}(s_j)$ *and* $\mathcal{N}(s_0^*) \subseteq \bigcap_{j=0}^{\kappa} \mathcal{N}(s_j^*)$.
(iii) $\bigcup_{j=0}^{\kappa} \mathcal{R}(s_j) \subseteq \mathcal{R}(s_0)$ *and* $\bigcup_{j=0}^{\kappa} \mathcal{R}(s_j^*) \subseteq \mathcal{R}(s_0^*)$.
(iv) $(s_j^*)_{j=0}^{\kappa} \in \mathcal{D}_{q \times p, \kappa}$.
(v) $s_j s_0^{\dagger} s_0 = s_j$ *and* $s_0 s_0^{\dagger} s_j = s_j$ *for all* $j \in \mathbb{Z}_{0, \kappa}$.
(vi) $s_0 s_0^{\dagger} s_j s_0^{\dagger} s_0 = s_j$ *for all* $j \in \mathbb{Z}_{0, \kappa}$.

We will next recognize that constructing the sequence reciprocal to a sequence $(s_j)_{j=0}^{\kappa}$ always yields a member of $\mathcal{D}_{p \times q, \kappa}$ (see [16, Proposition 5.10]).

Proposition 4.9. *If $\kappa \in \mathbb{N}_0 \cup \{+\infty\}$ and $(s_j)_{j=0}^{\kappa}$ is a sequence in $\mathbb{C}^{p \times q}$, then*

(a) $(s_j^{\sharp})_{j=0}^{\kappa} \in \mathcal{D}_{q \times p, \kappa}$.
(b) *For any $j \in \mathbb{Z}_{0, \kappa}$ both of the following identities hold:*

$$s_0^{\dagger} s_0 s_j^{\sharp} = s_j^{\sharp} \qquad \text{and} \qquad s_j^{\sharp} s_0 s_0^{\dagger} = s_j^{\sharp}$$

(c) $s_0^{\sharp} \left(\sum_{l=0}^{k} s_{k-l} s_l^{\sharp} \right) = 0_{p \times q}$ *for each $k \in \mathbb{Z}_{1, \kappa}$.*

(d) *If $\bigcup_{j=1}^{\kappa} \mathcal{R}(s_j) \subseteq \mathcal{R}(s_0)$, then $\sum_{l=0}^{k} s_{l-k} s_l^{\sharp} = 0_{p \times p}$ for each $k \in \mathbb{Z}_{1, \kappa}$.*

Now we compute the sequence $((s^\sharp)^\sharp_j)^\kappa_{j=0}$.

Proposition 4.10. *Let $\kappa \in \mathbb{N}_0 \cup \{+\infty\}$. If $(s_j)^\kappa_{j=0}$ is a sequence in $\mathbb{C}^{p \times q}$ and if we define the sequence $(\tilde{s}_j)^\kappa_{j=0}$ via $\tilde{s}_j := s_0 s_0^\dagger s_j s_0^\dagger s_0$, for each $j \in \mathbb{Z}_{0,\kappa}$, then:*

(a) *Both of these sequences have the same reciprocal sequence, i.e.:*

$$\left(s^\sharp_j\right)^\kappa_{j=0} = \left(\tilde{s}^\sharp_j\right)^\kappa_{j=0}.$$

(b) $\left(\tilde{s}_j\right)^\kappa_{j=0} \in \mathcal{D}_{p \times q, \kappa}$.

(c) *The reciprocal sequence $\left(\left(s^\sharp_j\right)^\sharp\right)^\kappa_{j=0}$ for the sequence $\left(s^\sharp_j\right)^\kappa_{j=0}$ satisfies*

$$\left(\left(s^\sharp_j\right)^\sharp\right)^\kappa_{j=0} = \left(\tilde{s}_j\right)^\kappa_{j=0}.$$

(d) *The following three statements are all equivalent:*
   (i) $(s_j)^\kappa_{j=0}$ *is the reciprocal sequence for $\left(s^\sharp_j\right)^\kappa_{j=0}$.*
   (ii) $(s_j)^\kappa_{j=0} = \left(\tilde{s}_j\right)^\kappa_{j=0}$.
   (iii) $(s_j)^\kappa_{j=0} \in \mathcal{D}_{p \times q, \kappa}$.

For a proof of Proposition 4.10, see [16, Proposition 5.13, Remark 4.7].

In order to formulate a particularly important result from [16], which we will later need, we first introduce two specific types of block Toeplitz matrices.

Notation 4.11. Given a $\kappa \in \mathbb{N}_0 \cup \{+\infty\}$ and a sequence $(s_j)^\kappa_{j=0}$ in $\mathbb{C}^{p \times q}$, we define, for each $m \in \mathbb{Z}_{0,\kappa}$, the triangular block Toeplitz matrices $\mathbf{S}^{(s)}_m$ and $\mathbb{S}^{(s)}_m$ as

$$\mathbf{S}^{(s)}_m := \begin{bmatrix} s_0 & 0 & 0 & \cdots & 0 \\ s_1 & s_0 & 0 & \cdots & 0 \\ s_2 & s_1 & s_0 & \cdots & 0 \\ \vdots & \vdots & \vdots & \ddots & \vdots \\ s_m & s_{m-1} & s_{m-2} & \cdots & s_0 \end{bmatrix} \tag{4.3}$$

and

$$\mathbb{S}^{(s)}_m := \begin{bmatrix} s_0 & s_1 & s_2 & \cdots & s_m \\ 0 & s_0 & s_1 & \cdots & s_{m-1} \\ 0 & 0 & s_0 & \cdots & s_{m-2} \\ \vdots & \vdots & \vdots & \ddots & \vdots \\ 0 & 0 & 0 & \cdots & s_0 \end{bmatrix}, \tag{4.4}$$

respectively. Whenever it is clear which sequence is meant, we will, respectively, simply write $\mathbf{S}_m$ and $\mathbb{S}_m$ instead of $\mathbf{S}^{(s)}_m$ and $\mathbb{S}^{(s)}_m$.

Remark 4.12. Let $\kappa \in \mathbb{N}_0 \cup \{+\infty\}$ and let $(s_j)^\kappa_{j=0}$ be a sequence from $\mathbb{C}^{p \times q}$. For each $m \in \mathbb{N}_0$, then

$$\mathbb{S}^{(s)}_m = [\mathbf{S}^{(s^*)}_m]^* \tag{4.5}$$

where $\mathbf{S}_m^{(s^*)}$ is the lower triangular block Toeplitz matrix (see (4.3)) corresponding to the sequence $(s_j^*)_{j=0}^{\kappa}$. If $(s_j)_{j=0}^{\kappa}$ is a sequence from $\mathbb{C}_{\mathrm{H}}^{q \times q}$ then, for each $m \in \mathbb{N}_0$, we have

$$\mathbb{S}_m^{(s)} = [\mathbf{S}_m^{(s)}]^*. \tag{4.6}$$

Notation 4.13. For each $\kappa \in \mathbb{N}_0 \cup \{+\infty\}$, for each sequence $(s_j)_{j=0}^{\kappa}$ of complex $p \times q$-matrices, and for each $m \in \mathbb{Z}_{0,\kappa}$, let the block Toeplitz matrices $\mathbf{S}_m^{\sharp}$ and $\mathbb{S}_m^{\sharp}$ be given, respectively, (using Notation 4.11, with $(s_j^{\sharp})_{j=0}^{\kappa}$ instead of $(s_j)_{j=0}^{\kappa}$) by

$$\mathbf{S}_m^{\sharp} := \mathbf{S}_m^{(s^{\sharp})} \qquad \text{and} \qquad \mathbb{S}_m^{\sharp} := \mathbb{S}_m^{(s^{\sharp})}.$$

The following result (see [16, Proposition 4.20]) is of central importance to this paper.

Proposition 4.14. *If $\kappa \in \mathbb{N}_0 \cup \{+\infty\}$ and $(s_j)_{j=0}^{\kappa} \in \mathcal{D}_{p \times q, \kappa}$, then $\mathbf{S}_m^{\dagger} = \mathbf{S}_m^{\sharp}$ and $\mathbb{S}_m^{\dagger} = \mathbb{S}_m^{\sharp}$ for each $m \in \mathbb{Z}_{0,\kappa}$.*

If $r, s \in \mathbb{N}$, $A = (a_{jk})_{\substack{j=1,\ldots,p \\ k=1,\ldots,q}} \in \mathbb{C}^{p \times q}$ and $B \in \mathbb{C}^{r \times s}$, then the *Kronecker product* $A \otimes B$ of the matrices A and B is given by $A \otimes B := (a_{jk}B)_{\substack{j=1,\ldots,p \\ k=1,\ldots,q}}$.

It should be noted that, if $s_0 \in \mathbb{C}^{p \times q}$ and if $m \in \mathbb{N}$, then the complex $(m+1)p \times (m+1)q$ matrix $\mathrm{diag}\,(s_0, s_0, \ldots, s_0)$ can be expressed as $I_{m+1} \otimes s_0$.

Lemma 4.15. *If $\kappa \in \mathbb{N}_0 \cup \{+\infty\}$ and $(s_j)_{j=0}^{\kappa} \in \mathcal{D}_{p \times q, \kappa}$, then, for each $m \in \mathbb{Z}_{0,\kappa}$:*

$$\mathbf{S}_m \mathbf{S}_m^{\dagger} = I_{m+1} \otimes (s_0 s_0^{\dagger}), \qquad \mathbf{S}_m^{\dagger} \mathbf{S}_m = I_{m+1} \otimes (s_0^{\dagger} s_0) \tag{4.7}$$

and

$$\mathbb{S}_m \mathbb{S}_m^{\dagger} = I_{m+1} \otimes (s_0 s_0^{\dagger}), \qquad \mathbb{S}_m^{\dagger} \mathbb{S}_m = I_{m+1} \otimes (s_0^{\dagger} s_0). \tag{4.8}$$

If, furthermore, $p = q$ and if $\mathcal{R}(s_0) = \mathcal{R}(s_0^*)$, then

$$\mathbf{S}_m \mathbf{S}_m^{\dagger} = \mathbf{S}_m^{\dagger} \mathbf{S}_m \qquad \text{and} \qquad \mathbb{S}_m \mathbb{S}_m^{\dagger} = \mathbb{S}_m^{\dagger} \mathbb{S}_m,$$

for every $m \in \mathbb{Z}_{0,\kappa}$.

Proof. Combine part (a) of Theorem 4.21 in [16] with Proposition 3.6 in [16].   □

In anticipation of later applications to Hankel non-negative definite sequences of matrices, we next consider a slight modification of first term dominancy.

Definition 4.16. Let $m \in \mathbb{N}$. A sequence $(s_j)_{j=0}^{m}$ of complex $p \times q$-matrices for which $(s_j)_{j=0}^{m-1} \in \mathcal{D}_{p \times q, m-1}$ is called a *nearly first term dominant* sequence. The set of all nearly first term dominant sequences $(s_j)_{j=0}^{m}$ will be denoted by $\tilde{\mathcal{D}}_{p \times q, m}$. We also set $\tilde{\mathcal{D}}_{p \times q, 0} := \mathcal{D}_{p \times q, 0}$.

Remark 4.17. For each $m \in \mathbb{N}$, the inclusion $\mathcal{D}_{p \times q, m} \subseteq \tilde{\mathcal{D}}_{p \times q, m}$ holds.

Remark 4.18. Let $m \in \mathbb{N}$ and $(s_j)_{j=0}^{m}$ be a sequence from $\mathbb{C}^{p \times q}$. Then the combination of part (a) of Proposition 4.9 and Remark 4.17 shows that $(s_j^{\sharp})_{j=0}^{m} \in \tilde{\mathcal{D}}_{q \times p, m}$.

Lemma 4.19. *Let $m \in \mathbb{N}$ and let $(s_j)_{j=0}^m \in \tilde{\mathcal{D}}_{p \times q, m}$. Let $t_j := s_j$ for each $j \in \mathbb{Z}_{0, m-1}$ and let $t_m := s_0 s_0^\dagger s_m s_0^\dagger s_0$. Furthermore, let $\tilde{s}_j := s_0 s_0^\dagger s_j s_0^\dagger s_0$ for each $j \in \mathbb{Z}_{0, m}$. Then:*

(a) $(\tilde{s}_j)_{j=0}^m = (t_j)_{j=0}^m$.
(b) $(s_j^\sharp)_{j=0}^m = (t_j^\sharp)_{j=0}^m$.
(c) $(t_j)_{j=0}^m \in \mathcal{D}_{p \times q, m}$.
(d) $((t^\sharp)_j^\sharp)_{j=0}^m = (t_j)_{j=0}^m$.
(e) *Let $m = 2n$, for some positive integer n. Then*

$$H_n^{(t)} = H_n^{(s)} + \operatorname{diag}(0_{np \times nq}, s_0 s_0^\dagger s_{2n} s_0^\dagger s_0 - s_{2n})$$

and

$$H_n^{(t)} = \left[I_{n+1} \otimes (s_0 s_0^\dagger) \right] H_n^{(s)} \left[I_{n+1} \otimes (s_0^\dagger s_0) \right].$$

Proof. (a) This follows from Proposition 4.8.

(b) Proposition 4.10 yields $(s_j^\sharp)_{j=0}^m = (\tilde{s}_j^\sharp)_{j=0}^m$. Thus, (a) implies (b).

(c) Proposition 4.10 yields $(\tilde{s}_j)_{j=0}^m \in \mathcal{D}_{p \times q, m}$. Hence, (a) implies (c).

(d) Apply (c) and part (d) of Proposition 4.10.

(e) The first identity is obvious. Using (a) we obtain

$$H_n^{(t)} = H_n^{(\tilde{s})} = [\tilde{s}_{j+k}]_{j,k=0}^n = [s_0 s_0^\dagger s_{j+k} s_0^\dagger s_0]_{j,k=0}^n$$
$$= \left[I_{n+1} \otimes (s_0 s_0^\dagger) \right] H_n^{(s)} \left[I_{n+1} \otimes (s_0^\dagger s_0) \right]. \qquad \square$$

The following result is the analogue of Lemma 4.15 for the class $\tilde{\mathcal{D}}_{p \times q, \kappa}$.

Lemma 4.20. *Let $m \in \mathbb{N}$ and $(s_j)_{j=0}^m \in \tilde{\mathcal{D}}_{p \times q, m}$. Then*

$$\mathbf{S}_m^{(s)} \mathbf{S}_m^{(s^\sharp)} = \left[I_{m+1} \otimes (s_0 s_0^\dagger) \right] + \left[\begin{array}{c|c} 0_{mp \times p} & 0_{mp \times mp} \\ \hline (I_p - s_0 s_0^\dagger) s_m s_0^\dagger & 0_{p \times mp} \end{array} \right]$$

$$\mathbf{S}_m^{(s^\sharp)} \mathbf{S}_m^{(s)} = \left[I_{m+1} \otimes (s_0^\dagger s_0) \right] + \left[\begin{array}{c|c} 0_{mq \times q} & 0_{mq \times mq} \\ \hline s_0^\dagger s_m (I_q - s_0^\dagger s_0) & 0_{q \times mq} \end{array} \right]. \qquad (4.9)$$

Proof. Let $t_j := s_j$ for all $j \in \mathbb{Z}_{0, m-1}$ and $t_m := s_0 s_0^\dagger s_m s_0^\dagger s_0$. From part (b) of Lemma 4.19, we get $t_j^\sharp = s_j^\sharp$ for all $j \in \mathbb{Z}_{0, m}$ and hence $\mathbf{S}_m^{(s^\sharp)} = \mathbf{S}_m^{(t^\sharp)}$. From $(s_j)_{j=0}^m \in \tilde{\mathcal{D}}_{p \times q, m}$ and part (c) of Lemma 4.19 we get $(t_j)_{j=0}^m \in \mathcal{D}_{p \times q, m}$ which, in view of [16, Proposition 4.20], implies $(\mathbf{S}_m^{(t)})^\dagger = \mathbf{S}_m^{(t^\sharp)}$. Because of $(t_j)_{j=0}^m \in \mathcal{D}_{p \times q, m}$ and Lemma 4.15, we have

$$\mathbf{S}_m^{(t)} (\mathbf{S}_m^{(t)})^\dagger = I_{m+1} \otimes (s_0 s_0^\dagger) \qquad \text{and} \qquad (\mathbf{S}_m^{(t)})^\dagger \mathbf{S}_m^{(t)} = I_{m+1} \otimes (s_0^\dagger s_0).$$

Hence, taking into account that Definition 4.1 implies $s_0^\sharp = s_0^\dagger$, we get

$$
\begin{aligned}
\mathbf{S}_m^{(s)}\mathbf{S}_m^{(s^\sharp)} &= \left(\mathbf{S}_m^{(t)} + \left[\begin{array}{c|c} 0_{mp\times q} & 0_{mp\times mq} \\ \hline s_m - t_m & 0_{p\times mq} \end{array}\right]\right)\mathbf{S}_m^{(s^\sharp)} \\
&= \mathbf{S}_m^{(t)}\mathbf{S}_m^{(s^\sharp)} + \left[\begin{array}{c|c} 0_{mp\times q} & 0_{mp\times mq} \\ \hline s_m - t_m & 0_{p\times mq} \end{array}\right]\mathbf{S}_m^{(s^\sharp)} \\
&= \mathbf{S}_m^{(t)}\mathbf{S}_m^{(t^\sharp)} + \left[\begin{array}{c|c} 0_{mp\times q} & 0_{mp\times mq} \\ \hline s_m - s_0 s_0^\dagger s_m s_0^\dagger s_0 & 0_{p\times mq} \end{array}\right]\left[\begin{array}{cc} s_0^\sharp & 0_{q\times mp} \\ * & * \end{array}\right] \\
&= \mathbf{S}_m^{(t)}(\mathbf{S}_m^{(t)})^\dagger + \left[\begin{array}{c|c} 0_{mp\times q} & 0_{mp\times mq} \\ \hline s_m - s_0 s_0^\dagger s_m s_0^\dagger s_0 & 0_{p\times mq} \end{array}\right]\left[\begin{array}{cc} s_0^\dagger & 0_{q\times mp} \\ * & * \end{array}\right] \\
&= \left[I_{m+1}\otimes(s_0 s_0^\dagger)\right] + \left[\begin{array}{c|c} 0_{mp\times p} & 0_{mp\times mp} \\ \hline (I_p - s_0 s_0^\dagger)s_m s_0^\dagger & 0_{p\times mp} \end{array}\right]
\end{aligned}
$$

and, similarly, we obtain (4.9). $\square$

Now we want to reformulate Lemma 4.20 in terms of Cauchy products.

Notation 4.21. Let $\kappa \in \mathbb{N}_0 \cup \{+\infty\}$. For all sequences $(s_j)_{j=0}^\kappa$ and $(t_j)_{j=0}^\kappa$ in $\mathbb{C}^{p\times q}$ and $\mathbb{C}^{q\times r}$, respectively, we will denote the *Cauchy product* of these two sequences by $((s\odot t)_j)_{j=0}^\kappa$, i.e., for all $j \in \mathbb{Z}_{0,\kappa}$:

$$
(s\odot t)_j = \sum_{l=0}^j s_l t_{j-l}.
$$

Remark 4.22. Let $\kappa \in \mathbb{N}_0 \cup \{+\infty\}$. Suppose that $(s_j)_{j=0}^\kappa$ and $(t_j)_{j=0}^\kappa$ are, respectively, sequences in $\mathbb{C}^{p\times q}$ and $\mathbb{C}^{q\times r}$. If $(u_j)_{j=0}^\kappa$ is a sequence in $\mathbb{C}^{p\times r}$, then

$$
(u_j)_{j=0}^\kappa = ((s\odot t)_j)_{j=0}^\kappa \qquad \text{if and only if} \qquad \mathbf{S}_m^{(s)}\mathbf{S}_m^{(t)} = \mathbf{S}_m^{(u)} \text{ for all } m \in \mathbb{Z}_{0,\kappa}.
$$

Proposition 4.23. *Let* $m \in \mathbb{N}$ *and* $(s_j)_{j=0}^m \in \tilde{\mathcal{D}}_{p\times q,m}$. *Then*

$$
(s\odot s^\sharp)_j = \begin{cases} s_0 s_0^\dagger & \text{if } j = 0 \\ 0_{p\times p} & \text{if } 1 \leq j \leq m-1 \\ (I_p - s_0 s_0^\dagger)s_m s_0^\dagger & \text{if } j = m \end{cases}
$$

and

$$
(s^\sharp \odot s)_j = \begin{cases} s_0^\dagger s_0 & \text{if } j = 0 \\ 0_{q\times q} & \text{if } 1 \leq j \leq m-1 \\ s_0^\dagger s_m(I_q - s_0^\dagger s_0) & \text{if } j = m \end{cases}.
$$

Proof. Use Lemma 4.20 and Remark 4.22. $\square$

Now we consider the classes of Hankel non-negative definite extendable sequences and Hankel non-negative definite sequences against the background of this section.

Proposition 4.24. *For each* $\kappa \in \mathbb{N}_0 \cup \{+\infty\}$, *we have* $\mathcal{H}_{q,\kappa}^{\geq,e} \subseteq \mathcal{D}_{q\times q,\kappa}$.

Proof. Use Lemma 3.1. □

Proposition 4.25. *For each $n \in \mathbb{N}$, the inclusion $\mathcal{H}_{q,2n}^{\geq} \subseteq \tilde{\mathcal{D}}_{q \times q, 2n}$ holds.*

Proof. Apply Lemma 3.2. □

5. Some identities for block Hankel matrices associated with Cauchy products

In this section, we study the interplay between the operation of forming Cauchy products of sequences of complex matrices and the construction of block Hankel matrices.

Let $n \in \mathbb{N}_0$ and let $(s_j)_{j=0}^{2n}$ be a sequence from $\mathbb{C}^{p \times q}$. Then we consider the block Hankel matrix

$$H_n^{(s)} := [s_{j+k}]_{j,k=0}^n. \tag{5.1}$$

Now we want to derive a first formula for the block Hankel matrix associated with the Cauchy product of two sequences of matrices. To prepare this we introduce a particular block matrix.

For each $m \in \mathbb{N}$, let the complex $mq \times mq$ matrix $U_{q,m}$ be given by

$$U_{q,m} := \begin{bmatrix} 0_{q \times q} & 0_{q \times q} & \cdots & 0_{q \times q} & I_q \\ 0_{q \times q} & 0_{q \times q} & \cdots & I_q & 0_{q \times q} \\ \vdots & \vdots & & \vdots & \vdots \\ I_q & 0_{q \times q} & \cdots & 0_{q \times q} & 0_{q \times q} \end{bmatrix}, \tag{5.2}$$

i.e., we set $U_{q,m} := [\delta_{j,m-k+1} I_q]_{j,k=1}^m$, where $\delta_{\ell,m}$ is the Kronecker symbol: $\delta_{\ell,m} := 1$ if $m = \ell$ and $\delta_{\ell,m} := 0$ otherwise. Clearly,

$$U_{q,m}^2 = I_{mq}. \tag{5.3}$$

Remark 5.1. Let $\kappa \in \mathbb{N}_0 \cup \{+\infty\}$, $(s_j)_{j=0}^{\kappa}$ be a sequence from $\mathbb{C}^{p \times q}$ and $n \in \mathbb{N}$ with $2n \leq \kappa$. Then:

(a) The matrix $\mathbf{S}_{2n}^{(s)}$ admits the block representations

$$\mathbf{S}_{2n}^{(s)} = \left[\begin{array}{c|c} [\mathbf{S}_{n-1}^{(s)}, 0_{np \times q}] & 0_{np \times nq} \\ \hline H_n^{(s)} U_{q,n+1} & \begin{bmatrix} 0_{p \times nq} \\ \mathbf{S}_{n-1}^{(s)} \end{bmatrix} \end{array} \right], \tag{5.4}$$

$$\mathbf{S}_{2n}^{(s)} = \left[\begin{array}{c|c} \mathbf{S}_n^{(s)} & 0_{(n+1)p \times nq} \\ \hline [0_{np \times p}, I_{np}] H_n^{(s)} U_{q,n+1} & \mathbf{S}_{n-1}^{(s)} \end{array} \right] \tag{5.5}$$

and

$$\mathbf{S}_{2n}^{(s)} = \left[\begin{array}{c|c} \mathbf{S}_{n-1}^{(s)} & 0_{np \times (n+1)q} \\ \hline H_n^{(s)} U_{q,n+1} \begin{bmatrix} I_{nq} \\ 0_{q \times nq} \end{bmatrix} & \mathbf{S}_n^{(s)} \end{array} \right]. \tag{5.6}$$

(b)
$$H_n^{(s)} = [0_{(n+1)p \times np}, I_{(n+1)p}]\mathbf{S}_{2n}^{(s)} \begin{bmatrix} U_{q,n+1} \\ 0_{nq \times (n+1)q} \end{bmatrix}.$$

(Apply part (a) and $U_{q,n+1}^2 = I_{(n+1)q}$.)

Proposition 5.2. *Let* $\kappa \in \mathbb{N}_0 \cup \{+\infty\}$, $(s_j)_{j=0}^{\kappa}$ *be a sequence from* $\mathbb{C}^{p \times q}$, $(t_j)_{j=0}^{\kappa}$ *be a sequence from* $\mathbb{C}^{q \times r}$ *and* $n \in \mathbb{N}$ *with* $2n \le \kappa$. *Then*

$$H_n^{(s \odot t)} = H_n^{(s)} \mathbb{S}_n^{(t)} + \mathrm{diag}(0_{p \times q}, \mathbb{S}_{n-1}^{(s)}) H_n^{(t)} \tag{5.7}$$

and

$$H_n^{(s \odot t)} = H_n^{(s)} \mathrm{diag}(0_{p \times q}, \mathbb{S}_{n-1}^{(t)}) + \mathbb{S}_n^{(s)} H_n^{(t)}. \tag{5.8}$$

Proof. We have

$$H_n^{(s \odot t)} = [0_{(n+1)p \times np}, I_{(n+1)p}]\mathbf{S}_{2n}^{(s \odot t)} \begin{bmatrix} U_{r,n+1} \\ 0_{nr \times (n+1)r} \end{bmatrix}$$

$$= [0_{(n+1)p \times np}, I_{(n+1)p}]\mathbf{S}_{2n}^{(s)}\mathbf{S}_{2n}^{(t)} \begin{bmatrix} U_{r,n+1} \\ 0_{nr \times (n+1)r} \end{bmatrix}$$

$$= [0_{(n+1)p \times np}, I_{(n+1)p}] \left[\begin{array}{c|c} * & * \\ \hline H_n^{(s)} U_{q,n+1} & \begin{bmatrix} 0_{p \times nq} \\ \mathbb{S}_{n-1}^{(s)} \end{bmatrix} \end{array} \right]$$

$$\times \left[\begin{array}{c|c} \mathbb{S}_n^{(t)} & * \\ \hline [0_{nq \times q}, I_{nq}]H_n^{(t)} U_{r,n+1} & * \end{array} \right] \begin{bmatrix} U_{r,n+1} \\ 0_{nr \times (n+1)r} \end{bmatrix}$$

$$= H_n^{(s)} U_{q,n+1} \mathbb{S}_n^{(t)} U_{r,n+1} + \begin{bmatrix} 0_{p \times nq} \\ \mathbb{S}_{n-1}^{(s)} \end{bmatrix} [0_{nq \times q}, I_{nq}]H_n^{(t)} U_{r,n+1}^2$$

$$= H_n^{(s)} \mathbb{S}_n^{(t)} + \mathrm{diag}(0_{p \times q}, \mathbb{S}_{n-1}^{(s)}) H_n^{(t)},$$

where the first equality is due to part (b) of Remark 5.1, the second to Remark 4.22, the third to part (a) of Remark 5.1, the fourth to $U_{r,n+1}^2 = I_{(n+1)r}$ and the fifth to [16, Remark 2.16]. Thus, (5.7) is checked. Similarly, one can prove that (5.8) holds true. □

Now let $n \in \mathbb{N}_0$ and let $(s_j)_{j=0}^{2n+1}$ be a sequence from $\mathbb{C}^{p \times q}$. Then we will consider the block Hankel matrix

$$K_n^{(s)} := [s_{j+k+1}]_{j,k=0}^n.$$

Remark 5.3. Let $\kappa \in \mathbb{N}_0 \cup \{+\infty\}$, $(s_j)_{j=0}^{\kappa}$ be a sequence from $\mathbb{C}^{p \times q}$ and $n \in \mathbb{N}_0$ with $2n + 1 \le \kappa$. Then:

(a) The matrix $\mathbf{S}_{2n+1}^{(s)}$ admits the block representation

$$\mathbf{S}_{2n+1}^{(s)} = \left[\begin{array}{c|c} \mathbb{S}_n^{(s)} & 0_{(n+1)p \times (n+1)q} \\ \hline K_n^{(s)} U_{q,n+1} & \mathbb{S}_n^{(s)} \end{array} \right].$$

(b)
$$K_n^{(s)} = [0_{(n+1)p \times (n+1)p}, I_{(n+1)p}] \mathbb{S}_{2n+1}^{(s)} \begin{bmatrix} U_{q,n+1} \\ 0_{(n+1)q \times (n+1)q} \end{bmatrix}.$$

(Apply part (a) and $U_{q,n+1}^2 = I_{(n+1)q}$.)

Proposition 5.4. *Let* $\kappa \in \mathbb{N}_0 \cup \{+\infty\}$, $(s_j)_{j=0}^{\kappa}$ *be a sequence from* $\mathbb{C}^{p \times q}$, $(t_j)_{j=0}^{\kappa}$ *be a sequence from* $\mathbb{C}^{q \times r}$ *and* $n \in \mathbb{N}_0$ *with* $2n + 1 \leq \kappa$. *Then*

$$K_n^{(s \odot t)} = K_n^{(s)} \mathbb{S}_n^{(t)} + \mathbb{S}_n^{(s)} K_n^{(t)}.$$

Proof. We have

$$K_n^{(s \odot t)} = [0_{(n+1)p \times (n+1)p}, I_{(n+1)p}] \mathbb{S}_{2n+1}^{(s \odot t)} \begin{bmatrix} U_{r,n+1} \\ 0_{(n+1)r \times (n+1)r} \end{bmatrix}$$

$$= [0_{(n+1)p \times (n+1)p}, I_{(n+1)p}] \mathbb{S}_{2n+1}^{(s)} \mathbb{S}_{2n+1}^{(t)} \begin{bmatrix} U_{r,n+1} \\ 0_{(n+1)r \times (n+1)r} \end{bmatrix}$$

$$= [0_{(n+1)p \times (n+1)p}, I_{(n+1)p}] \begin{bmatrix} * & * \\ \hline K_n^{(s)} U_{q,n+1} & \mathbb{S}_n^{(s)} \end{bmatrix}$$

$$\times \begin{bmatrix} \mathbb{S}_n^{(t)} & * \\ \hline K_n^{(t)} U_{r,n+1} & * \end{bmatrix} \begin{bmatrix} U_{r,n+1} \\ 0_{(n+1)r \times (n+1)r} \end{bmatrix}$$

$$= K_n^{(s)} U_{q,n+1} \mathbb{S}_n^{(t)} U_{r,n+1} + \mathbb{S}_n^{(s)} K_n^{(t)} U_{r,n+1}^2 = K_n^{(s)} \mathbb{S}_n^{(t)} + \mathbb{S}_n^{(s)} K_n^{(t)}$$

where the first equality is due to part (b) of Remark 5.3, the second to Remark 4.22, the third to part (a) of Remark 5.3, the fourth to $U_{r,n+1}^2 = I_{(n+1)r}$ and the fifth to [16, Remark 2.16]. $\square$

Let $n \in \mathbb{N}_0$ and let $(s_j)_{j=0}^{2n+2}$ be a sequence from $\mathbb{C}^{p \times q}$. We now consider the block Hankel matrix

$$G_n^{(s)} := [s_{j+k+2}]_{j,k=0}^n.$$

Proposition 5.5. *Let* $\kappa \in \mathbb{N}_0 \cup \{+\infty\}$, $(s_j)_{j=0}^{\kappa}$ *be a sequence from* $\mathbb{C}^{p \times q}$, $(t_j)_{j=0}^{\kappa}$ *be a sequence from* $\mathbb{C}^{q \times r}$ *and* $n \in \mathbb{N}_0$ *with* $2n + 2 \leq \kappa$. *Then*

$$G_n^{(s \odot t)} = G_n^{(s)} \mathbb{S}_n^{(t)} + y_{1,n+1}^{(s)} z_{1,n+1}^{(t)} + \mathbb{S}_n^{(s)} G_n^{(t)}.$$

Proof. From Proposition 5.2, we have

$$G_n^{(s \odot t)} = [0_{(n+1)p \times p}, I_{(n+1)p}] H_{n+1}^{(s \odot t)} \begin{bmatrix} 0_{r \times (n+1)r} \\ I_{(n+1)r} \end{bmatrix}$$

$$= [0_{(n+1)p \times p}, I_{(n+1)p}] \left[H_{n+1}^{(s)} \mathbb{S}_{n+1}^{(t)} + \text{diag}(0_{p \times q}, \mathbb{S}_n^{(s)}) H_{n+1}^{(t)} \right] \begin{bmatrix} 0_{r \times (n+1)r} \\ I_{(n+1)r} \end{bmatrix}$$

$$= [0_{(n+1)p \times p}, I_{(n+1)p}] H_{n+1}^{(s)} \mathbb{S}_{n+1}^{(t)} \begin{bmatrix} 0_{r \times (n+1)r} \\ I_{(n+1)r} \end{bmatrix}$$

$$+ [0_{(n+1)p \times p}, I_{(n+1)p}] \text{diag}(0_{p \times q}, \mathbb{S}_n^{(s)}) H_{n+1}^{(t)} \begin{bmatrix} 0_{r \times (n+1)r} \\ I_{(n+1)r} \end{bmatrix}$$

$$= [0_{(n+1)p\times p}, I_{(n+1)p}] \begin{bmatrix} * & * \\ y_{1,n+1}^{(s)} & G_n^{(s)} \end{bmatrix} \begin{bmatrix} * & z_{1,n+1}^{(t)} \\ * & \mathbb{S}_n^{(t)} \end{bmatrix} \begin{bmatrix} 0_{r\times(n+1)r} \\ I_{(n+1)r} \end{bmatrix}$$

$$+ [0_{(n+1)p\times p}, I_{(n+1)p}] \begin{bmatrix} * & * \\ 0_{(n+1)p\times q} & \mathbb{S}_n^{(s)} \end{bmatrix} \begin{bmatrix} * & * \\ * & G_n^{(t)} \end{bmatrix} \begin{bmatrix} 0_{r\times(n+1)r} \\ I_{(n+1)r} \end{bmatrix}$$

$$= [y_{1,n+1}^{(s)}, G_n^{(s)}] \begin{bmatrix} z_{1,n+1}^{(t)} \\ \mathbb{S}_n^{(t)} \end{bmatrix} + [0_{(n+1)p\times q}, \mathbb{S}_n^{(s)}] \begin{bmatrix} * \\ G_n^{(t)} \end{bmatrix}$$

$$= G_n^{(s)} \mathbb{S}_n^{(t)} + y_{1,n+1}^{(s)} z_{1,n+1}^{(t)} + \mathbb{S}_n^{(s)} G_n^{(t)}. \qquad \square$$

6. Some identities for block Hankel matrices formed by a sequence and its reciprocal

The main theme of this section is the investigation of the interplay between various block Hankel matrices. Let $n \in \mathbb{N}$ and let $(s_j)_{j=0}^{2n}$ be a sequence from $\mathbb{C}^{p\times q}$. Further, let $(s_j^\sharp)_{j=0}^{2n}$ be the reciprocal sequence corresponding to $(s_j)_{j=0}^{2n}$. Then we will study the interplay between the block Hankel matrices $H_n^{(s)}$ given by (5.1) and

$$H_n^{(s^\sharp)} := [s_{j+k}^\sharp]_{j,k=0}^n. \qquad (6.1)$$

Whenever it is clear which sequence is meant, we will, respectively, write H_n and $H_n^\sharp$ instead of $H_n^{(s)}$ and $H_n^{(s^\sharp)}$.

To prepare the formulation of the first main result of this section we introduce some notation. For each $n \in \mathbb{N}_0$, let

$$v_{q,n} := \mathrm{col}(\delta_{j0} I_q)_{j=0}^n. \qquad (6.2)$$

Let $\kappa \in \mathbb{N}_0 \cup \{+\infty\}$ and let $(s_j)_{j=0}^\kappa$ be a sequence from $\mathbb{C}^{p\times q}$ with reciprocal sequence $(s_j^\sharp)_{j=0}^\kappa$. Then, for all integers j and k satisfying $0 \le j \le k \le \kappa$, we set

$$y_{jk}^\sharp := \mathrm{col}(s_l^\sharp)_{l=j}^k \qquad \text{and} \qquad z_{jk}^\sharp := \mathrm{row}(s_l^\sharp)_{l=j}^k. \qquad (6.3)$$

Theorem 6.1. *Let* $n \in \mathbb{N}_0$ *and let* $(s_j)_{j=0}^{2n} \in \tilde{\mathcal{D}}_{p\times q, 2n}$. *Then:*

(a) *The equations*

$$H_n^\sharp + \mathbb{S}_n^\sharp H_n \mathbb{S}_n^\sharp = y_{0,n}^\sharp v_{p,n}^* + v_{q,n} z_{0,n}^\sharp \qquad (6.4)$$

and

$$H_n^\sharp + \mathbb{S}_n^\dagger H_n \mathbb{S}_n^\dagger = y_{0,n}^\sharp v_{p,n}^* + v_{q,n} z_{0,n}^\sharp \qquad (6.5)$$

hold.

(b) *Let*

$$\Theta_{n;p,q} := \begin{cases} 0_{p\times q} & \text{if } n = 0 \\ \mathrm{diag}(0_{np\times nq}, s_0 s_0^\dagger s_{2n} s_0^\dagger s_0 - s_{2n}) & \text{if } n \in \mathbb{N}. \end{cases} \qquad (6.6)$$

Then

$$H_n + \mathbb{S}_n H_n^\sharp \mathbb{S}_n = y_{0,n} v_{q,n}^* + v_{p,n} z_{0,n} - \Theta_{n;p,q}.$$

Proof. (a) In the case $n = 0$ both assertions are an immediate consequence of (5.1), (6.1), (6.2), (6.3) and Definition 4.1. Suppose now $n \in \mathbb{N}$. From Proposition 5.2 we get then

$$H_n \mathbb{S}_n^{\sharp} = H_n^{(s \odot s^{\sharp})} - \operatorname{diag}(0_{p \times q}, \mathbf{S}_{n-1}) H_n^{\sharp}. \tag{6.7}$$

Because of $(s_j)_{j=0}^{2n} \in \tilde{\mathcal{D}}_{p \times q, 2n}$, it follows from Proposition 4.23 that

$$H_n^{(s \odot s^{\sharp})} = \operatorname{diag}(s_0 s_0^{\dagger}, 0_{np \times np}) + \operatorname{diag}\left(0_{np \times np}, (I_p - s_0 s_0^{\dagger}) s_{2n} s_0^{\dagger}\right). \tag{6.8}$$

Since $(s_j)_{j=0}^{2n} \in \tilde{\mathcal{D}}_{p \times q, 2n}$, we see from Remark 4.7 that $(s_j)_{j=0}^{n} \in \mathcal{D}_{p \times q, n}$ which in view of Proposition 4.14 implies

$$\mathbf{S}_n^{\dagger} = \mathbf{S}_n^{\sharp}, \tag{6.9}$$
$$\mathbb{S}_n^{\dagger} = \mathbb{S}_n^{\sharp} \tag{6.10}$$

and, in view of [16, Lemma 4.18], moreover

$$\mathbf{S}_n^{\dagger} = \begin{bmatrix} * & 0_{q \times np} \\ * & \mathbf{S}_{n-1}^{\dagger} \end{bmatrix}. \tag{6.11}$$

From Definition 4.16 and Remark 4.7 we get $(s_j)_{j=0}^{n-1} \in \mathcal{D}_{p \times q, n-1}$. Hence, Lemma 4.15 yields

$$\mathbf{S}_{n-1}^{\dagger} \mathbf{S}_{n-1} = I_n \otimes (s_0^{\dagger} s_0). \tag{6.12}$$

Thus, using (6.9), (6.11) and (6.12), we obtain

$$\begin{aligned}
I_{(n+1)q} - \mathbf{S}_n^{\sharp} \operatorname{diag}(0_{p \times q}, \mathbf{S}_{n-1}) &= I_{(n+1)q} - \mathbf{S}_n^{\dagger} \operatorname{diag}(0_{p \times q}, \mathbf{S}_{n-1}) \\
&= I_{(n+1)q} - \begin{bmatrix} * & 0_{q \times np} \\ * & \mathbf{S}_{n-1}^{\dagger} \end{bmatrix} \operatorname{diag}(0_{p \times q}, \mathbf{S}_{n-1}) \\
&= \operatorname{diag}(I_q, I_{nq}) - \operatorname{diag}(0_{q \times q}, \mathbf{S}_{n-1}^{\dagger} \mathbf{S}_{n-1}) \\
&= \operatorname{diag}(I_q, I_n \otimes I_q) - \operatorname{diag}\left(0_{q \times q}, I_n \otimes (s_0^{\dagger} s_0)\right) \\
&= \operatorname{diag}\left(I_q, I_n \otimes (I_q - s_0^{\dagger} s_0)\right). \tag{6.13}
\end{aligned}$$

From part (b) of Proposition 4.9 we get $s_j^{\sharp} s_0 s_0^{\dagger} = s_j^{\sharp}$ for all $j \in \mathbb{Z}_{0,n}$ and hence

$$\begin{aligned}
\mathbf{S}_n^{\sharp} \operatorname{diag}(s_0 s_0^{\dagger}, 0_{np \times np}) &= [y_{0,n}^{\sharp}, *] \operatorname{diag}(s_0 s_0^{\dagger}, 0_{np \times np}) \\
&= [y_{0,n}^{\sharp} s_0 s_0^{\dagger}, 0_{(n+1)q \times np}] = [y_{0,n}^{\sharp}, 0_{(n+1)q \times np}] \tag{6.14} \\
&= y_{0,n}^{\sharp} v_{p,n}^{*}.
\end{aligned}$$

In view of $s_0^\sharp = s_0^\dagger$, we have

$$\mathbf{S}_n^\sharp \operatorname{diag}\left(0_{np \times np}, (I_p - s_0 s_0^\dagger) s_{2n} s_0^\dagger\right) = \begin{bmatrix} * & 0_{nq \times p} \\ * & s_0^\sharp \end{bmatrix} \operatorname{diag}\left(0_{np \times np}, (I_p - s_0 s_0^\dagger) s_{2n} s_0^\dagger\right)$$

$$= \operatorname{diag}\left(0_{nq \times np}, s_0^\sharp (I_p - s_0 s_0^\dagger) s_{2n} s_0^\dagger\right)$$

$$= 0_{(n+1)q \times (n+1)p}. \tag{6.15}$$

From part (b) of Proposition 4.9 we get $s_0^\dagger s_0 s_j^\sharp = s_j^\sharp$ for all $j \in \mathbb{Z}_{1,2n}$ and hence

$$\operatorname{diag}\left(I_q, I_n \otimes (I_q - s_0^\dagger s_0)\right) H_n^\sharp = \operatorname{diag}\left(I_q, I_n \otimes (I_q - s_0^\dagger s_0)\right) \begin{bmatrix} z_{0,n}^\sharp \\ [y_{1,n}^\sharp, G_{n-1}^\sharp] \end{bmatrix}$$

$$= \begin{bmatrix} I_q z_{0,n}^\sharp \\ [I_n \otimes (I_q - s_0^\dagger s_0)][y_{1,n}^\sharp, G_{n-1}^\sharp] \end{bmatrix}$$

$$= \begin{bmatrix} z_{0,n}^\sharp \\ [[I_n \otimes (I_q - s_0^\dagger s_0)]y_{1,n}^\sharp, [I_n \otimes (I_q - s_0^\dagger s_0)]G_{n-1}^\sharp] \end{bmatrix} = \begin{bmatrix} z_{0,n}^\sharp \\ [0_{nq \times p}, 0_{nq \times np}] \end{bmatrix}$$

$$= \begin{bmatrix} z_{0,n}^\sharp \\ 0_{nq \times (n+1)p} \end{bmatrix} = v_{q,n} z_{0,n}^\sharp. \tag{6.16}$$

Thus, using (6.7), (6.8), (6.13), (6.14), (6.15) and (6.16), we finally obtain

$$H_n^\sharp + \mathbf{S}_n^\sharp H_n \mathbf{S}_n^\sharp = H_n^\sharp + \mathbf{S}_n^\sharp \left[H_n^{(s \odot s^\sharp)} - \operatorname{diag}(0_{p \times q}, \mathbf{S}_{n-1}) H_n^\sharp \right]$$

$$= \mathbf{S}_n^\sharp H_n^{(s \odot s^\sharp)} + \left[I_{(n+1)q} - \mathbf{S}_n^\sharp \operatorname{diag}(0_{p \times q}, \mathbf{S}_{n-1}) \right] H_n^\sharp$$

$$= \mathbf{S}_n^\sharp \left[\operatorname{diag}(s_0 s_0^\sharp, 0_{np \times np}) + \operatorname{diag}\left(0_{np \times np}, (I_p - s_0 s_0^\dagger) s_{2n} s_0^\dagger\right) \right]$$

$$+ \operatorname{diag}\left(I_q, I_n \otimes (I_q - s_0^\dagger s_0)\right) H_n^\sharp$$

$$= y_{0,n}^\sharp v_{p,n}^* + v_{q,n} z_{0,n}^\sharp.$$

Hence (6.4) is proved. Taking (6.9) and (6.10) into account, then (6.5) follows from (6.4). Thus, (a) is proved.

(b) From (5.1), (4.3), (6.1), (4.4), Definition 4.1, (2.2) and (6.2), we see that

$$H_0 + \mathbf{S}_0 H_0^\sharp \mathbf{S}_0 = s_0 + s_0 s_0^\sharp s_0 = s_0 + s_0 s_0^\dagger s_0 = s_0 + s_0 = y_{0,0} v_{q,0}^* + v_{p,0} z_{0,0}.$$

Thus, the assertion is proved in the case $n = 0$. Now suppose $n \in \mathbb{N}$. Let $t_j := s_j$ for each $j \in \mathbb{Z}_{0,2n-1}$ and $t_{2n} := s_0 s_0^\dagger s_{2n} s_0^\dagger s_0$. Then parts (b) and (d) of Lemma 4.19 yield

$$(s_j^\sharp)_{j=0}^{2n} = (t_j^\sharp)_{j=0}^{2n} \tag{6.17}$$

and

$$\left((t^\sharp)_j^\sharp\right)_{j=0}^{2n} = (t_j)_{j=0}^{2n}, \tag{6.18}$$

respectively. In view of Remark 4.18, we have $(t_j^\sharp)_{j=0}^{2n} \in \tilde{\mathcal{D}}_{q \times p, 2n}$. Thus, applying Theorem 6.1 to the sequence $(t_j^\sharp)_{j=0}^{2n}$ and taking (6.18) into account, we obtain

$$H_n^{(t)} + \mathbf{S}_n^{(t)} H_n^{(t^\sharp)} \mathbb{S}_n^{(t)} = y_{0,n}^{(t)} v_{p,n}^* + v_{q,n} z_{0,n}^{(t)}. \tag{6.19}$$

In view of part (e) of Lemma 4.19, we have

$$H_n^{(t)} = H_n + \Theta_{n;p,q}. \tag{6.20}$$

From (6.17) we see

$$H_n^\sharp = H_n^{(t^\sharp)}. \tag{6.21}$$

From the definition of the sequence $(t_j)_{j=0}^{2n}$ the equations

$$\mathbf{S}_n^{(t)} = \mathbf{S}_n, \qquad \mathbb{S}_n^{(t)} = \mathbb{S}_n, \qquad y_{0,n}^{(t)} = y_{0,n} \qquad \text{and} \qquad z_{0,n}^{(t)} = z_{0,n}$$

are obvious. Thus, taking (6.20) and (6.21) into account the assertion is an immediate consequence of (6.19). □

Corollary 6.2. *Let* $\kappa \in \mathbb{Z}_{2,\infty}$ *and let* $(s_j)_{j=0}^{\kappa} \in \tilde{\mathcal{D}}_{p \times q, \kappa}$. *For every choice of integers* $m \in \mathbb{Z}_{1,\kappa-1}$ *and* $n \in \mathbb{Z}_{1,\kappa-m}$, *then*

$$\sum_{k=0}^{m} \sum_{l=0}^{n} s_{m-k}^\sharp s_{k+l} s_{n-l}^\sharp = -s_{m+n}^\sharp.$$

Proof. Compare the $p \times q$ blocks in the $p \times q$ block partition of the lower right $np \times nq$ block of the left- and right-hand side of equation (6.4) in Theorem 6.1. □

Corollary 6.3. *Let* $n \in \mathbb{N}_0$ *and let* $(s_j)_{j=0}^{2n} \in \mathcal{D}_{p \times q, 2n}$. *Then* (6.4), (6.5), *and*

$$H_n + \mathbf{S}_n H_n^\sharp \mathbb{S}_n = y_{0,n} v_{q,n}^* + v_{p,n} z_{0,n} \tag{6.22}$$

hold true.

Proof. In the case $n = 0$ equation (6.22) is trivial. Now we consider the case $n \in \mathbb{N}$. From Remark 4.17 we obtain $(s_j)_{j=0}^{2n} \in \tilde{\mathcal{D}}_{p \times q, 2n}$. Hence, part (a) of Theorem 6.1 yields (6.4) and (6.5). Because of $(s_j)_{j=0}^{2n} \in \mathcal{D}_{p \times q, 2n}$, Proposition 4.8 implies $s_0 s_0^\dagger s_{2n} s_0^\dagger s_0 = s_{2n}$. Consequently, applying part (b) of Theorem 6.1 gives us (6.22). □

In view of Proposition 4.25 and Proposition 4.24, we are now able to apply the preceding results to the classes of Hankel non-negative definite sequences and Hankel non-negative definite extendable sequences.

Proposition 6.4. *Let* $\kappa \in \mathbb{N}_0 \cup \{+\infty\}$ *and let* $(s_j)_{j=0}^{2\kappa} \in \mathcal{H}_{q,2\kappa}^{\geq}$. *For each* $n \in \mathbb{Z}_{0,\kappa}$ *then*

$$H_n^\sharp + \mathbf{S}_n^\sharp H_n (\mathbf{S}_n^\sharp)^* = y_{0,n}^\sharp v_{q,n}^* + v_{p,n} z_{0,n}^\sharp, \tag{6.23}$$

$$H_n^\sharp + \mathbf{S}_n^\dagger H_n (\mathbf{S}_n^\dagger)^* = y_{0,n}^\sharp v_{q,n}^* + v_{p,n} z_{0,n}^\sharp, \tag{6.24}$$

and

$$H_n + \mathbf{S}_n H_n^\sharp \mathbf{S}_n^* = y_{0,n} v_{q,n}^* + v_{p,n} z_{0,n} - \Theta_{n;q,q}, \tag{6.25}$$

where $\Theta_{n;q,q}$ *is given by* (6.6).

Proof. Let $n \in \mathbb{Z}_{0,\kappa}$. Then $(s_j)_{j=0}^{2n}$ belongs to $\mathcal{H}_{q,2n}^{\geq}$. From part (b) of Lemma 3.2 we infer then, $s_j^* = s_j$ for each $j \in \mathbb{Z}_{0,n}$. Hence, (4.6) implies $\mathbb{S}_n = \mathbf{S}_n^*$. Because of [16, Corollary 5.17], we get $(s_j^\sharp)^* = s_j^\sharp$ for each $j \in \mathbb{Z}_{0,n}$. Consequently, $\mathbb{S}_n^\sharp = (\mathbf{S}_n^\sharp)^*$. Thus, in the case $n = 0$, the assertion follows immediately. Now suppose $n \geq 1$. From Proposition 4.25 we see that $(s_j)_{j=0}^{2n} \in \tilde{\mathcal{D}}_{q \times q, 2n}$ is valid. Hence, the application of Theorem 6.1 and part (a) of Lemma A.1 yields (6.23) and (6.24), whereas part (b) of Theorem 6.1 provides us (6.25). $\qquad\square$

Proposition 6.5. *Let $m \in \mathbb{N}_0$ and let $(s_j)_{j=0}^{2m} \in \mathcal{H}_{q,2m}^{\geq,\mathrm{e}}$. For each $n \in \mathbb{Z}_{0,m}$, then the equations (6.23), (6.24), and*

$$H_n + \mathbf{S}_n H_n^\sharp \mathbf{S}_n^* = y_{0,n} v_{q,n}^* + v_{q,n} z_{0,n}$$

hold true.

Proof. From Proposition 4.24 we see that $(s_j)_{j=0}^{2m}$ belongs to $\mathcal{D}_{q \times q, m}$. Consequently, for each $n \in \mathbb{Z}_{0,m}$, from Proposition 4.8 we get $s_0 s_0^\dagger s_{2n} s_0^\dagger s_0 = s_{2n}$. Taking into account that $(s_j)_{j=0}^{2m} \in \mathcal{H}_{q,2m}^{\geq}$ the application of Proposition 6.4 completes the proof. $\qquad\square$

Our next goal can be described as follows. Let $n \in \mathbb{N}_0$ and let $(s_j)_{j=0}^{2n} \in \tilde{\mathcal{D}}_{p \times q, 2n}$. Then starting from Theorem 6.1 we are looking for further relations between the block Hankel matrices H_n and $H_n^\sharp$. The following lemma contains the key to reach this goal.

Lemma 6.6. *Let $n \in \mathbb{N}_0$ and $(s_j)_{j=0}^n \in \mathcal{D}_{p \times q, n}$. Then*

$$\mathbf{S}_n^\sharp (y_{0,n} v_{q,n}^* + v_{p,n} z_{0,n}) \mathbb{S}_n^\sharp = y_{0,n}^\sharp v_{p,n}^* + v_{q,n} z_{0,n}^\sharp \tag{6.26}$$

and

$$\mathbf{S}_n (y_{0,n}^\sharp v_{p,n}^* + v_{q,n} z_{0,n}^\sharp) \mathbb{S}_n = y_{0,n} v_{q,n}^* + v_{p,n} z_{0,n}. \tag{6.27}$$

Proof. Using (4.3), (2.2), (6.2), $s_0^\sharp = s_0^\dagger$ and (6.3), we easily see that (6.26) is true for $n = 0$.

Now suppose $n \in \mathbb{N}$. From (4.3), (4.4) and (2.2) we see that $y_{0,n}$ and $z_{0,n}$ are the first $(n+1)p \times q$-block column of $\mathbf{S}_n$ and the first $p \times (n+1)q$-block row of $\mathbb{S}_n$, respectively. Thus, the application of Proposition 4.14 and Lemma 4.15 yields

$$\mathbf{S}_n^\sharp y_{0,n} = \mathbf{S}_n^\dagger y_{0,n} = \mathbf{S}_n^\dagger \mathbf{S}_n \begin{bmatrix} I_q \\ 0_{q \times q} \end{bmatrix} = \begin{bmatrix} s_0^\dagger s_0 \\ 0_{nq \times q} \end{bmatrix} \tag{6.28}$$

and

$$z_{0,n} \mathbb{S}_n^\sharp = z_{0,n} \mathbb{S}_n^\dagger = \begin{bmatrix} I_p & 0_{p \times np} \end{bmatrix} \mathbb{S}_n \mathbb{S}_n^\dagger = [s_0 s_0^\dagger, 0_{p \times np}]. \tag{6.29}$$

Using (4.3), (4.4) and (6.2), we get

$$v_{q,n}^* \mathbb{S}_n^\sharp = [s_0^\sharp, \dots, s_n^\sharp] \tag{6.30}$$

and

$$\mathbf{S}_n^\sharp v_{p,n} = \begin{bmatrix} s_0^\sharp \\ \vdots \\ s_n^\sharp \end{bmatrix}. \tag{6.31}$$

Applying (6.28), (6.29), (6.30), (6.31), part (b) of Proposition 4.9, (6.2) and (6.3), we obtain

$$\mathbf{S}_n^\sharp (y_{0,n} v_{q,n}^* + v_{p,n} z_{0,n}) \mathbb{S}_n^\sharp = \mathbf{S}_n^\sharp y_{0,n} v_{q,n}^* \mathbb{S}_n^\sharp + \mathbf{S}_n^\sharp v_{p,n} z_{0,n} \mathbb{S}_n^\sharp$$

$$= \begin{bmatrix} s_0^\dagger s_0 \\ 0_{nq \times q} \end{bmatrix} [s_0^\sharp, \dots, s_n^\sharp] + \begin{bmatrix} s_0^\sharp \\ \vdots \\ s_n^\sharp \end{bmatrix} [s_0 s_0^\dagger, 0_{p \times np}]$$

$$= \begin{bmatrix} [s_0^\dagger s_0 s_0^\sharp, \dots, s_0^\dagger s_0 s_n^\sharp] \\ 0_{nq \times (n+1)q} \end{bmatrix} + \left[\begin{bmatrix} s_0^\sharp s_0 s_0^\dagger \\ \vdots \\ s_n^\sharp s_0 s_0^\dagger \end{bmatrix}, 0_{(n+1)p \times np} \right]$$

$$= \begin{bmatrix} [s_0^\sharp, \dots, s_n^\sharp] \\ 0_{nq \times (n+1)q} \end{bmatrix} + \left[\begin{bmatrix} s_0^\sharp \\ \vdots \\ s_n^\sharp \end{bmatrix}, 0_{(n+1)p \times np} \right]$$

$$= v_{q,n} z_{0,n}^\sharp + y_{0,n}^\sharp v_{p,n}^*.$$

Thus, (6.26) is also proved for each $n \in \mathbb{N}$. Part (a) of Proposition 4.9 shows that

$$(s_j^\sharp)_{j=0}^n \in \mathcal{D}_{q \times p, n}. \tag{6.32}$$

Since $(s_j)_{j=0}^n$ belongs to $\mathcal{D}_{p \times q, n}$, we see from part (d) of Proposition 4.10 that

$$\left((s_j^\sharp)^\sharp \right)_{j=0}^n = (s_j)_{j=0}^n. \tag{6.33}$$

Because of (6.32), we can apply (6.26) to the sequence $(s_j^\sharp)_{j=0}^n$. Thus, taking (6.33) into account, we obtain (6.27). $\qquad\square$

Lemma 6.7. *Let $n \in \mathbb{N}_0$ and let $(s_j)_{j=0}^{2n}$ be a sequence from $\mathbb{C}^{p \times q}$. Then*

$$\mathbf{S}_n^\sharp \Theta_{n;p,q} \mathbb{S}_n^\sharp = 0_{(n+1)q \times (n+1)p}, \tag{6.34}$$

where $\Theta_{n;p,q}$ is given (6.6).

Proof. In the case $n = 0$ the assertion is obvious. Now suppose $n \in \mathbb{N}$. Let

$$a := s_0^\sharp (s_0 s_0^\dagger s_{2n} s_0^\dagger s_0 - s_{2n}) s_0^\sharp. \tag{6.35}$$

Then a closer look at the matrices $\mathbf{S}_n^\sharp$, $\Theta_{n;p,q}$ and $\mathbb{S}_n^\sharp$ shows that

$$\mathbf{S}_n^\sharp \Theta_{n;p,q} \mathbb{S}_n^\sharp = \mathrm{diag}(0_{nq \times np}, a). \tag{6.36}$$

In view of Definition 4.1, we have $s_0^\sharp = s_0^\dagger$. Thus, from (6.35) we infer $a = 0_{q \times p}$. Now, combining this and (6.36) yields (6.34). $\qquad\square$

Theorem 6.8. *Let* $n \in \mathbb{N}_0$ *and let* $(s_j)_{j=0}^{2n} \in \tilde{\mathcal{D}}_{p \times q, 2n}$. *Then:*

(a) *The equations*

$$H_n^\sharp = -\mathbb{S}_n^\sharp \left[H_n - (y_{0,n} v_{q,n}^* + v_{p,n} z_{0,n}) \right] \mathbb{S}_n^\sharp \tag{6.37}$$

and

$$H_n^\sharp = -\mathbb{S}_n^\dagger \left[H_n - (y_{0,n} v_{q,n}^* + v_{p,n} z_{0,n}) \right] \mathbb{S}_n^\dagger \tag{6.38}$$

hold.

(b)

$$H_n = -\Theta_{n;p,q} - \mathbb{S}_n \left[H_n^\sharp - (y_{0,n}^\sharp v_{q,n}^* + v_{p,n} z_{0,n}^\sharp) \right] \mathbb{S}_n,$$

where $\Theta_{n;p,q}$ *is given by* (6.6).

(c) $\operatorname{rank}(H_n^\sharp) = \operatorname{rank}[H_n + \Theta_{n;p,q} - (y_{0,n} v_{q,n}^* + v_{p,n} z_{0,n})]$.

(d) *If* $p = q$, *then* $\det(H_n^\sharp) = (-1)^{(n+1)q} [(\det s_0)^\dagger]^{2n+2} \det(H_n - (y_{0,n} v_{q,n}^* + v_{p,n} z_{0,n}))$.

Proof. In view of $(s_j)_{j=0}^{2n} \in \tilde{\mathcal{D}}_{p \times q, 2n}$ we see from Definition 4.16 and Remark 4.7 that

$$(s_j)_{j=0}^n \in \mathcal{D}_{p \times q, n}. \tag{6.39}$$

(a) Equation (6.37) is an immediate consequence of (6.4), (6.39) and Lemma 6.6. The identity (6.38) follows, in view of (6.39) and Proposition 4.14, from (6.37).

(b) Taking (6.39) into account and combining part (b) of Theorem 6.1 and Lemma 6.6 yields (b).

(c) From part (b) of Theorem 6.1 we obtain

$$H_n + \Theta_{n;p,q} - (y_{0,n} v_{q,n}^* + v_{p,n} z_{0,n}) = -\mathbb{S}_n H_n^\sharp \mathbb{S}_n.$$

Thus,

$$\operatorname{rank} \left[H_n + \Theta_{n;p,q} - (y_{0,n} v_{q,n}^* + v_{p,n} z_{0,n}) \right] \leq \operatorname{rank} H_n^\sharp. \tag{6.40}$$

The combination of (6.37) and Lemma 6.7 yields

$$H_n^\sharp = -\mathbb{S}_n^\sharp \left[H_n - (y_{0,n} v_{q,n}^* + v_{p,n} z_{0,n}) \right] \mathbb{S}_n^\sharp - \mathbb{S}_n^\sharp \Theta_{n;p,q} \mathbb{S}_n^\sharp$$
$$= -\mathbb{S}_n^\sharp \left[H_n + \Theta_{n;p,q} - (y_{0,n} v_{q,n}^* + v_{p,n} z_{0,n}) \right] \mathbb{S}_n^\sharp.$$

Hence,

$$\operatorname{rank} H_n^\sharp \leq \operatorname{rank} \left[H_n + \Theta_{n;p,q} - (y_{0,n} v_{q,n}^* + v_{p,n} z_{0,n}) \right]. \tag{6.41}$$

Applying (6.40) and (6.41) it follows (c).

(d) Part (d) is an immediate consequence of (6.37) and $s_0^\sharp = s_0^\dagger$. □

Now we study the interplay between slightly modified block Hankel matrices. Let $n \in \mathbb{N}_0$ and let $(s_j)_{j=0}^{2n+1}$ be a sequence from $\mathbb{C}^{p \times q}$ with reciprocal sequence $(s_j^\sharp)_{j=0}^{2n+1}$. Then we will study the interplay between the block Hankel matrices

$$K_n^{(s)} := [s_{j+k+1}]_{j,k=0}^n \tag{6.42}$$

and

$$K_n^{(s^\sharp)} := [s_{j+k+1}^\sharp]_{j,k=0}^n. \tag{6.43}$$

Whenever it is clear which sequence is meant we will, respectively, write K_n and $K_n^\sharp$ instead of $K_n^{(s)}$ and $K_n^{(s^\sharp)}$.

Now we are able to formulate the third main result of this section.

Theorem 6.9. *Let $n \in \mathbb{N}_0$ and $(s_j)_{j=0}^{2n+1} \in \tilde{\mathcal{D}}_{p\times q,2n+1}$. Then:*

(a) $K_n^\sharp = -\mathbf{S}_n^\sharp K_n \mathbb{S}_n^\sharp$ *and* $K_n^\sharp = -\mathbf{S}_n^\dagger K_n \mathbb{S}_n^\dagger$.

(b) $K_n = \Xi_n - \mathbf{S}_n K_n^\sharp \mathbb{S}_n$, *where*

$$
\Xi_n := \begin{cases} s_{2n+1} - s_0 s_0^\dagger s_{2n+1} s_0^\dagger s_0 & \text{if } n = 0 \\ \operatorname{diag}(0_{np\times nq}, s_{2n+1} - s_0 s_0^\dagger s_{2n+1} s_0^\dagger s_0) & \text{if } n \in \mathbb{N}. \end{cases}
$$

(c) $\operatorname{rank} K_n^\sharp = \operatorname{rank}(K_n - \Xi_n)$.

(d) *If $p = q$, then $\det(K_n^\sharp) = (-1)^{(n+1)q}[(\det s_0)^\dagger]^{2n+2} \det K_n$.*

Proof. (a) In the case $n = 0$, the equations of (a) are immediate consequences of Definition 4.1. Now suppose $n \in \mathbb{N}$. From Proposition 5.4 we get

$$
K_n^{(s\odot s^\sharp)} = \mathbf{S}_n K_n^\sharp + K_n \mathbb{S}_n^\sharp, \tag{6.44}
$$

whereas Proposition 4.23 implies

$$
K_n^{(s\odot s^\sharp)} = \operatorname{diag}\left(0_{np\times np}, (I_p - s_0 s_0^\dagger) s_{2n+1} s_0^\dagger\right). \tag{6.45}
$$

From Definition 4.16 and Remark 4.7, we get (6.39). Thus, from Proposition 4.14 we infer that (6.9) and (6.10) hold true, whereas (6.9) and Lemma 4.15 imply

$$
\mathbf{S}_n^\sharp \mathbf{S}_n = \mathbf{S}_n^\dagger \mathbf{S}_n = I_{n+1} \otimes (s_0^\dagger s_0). \tag{6.46}
$$

Using (6.44) and (6.46), we conclude

$$
-\mathbf{S}_n^\sharp K_n \mathbb{S}_n^\sharp = \mathbf{S}_n^\sharp(\mathbf{S}_n K_n^\sharp - K_n^{(s\odot s^\sharp)}) = \left[I_{n+1} \otimes (s_0^\dagger s_0)\right] K_n^\sharp - \mathbf{S}_n^\sharp K_n^{(s\odot s^\sharp)}. \tag{6.47}
$$

From part (b) of Proposition 4.9 we get $s_0^\dagger s_0 s_j^\sharp = s_j^\sharp$ for all $j \in \mathbb{Z}_{1,2n+1}$. Thus,

$$
\left[I_{n+1} \otimes (s_0^\dagger s_0)\right] K_n^\sharp = K_n^\sharp. \tag{6.48}
$$

In view of Definition 4.1, we have $s_0^\sharp = s_0^\dagger$. Thus, we get

$$
\mathbf{S}_n^\sharp \operatorname{diag}\left(0_{np\times np}, (I_p - s_0 s_0^\dagger) s_{2n+1} s_0^\dagger\right)
$$
$$
= \begin{bmatrix} * & 0_{nq\times p} \\ * & s_0^\sharp \end{bmatrix} \operatorname{diag}\left(0_{np\times np}, (I_p - s_0 s_0^\dagger) s_{2n+1} s_0^\dagger\right)
$$
$$
= \operatorname{diag}\left(0_{nq\times np}, s_0^\sharp(I_p - s_0 s_0^\dagger) s_{2n+1} s_0^\dagger\right) = \operatorname{diag}\left(0_{nq\times np}, s_0^\dagger(I_p - s_0 s_0^\dagger) s_{2n+1} s_0^\dagger\right)
$$
$$
= 0_{(n+1)q\times(n+1)p}. \tag{6.49}
$$

From (6.45) and (6.49) we infer

$$
\mathbf{S}_n^\sharp K_n^{(s\odot s^\sharp)} = 0_{(n+q)q\times(n+1)p}. \tag{6.50}
$$

Combining (6.47), (6.48) and (6.50), we obtain the first equation in part (a). In view of (6.9) and (6.10), the second equation holds as well. Thus, (a) is proved.

(b) Let $t_j := s_j$ for each $j \in \mathbb{Z}_{0,2n}$ and $t_{2n+1} := s_0 s_0^\dagger s_{2n+1} s_0^\dagger s_0$. Then part (b) and part (d) of Lemma 4.19 yield

$$(s_j^\sharp)_{j=0}^{2n+1} = (t_j^\sharp)_{j=0}^{2n+1} \tag{6.51}$$

and

$$\left((t^\sharp)_j^\sharp\right)_{j=0}^{2n+1} = (t_j)_{j=0}^{2n+1}. \tag{6.52}$$

In view of Remark 4.18, we have $(t_j^\sharp)_{j=0}^{2n+1} \in \tilde{\mathcal{D}}_{q\times p,2n+1}$. Thus applying part (a) to the sequence $(t_j^\sharp)_{j=0}^{2n+1}$ and taking (6.52) into account, we obtain

$$K_n^{(t)} = -\mathbf{S}_n^{(t)} K_n^{(t^\sharp)} \mathbb{S}_n^{(t)}. \tag{6.53}$$

From the construction of the sequence $(t_j)_{j=0}^{2n+1}$ we see

$$K_n^{(t)} = K_n - \Xi_n, \qquad \mathbf{S}_n^{(t)} = \mathbf{S}_n \qquad \text{and} \qquad \mathbb{S}_n^{(t)} = \mathbb{S}_n. \tag{6.54}$$

From (6.51) we infer

$$K_n^{(t^\sharp)} = K_n^\sharp. \tag{6.55}$$

The combination of (6.53), (6.54) and (6.55) yields now the assertion of (b).

(c) In view of (b), we have $K_n = \Xi_n - \mathbf{S}_n K_n^\sharp \mathbb{S}_n$. Thus,

$$\operatorname{rank}(K_n - \Xi_n) \le \operatorname{rank} K_n^\sharp. \tag{6.56}$$

Because of part (c) of Lemma 4.19, we have $(t_j)_{j=0}^{2n+1} \in \mathcal{D}_{p\times q,2n+1}$. Thus, Remark 4.17 implies $(t_j)_{j=0}^{2n+1} \in \tilde{\mathcal{D}}_{p\times q,2n+1}$. Hence, applying part (a) to the sequence $(t_j)_{j=0}^{2n+1}$ gives

$$K_n^{(t^\sharp)} = -\mathbf{S}_n^{(t^\sharp)} K_n^{(t)} \mathbb{S}_n^{(t^\sharp)}. \tag{6.57}$$

Taking (6.51) and the first equation in (6.54) into account, formula (6.57) can be rewritten as

$$K_n^\sharp = -\mathbf{S}_n^\sharp (K_n - \Xi_n) \mathbb{S}_n^\sharp.$$

Consequently,

$$\operatorname{rank} K_n^\sharp \le \operatorname{rank}(K_n - \Xi_n). \tag{6.58}$$

Now the combination of (6.54) and (6.58) yields the assertion of (c).

(d) This is a consequence of (a). $\qquad\square$

Corollary 6.10. *Let $n \in \mathbb{N}_0$ and $(s_j)_{j=0}^{2n+1} \in \mathcal{D}_{p\times q,2n+1}$. Then*

$$K_n^\sharp = -\mathbf{S}_n^\sharp K_n \mathbb{S}_n^\sharp, \qquad\qquad K_n^\sharp = -\mathbf{S}_n^\dagger K_n \mathbb{S}_n^\dagger, \tag{6.59}$$

$$K_n = -\mathbf{S}_n K_n^\sharp \mathbb{S}_n \qquad and \qquad \operatorname{rank} K_n^\sharp = \operatorname{rank} K_n. \tag{6.60}$$

Proof. In view of Remark 4.17, the equations in (6.59) follow from part (a) of Theorem 6.9. Because of Proposition 4.8, we have $s_{2n+1} = s_0 s_0^\dagger s_{2n+1} s_0^\dagger s_0$. Thus, $\Xi_n = 0_{(n+1)p\times(n+1)q}$. Now (6.60) is a direct consequence of part (b) and part (c) of Theorem 6.9. $\qquad\square$

Corollary 6.11. *Let $n \in \mathbb{N}_0$ and $(s_j)_{j=0}^{2n+1} \in \mathcal{H}_{q,2n+1}^{\geq,e}$. Then*

$$K_n^\sharp = -\mathbf{S}_n^\sharp K_n(\mathbf{S}_n^\sharp)^*, \qquad\qquad\qquad K_n^\sharp = -\mathbf{S}_n^\dagger K_n(\mathbf{S}_n^\dagger)^*,$$
$$K_n = -\mathbf{S}_n K_n^\sharp \mathbf{S}_n^* \qquad and \qquad \operatorname{rank} K_n^\sharp = \operatorname{rank} K_n.$$

Proof. Part (b) of Lemma 3.1 shows that $(s_j)_{j=0}^{2n+1}$ is a sequence from $\mathbb{C}_H^{q \times q}$. Hence, (4.6) in Remark 4.12 implies $\mathbb{S}_n = \mathbf{S}_n^*$. Thus, the assertions follow by combining Proposition 4.24 and Corollary 6.10. $\square$

In the rest of this section we will study the interplay between block Hankel matrices of a third type. Let $n \in \mathbb{N}$ and let $(s_j)_{j=0}^{2n+2}$ be a sequence from $\mathbb{C}^{p \times q}$ with reciprocal sequence $(s_j^\sharp)_{j=0}^{2n+2}$. Then we will study some crossrelations between the block Hankel matrices

$$G_n^{(s)} := [s_{j+k+2}]_{j,k=0}^n \qquad and \qquad G_n^{(s^\sharp)} := [s_{j+k+2}^\sharp]_{j,k=0}^n.$$

Whenever it is clear which sequence is meant we will, respectively, write G_n and $G_n^\sharp$ instead of $G_n^{(s)}$ and $G_n^{(s^\sharp)}$.

Lemma 6.12. *Let $n \in \mathbb{N}_0$ and let $(s_j)_{j=0}^{2n+2}$ be a sequence from $\mathbb{C}^{p \times q}$. Then*

$$\mathbf{S}_n^\sharp \Gamma_{n;p,q} \mathbb{S}_n^\sharp = 0_{(n+1)q \times (n+1)p},$$

where

$$\Gamma_{n;p,q} := \begin{cases} s_0 s_0^\dagger s_2 s_0^\dagger s_0 - s_2 & \text{if } n = 0 \\ \operatorname{diag}(0_{np \times nq}, s_0 s_0^\dagger s_{2n+2} s_0^\dagger s_0 - s_{2n+2}) & \text{if } n \in \mathbb{N}. \end{cases} \qquad (6.61)$$

Lemma 6.12 can be proved analogously to Lemma 6.7. We omit the details.

Theorem 6.13. *Let $n \in \mathbb{N}_0$ and let $(s_j)_{j=0}^{2n+2} \in \tilde{\mathcal{D}}_{p \times q, 2n+2}$. Then:*

(a) *The equations*

$$G_n^\sharp = -\mathbf{S}_n^\sharp (G_n - y_{1,n+1} s_0^\dagger z_{1,n+1}) \mathbb{S}_n^\sharp$$

and

$$G_n^\sharp = -\mathbf{S}_n^\dagger (G_n - y_{1,n+1} s_0^\dagger z_{1,n+1}) \mathbb{S}_n^\dagger$$

hold true.

(b) *Let $\Gamma_{n;p,q}$ be defined via (6.61). Then*

$$G_n = -\Gamma_{n;p,q} - \mathbf{S}_n \left[G_n^\sharp - y_{1,n+1}^\sharp (s_0^\sharp)^\dagger z_{1,n+1}^\sharp \right] \mathbb{S}_n.$$

(c) $\operatorname{rank} G_n^\sharp = \operatorname{rank}(G_n + \Gamma_{n;p,q} - y_{1,n+1} s_0^\dagger z_{1,n+1}).$

(d) *If $p = q$, then $\det(G_n^\sharp) = (-1)^{(n+1)q}[(\det s_0)^\dagger]^{2n+2} \det(G_n - y_{1,n+1} s_0^\dagger z_{1,n+1}).$*

Proof. (a) From Proposition 5.5 we get

$$G_n^{(s \odot s^\sharp)} = G_n \mathbb{S}_n^\sharp + y_{1,n+1} z_{1,n+1}^\sharp + \mathbf{S}_n G_n^\sharp, \qquad (6.62)$$

whereas Proposition 4.23 implies

$$G_n^{(s\odot s^\sharp)} = \begin{cases} (I_p - s_0 s_0^\dagger) s_2 s_0^\dagger & \text{if } n = 0 \\ \operatorname{diag}(0_{np\times np}, (I_p - s_0 s_0^\dagger) s_{2n+2} s_0^\dagger) & \text{if } n \in \mathbb{N}. \end{cases} \tag{6.63}$$

From Definition 4.16 we get

$$(s_j)_{j=0}^{2n+1} \in \mathcal{D}_{p\times q, 2n+1}. \tag{6.64}$$

Let us consider an arbitrary $m \in \mathbb{Z}_{0,2n+1}$. Taking (6.64) and $m \in \mathbb{Z}_{0,2n+1}$ into account, we infer from Proposition 4.14 that

$$\mathbf{S}_m^\dagger = \mathbf{S}_m^\sharp, \qquad\qquad \mathbb{S}_m^\dagger = \mathbb{S}_m^\sharp \tag{6.65}$$

whereas Lemma 4.15 implies (4.7) and (4.8). In view of (6.65) we get from (4.7) and (4.8) that

$$\mathbf{S}_m^\sharp \mathbf{S}_m = I_{m+1} \otimes (s_0^\dagger s_0) \qquad \text{and} \qquad \mathbb{S}_m^\sharp \mathbb{S}_m = I_{m+1} \otimes (s_0^\dagger s_0) \tag{6.66}$$

and

$$\mathbf{S}_m \mathbf{S}_m^\sharp = I_{m+1} \otimes (s_0 s_0^\dagger) \qquad \text{and} \qquad \mathbb{S}_m \mathbb{S}_m^\sharp = I_{m+1} \otimes (s_0 s_0^\dagger). \tag{6.67}$$

Using (6.62), (6.66) and (6.63), we get

$$\begin{aligned}
-\mathbf{S}_n^\sharp (G_n - y_{1,n+1} s_0^\dagger z_{1,n+1}) \mathbb{S}_n^\sharp &= \mathbf{S}_n^\sharp (-G_n \mathbb{S}_n^\sharp + y_{1,n+1} s_0^\dagger z_{1,n+1} \mathbb{S}_n^\sharp) \\
&= \mathbf{S}_n^\sharp (y_{1,n+1} z_{1,n+1}^\sharp + \mathbf{S}_n G_n^\sharp - G_n^{(s\odot s^\sharp)} + y_{1,n+1} s_0^\dagger z_{1,n+1} \mathbb{S}_n^\sharp) \\
&= \mathbf{S}_n^\sharp \mathbf{S}_n G_n^\sharp - \mathbf{S}_n^\sharp G_n^{(s\odot s^\sharp)} + \mathbf{S}_n^\sharp y_{1,n+1} (z_{1,n+1}^\sharp + s_0^\dagger z_{1,n+1} \mathbb{S}_n^\sharp) \\
&= \left[I_{n+1} \otimes (s_0^\dagger s_0) \right] G_n^\sharp - \mathbf{S}_n^\sharp G_n^{(s\odot s^\sharp)} + \mathbf{S}_n^\sharp y_{1,n+1} (z_{1,n+1}^\sharp + s_0^\dagger z_{1,n+1} \mathbb{S}_n^\sharp).
\end{aligned} \tag{6.68}$$

From part (b) of Proposition 4.9 we get $s_0^\dagger s_0 s_j^\sharp = s_j^\sharp$ for all $j \in \mathbb{Z}_{1,2n+2}$. Thus,

$$\left[I_{n+1} \otimes (s_0^\dagger s_0) \right] G_n^\sharp = G_n^\sharp \tag{6.69}$$

and

$$s_0^\dagger s_0 z_{1,n+1}^\sharp = z_{1,n+1}^\sharp. \tag{6.70}$$

From (6.70) and (6.67), we obtain

$$\begin{aligned}
z_{1,n+1}^\sharp + s_0^\dagger z_{1,n+1} \mathbb{S}_n^\sharp &= s_0^\dagger s_0 z_{1,n+1}^\sharp + s_0^\dagger z_{1,n+1} \mathbb{S}_n^\sharp = s_0^\dagger [s_0, z_{1,n+1}] \begin{bmatrix} z_{1,n+1}^\sharp \\ \mathbb{S}_n^\sharp \end{bmatrix} \\
&= s_0^\dagger [I_p, 0_{p\times(n+1)p}] \begin{bmatrix} s_0 & z_{1,n+1} \\ 0_{(n+1)p\times q} & \mathbb{S}_n \end{bmatrix} \begin{bmatrix} s_0^\sharp & z_{1,n+1}^\sharp \\ 0_{(n+1)q\times p} & \mathbb{S}_n^\sharp \end{bmatrix} \begin{bmatrix} 0_{p\times(n+1)p} \\ I_{(n+1)p} \end{bmatrix} \\
&= s_0^\dagger [I_p, 0_{p\times(n+1)p}] \mathbb{S}_{n+1} \mathbb{S}_{n+1}^\sharp \begin{bmatrix} 0_{p\times(n+1)p} \\ I_{(n+1)p} \end{bmatrix} \\
&= s_0^\dagger [I_p, 0_{p\times(n+1)p}] \left[I_{n+2} \otimes (s_0 s_0^\dagger) \right] \begin{bmatrix} 0_{p\times(n+1)p} \\ I_{(n+1)p} \end{bmatrix} = 0_{q\times(n+1)p},
\end{aligned}$$

which implies

$$\mathbf{S}_n^\sharp y_{1,n+1} (z_{1,n+1}^\sharp + s_0^\dagger z_{1,n+1} \mathbb{S}_n^\sharp) = 0_{(n+1)q\times(n+1)p}. \tag{6.71}$$

In view of Definition 4.1, we have $s_0^\sharp = s_0^\dagger$. Taking this into account, we get

$$\mathbf{S}_0^\sharp (I_p - s_0 s_0^\dagger) s_2 s_0^\dagger = s_0^\sharp (I_p - s_0 s_0^\dagger) s_2 s_0^\dagger = s_0^\dagger (I_p - s_0 s_0^\dagger) s_2 s_0^\dagger = 0_{q \times p} \qquad (6.72)$$

and in the case $n \in \mathbb{N}$ furthermore

$$
\begin{aligned}
& \mathbf{S}_n^\sharp \operatorname{diag}\left(0_{np \times np}, (I_p - s_0 s_0^\dagger) s_{2n+2} s_0^\dagger\right) \\
&= \begin{bmatrix} * & 0_{nq \times p} \\ * & s_0^\sharp \end{bmatrix} \operatorname{diag}\left(0_{np \times np}, (I_p - s_0 s_0^\dagger) s_{2n+2} s_0^\dagger\right) \\
&= \operatorname{diag}\left(0_{nq \times np}, s_0^\sharp (I_p - s_0 s_0^\dagger) s_{2n+2} s_0^\dagger\right) \\
&= \operatorname{diag}\left(0_{nq \times np}, s_0^\dagger (I_p - s_0 s_0^\dagger) s_{2n+2} s_0^\dagger\right) = 0_{(n+1)q \times (n+1)p}.
\end{aligned}
\qquad (6.73)
$$

Combining (6.63), (6.72) and (6.73), we infer

$$\mathbf{S}_n^\sharp G_n^{(s \odot s^\sharp)} = 0_{(n+1)q \times (n+1)p}. \qquad (6.74)$$

Applying (6.68), (6.69), (6.74) and (6.71), we obtain

$$-\mathbf{S}_n^\sharp (G_n - y_{1,n+1} s_0^\dagger z_{1,n+1}) \mathbb{S}_n^\sharp = G_n^\sharp. \qquad (6.75)$$

Using $m \in \mathbb{Z}_{0,2n+1}$ and (6.65), the second assertion of (a) follows immediately from (6.75). Thus (a) is proved.

(b) Let $t_j := s_j$ for each $j \in \mathbb{Z}_{0,2n+1}$ and $t_{2n+2} := s_0 s_0^\dagger s_{2n+2} s_0^\dagger s_0$. Then part (b) and part (d) of Lemma 4.19 yield

$$(s_j^\sharp)_{j=0}^{2n+2} = (t_j^\sharp)_{j=0}^{2n+2} \qquad (6.76)$$

and

$$\left((t^\sharp)_j^\sharp\right)_{j=0}^{2n+2} = (t_j)_{j=0}^{2n+2}. \qquad (6.77)$$

In view of Remark 4.18, we have $(t_j^\sharp)_{j=0}^{2n+2} \in \tilde{\mathcal{D}}_{q \times p, 2n+2}$. Thus applying part (a) to the sequence $(t_j^\sharp)_{j=0}^{2n+2}$ and taking (6.77) into account, we obtain

$$G_n^{(t)} = -\mathbf{S}_n^{(t)} \left[G_n^{(t^\sharp)} - y_{1,n+1}^{(t^\sharp)} (t_0^\sharp)^\dagger z_{1,n+1}^{(t^\sharp)} \right] \mathbb{S}_n^{(t)}. \qquad (6.78)$$

From the construction of the sequence $(t_j)_{j=0}^{2n+2}$ we see

$$G_n^{(t)} = G_n + \Gamma_{n;p,q}, \qquad \mathbf{S}_n^{(t)} = \mathbf{S}_n \qquad \text{and} \qquad \mathbb{S}_n^{(t)} = \mathbb{S}_n. \qquad (6.79)$$

From (6.76) we infer

$$G_n^{(t^\sharp)} = G_n^\sharp, \qquad y_{1,n+1}^{(t^\sharp)} = y_{1,n+1}^\sharp \qquad \text{and} \qquad z_{1,n+1}^{(t^\sharp)} = z_{1,n+1}^\sharp. \qquad (6.80)$$

The combination of (6.78), (6.79) and (6.80) yields now the assertion of (b).

(c) Lemma 6.12 implies

$$\mathbf{S}_n^\sharp \Gamma_{n;p,q} \mathbb{S}_n^\sharp = 0_{(n+1)q \times (n+1)p}. \qquad (6.81)$$

Applying (6.75) and (6.79) gives us

$$G_n^\sharp = -\mathbf{S}_n^\sharp(G_n - y_{1,n+1}s_0^\dagger z_{1,n+1})\mathbb{S}_n^\sharp - \mathbf{S}_n^\sharp \Gamma_{n;p,q}\mathbb{S}_n^\sharp$$
$$= -\mathbf{S}_n^\sharp(G_n - y_{1,n+1}s_0^\dagger z_{1,n+1} + \Gamma_{n;p,q})\mathbb{S}_n^\sharp.$$

Thus

$$\operatorname{rank}G_n^\sharp \le \operatorname{rank}(G_n - y_{1,n+1}s_0^\dagger z_{1,n+1} + \Gamma_{n;p,q}). \tag{6.82}$$

Using $s_0^\sharp = s_0^\dagger$, we get

$$\mathbf{S}_0\mathbf{S}_0^\sharp G_0\mathbb{S}_0^\sharp\mathbb{S}_0 = s_0 s_0^\sharp s_2 s_0^\sharp s_0 = s_2 + (s_0 s_0^\dagger s_2 s_0^\dagger s_0 - s_2) = G_0 + \Gamma_0. \tag{6.83}$$

Now we consider the case $n \in \mathbb{N}$. In view of (6.64) we see from Proposition 4.8 that $s_0 s_0^\dagger s_j = s_j$ and $s_j s_0 s_0^\dagger = s_j$ for all $j \in \mathbb{Z}_{0,2n+1}$. Thus, taking $n \in \mathbb{N}$ into account, we get

$$\left[I_{n+1} \otimes (s_0 s_0^\dagger)\right] G_n = G_n + \operatorname{diag}\left(0_{np \times nq}, (s_0 s_0^\dagger - I_p)s_{2n+2}\right) \tag{6.84}$$

and

$$G_n \left[I_{n+1} \otimes (s_0^\dagger s_0)\right] = G_n + \operatorname{diag}\left(0_{np \times nq}, (s_0^\dagger s_0 - I_q)s_{2n+2}\right). \tag{6.85}$$

Using $m \in \mathbb{Z}_{0,2n+1}$, (6.66), (6.67), (6.84) and (6.61), we obtain

$$\mathbf{S}_n\mathbf{S}_n^\sharp G_n \mathbb{S}_n^\sharp \mathbb{S}_n = \left[I_{n+1} \otimes (s_0 s_0^\dagger)\right] G_n \left[I_{n+1} \otimes (s_0^\dagger s_0)\right]$$
$$= \left[G_n + \operatorname{diag}\left(0_{np \times nq}, (s_0 s_0^\dagger - I_p)s_{2n+2}\right)\right]\left[I_{n+1} \otimes (s_0^\dagger s_0)\right]$$
$$= G_n\left[I_{n+1} \otimes (s_0^\dagger s_0)\right] + \operatorname{diag}\left(0_{np \times nq}, (s_0 s_0^\dagger - I_p)s_{2n+2}s_0^\dagger s_0\right)$$
$$= G_n + \operatorname{diag}\left(0_{np \times nq}, s_{2n+2}(s_0^\dagger s_0 - I_q)\right) + \operatorname{diag}\left(0_{np \times nq}, (s_0 s_0^\dagger - I_p)s_{2n+2}s_0^\dagger s_0\right)$$
$$= G_n + \operatorname{diag}(0_{np \times nq}, s_0 s_0^\dagger s_{2n+2}s_0^\dagger s_0 - s_{2n+2}) = G_n + \Gamma_{n;p,q}. \tag{6.86}$$

In view of (6.64) we have $s_0 s_0^\dagger s_j = s_j$ and $s_j s_0^\dagger s_0 = s_j$ for all $j \in \mathbb{Z}_{1,n+1}$. Thus

$$\left[I_{n+1} \otimes (s_0 s_0^\dagger)\right] y_{1,n+1} = y_{1,n+1} \tag{6.87}$$

and

$$z_{1,n+1}\left[I_{n+1} \otimes (s_0^\dagger s_0)\right] = z_{1,n+1}. \tag{6.88}$$

From $m \in \mathbb{Z}_{0,2n+1}$, (6.66), (6.67), (6.87) and (6.88) we get

$$\mathbf{S}_n\mathbf{S}_n^\sharp y_{1,n+1}s_0^\dagger z_{1,n+1}\mathbb{S}_n^\sharp \mathbb{S}_n = \left[I_{n+1} \otimes (s_0 s_0^\dagger)\right] y_{1,n+1}s_0^\dagger z_{1,n+1}\left[I_{n+1} \otimes (s_0^\dagger s_0)\right]$$
$$= y_{1,n+1}s_0^\dagger z_{1,n+1}. \tag{6.89}$$

Using (6.83), $n \in \mathbb{N}$, (6.86), (6.89) and (6.75) we infer

$$G_n - y_{1,n+1}s_0^\dagger z_{1,n+1} + \Gamma_{n;p,q} = (G_n + \Gamma_{n;p,q}) - y_{1,n+1}s_0^\dagger z_{1,n+1}$$
$$= \mathbf{S}_n\mathbf{S}_n^\sharp(G_n - y_{1,n+1}s_0^\dagger z_{1,n+1})\mathbb{S}_n^\sharp \mathbb{S}_n = -\mathbf{S}_n G_n^\sharp \mathbb{S}_n.$$

Thus,

$$\mathrm{rank}(G_n - y_{1,n+1}s_0^\dagger z_{1,n+1} + \Gamma_{n;p,q}) \le \mathrm{rank}\, G_n^\sharp. \tag{6.90}$$

The combination of (6.82) and (6.90) yields the assertion of (c).

(d) This is an immediate consequence of (a). □

Corollary 6.14. *Let* $n \in \mathbb{N}_0$ *and let* $(s_j)_{j=0}^{2n+2} \in \tilde{\mathcal{D}}_{q \times q, 2n+2}$ *be a sequence of Hermitian matrices. Then*

$$G_n^\sharp = -\mathbf{S}_n^\dagger(G_n - y_{1,n+1}s_0^\dagger z_{1,n+1})(\mathbf{S}_n^\dagger)^*. \tag{6.91}$$

Proof. In view of (4.6) we have $\mathbf{S}_n^* = \mathbb{S}_n$. Thus, $\mathbb{S}_n^\dagger = (\mathbf{S}_n^*)^\dagger = (\mathbf{S}_n^\dagger)^*$. Now (6.91) follows from part (a) of Theorem 6.13. □

Corollary 6.15. *Let* $n \in \mathbb{N}_0$ *and* $(s_j)_{j=0}^{2n+2} \in \mathcal{D}_{p \times q, 2n+2}$. *Then:*

(a) *The block Hankel matrices* $G_n^\sharp$ *and* G_n *admit the multiplicative decompositions*

$$G_n^\sharp = -\mathbf{S}_n^\sharp(G_n - y_{1,n+1}s_0^\dagger z_{1,n+1})\mathbb{S}_n^\sharp, \tag{6.92}$$

$$G_n^\sharp = -\mathbf{S}_n^\dagger(G_n - y_{1,n+1}s_0^\dagger z_{1,n+1})\mathbb{S}_n^\dagger \tag{6.93}$$

and

$$G_n = -\mathbf{S}_n \left[G_n^\sharp - y_{1,n+1}^\sharp(s_0^\sharp)^\dagger z_{1,n+1}^\sharp \right] \mathbb{S}_n, \tag{6.94}$$

respectively.

(b) $\mathrm{rank}\, G_n^\sharp = \mathrm{rank}(G_n - y_{1,n+1}s_0^\dagger z_{1,n+1})$.

Proof. In view of Remark 4.17, the equations (6.92) and (6.93) follow from part (a) of Theorem 6.13. Because of Proposition 4.8, we have $s_{2n+2} = s_0 s_0^\dagger s_{2n+2} s_0^\dagger s_0$. Thus, (6.94) and (b) follow from part (b) and part (c) of Theorem 6.13, respectively. □

Corollary 6.16. *Let* $n \in \mathbb{N}_0$ *and* $(s_j)_{j=0}^{2n+2} \in \mathcal{H}_{q,2n+2}^{\ge,\mathrm{e}}$. *Then:*

(a) *The block Hankel matrices* $G_n^\sharp$ *and* G_n *admit the multiplicative decompositions*

$$G_n^\sharp = -\mathbf{S}_n^\sharp(G_n - y_{1,n+1}s_0^\dagger y_{1,n+1}^*)(\mathbf{S}_n^\sharp)^*,$$

$$G_n^\sharp = -\mathbf{S}_n^\dagger(G_n - y_{1,n+1}s_0^\dagger y_{1,n+1}^*)(\mathbf{S}_n^\dagger)^*$$

and

$$G_n = -\mathbf{S}_n \left[G_n^\sharp - y_{1,n+1}^\sharp(s_0^\sharp)^\dagger(y_{1,n+1}^\sharp)^* \right] \mathbf{S}_n^*,$$

respectively.

(b) $\mathrm{rank}\, G_n^\sharp = \mathrm{rank}(G_n - y_{1,n+1}s_0^\dagger z_{1,n+1})$.

Proof. Part (b) of Lemma 3.1 shows that $(s_j)_{j=0}^{2n+2}$ is a sequence from $\mathbb{C}_{\mathrm{H}}^{q \times q}$. Thus, we infer from (4.6) that $\mathbb{S}_n = \mathbf{S}_n^*$. Hence, the assertions follow by combining Proposition 4.24 and Corollary 6.15. □

7. The shortened negative reciprocal sequence corresponding to a sequence from $\tilde{\mathcal{D}}_{p \times q, \kappa}$

The investigations of this section are inspired by Theorem 6.13 and its Corollaries. This will be explained now. Let $n \in \mathbb{N}$ and let $(s_j)_{j=0}^{2n} \in \tilde{\mathcal{D}}_{p \times q, 2n}$. We denote by $(s_j^\sharp)_{j=0}^{2n}$ the reciprocal sequence corresponding to $(s_j)_{j=0}^{2n}$. Then our interest will be concentrated on the shortened negative reciprocal sequence $(-s_{j+2}^\sharp)_{j=0}^{2(n-1)}$. The starting point of our studies are the multiplicative decompositions for the block Hankel matrix

$$G_{n-1}^\sharp := [s_{j+k+2}^\sharp]_{j,k=0}^{n-1} \tag{7.1}$$

which were obtained in part (a) of Theorem 6.13.

Now we are going to consider these multiplicative decompositions under a special aspect. In order to explain this more precisely, we start from the block Hankel matrix

$$H_n := [s_{j+k}]_{j,k=0}^n,$$

and consider its block decomposition

$$H_n = \begin{bmatrix} s_0 & z_{1,n} \\ y_{1,n} & G_{n-1} \end{bmatrix}, \tag{7.2}$$

where $y_{1,n}$ and $z_{1,n}$ are given in (2.2) and where

$$G_{n-1} := [s_{j+k+2}]_{j,k=0}^{n-1}. \tag{7.3}$$

Then we see that the matrix

$$G_{n-1} - y_{1,n} s_0^\dagger z_{1,n} \tag{7.4}$$

(which plays an important role in Theorem 6.13) is a Schur complement in the matrix H_n given via (7.2). Because several important properties of the sequence $(s_j)_{j=0}^{2n}$ have interesting consequences for the Schur complement $G_{n-1} - y_{1,n} s_0^\dagger z_{1,n}$ the application of Theorem 6.13 and its Corollaries promises interesting observations for sequences belonging to various subclasses of $\tilde{\mathcal{D}}_{p \times q, 2n}$. The central goal of this section is a detailed study of this topic. Our main interest is concentrated on classes of sequences with particular Hankel properties which were already considered in Section 3.

Now we are going to investigate the shortened negative reciprocal sequence in the case of sequences of matrices with prescribed Hankel properties.

First we investigate the sequence $(t_j)_{j=0}^{2n}$ introduced in Lemma 4.19 for the case of a given Hankel non-negative definite sequence $(s_j)_{j=0}^{2n}$.

Lemma 7.1. Let $n \in \mathbb{N}$ and let $(s_j)_{j=0}^{2n} \in \mathcal{H}_{q,2n}^{\geq}$. Let $t_j := s_j$ for each $j \in \mathbb{Z}_{0,2n-1}$ and let $t_{2n} := s_0 s_0^\dagger s_{2n} s_0^\dagger s_0$. Then:

(a) $(t_j)_{j=0}^{2n} \in \mathcal{D}_{q \times q, 2n}$.

(b) $t_j = s_0 s_0^\dagger s_j (s_0 s_0^\dagger)^*$ for each $j \in \mathbb{Z}_{0,2n}$.

(c) $(t_j)_{j=0}^{2n} \in \mathcal{H}_{q,2n}^{\geq}$.

(d) *The matrix $H_n^{(t)} := (t_{j+k})_{j,\,k=1}^n$ admits the block decomposition*

$$H_n^{(t)} = \begin{bmatrix} s_0 & z_{1,n} \\ y_{1,n} & G_{n-1} + \Gamma_{n-1;q,\,q} \end{bmatrix},$$

where $y_{1,n}$ and $z_{1,n}$ are given by (2.2) and where $\Gamma_{n-1;q,\,q}$ is defined in (6.61).

(e) $\operatorname{rank} H_n^{(t)} = \operatorname{rank} s_0 + \operatorname{rank}(G_{n-1} + \Gamma_{n-1;q,\,q} - y_{1,n} s_0^\dagger z_{1,n})$.

(f) *If $n = 1$, then $(t_j)_{j=0}^{2n} \in \mathcal{H}_{q,2n}^{\geq,\mathrm{e}}$.*

Proof. (a) Apply part (c) of Lemma 4.19.

(b) In view of part (b) of Lemma 3.2 we have $s_0^* = s_0$. Thus, applying part (a) of Lemma A.1 yields $s_0^\dagger s_0 = s_0 s_0^\dagger$. Hence, taking part (a) of Lemma 4.19 into account, for each $j \in \mathbb{Z}_{0,2n}$, we have

$$t_j = s_0 s_0^\dagger s_j s_0^\dagger s_0 = s_0 s_0^\dagger s_j s_0 s_0^\dagger = s_0 s_0^\dagger s_j (s_0 s_0^\dagger)^*.$$

(c) Because of (b), from Remark 3.3, we see that $(t_j)_{j=0}^{2n} \in \mathcal{H}_{q,2n}^\geq$.

(d) Using part (e) of Lemma 4.19, (6.61) and (7.2), we get

$$H_n^{(t)} = H_n^{(s)} + \operatorname{diag}(0_{nq \times nq}, s_0 s_0^\dagger s_{2n} s_0^\dagger s_0 - s_{2n}) = H_n^{(s)} + \Gamma_{n;q,\,q}$$

$$= \begin{bmatrix} s_0 & z_{1,n} \\ y_{1,n} & G_{n-1} \end{bmatrix} + \operatorname{diag}(0_{q \times q}, \Gamma_{n-1;q,\,q}) = \begin{bmatrix} s_0 & z_{1,n} \\ y_{1,n} & G_{n-1} + \Gamma_{n-1;q,\,q} \end{bmatrix}.$$

(e) In view of (c), we have $H_n^{(t)} \in \mathbb{C}_\geq^{(n+1)q \times (n+1)q}$. Thus, taking (d) into account, we infer the assertion of (e) from a well-known result on non-negative hermitian block matrices (see, e.g., [10, Lemma 1.1.7 and Lemma 1.1.9]).

(f) In view of (c) and

$$\mathcal{R}(t_2) = \mathcal{R}(s_0 s_0^\dagger s_2 s_0^\dagger s_0) \subseteq \mathcal{R}(s_0),$$

the assertion of (f) follows from [12, Lemma 2.7]. □

The following example shows that the assertion of part (f) of Lemma 7.1 is not true for $n = 2$.

Example 7.2. Let $s_0 := I_q$, $s_j := 0_{q \times q}$ for $j \in \{1,2,3\}$ and $s_4 := I_q$. Then $H_2 = \operatorname{diag}(I_q, 0_{q \times q}, I_q)$. Thus, $(s_j)_{j=0}^4 \in \mathcal{H}_{q,4}^\geq \cap \mathcal{D}_{q \times q,4}$, but $(s_j)_{j=0}^4 \notin \mathcal{H}_{q,4}^{\geq,\mathrm{e}}$. Furthermore, it is obvious that using the notation of Lemma 7.1 we have $(s_j)_{j=0}^4 = (t_j)_{j=0}^4$.

Proposition 7.3. *Let $n \in \mathbb{N}$ and let $(s_j)_{j=0}^{2n} \in \mathcal{H}_{q,2n}^\geq$. Then:*

(a) *The block Hankel matrix $-G_{n-1}^\sharp$ is non-negative Hermitian.*

(b) $\operatorname{rank}\left[H_n + \operatorname{diag}(0_{nq \times nq}, s_0 s_0^\dagger s_{2n} s_0^\dagger s_0 - s_{2n})\right] = \operatorname{rank} s_0 + \operatorname{rank}(G_{n-1}^\sharp)$.

Proof. (a) In view of $(s_j)_{j=0}^{2n} \in \mathcal{H}_{q,2n}^\geq$ we see from Proposition 4.25 that $(s_j)_{j=0}^{2n} \in \tilde{\mathcal{D}}_{q \times q,2n}$, whereas part (b) of Lemma 3.2 shows that $(s_j)_{j=0}^{2n}$ is a sequence from $\mathbb{C}_\mathrm{H}^{q \times q}$. Thus, Corollary 6.14 yields

$$-G_{n-1}^\sharp = \mathbf{S}_{n-1}^\dagger (G_{n-1} - y_{1,n} s_0^\dagger z_{1,n})(\mathbf{S}_{n-1}^\dagger)^*. \tag{7.5}$$

From $H_n \in \mathbb{C}_{\geq}^{(n+1)q \times (n+1)q}$ we infer from (7.2) that

$$G_{n-1} - y_{1,n} s_0^\dagger z_{1,n} \in \mathbb{C}_{\geq}^{nq \times nq}. \tag{7.6}$$

Applying (7.5) and (7.6) we get $-G_{n-1}^\sharp \in \mathbb{C}_{\geq}^{nq \times nq}$.

(b) Because of $(s_j)_{j=0}^{2n} \in \tilde{\mathcal{D}}_{q \times q, 2n}$ we see from part (b) of Theorem 6.13 that

$$\operatorname{rank} G_{n-1}^\sharp = \operatorname{rank}(G_{n-1} - y_{1,n} s_0^\dagger z_{1,n} + \Gamma_{n-1;q,q}). \tag{7.7}$$

We set $t_j := s_j$ for each $j \in \mathbb{Z}_{0,2n-1}$ and $t_{2n} := s_0 s_0^\dagger s_{2n} s_0^\dagger s_0$. Then part (d) of Lemma 4.19 yields

$$H_n^{(t)} = H_n + \operatorname{diag}(0_{nq \times nq}, s_0 s_0^\dagger s_{2n} s_0^\dagger s_0 - s_{2n}), \tag{7.8}$$

whereas part (e) of Lemma 7.1 implies

$$\operatorname{rank} H_n^{(t)} = \operatorname{rank} s_0 + \operatorname{rank}(G_{n-1} - y_{1,n} s_0^\dagger z_{1,n} + \Gamma_{n-1;q,q}). \tag{7.9}$$

The combination of (7.8), (7.9) and (7.7) yields now the assertion of (b). $\qquad\square$

Corollary 7.4. *Let $n \in \mathbb{N}$ and let $(s_j)_{j=0}^{2n} \in \mathcal{H}_{q,2n}^{\geq} \cap \mathcal{D}_{q \times q, 2n}$. Then*

$$\operatorname{rank} H_n = \operatorname{rank} s_0 + \operatorname{rank} G_{n-1}^\sharp.$$

Proof. In view of $(s_j)_{j=0}^{2n} \in \mathcal{D}_{q \times q, 2n}$, the application of Proposition 4.8 provides $s_0 s_0^\dagger s_{2n} s_0^\dagger s_0 = s_{2n}$. Thus, the assertion follows from part (b) of Proposition 7.3. $\qquad\square$

Corollary 7.5. *Let $n \in \mathbb{N}$ and let $(s_j)_{j=0}^{2n} \in \mathcal{H}_{q,2n}^{\geq,\mathrm{e}}$. Then:*

(a) *The block Hankel matrix $-G_{n-1}^\sharp$ is non-negative Hermitian.*

(b) $\operatorname{rank} H_n = \operatorname{rank} s_0 + \operatorname{rank} G_{n-1}^\sharp.$

Proof. Clearly, $(s_j)_{j=0}^{2n} \in \mathcal{H}_{q,2n}^{\geq}$, whereas Proposition 4.24 implies that $(s_j)_{j=0}^{2n}$ belongs to $\mathcal{D}_{q \times q, 2n}$. Thus, part (a) of Proposition 7.3 and Corollary 7.4 yield the assertions of (a) and (b), respectively. $\qquad\square$

Proposition 7.6. *Let $\kappa \in \mathbb{N} \cup \{+\infty\}$ and let $(s_j)_{j=0}^{2\kappa} \in \mathcal{H}_{q,2\kappa}^{\geq}$. Then $(-s_{j+2}^\sharp)_{j=0}^{2(\kappa-1)}$ belongs to $\mathcal{H}_{q,2(\kappa-1)}^{\geq}$.*

Proof. Apply part (a) of Proposition 7.3. $\qquad\square$

Proposition 7.7. *Let $m \in \mathbb{Z}_{2,\infty}$ and let $(s_j)_{j=0}^{m} \in \mathcal{H}_{q,m}^{\geq,\mathrm{e}}$. Then $(-s_{j+2}^\sharp)_{j=0}^{m-2} \in \mathcal{H}_{q,m-2}^{\geq,\mathrm{e}}$.*

Proof. We first consider the case that $m = 2n$ with some $n \in \mathbb{N}$. Because of $(s_j)_{j=0}^{2n} \in \mathcal{H}_{q,2n}^{\geq,\mathrm{e}}$, there are complex $q \times q$-matrices s_{2n+1} and s_{2n+2} such that $(s_j)_{j=0}^{2(n+1)} \in \mathcal{H}_{q,2(n+1)}^{\geq}$. Then Proposition 7.6 yields that the reciprocal sequence $(\tilde{s}_j^\sharp)_{j=0}^{2(n+1)}$ corresponding to $(s_j)_{j=0}^{2(n+1)}$ fulfills $(-\tilde{s}_{j+2}^\sharp)_{j=0}^{2n} \in \mathcal{H}_{q,2n}^{\geq}$. Thus, the sequence $(-\tilde{s}_{j+2}^\sharp)_{j=0}^{2(n-1)}$ belongs to $\mathcal{H}_{q,2(n-1)}^{\geq,\mathrm{e}}$. In view of Remark 4.2, we have $\tilde{s}_j^\sharp = s_j^\sharp$

for $j \in \mathbb{Z}_{0,2n}$. This completes the proof in this case. The case $m = 2n + 1$ with some $n \in \mathbb{N}$ can be treated similarly. $\square$

Proposition 7.8. *Let* $\kappa \in \mathbb{N} \cup \{+\infty\}$ *and let* $(s_j)_{j=0}^{2\kappa} \in \mathcal{H}_{q,2\kappa}^{>}$. *Then* $(-s_{j+2}^{\sharp})_{j=0}^{2(\kappa-1)}$ *belongs to* $\mathcal{H}_{q,2(\kappa-1)}^{>}$.

Proof. Let $n \in \mathbb{Z}_{1,\kappa}$. Because $(s_j)_{j=0}^{2n}$ belongs to $\mathcal{H}_{q,2n}^{>}$ and [12, Remark 2.8], we have $(s_j)_{j=0}^{2n} \in \mathcal{H}_{q,2n}^{\geq,e}$. Thus, Corollary 7.5 implies

$$\operatorname{rank} H_n = \operatorname{rank} s_0 + \operatorname{rank} G_{n-1}^{\sharp}. \tag{7.10}$$

In view of $(s_j)_{j=0}^{2n} \in \mathcal{H}_{q,2n}^{>}$, the matrix H_n is positive Hermitian and, consequently, non-singular. Thus, (7.10) implies $\det G_{n-1}^{\sharp} \neq 0$. Because of part (a) of Proposition 7.3, the matrix $-G_{n-1}^{\sharp}$ is non-negative Hermitian we get $-G_{n-1}^{\sharp} \in \mathbb{C}_{>}^{nq \times nq}$. This completes the proof. $\square$

Proposition 7.9. *Let* $n \in \mathbb{N}$, *and let* $(s_j)_{j=0}^{2n} \in \mathcal{H}_{q,2n}^{\geq,\mathrm{cd}}$. *Then* $(-s_{j+2}^{\sharp})_{j=0}^{2(n-1)}$ *belongs to* $\mathcal{H}_{q,2(n-1)}^{\geq,\mathrm{cd}}$.

Proof. Because of $(s_j)_{j=0}^{2n} \in \mathcal{H}_{q,2n}^{\geq,\mathrm{cd}}$ and [12, Corollary 2.14], we have $(s_j)_{j=0}^{2n} \in \mathcal{H}_{q,2n}^{\geq,e}$. Part (b) of Corollary 7.5 provides us then

$$\operatorname{rank} H_n = \operatorname{rank} s_0 + \operatorname{rank}(-G_{n-1}^{\sharp}). \tag{7.11}$$

Since $(s_j)_{j=0}^{2n}$ belongs to $\mathcal{H}_{q,2n}^{\geq,\mathrm{cd}}$, the application of [12, Remark 2.1 (a)] in combination with (1.2) yields $\operatorname{rank} H_n = \operatorname{rank} H_{n-1}$. Taking additionally (7.11) into account, we get

$$\operatorname{rank}(-G_{n-1}^{\sharp}) = \operatorname{rank} H_n - \operatorname{rank} s_0 = \operatorname{rank} H_{n-1} - \operatorname{rank} s_0. \tag{7.12}$$

In the case $n = 1$ from (7.12) we obtain $\operatorname{rank}(-s_2^{\sharp}) = \operatorname{rank} H_0 - \operatorname{rank} s_0 = 0$, i.e., $-s_2^{\sharp} = 0_{q \times q}$, and consequently $(-s_{j+2}^{\sharp})_{j=0}^{0} \in \mathcal{H}_{q,0}^{\geq,\mathrm{cd}}$.

We now consider the case $n > 1$. Because of $(s_j)_{j=0}^{2n} \in \mathcal{H}_{q,2n}^{\geq}$ and part (a) of Proposition 7.3, we get $-G_{n-1}^{\sharp} \in \mathbb{C}_{\geq}^{nq \times nq}$. In view of $(s_j)_{j=0}^{2n} \in \mathcal{H}_{q,2n}^{\geq}$, we have $(s_j)_{j=0}^{2(n-1)} \in \mathcal{H}_{q,2(n-1)}^{\geq,e}$. Part (b) of Corollary 7.5 provides us then $\operatorname{rank} G_{n-2}^{\sharp} = \operatorname{rank} H_{n-1} - \operatorname{rank} s_0$. This equation and (7.12) yield $\operatorname{rank}(-G_{n-1}^{\sharp}) - \operatorname{rank}(-G_{n-2}^{\sharp}) = 0$. Since $-G_{n-1}^{\sharp} \in \mathbb{C}_{\geq}^{nq \times nq}$ and [12, Remark 2.1 (a)], this implies $(-s_{j+2}^{\sharp})_{j=0}^{2(n-1)} \in \mathcal{H}_{q,2(n-1)}^{\geq,\mathrm{cd}}$. $\square$

Proposition 7.10. *Let* $n \in \mathbb{N}_0$ *and* $(s_j)_{j=0}^{\infty} \in \mathcal{H}_{q,\infty}^{\geq,\mathrm{cd},n}$. *Then the sequence* $(-s_{j+2}^{\sharp})_{j=0}^{\infty}$ *belongs to* $\mathcal{H}_{q,\infty}^{\geq,\mathrm{cd},\max\{0,n-1\}}$.

Proof. We first consider the case $n = 0$. Because of (2.1) we have then $s_0 = 0_{q \times q}$ which, in view of Definition 4.1, implies $-s_k^{\sharp} = 0_{q \times q}$ for every $k \in \mathbb{Z}_{2,+\infty}$. Thus, $(-s_{j+2}^{\sharp})_{j=0}^{\infty}$ belongs to $\mathcal{H}_{q,\infty}^{\geq}$ and is completely degenerate of order 0. In the case $n > 0$ the assertion follows from Proposition 7.6, Remark 4.2 and Proposition 7.9. □

The following example shows that in the case of $m \in \mathbb{Z}_{2,\infty}$ and a sequence $(s_j)_{j=0}^m \in \mathcal{D}_{p \times q,m}$ the sequence $(-s_{j+2}^{\sharp})_{j=0}^{m-2}$ does not necessarily belong to one of the classes $\mathcal{D}_{p \times q,m-2}$ or $\tilde{\mathcal{D}}_{p \times q,m-2}$.

Example 7.11. Let $s_0 := I_q$, $s_1 := 0_{q \times q}$, $s_2 := 0_{q \times q}$, $s_3 := I_q$, and $s_4 := 0_{q \times q}$. In view of Example 4.6 we have then $(s_j)_{j=0}^4 \in \mathcal{D}_{q,4}$. Using Definition 4.1 we obtain $-s_2^{\sharp} = 0_{q \times q}$, $-s_3^{\sharp} = I_q$, and $-s_4^{\sharp} = 0_{q \times q}$ and, in view of Definition 4.5 and Definition 4.16, in particular $(-s_{j+2}^{\sharp})_{j=0}^2 \notin \tilde{\mathcal{D}}_{q,2}$.

Now we turn our attention to a further class of sequences of complex matrices, which is characterized by a remarkable behaviour of the associated block Hankel matrices.

Definition 7.12. Let $\kappa \in \mathbb{N}_0 \cup \{+\infty\}$ and let $(s_j)_{j=0}^{2\kappa}$ be a sequence from $\mathbb{C}^{q \times q}$. Then $(s_j)_{j=0}^{2\kappa}$ is called *strictly regular* if $\det H_n \neq 0$ for each $n \in \mathbb{Z}_{0,\kappa}$. The notation $\mathcal{H}_{q,2\kappa}^{\mathrm{sr}}$ stands for the set of all strictly regular sequences $(s_j)_{j=0}^{2\kappa}$ from $\mathbb{C}^{q \times q}$.

Remark 7.13. Let $\kappa \in \mathbb{N}_0 \cup \{+\infty\}$ and let $(s_j)_{j=0}^{2\kappa} \in \mathcal{H}_{q,2\kappa}^{\mathrm{sr}}$. Then $(s_j)_{j=0}^{2n} \in \mathcal{H}_{q,2n}^{\mathrm{sr}}$ for all $n \in \mathbb{Z}_{0,\kappa}$.

Remark 7.14. Let $\kappa \in \mathbb{N}_0 \cup \{+\infty\}$. Then Example 4.6 shows that $\mathcal{H}_{q,2\kappa}^{\mathrm{sr}} \subseteq \mathcal{D}_{q \times q,2\kappa}$.

Remark 7.15. For each $\kappa \in \mathbb{N}_0 \cup \{+\infty\}$, the inclusion $\mathcal{H}_{q,2\kappa}^{>} \subseteq \mathcal{H}_{q,2\kappa}^{\mathrm{sr}}$ holds.

Lemma 7.16. *Let $n \in \mathbb{N}$ and let $(s_j)_{j=0}^{2n} \in \mathcal{H}_{q,2n}^{\mathrm{sr}}$. Then:*

(a) $\det H_n = \det s_0 \cdot \det(G_{n-1} - y_{1,n} s_0^{\dagger} z_{1,n})$.

(b) $\det(G_{n-1}^{\sharp}) \neq 0$.

Proof. (a) By assumption, we have $\det s_0 \neq 0$. Thus, part (a) is an immediate consequence of the Schur-Frobenius formula (see, e.g., [10, Lemma 1.1.7]).

(b) By assumption we have $\det H_n \neq 0$. Thus, part (a) implies

$$\det(G_{n-1} - y_{1,n} s_0^{\dagger} z_{1,n}) \neq 0. \tag{7.13}$$

Remark 7.14 gives $(s_j)_{j=0}^{2n} \in \mathcal{D}_{q \times q,2n}$. Hence, part (b) of Corollary 6.16 and (7.13) yield

$$\operatorname{rank} G_{n-1}^{\sharp} = \operatorname{rank}(G_{n-1} - y_{1,n} s_0^{\dagger} z_{1,n}). \tag{7.14}$$

This implies the assertion of (b). □

Proposition 7.17. *Let $\kappa \in \mathbb{N} \cup \{+\infty\}$ and let $(s_j)_{j=0}^{2\kappa} \in \mathcal{H}_{q,2\kappa}^{\mathrm{sr}}$. Then $(-s_{j+2}^{\sharp})_{j=0}^{2(\kappa-1)}$ belongs to $\mathcal{H}_{q,2(\kappa-1)}^{\mathrm{sr}}$.*

Proof. Use part (b) of Lemma 7.16. □

In our subsequent considerations it will be advantageous to replace the sequence $(-s_{j+2}^{\sharp})_{j=0}^{2(n-1)}$ by a slightly modified sequence. This will be done in the next section.

8. The first Schur transform of a sequence of $p \times q$-matrices

In [9], Chen and Hu presented a Schur-type algorithm to parametrize the solution set of the matricial truncated Hamburger moment problem. In this way, starting with an arbitrary sequence $(s_j)_{j=0}^{2n} \in \mathcal{H}_{q,2n}^{\geq}$ containing the given power moments of the moment problem in question, they constructed a particular sequence $(s_j^{(1)})_{j=0}^{2(n-1)}$ belonging to $\mathcal{H}_{q,2(n-1)}^{\geq,e}$. This sequence corresponds to the given data of a truncated Hamburger moment problem which is connected by a particular explicit interrelation with the original moment problem. In this section, we are going to discuss in detail this sequence $(s_j^{(1)})_{j=0}^{2(n-1)}$ in an alternative and more general setting. More precisely, we will construct the sequence $(s_j^{(1)})_{j=0}^{2(n-1)}$ by an appropriate normalization of the sequence $(-s_{j+2}^{\sharp})_{j=0}^{2(n-1)}$, which was studied in Section 7. We will verify that, in the special case of a given sequence $(s_j)_{j=0}^{2n} \in \mathcal{H}_{q,2n}^{\geq,e}$, our Schur-type procedure produces exactly the sequence $(s_j^{(1)})_{j=0}^{2(n-1)}$ which was obtained by Chen/Hu [9] in a different way.

The following notion is one of the central objects in this paper. It describes the basic step of the Schur-type algorithm which will be developed in Section 9.

Definition 8.1. Let $\kappa \in \mathbb{Z}_{2,\infty} \cup \{+\infty\}$, let $(s_j)_{j=0}^{\kappa}$ be a sequence of complex $p \times q$ matrices, and let $(s_j^{\sharp})_{j=0}^{\kappa}$ be the reciprocal sequence corresponding to $(s_j)_{j=0}^{\kappa}$. Then the sequence $(s_j^{(1)})_{j=0}^{\kappa-2}$ defined for each $j \in \mathbb{Z}_{0,\kappa-2}$ by

$$s_j^{(1)} := -s_0 s_{j+2}^{\sharp} s_0$$

is called *the first Schur transform of* $(s_j)_{j=0}^{\kappa}$.

The main theme of this section is the investigation of the first Schur transform of a (finite or infinite) sequence of complex matrices.

Remark 8.2. Let $\kappa \in \mathbb{Z}_{2,\infty} \cup \{+\infty\}$, let $(s_j)_{j=0}^{\kappa}$ be a sequence of complex $p \times q$-matrices, and let $(s_j^{(1)})_{k=0}^{\kappa-2}$ be the first Schur transform of $(s_j)_{j=0}^{\kappa}$. In view of Definition 8.1 and Remark 4.2, for each $l \in \mathbb{Z}_{2,\kappa}$, then $(s_j^{(1)})_{j=0}^{l-2}$ is the first Schur transform of $(s_j)_{j=0}^{l}$.

If $\kappa \in \mathbb{Z}_{2,\infty} \cup \{+\infty\}$ and if $(s_j)_{j=0}^{\kappa}$ is a sequence of complex $p \times q$-matrices, then we will continue to use the notations $(s_j^{(1)})_{j=0}^{\kappa-2}$ and $(s_j^{\sharp})_{j=0}^{\kappa}$ to designate the first Schur transform of $(s_j)_{j=0}^{\kappa}$ and the reciprocal sequence corresponding to $(s_j)_{j=0}^{\kappa}$, respectively.

Example 8.3. Let $\kappa \in \mathbb{Z}_{2,\infty} \cup \{+\infty\}$, and let $(s_j)_{j=0}^{\kappa}$ be a sequence of complex $p \times q$-matrices satisfying $s_j = 0_{p \times q}$ for each $j \in \mathbb{Z}_{1,\kappa}$. Then Example 4.3 shows that $s_j^{(1)} = 0_{p \times q}$ for each $j \in \mathbb{Z}_{0,\kappa-2}$.

Remark 8.4. Let $\kappa \in \mathbb{Z}_{2,\infty} \cup \{+\infty\}$ and let $(s_j)_{j=0}^{\kappa}$ be a sequence of complex $p \times q$-matrices. In view of Definition 8.1, for every $j \in \mathbb{Z}_{0,\kappa-2}$, then $\mathcal{N}(s_0) \subseteq \mathcal{N}(s_j^{(1)})$ and $\mathcal{R}(s_j^{(1)}) \subseteq \mathcal{R}(s_0)$.

Remark 8.5. Let $\kappa \in \mathbb{Z}_{2,\infty} \cup \{+\infty\}$ and let $(s_j)_{j=0}^{\kappa}$ be a sequence of complex $p \times q$-matrices.

(a) From Definition 8.1 one can easily see that

$$s_j^{(1)} s_0^{\dagger} s_0 = s_j^{(1)} \qquad \text{and} \qquad s_0 s_0^{\dagger} s_j^{(1)} = s_j^{(1)}$$

hold for each $j \in \mathbb{Z}_{0,\kappa-2}$.

(b) Part (b) of Proposition 4.9 implies $s_{j+2}^{\sharp} = -s_0^{\dagger} s_j^{(1)} s_0^{\dagger}$ for each $j \in \mathbb{Z}_{0,\kappa-2}$.

Remark 8.6. Let $\kappa \in \mathbb{N} \cup \{+\infty\}$ with $\kappa \geq 2$ and let $(s_j)_{j=0}^{\kappa}$ be a sequence of complex $p \times q$-matrices. Further, let $o, r \in \mathbb{N}$, $U \in \mathbb{C}^{o \times p}$ with $U^*U = I_p$ and $V \in \mathbb{C}^{q \times r}$ with $VV^* = I_q$. Then, [16, Lemma 5.19] implies $(Us_jV)^{(1)} = Us_j^{(1)}V$ for every $j \in \mathbb{Z}_{0,\kappa-2}$.

Remark 8.7. Let $l \in \mathbb{N}$, $(p_k)_{k=1}^{l}$ and $(q_k)_{k=1}^{l}$ be sequences from $\mathbb{N}$, $\kappa \in \mathbb{Z}_{2,+\infty} \cup \{+\infty\}$ and for all $k \in \mathbb{Z}_{1,l}$ let $(s_{k;j})_{j=0}^{\kappa}$ be a sequence from $\mathbb{C}^{p_k \times q_k}$. Then [16, Remark 5.20] implies $[\text{diag}\,(s_{k;j})_{k=1}^{l}]^{(1)} = \text{diag}\,(s_{k;j}^{(1)})_{k=1}^{l}$ for every $j \in \mathbb{Z}_{0,\kappa-2}$.

Remark 8.8. Let $n \in \mathbb{N}$ and let $(s_j)_{j=0}^{2n} \in \mathcal{H}_{q,2n}^{\geq,e}$ be such that $1 \leq \text{rank}\,s_0 \leq q - 1$. Let $r := \text{rank}\,s_0$. According to part (c) of Corollary 3.16, let the unitary $q \times q$ matrix U and the sequence $(\tilde{s}_j)_{j=0}^{2n} \in \mathcal{H}_{r,2n}^{\geq,e}$ be such that for each $j \in \mathbb{Z}_{0,2n}$ the relation $s_j = U^* \cdot \text{diag}(\tilde{s}_j, 0_{(q-r)\times(q-r)}) \cdot U$ holds. Then the application of Remark 8.6, Remark 8.7, and Example 8.3 shows that $s_j^{(1)} = U^* \cdot \text{diag}(\tilde{s}_j^{(1)}, 0_{(q-r)\times(q-r)}) \cdot U$ for every $j \in \mathbb{Z}_{0,2(n-1)}$.

Remark 8.9. Let $\kappa \in \mathbb{Z}_{2,\infty} \cup \{+\infty\}$ and let $(s_j)_{j=0}^{\kappa}$ and $(t_j)_{j=0}^{\kappa}$ be sequences from $\mathbb{C}^{p \times q}$ such that $(s_j^{\sharp})_{j=0}^{\kappa} = (t_j^{\sharp})_{j=0}^{\kappa}$ is satisfied. Then Definition 4.1 and part (a) of Lemma A.1 show that $s_0 = t_0$. Thus, the identity $(s_j^{(1)})_{j=0}^{\kappa-2} = (t_j^{(1)})_{j=0}^{\kappa-2}$ holds for the first Schur transforms of $(s_j)_{j=0}^{\kappa}$ and $(t_j)_{j=0}^{\kappa}$.

Remark 8.10. Let $\kappa \in \mathbb{Z}_{2,\infty} \cup \{+\infty\}$ and let $(s_j)_{j=0}^{\kappa}$ be a sequence from $\mathbb{C}^{p \times q}$. For each $j \in \mathbb{Z}_{0,\kappa}$ let $\tilde{s}_j := s_0 s_0^{\dagger} s_j s_0^{\dagger} s_0$. Then the combination of part (a) of Proposition 4.10 and Remark 8.9 yields the identity $(s_j^{(1)})_{j=0}^{\kappa-2} = (\tilde{s}_j^{(1)})_{j=0}^{\kappa-2}$ for the first Schur transforms of $(s_j)_{j=0}^{\kappa}$ and $(\tilde{s}_j)_{j=0}^{\kappa}$.

Remark 8.11. Let $m \in \mathbb{Z}_{2,\infty}$ and let $(s_j)_{j=0}^{m} \in \tilde{\mathcal{D}}_{p \times q,m}$. Let $t_j := s_j$ for each $j \in \mathbb{Z}_{0,m-1}$ and let $t_m := s_0 s_0^{\dagger} s_m s_0^{\dagger} s_0$. Then the combination of part (b) of Lemma 4.19 and Remark 8.9 provides us the identity $(s_j^{(1)})_{j=0}^{m-2} = (t_j^{(1)})_{j=0}^{m-2}$ for the first Schur transforms of $(s_j)_{j=0}^{m}$ and $(t_j)_{j=0}^{m}$.

Now we modify the results stated in Proposition 7.6, Proposition 7.7, Proposition 7.8, and Proposition 7.9.

Proposition 8.12. *Let* $n \in \mathbb{N}$ *and let* $(s_j)_{j=0}^{2n} \in \mathcal{H}_{q,2n}^{\geq}$. *Then* $(s_j^{(1)})_{j=0}^{2(n-1)} \in \mathcal{H}_{q,2(n-1)}^{\geq}$.

Proof. Part (b) of Lemma 3.2 implies $s_0^* = s_0$. Thus, in view of Definition 8.1 and Proposition 7.6, the application of Remark 3.3 yields the assertion. □

Proposition 8.13. *Let* $m \in \mathbb{N}$ *with* $m \geq 2$, *and let* $(s_j)_{j=0}^m \in \mathcal{H}_{q,m}^{\geq,e}$. *Then* $(s_j^{(1)})_{j=0}^{m-2} \in \mathcal{H}_{q,m-2}^{\geq,e}$.

Proof. From part (b) of Lemma 3.1 we see that $s_0^* = s_0$. Hence, in view of Definition 8.1 and Proposition 7.7, the application of Remark 3.4 yields the assertion. □

Proposition 8.14. *Let* $n \in \mathbb{N}$ *and* $(s_j)_{j=0}^{2n} \in \mathcal{H}_{q,2n}^{>}$. *Then* $(s_j^{(1)})_{j=0}^{2(n-1)} \in \mathcal{H}_{q,2(n-1)}^{>}$.

Proof. The choice of $(s_j)_{j=0}^{2n}$ particularly implies that s_0 is a non-singular and Hermitian matrix. Thus, in view of Definition 8.1 and Proposition 7.8, the application of Remark 3.5 completes the proof. □

Proposition 8.15. *Let* $n \in \mathbb{N}$ *and* $(s_j)_{j=0}^{2n} \in \mathcal{H}_{q,2n}^{\geq,cd}$. *Then* $(s_j^{(1)})_{j=0}^{2(n-1)} \in \mathcal{H}_{q,2(n-1)}^{\geq,cd}$.

Proof. Since $(s_j)_{j=0}^{2n} \in \mathcal{H}_{q,2n}^{\geq}$ holds, the application of part (b) of Lemma 3.2 gives $s_0^* = s_0$. Hence, in view of part (a) of Lemma A.1 and Definition 4.1 we get

$$\mathcal{N}(s_0) = \mathcal{N}(s_0^\dagger) = \mathcal{N}(s_0^\sharp).$$

Because of part (a) of Proposition 4.9 and Definition 4.5 we obtain in the case $n - 1 > 0$ that

$$\mathcal{N}(s_0) = \mathcal{N}(s_0^\sharp) \subseteq \bigcap_{k=0}^{2n} \mathcal{N}(s_k^\sharp) \subseteq \bigcap_{j=0}^{2(n-1)-1} \mathcal{N}(s_{j+2}^\sharp) = \bigcap_{j=0}^{2(n-1)-1} \mathcal{N}(-s_{j+2}^\sharp).$$

Thus, in view of Definition 8.1 and Proposition 7.9, the application of Proposition 3.12 yields the assertion. □

In view of Definition 4.16 and Definition 4.5, the following example shows that the first Schur transform of a sequence belonging to $\tilde{\mathcal{D}}_{q,m}$ or $\mathcal{D}_{q,m}$ does not necessarily belong to $\tilde{\mathcal{D}}_{q,m-1}$ or $\mathcal{D}_{q,m-1}$.

Example 8.16. Let $s_0 := I_q$, $s_1 := 0_{q \times q}$, $s_2 := 0_{q \times q}$, $s_3 := I_q$, and $s_4 := 0_{q \times q}$. From Example 4.6 we know that $(s_j)_{j=0}^4 \in \mathcal{D}_{q,4}$. Furthermore, in view of Definition 8.1 and Example 7.11, we have $s_0^{(1)} = 0_{q \times q}$, $s_1^{(1)} = I_q$, and $s_2^{(1)} = 0_{q \times q}$. In particular, taking Definition 4.5 and Definition 4.16 into account we see that $(s_j^{(1)})_{j=0}^2 \notin \tilde{\mathcal{D}}_{q,2}$.

Now we will study the block Hankel matrices associated with the first Schur transform of a sequence from $\mathbb{C}^{p \times q}$.

Lemma 8.17. *Let $n \in \mathbb{N}$, and let $(s_j)_{j=0}^{2n}$ be a sequence of complex $p \times q$-matrices. Let*

$$H_{n-1}^{(1)} := [s_{j+k}^{(1)}]_{j,k=0}^{n-1} \tag{8.1}$$

and let $G_{n-1}^{\sharp}$ be defined in (7.1). Then

(a) $H_{n-1}^{(1)} = -(I_n \otimes s_0)G_{n-1}^{\sharp}(I_n \otimes s_0).$
(b) $G_{n-1}^{\sharp} = -(I_n \otimes s_0^{\dagger})H_{n-1}^{(1)}(I_n \otimes s_0^{\dagger}).$
(c) $\operatorname{rank} H_{n-1}^{(1)} = \operatorname{rank} G_{n-1}^{\sharp}.$

Proof. The assertion of (a) is clear from Definition 8.1. Taking part (b) of Proposition 4.9 into account part (b) follows from (a). Finally, part (c) follows by combining (a) and (b). □

Now, for an arbitrary $n \in \mathbb{N}$, we consider a sequence $(s_j)_{j=0}^{2n} \in \tilde{\mathcal{D}}_{p \times q, 2n}$. We are going to establish various multiplicative representations for the block Hankel matrix $H_{n-1}^{(1)}$ defined in (8.1). Before doing this we still need some preparations.

For each $m \in \mathbb{N}_0$ and each sequence $(s_j)_{j=0}^{m}$ of complex $p \times q$-matrices, let

$$\Delta_m := (I_{m+1} \otimes s_0)\mathbf{S}_m^{\dagger} + \left[I_{m+1} \otimes (I_p - s_0 s_0^{\dagger})\right] \tag{8.2}$$

and let

$$\nabla_m := \mathbb{S}_m^{\dagger}(I_{m+1} \otimes s_0) + \left[I_{m+1} \otimes (I_q - s_0^{\dagger} s_0)\right], \tag{8.3}$$

where $\mathbf{S}_m$ and $\mathbb{S}_m$ are the block Toeplitz matrices given by (4.3) and (4.4), respectively. The following result shows that under certain conditions the matrices Δ_m and ∇_m belong to the sets $\mathfrak{L}_{p,m}$ and $\mathfrak{U}_{q,m}$, respectively, which are introduced in Definition B.1.

Lemma 8.18. *Let $\kappa \in \mathbb{N}_0 \cup \{+\infty\}$ and let $(s_j)_{j=0}^{\kappa} \in \mathcal{D}_{p \times q, \kappa}$. For each $m \in \mathbb{Z}_{0,\kappa}$, then the matrix Δ_m defined by (8.2) is a non-singular block Toeplitz matrix which admits the $p \times p$ block representation*

$$\Delta_m = \begin{cases} I_p & \text{if } m = 0 \\[2mm] \begin{bmatrix} I_p & 0 & 0 & \cdots & 0 & 0 \\ s_0 s_1^{\sharp} & I_p & 0 & \cdots & 0 & 0 \\ s_0 s_2^{\sharp} & s_0 s_1^{\sharp} & I_p & \cdots & 0 & 0 \\ \vdots & \vdots & \vdots & \ddots & \vdots & \vdots \\ s_0 s_{m-1}^{\sharp} & s_0 s_{m-2}^{\sharp} & s_0 s_{m-3}^{\sharp} & \cdots & I_p & 0 \\ s_0 s_m^{\sharp} & s_0 s_{m-1}^{\sharp} & s_0 s_{m-2}^{\sharp} & \cdots & s_0 s_1^{\sharp} & I_p \end{bmatrix} & \text{if } m > 0 \end{cases}$$

and which in particular belongs to $\mathfrak{L}_{p,m}$, whereas the matrix ∇_m defined by (8.3) is a non-singular block Toeplitz matrix which admits the $q \times q$ block representation

$$
\nabla_m = \begin{cases}
I_q & \text{if } m = 0 \\[2mm]
\begin{bmatrix}
I_q & s_1^\sharp s_0 & s_2^\sharp s_0 & \cdots & s_{m-1}^\sharp s_0 & s_m^\sharp s_0 \\
0 & I_q & s_1^\sharp s_0 & \cdots & s_{m-2}^\sharp s_0 & s_{m-1}^\sharp s_0 \\
0 & 0 & I_q & \cdots & s_{m-3}^\sharp s_0 & s_{m-2}^\sharp s_0 \\
\vdots & \vdots & \vdots & \ddots & \vdots & \vdots \\
0 & 0 & 0 & \cdots & I_q & s_1^\sharp s_0 \\
0 & 0 & 0 & \cdots & 0 & I_q
\end{bmatrix} & \text{if } m > 0
\end{cases}
$$

and which in particular belongs to $\mathfrak{U}_{q,m}$.

Proof. Use Proposition 4.14 and $s_0^\sharp = s_0^\dagger$ which follows from Definition 4.1. $\qquad\square$

Proposition 8.19. *Let $n \in \mathbb{N}$ and let $(s_j)_{j=0}^{2n} \in \tilde{\mathcal{D}}_{p\times q,2n}$. Then the block Hankel matrix $H_{n-1}^{(1)}$ defined in (8.1) admits the representations*

$$
H_{n-1}^{(1)} = (I_n \otimes s_0)\mathbf{S}_{n-1}^\dagger(G_{n-1} - y_{1,n}s_0^\dagger z_{1,n})\mathbb{S}_{n-1}^\dagger(I_n \otimes s_0) \tag{8.4}
$$

and

$$
H_{n-1}^{(1)} = \Delta_{n-1}(G_{n-1} - y_{1,n}s_0^\dagger z_{1,n} + \Gamma_{n-1;q,q})\nabla_{n-1}. \tag{8.5}
$$

Furthermore,

$$
\operatorname{rank} H_{n-1}^{(1)} = \operatorname{rank}(G_{n-1} - y_{1,n}s_0^\dagger z_{1,n} + \Gamma_{n-1;q,q}). \tag{8.6}
$$

Proof. Combining part (a) of Lemma 8.17 and part (a) of Theorem 6.13 we get (8.4). Because of $(s_j)_{j=0}^{2n} \in \tilde{\mathcal{D}}_{p\times q,2n}$ and Definition 4.16, we have

$$
(s_j)_{j=0}^{2n-1} \in \mathcal{D}_{p\times q,2n-1}. \tag{8.7}
$$

From (8.7) and Proposition 4.14 we obtain then $\mathbf{S}_{n-1}^\dagger = \mathbf{S}_{n-1}^\sharp$ and $\mathbb{S}_{n-1}^\dagger = \mathbb{S}_{n-1}^\sharp$. Thus, taking Lemma 6.12 into account, we obtain

$$
\mathbf{S}_{n-1}^\dagger \Gamma_{n-1;q,q}\mathbb{S}_{n-1}^\dagger = 0_{nq\times np}.
$$

Hence, from (8.4) we get

$$
H_{n-1}^{(1)} = (I_n \otimes s_0)\mathbf{S}_{n-1}^\dagger(G_{n-1} - y_{1,n}s_0^\dagger z_{1,n} + \Gamma_{n-1;q,q})\mathbb{S}_{n-1}^\dagger(I_n \otimes s_0). \tag{8.8}
$$

Because of (8.7), the application of Proposition 4.8 yields $(I_p - s_0 s_0^\dagger)s_j = 0_{p\times q}$ and $s_j(I_q - s_0^\dagger s_0) = 0_{p\times q}$ for each $j \in \mathbb{Z}_{0,2n-1}$. Thus,

$$
\left[I_n \otimes (I_p - s_0 s_0^\dagger)\right](G_{n-1} - y_{1,n}s_0^\dagger z_{1,n} + \Gamma_{n-1;q,q}) = 0_{np\times nq}
$$

and

$$
(G_{n-1} - y_{1,n}s_0^\dagger z_{1,n} + \Gamma_{n-1;q,q})\left[I_n \otimes (I_q - s_0^\dagger s_0)\right] = 0_{np\times nq}.
$$

Hence, using (8.8), (8.2), and (8.3) we obtain (8.5). Because of (8.7) and Lemma 8.18 we see that the matrices Δ_{n-1} and ∇_{n-1} are both non-singular. Thus, from (8.5) we infer (8.6). $\qquad\square$

Corollary 8.20. *Let $n \in \mathbb{N}$ and let $(s_j)_{j=0}^{2n} \in \mathcal{D}_{p \times q, 2n}$. Then the matrix $H_{n-1}^{(1)}$ admits the representations (8.4) and $H_{n-1}^{(1)} = \Delta_{n-1}(G_{n-1} - y_{1,n}s_0^\dagger z_{1,n})\nabla_{n-1}$. Moreover, $\operatorname{rank} H_{n-1}^{(1)} = \operatorname{rank}(G_{n-1} - y_{1,n}s_0^\dagger z_{1,n})$.*

Proof. Because of $(s_j)_{j=0}^{2n} \in \mathcal{D}_{p \times q, 2n}$, we obtain from Remark 4.17 that $(s_j)_{j=0}^{2n} \in \tilde{\mathcal{D}}_{p \times q, 2n}$, whereas Proposition 4.8 shows that $\Gamma_{n-1;q,q} = 0_{np \times nq}$. Hence, Proposition 8.19 yields the assertion. $\qquad\square$

Corollary 8.21. *Let $n \in \mathbb{N}$ and let $(s_j)_{j=0}^{2n} \in \mathcal{H}_{q,2n}^{\geq}$. Then*

$$H_{n-1}^{(1)} = (I_n \otimes s_0)S_{n-1}^\dagger(G_{n-1} - y_{1,n}s_0^\dagger z_{1,n})(S_{n-1}^\dagger)^*(I_n \otimes s_0), \tag{8.9}$$

$$H_{n-1}^{(1)} = \Delta_{n-1}(G_{n-1} - y_{1,n}s_0^\dagger z_{1,n} + \Gamma_{n-1;q,q})\Delta_{n-1}^*, \tag{8.10}$$

and

$$\operatorname{rank} H_{n-1}^{(1)} = \operatorname{rank}\left[H_n + \operatorname{diag}(0_{nq \times nq}, s_0 s_0^\dagger s_{2n} s_0^\dagger s_0 - s_{2n})\right] - \operatorname{rank} s_0. \tag{8.11}$$

Proof. From Proposition 4.25 we see that $(s_j)_{j=0}^{2n} \in \tilde{\mathcal{D}}_{q,2n}$ is valid. Part (b) of Lemma 3.2 yields $s_j^* = s_j$ for each $j \in \mathbb{Z}_{0,n-1}$. Thus (4.6) implies $\mathbb{S}_{n-1} = \mathbf{S}_{n-1}^*$. Hence, using part (a) of Lemma A.1 we get $(\mathbf{S}_{n-1}^*)^* = (\mathbf{S}_{n-1}^*)^\dagger = \mathbb{S}_{n-1}^\dagger$ and $(s_0 s_0^\dagger)^* = (s_0^\dagger)^* s_0^* = (s_0^*)^\dagger s_0^* = s_0^\dagger s_0$. Taking additionally into account (8.2) and (8.3), we can conclude $\Delta_{n-1}^* = \nabla_{n-1}$. Hence, application of Proposition 8.19 yields (8.9), (8.10), and (8.6). Thus, using part (c) of Theorem 6.13 and Proposition 7.3, we finally obtain (8.11). $\qquad\square$

Corollary 8.22. *Let $n \in \mathbb{N}$ and let $(s_j)_{j=0}^{2n} \in \mathcal{H}_{q,2n}^{\geq,e}$. Then equations (8.9),*

$$H_{n-1}^{(1)} = \Delta_{n-1}(G_{n-1} - y_{1,n}s_0^\dagger z_{1,n})\Delta_{n-1}^*,$$

and $\operatorname{rank} H_{n-1}^{(1)} = \operatorname{rank} H_n - \operatorname{rank} s_0$ hold true.

Proof. Because of Proposition 4.24 we have $(s_j)_{j=0}^{2n} \in \mathcal{D}_{q \times q, 2n}$. Thus, Proposition 4.8 provides us $s_0 s_0^\dagger s_{2n} s_0^\dagger s_0 = s_{2n}$ and consequently $\Gamma_{n-1;q,q} = 0_{nq \times nq}$. Hence, the application of Corollary 8.21 yields all assertions. $\qquad\square$

Now we prove an interesting recursion formula for the Schur transform of a sequence $(s_j)_{j=0}^{\kappa} \in \mathcal{D}_{p \times q, \kappa}$.

Proposition 8.23. *Let* $\kappa \in \mathbb{N} \cup \{+\infty\}$ *with* $\kappa \geq 2$, *let* $(s_j)_{j=0}^{\kappa} \in \mathcal{D}_{p \times q, \kappa}$, *and let* $(s_j^{(1)})_{k=0}^{\kappa-2}$ *be the first Schur transform of* $(s_j)_{j=0}^{\kappa}$. *For every* $j \in \mathbb{Z}_{0,\kappa-2}$, *then*

$$
s_j^{(1)} = \begin{cases}
s_2 - s_1 s_0^{\dagger} s_1 & \text{if } j = 0 \\
s_{j+2} - s_{j+1} s_0^{\dagger} s_1 - \displaystyle\sum_{l=0}^{j-1} s_{j-l} s_0^{\dagger} s_l^{(1)} & \text{if } j > 0
\end{cases}.
$$

Proof. If $\kappa < +\infty$, then, according to [16, Remark 4.8], let $(s_j)_{j=\kappa+1}^{\infty}$ be a sequence of complex $p \times q$-matrices such that $(s_j)_{j=0}^{\infty}$ belongs to $\mathcal{D}_{p \times q, \infty}$. Moreover, in view of Remark 8.2, if $\kappa < +\infty$ and if $(s_j^{(1)})_{j=0}^{\infty}$ is the first Schur transform of $(s_j)_{j=0}^{\infty}$, then $(s_j^{(1)})_{j=0}^{\kappa-2}$ is the first Schur transform of $(s_j)_{j=0}^{\kappa}$. Thus, it is sufficient to consider the case $\kappa = +\infty$. Because of $(s_j)_{j=0}^{\infty} \in \mathcal{D}_{p \times q, \infty}$, the application of Proposition 4.8 yields

$$
s_0 s_0^{\dagger} s_j = s_j \qquad \text{and} \qquad s_j s_0^{\dagger} s_0 = s_j \qquad \text{for each } j \in \mathbb{N}_0. \tag{8.12}
$$

From Definition 8.1, Definition 4.1, and (8.12) we get immediately $s_0^{(1)} = s_2 - s_1 s_0^{\dagger} s_1$ and, for each $j \in \mathbb{N}$, in view of part (b) of Remark 8.5, furthermore

$$
s_j^{(1)} = -s_0 \left(-s_0^{\dagger} \sum_{l=0}^{j+1} s_{j+2-l} s_l^{\#} \right) s_0 = \sum_{l=0}^{j+1} s_{j+2-l} s_l^{\#} s_0
$$

$$
= s_{j+2} s_0^{\dagger} s_0 + s_{j+1} (-s_0^{\dagger} s_1 s_0^{\#}) s_0 + \sum_{l=2}^{j+1} s_{j+2-l} s_l^{\#} s_0
$$

$$
= s_{j+2} - s_{j+1} s_0^{\dagger} s_1 + \sum_{l=2}^{j+1} s_{j+2-l} (-s_0^{\dagger} s_{l-2}^{(1)} s_0^{\dagger}) s_0.
$$

Applying part (a) of Remark 8.5 completes the proof. ☐

Corollary 8.24. *Let* $m \in \mathbb{Z}_{2,\infty}$ *and let* $(s_j)_{j=0}^{m} \in \tilde{\mathcal{D}}_{p \times q, m}$. *Then*

$$
s_{m-1}^{(1)} = \begin{cases}
s_0 s_0^{\dagger} s_2 s_0^{\dagger} s_0 - s_1 s_0^{\dagger} s_1 & \text{if } m = 2 \\
s_0 s_0^{\dagger} s_m s_0^{\dagger} s_0 - s_{m-1} s_0^{\dagger} s_1 - \displaystyle\sum_{l=0}^{m-3} s_{m-2-l} s_0^{\dagger} s_l^{(1)} & \text{if } m > 2
\end{cases}.
$$

Proof. Let $t_j := s_j$ for each $j \in \mathbb{Z}_{0,m-1}$ and let $t_j := s_0 s_0^{\dagger} s_m s_0^{\dagger} s_0$ for $j = m$. Then, because of $(s_j)_{j=0}^{m} \in \tilde{\mathcal{D}}_{p \times q, m}$, part (c) of Lemma 4.19 yields $(t_j)_{j=0}^{m} \in \mathcal{D}_{p \times q, m}$. Thus, applying Proposition 8.23 yields the assertion. ☐

We continue with some statements about the first Schur transform of infinite sequences belonging to particular subclasses of $\mathcal{H}_{q,\infty}^{\geq}$.

Proposition 8.25. *Let* $(s_j)_{j=0}^{\infty}$ *be a sequence from* $\mathbb{C}^{q \times q}$. *Let* $(s_j^{(1)})_{j=0}^{\infty}$ *be the first Schur transform of* $(s_j)_{j=0}^{\infty}$. *Then:*

(a) If $(s_j)_{j=0}^{\infty} \in \mathcal{H}_{q,\infty}^{\geq}$, then $(s_j^{(1)})_{j=0}^{\infty} \in \mathcal{H}_{q,\infty}^{\geq}$.

(b) If $(s_j)_{j=0}^{\infty} \in \mathcal{H}_{q,\infty}^{>}$, then $(s_j^{(1)})_{j=0}^{\infty} \in \mathcal{H}_{q,\infty}^{>}$.

(c) Let $n \in \mathbb{N}_0$ and let $(s_j)_{j=0}^{\infty} \in \mathcal{H}_{q,\infty}^{\geq}$ be completely degenerate of order n. Then $(s_j^{(1)})_{j=0}^{\infty}$ belongs to $\mathcal{H}_{q,\infty}^{\geq}$ and is completely degenerate of order $\max\{0, n-1\}$.

Proof. (a) This follows from Remark 8.2 and Proposition 8.12.

(b) This follows from Remark 8.2 and Proposition 8.14.

(c) We first consider the case $n = 0$. Because of (2.1), we have then $s_0 = 0_{q \times q}$ which, in view of Definition 8.1, implies $s_j^{(1)} = 0_{q \times q}$ for every $j \in \mathbb{N}_0$. Thus, $(s_j^{(1)})_{j=0}^{\infty}$ belongs to $\mathcal{H}_{q,\infty}^{\geq}$ and is completely degenerate of order 0. In the case $n > 0$ the assertion follows from (a), Remark 8.2, and Proposition 8.15. □

Proposition 8.26. Let $\kappa \in \mathbb{N} \cup \{+\infty\}$ and let $(s_j)_{j=0}^{2\kappa} \in \mathcal{H}_{q,2\kappa}^{\mathrm{sr}}$. Then the first Schur transform $(s_j^{(1)})_{j=0}^{2(\kappa-1)}$ of $(s_j)_{j=0}^{2\kappa}$ belongs to $\mathcal{H}_{q,2(\kappa-1)}^{\mathrm{sr}}$.

Proof. Let $n \in \mathbb{Z}_{1,\kappa}$. Then Remark 7.13 shows that $(s_j)_{j=0}^{2n} \in \mathcal{H}_{q,2n}^{\mathrm{sr}}$. Thus, part (b) of Lemma 7.16 implies $\det(G_{n-1}^{\sharp}) \neq 0$. Now, part (c) of Lemma 8.17 provides us with $\det(H_{n-1}^{(1)}) \neq 0$. This completes the proof. □

Now we want to show that the first Schur transform $(s_j^{(1)})_{j=0}^{2(n-1)}$ of a sequence $(s_j)_{j=0}^{2n} \in \mathcal{H}_{q,2n}^{\geq,\mathrm{e}}$ coincides with that sequence which was obtained in Chen/Hu [9] by showing that a certain matrix has $q \times q$ block Hankel structure (see [9, p. 206–207, formula (4.9)]). To realize this goal, we have to show that for certain matrices built from $(s_j)_{j=0}^{2n}$ the Moore-Penrose inverse coincides with the Drazin inverse.

If $A \in \mathbb{C}^{s \times s}$ is a square matrix, then by A^{D} we denote the Drazin inverse of A (see also Appendix A).

Lemma 8.27. Let $n \in \mathbb{N}$, and let $(s_j)_{j=0}^{2n} \in \mathcal{H}_{q,2n}^{\geq,\mathrm{e}}$. Then $\mathbf{S}_{n-1}^{\dagger} = \mathbf{S}_{n-1}^{\mathrm{D}}$ and $\mathbb{S}_{n-1}^{\dagger} = \mathbb{S}_{n-1}^{\mathrm{D}}$ where $\mathbf{S}_{n-1}$ and $\mathbb{S}_{n-1}$ are defined in Notation 4.11.

Proof. From part (b) of Lemma 3.1 we see that $(s_j)_{j=0}^{2n}$ is a sequence from $\mathbb{C}_{\mathrm{H}}^{q \times q}$. Hence, [16, Remark 4.12, part (a) and part (c)] shows that $\mathcal{R}(\mathbf{S}_{n-1}) = \mathcal{R}(\mathbf{S}_{n-1}^{*})$ and $\mathcal{R}(\mathbb{S}_{n-1}) = \mathcal{R}(\mathbb{S}_{n-1}^{*})$ hold. Thus, Proposition A.2 yields both assertions. □

Let $n \in \mathbb{N}$, and let $(s_j)_{j=0}^{2n} \in \mathcal{H}_{q,2n}^{\geq,\mathrm{e}}$. According to Corollary 8.22, the matrix $H_{n-1}^{(1)}$ admits the representation (8.9), whereas Lemma 8.27 shows that $\mathbf{S}_{n-1}^{\dagger} = \mathbf{S}_{n-1}^{\mathrm{D}}$ and $\mathbb{S}_{n-1}^{\dagger} = \mathbb{S}_{n-1}^{\mathrm{D}}$ hold. In view of part (b) of Lemma 3.1, the sequence $(s_j)_{j=0}^{2n}$ consists of Hermitian matrices. Thus we see from (4.6) that $\mathbb{S}_{n-1} = \mathbf{S}_{n-1}^{*}$. Hence, from part (a) of Lemma A.1 we get

$$\mathbb{S}_{n-1}^{\mathrm{D}} = \mathbb{S}_{n-1}^{\dagger} = (\mathbf{S}_{n-1}^{*})^{\dagger} = (\mathbf{S}_{n-1}^{\dagger})^{*}$$

and, according to (8.9), then

$$H_{n-1}^{(1)} = (I_n \otimes s_0)\mathbb{S}_{n-1}^{\mathrm{D}}(G_{n-1} - y_{1,n}s_0^{\dagger}z_{1,n})\mathbb{S}_{n-1}^{\mathrm{D}}(I_n \otimes s_0). \tag{8.13}$$

Thus, we see that the block Hankel matrix $H_{n-1}^{(1)}$ coincides with the matrix which occurs in formula (4.9) at p. 206–207 in the paper of Chen/Hu [9]. This means that the sequence used by Chen and Hu is exactly the first Schur transform $(s_j^{(1)})_{j=0}^{2(n-1)}$ of $(s_j)_{j=0}^{2n}$, because Chen and Hu obtained the sequence $(s_j^{(1)})_{j=0}^{2(n-1)}$ by proving that the matrix defined by the right-hand side of (8.13) has $q \times q$ block Hankel structure. Their proof is based on a regularization procedure in connection with the application of results due to Gekhtman/Shmoĭsh [17] on generalized Bezoutiants.

9. A Schur-type algorithm for sequences of complex $p \times q$-matrices

The Schur-type transform introduced in Section 8 generates in a natural way a corresponding algorithm for (finite or infinite) sequences of complex $p \times q$-matrices. The investigation of this algorithm is the central point of this section.

First we are going to extend Definition 8.1. Before we do that, let us note the following. If $\kappa \in \mathbb{N}_0 \cup \{+\infty\}$ and if $(s_j)_{j=0}^{\kappa}$ is a sequence of complex $p \times q$ matrices, then let $s_j^{(0)} := s_j$ for each $j \in \mathbb{Z}_{0,\kappa}$, and we will call $(s_j^{(0)})_{j=0}^{\kappa}$ the 0th Schur transform of $(s_j)_{j=0}^{\kappa}$. Now we define, recursively, the kth Schur transform.

Definition 9.1. Let $\kappa \in \mathbb{N}_0 \cup \{+\infty\}$ and let $(s_j)_{j=0}^{\kappa}$ be a sequence of complex $p \times q$ matrices. For each $k \in \mathbb{N}$ with $2k \leq \kappa$, the first Schur transform $(s_j^{(k)})_{j=0}^{\kappa-2k}$ of $(s_j^{(k-1)})_{j=0}^{\kappa-2(k-1)}$ is called the kth Schur transform of $(s_j)_{j=0}^{\kappa}$.

Remark 9.2. Let $\kappa \in \mathbb{N}_0 \cup \{+\infty\}$ and let $(s_j)_{j=0}^{\kappa}$ be a sequence of complex $p \times q$-matrices. Let $k, l \in \mathbb{N}_0$ be such that $2(k+l) \leq \kappa$. Then from Definition 9.1 it is immediately obvious that the $(k+l)$th Schur transform of $(s_j)_{j=0}^{\kappa}$ is the lth Schur transform of the kth Schur transform $(s_j^{(k)})_{j=0}^{\kappa-2k}$ of $(s_j)_{j=0}^{\kappa}$.

Now we are going to study the Schur-type algorithm introduced in Definition 9.1 for sequences belonging to $\mathcal{H}_{q,2n}^{\geq}$ and its distinguished subclasses and furthermore for the class $\mathcal{H}_{q,2n}^{\mathrm{sr}}$.

Proposition 9.3. *Let $n \in \mathbb{N}_0$ and let $(s_j)_{j=0}^{2n}$ be a sequence from $\mathbb{C}^{q \times q}$. Let $k \in \mathbb{Z}_{0,n}$ and let $(s_j^{(k)})_{j=0}^{2(n-k)}$ be the kth Schur transform of $(s_j)_{j=0}^{2n}$. Then:*

(a) *If $(s_j)_{j=0}^{2n} \in \mathcal{H}_{q,2n}^{\geq}$, then $(s_j^{(k)})_{j=0}^{2(n-k)} \in \mathcal{H}_{q,2(n-k)}^{\geq}$.*

(b) *If $(s_j)_{j=0}^{2n} \in \mathcal{H}_{q,2n}^{>}$, then $(s_j^{(k)})_{j=0}^{2(n-k)} \in \mathcal{H}_{q,2(n-k)}^{>}$.*

(c) *If $(s_j)_{j=0}^{2n} \in \mathcal{H}_{q,2n}^{\geq,\mathrm{cd}}$, then $(s_j^{(k)})_{j=0}^{2(n-k)} \in \mathcal{H}_{q,2(n-k)}^{\geq,\mathrm{cd}}$.*

(d) *If $(s_j)_{j=0}^{2n} \in \mathcal{H}_{q,2n}^{\mathrm{sr}}$, then $(s_j^{(k)})_{j=0}^{2(n-k)} \in \mathcal{H}_{q,2(n-k)}^{\mathrm{sr}}$.*

Proof. The assertions of (a), (b), (c) and (d) follow by induction from Remark 9.2, Proposition 8.12, Proposition 8.14, Proposition 8.15, and Proposition 8.26, respectively. $\qquad \square$

Proposition 9.4. *Let* $m \in \mathbb{N}_0$, *let* $(s_j)_{j=0}^{m} \in \mathcal{H}_{q,m}^{\geq,\mathrm{e}}$, *let* $k \in \mathbb{N}_0$ *be such that* $2k \leq m$ *holds. Then the kth Schur transform* $(s_j^{(k)})_{j=0}^{m-2k}$ *of* $(s_j)_{j=0}^{m}$ *belongs to* $\mathcal{H}_{q,m-2k}^{\geq,\mathrm{e}}$.

Proof. In the case $m \in \{0,1\}$ the assertion is obvious and in the case $m \in \mathbb{Z}_{2,+\infty}$ it follows by induction from Remark 9.2 and Proposition 8.13. $\square$

Proposition 9.5. *Let* $(s_j)_{j=0}^{\infty}$ *be a sequence from* $\mathbb{C}^{q\times q}$. *Let* $k \in \mathbb{N}_0$ *and let* $(s_j^{(k)})_{j=0}^{\infty}$ *be the kth Schur transform of* $(s_j)_{j=0}^{\infty}$. *Then:*

(a) *Let* $(s_j)_{j=0}^{\infty} \in \mathcal{H}_{q,\infty}^{\geq}$. *Then* $(s_j^{(k)})_{j=0}^{\infty} \in \mathcal{H}_{q,\infty}^{\geq}$.

(b) *Let* $(s_j)_{j=0}^{\infty} \in \mathcal{H}_{q,\infty}^{>}$. *Then* $(s_j^{(k)})_{j=0}^{\infty} \in \mathcal{H}_{q,\infty}^{>}$.

(c) *Let* $n \in \mathbb{N}_0$ *and let* $(s_j)_{j=0}^{\infty} \in \mathcal{H}_{q,\infty}^{\geq}$ *be completely degenerate of order n. Then* $(s_j^{(k)})_{j=0}^{\infty}$ *belongs to* $\mathcal{H}_{q,\infty}^{\geq}$ *and is completely degenerate of order* $\max\{0, n-k\}$.

(d) *Let* $(s_j)_{j=0}^{\infty} \in \mathcal{H}_{q,\infty}^{\mathrm{sr}}$. *Then* $(s_j^{(k)})_{j=0}^{\infty} \in \mathcal{H}_{q,\infty}^{\mathrm{sr}}$.

Proof. The assertions of (a), (b), and (c) follow by induction from Remark 9.2 and parts (a), (b), and (c) of Proposition 8.25, respectively. Furthermore, the application of Proposition 8.26 yields (d) by induction. $\square$

The following result is some generalization of Remark 8.4.

Proposition 9.6. *Let* $\kappa \in \mathbb{N}_0 \cup \{+\infty\}$ *and let* $(s_j)_{j=0}^{\kappa}$ *be a sequence of complex* $p \times q$*-matrices. For every* $k \in \mathbb{N}_0$ *with* $2k \leq \kappa - 2$, *every* $l \in \mathbb{N}$ *with* $2l \leq \kappa - 2k$, *and every* $j \in \mathbb{Z}_{0,\kappa-2(k+l)}$, *then* $\mathcal{N}(s_0^{(k)}) \subseteq \mathcal{N}(s_j^{(k+l)})$ *and* $\mathcal{R}(s_j^{(k+l)}) \subseteq \mathcal{R}(s_0^{(k)})$.

Proof. Let $k \in \mathbb{N}_0$ with $2k \leq \kappa - 2$ and let $l \in \mathbb{N}$ with $2l \leq \kappa - 2k$. Then we have $\kappa - 2(k+l-1) \geq 2$. Hence, keeping Definition 9.1 in mind, the application of Remark 8.4 to the sequence $(s_j^{(k+l-1)})_{j=0}^{\kappa-2(k+l-1)}$ yields $\mathcal{N}(s_0^{(k+l-1)}) \subseteq \mathcal{N}(s_j^{(k+l)})$ and $\mathcal{R}(s_j^{(k+l)}) \subseteq \mathcal{R}(s_0^{(k+l-1)})$ for every $j \in \mathbb{Z}_{0,\kappa-2(k+l)}$. Thus, it is sufficient to show the inclusions $\mathcal{N}(s_0^{(k)}) \subseteq \mathcal{N}(s_0^{(k+l-1)})$ and $\mathcal{R}(s_0^{(k+l-1)}) \subseteq \mathcal{R}(s_0^{(k)})$. However, because the case $l = 1$ is trivial, these inclusions follow by induction using Definition 9.1 and Remark 8.4. $\square$

It should be mentioned that the following important special case of Proposition 9.6 was already obtained with different methods by Chen/Hu [9, Theorem 3.9].

Proposition 9.7. *Let* $n \in \mathbb{N}$ *and let* $(s_j)_{j=0}^{2n} \in \mathcal{H}_{q,2n}^{\geq,\mathrm{e}}$. *Then* $\mathcal{R}(s_0^{(k+1)}) \subseteq \mathcal{R}(s_0^{(k)})$ *and in particular* $\operatorname{rank} s_0^{(k+1)} \leq \operatorname{rank} s_0^{(k)}$ *for each* $k \in \mathbb{Z}_{0,n-1}$.

Proof. This is an immediate consequence of Proposition 9.6. $\square$

Note that the example $s_0 = 0, s_1 = 0, s_2 = 1$ shows that there is a sequence in $\mathcal{H}_{1,2}^{\geq} \setminus \mathcal{H}_{1,2}^{\geq,\mathrm{e}}$ which fulfills $\mathcal{R}(s_1^{(0)}) \subseteq \mathcal{R}(s_0^{(1)})$.

The focus of our next considerations can be described as follows. Let $n \in \mathbb{N}$, let $(s_j)_{j=0}^{2n} \in \mathcal{H}_{q,2n}^{\geq,\mathrm{e}}$, and, for each $k \in \mathbb{Z}_{0,n}$ let $(s_j^{(k)})_{j=0}^{2(n-k)}$ be the kth Schur transform of $(s_j)_{j=0}^{2n}$. Then we want to show that, for each $k \in \mathbb{Z}_{0,n}$, the matrix L_k

introduced in (2.1) fulfills $L_k = s_0^{(k)}$. The main idea to realize this plan is to apply Lemma B.5 in an appropriate way. More precisely, we will determine matrices $\mathbb{F}_n$ and $\mathbb{W}_n$ belonging to the multiplicative group $\mathfrak{U}_{q,n}$ introduced in Appendix B such that the block Hankel matrix H_n admits the block diagonalizations

$$\mathbb{F}_n^* H_n \mathbb{F}_n = \mathrm{diag}(L_0, L_1, \ldots, L_n) \tag{9.1}$$

and

$$\mathbb{W}_n^* H_n \mathbb{W}_n = \mathrm{diag}(s_0^{(0)}, s_0^{(1)}, \ldots, s_0^{(n)}). \tag{9.2}$$

After realizing this goal the application of Lemma B.5 will bring the desired result.

Concerning the first block diagonalization (9.1), the following result was obtained in [15, Proposition 4.17].

Proposition 9.8. *Let* $n \in \mathbb{N}$ *and let* $(s_j)_{j=0}^{2n} \in \mathcal{H}_{q,2n}^{\geq}$. *For each* $k \in \mathbb{Z}_{1,n}$, *let*

$$F_k := \begin{bmatrix} I_{kq} & -H_{k-1}^\dagger y_{k,2k-1} \\ 0_{q \times kq} & I_q \end{bmatrix}$$

and let

$$\mathbf{F}_{k,n} := \begin{cases} \mathrm{diag}(F_k, I_{(n-k)q}) & \text{if } k \leq n-1 \\ F_n & \text{if } k = n \end{cases}.$$

Furthermore, let

$$\mathbb{F}_n := \mathbf{F}_{n,n} \mathbf{F}_{n-1,n} \cdots \mathbf{F}_{1,n},$$

and let $\mathbb{F}_n = (\mathbb{F}_{jk}^{(n)})_{j,k=0}^n$ *be the* $q \times q$ *block partition of* $\mathbb{F}_n$. *Then:*

(a) *The matrix* $\mathbb{F}_n$ *belongs to the set* $\mathfrak{U}_{q,n}$ *which is defined in Appendix B.*

(b) $L_{n-1} \mathbb{F}_{n-1,n}^{(n)} = M_{n-1} - s_{2n-1}$.

(c) $\mathbb{F}_n^* H_n \mathbb{F}_n = \mathrm{diag}(L_0, L_1, \ldots, L_n)$.

After having determined some matrix $\mathbb{F}_n \in \mathfrak{U}_{q,n}$ satisfying (9.1) we will now construct a matrix $\mathbb{W}_n \in \mathfrak{L}_{q,n}$ which fulfills (9.2). More precisely, the desired factorization (9.2) can be obtained only for a sequence $(s_j)_{j=0}^{2n} \in \mathcal{H}_{q,2n}^{\geq,e}$. In the case $(s_j)_{j=0}^{2n} \in \mathcal{H}_{q,2n}^{\geq}$, the situation is more delicate as the following result shows.

Proposition 9.9. *Let* $n \in \mathbb{N}$ *and let* $(s_j)_{j=0}^{2n} \in \mathcal{H}_{q,2n}^{\geq}$. *For each* $l \in \mathbb{Z}_{0,n-1}$, *let* $(s_j^{(l)})_{j=0}^{2(n-l)}$ *be the* lth *Schur transform of* $(s_j)_{j=0}^{2n}$. *For each* $k \in \mathbb{Z}_{1,n}$, *let the block Toeplitz matrix* $S_{k-1}^{(n-k)}$ *be defined by*

$$S_{k-1}^{(n-k)} := \begin{bmatrix} s_0^{(n-k)} & 0_{q \times q} & 0_{q \times q} & \cdots & 0_{q \times q} \\ s_1^{(n-k)} & s_0^{(n-k)} & 0_{q \times q} & \cdots & 0_{q \times q} \\ s_2^{(n-k)} & s_1^{(n-k)} & s_0^{(n-k)} & \cdots & 0_{q \times q} \\ \vdots & \vdots & \vdots & \ddots & \vdots \\ s_{k-1}^{(n-k)} & s_{k-2}^{(n-k)} & s_{k-3}^{(n-k)} & \cdots & s_0^{(n-k)} \end{bmatrix}$$

and let

$$y_{1,k}^{(n-k)} := \mathrm{col}(s_j^{(n-k)})_{j=1}^k.$$

Further, for each $k \in \mathbb{Z}_{1,n}$, let

$$
V_{k,n} := \left(
\begin{array}{c|c}
I_q & 0_{q \times kq} \\
\hline
\begin{array}{c}
-(I_k \otimes s_0^{(n-k)})(S_{k-1}^{(n-k)})^\dagger \\
\times y_{1,k}^{(n-k)}(s_0^{(n-k)})^\dagger
\end{array}
&
\begin{array}{c}
(I_k \otimes s_0^{(n-k)})(S_{k-1}^{(n-k)})^\dagger \\
+(I_k \otimes [I_q - s_0^{(n-k)}(s_0^{(n-k)})^\dagger])
\end{array}
\end{array}
\right)
$$

and

$$
\mathbb{V}_{k,n} := \begin{cases} \operatorname{diag}(I_{(n-k)q}, V_{k,n}) & \text{if } k \leq n-1 \\ V_{n,n} & \text{if } k = n \end{cases}.
$$

Let $\mathbb{V}_n := \mathbb{V}_{1,n}\mathbb{V}_{2,n}\cdots\mathbb{V}_{n,n}$ and let $\mathbb{V}_n = (\mathbb{V}_{lm}^{(n)})_{l,m=0}^n$ be the $q \times q$ block partition of $\mathbb{V}_n$. Then:

(a) *The matrix $\mathbb{V}_n$ belongs to the set $\mathfrak{L}_{q,n}$ which is defined in Definition B.1.*

(b) $\mathbb{V}_{n,n-1}^{(n)} = \sum_{j=0}^{n-1} s_0^{(j)}(s_1^{(j)})^\sharp.$

(c)

$$
\mathbb{V}_n H_n \mathbb{V}_n^*
$$

$$
= \operatorname{diag}\left(s_0^{(0)}, s_0^{(1)}, \ldots, s_0^{(n)} + \sum_{j=0}^{n-1}\left[s_{2(n-j)}^{(j)} - s_0^{(j)}(s_0^{(j)})^\dagger s_{2(n-j)}^{(j)}(s_0^{(j)})^\dagger s_0^{(j)}\right]\right).
$$

(d) *Let $(s_j)_{j=0}^{2n} \in \mathcal{H}_{q,2n}^{\geq,e}$. Then $\mathbb{V}_n H_n \mathbb{V}_n^* = \operatorname{diag}(s_0^{(0)}, s_0^{(1)}, \ldots, s_0^{(n)})$.*

Proof. Because of part (a) of Proposition 9.3 we have $(s_j^{(n-k)})_{j=0}^{2k} \in \mathcal{H}_{q,2k}^{\geq}$ for every $k \in \mathbb{Z}_{1,n}$, which, in view of Proposition 4.25, implies $(s_j^{(n-k)})_{j=0}^{2k} \in \tilde{\mathcal{D}}_{q,2k}$ for every $k \in \mathbb{Z}_{1,n}$. Keeping Definition 4.16 in mind, we obtain

$$
(s_j^{(n-k)})_{j=0}^{2k-1} \in \mathcal{D}_{q,2k-1} \qquad\qquad \text{for every } k \in \mathbb{Z}_{1,n}. \tag{9.3}
$$

Hence, from Lemma 8.18 applied to the sequence $(s_j^{(n-k)})_{j=0}^{2k}$ we see that

$$
\mathbf{V}_{k,n} \in \mathfrak{L}_{q,n} \qquad\qquad \text{for each } k \in \mathbb{Z}_{1,n}. \tag{9.4}
$$

(a) This is an immediate consequence of (9.4) and part (a) of Remark B.2.

(b) For each $k \in \mathbb{Z}_{1,n}$, let $V_{k,n} = (V_{lm}^{(k,n)})_{l,m=0}^k$ be the $q \times q$ block partition of $V_{k,n}$. Recalling Definition 4.1, we get

$$
\begin{aligned}
V_{1,0}^{(1,n)} &= -(I_1 \otimes s_0^{(n-1)})(S_0^{(n-1)})^\dagger y_{1,1}^{(n-1)}(s_0^{(n-1)})^\dagger \\
&= -s_0^{(n-1)}(s_0^{(n-1)})^\dagger s_1^{(n-1)}(s_0^{(n-1)})^\dagger \\
&= s_0^{(n-1)}\left[-(s_0^{(n-1)})^\dagger s_1^{(n-1)}(s_0^{(n-1)})^\sharp\right] = s_0^{(n-1)}(s_1^{(n-1)})^\sharp.
\end{aligned} \tag{9.5}
$$

In the case $n = 1$ we can easily see that $(\mathbb{V}_{lm}^{(n)})_{l,m=0}^{n} = (V_{lm}^{(n,n)})_{l,m=0}^{n}$ and, in view of (9.5), consequently

$$\mathbb{V}_{n,n-1}^{(n)} = V_{n,n-1}^{(n,n)} = V_{1,0}^{(1,n)} = s_0^{(n-1)}(s_1^{(n-1)})^{\sharp} = \sum_{j=0}^{n-1} s_0^{(j)}(s_1^{(j)})^{\sharp}.$$

Thus, if $n = 1$, then (b) is verified. Now let $n \geq 2$. For each $k \in \mathbb{Z}_{1,n}$ let $\mathbf{V}_{k,n} = (\mathbf{V}_{lm}^{(k,n)})_{l,m=0}^{n}$ be the $q \times q$ block partition of $\mathbf{V}_{k,n}$. Because of (9.4) and part (a) of Remark B.3, we have

$$\mathbb{V}_{n,n-1}^{(n)} = \sum_{k=1}^{n} \mathbf{V}_{n,n-1}^{(k,n)} = \sum_{k=1}^{n} V_{k,k-1}^{(k,n)}. \tag{9.6}$$

In view of (9.3), the application of Lemma 8.18 for every $k \in \mathbb{Z}_{2,n}$ to the sequence $(s_j^{(n-k)})_{j=0}^{2k-1}$ yields

$$V_{k,k-1}^{(k,n)} = s_0^{(n-k)}(s_1^{(n-k)})^{\sharp} \qquad\qquad \text{for each } k \in \mathbb{Z}_{2,n}. \tag{9.7}$$

From (9.6), (9.5), and (9.7) we conclude that

$$\mathbb{V}_{n,n-1}^{(n)} = \sum_{k=1}^{n} V_{k,k-1}^{(k,n)} = \sum_{k=1}^{n} s_0^{(n-k)}(s_1^{(n-k)})^{\sharp} = \sum_{j=0}^{n-1} s_0^{(j)}(s_1^{(j)})^{\sharp}.$$

Thus part (b) is also checked for $n \geq 2$.

(c) Because of $(s_j)_{j=0}^{2n} \in \mathcal{H}_{q,2n}^{\geq}$, part (c) of Lemma 3.2 implies $\mathcal{N}(s_0) \subseteq \mathcal{N}(s_j)$ and $\mathcal{R}(s_j) \subseteq \mathcal{R}(s_0)$ for every $j \in \mathbb{Z}_{1,n}$. From (2.2) and parts (b) and (c) of Lemma A.1 we obtain that

$$\left[I_n \otimes (s_0 s_0^{\dagger}) \right] y_{1,n} = y_{1,n}, \tag{9.8}$$

$$y_{1,n} s_0^{\dagger} s_0 = y_{1,n} \qquad\qquad \text{and} \qquad\qquad s_0 s_0^{\dagger} z_{1,n} = z_{1,n}. \tag{9.9}$$

In view of (8.2) and (9.8) we get

$$\Delta_{n-1} y_{1,n} = \left((I_n \otimes s_0) \mathbf{S}_{n-1}^{\dagger} + \left[I_n \otimes (I_q - s_0 s_0^{\dagger}) \right] \right) y_{1,n} = (I_n \otimes s_0) \mathbf{S}_{n-1}^{\dagger} y_{1,n}.$$

Hence, taking additionally (8.2) into account once again, we can conclude

$$\mathbf{V}_{n,n} = V_{n,n}$$

$$= \begin{bmatrix} I_q & 0_{q \times nq} \\ -(I_n \otimes s_0) \mathbf{S}_{n-1}^{\dagger} y_{1,n} s_0^{\dagger} & (I_n \otimes s_0) \mathbf{S}_{n-1}^{\dagger} + [I_n \otimes (I_q - s_0 s_0^{\dagger})] \end{bmatrix} \tag{9.10}$$

$$= \begin{bmatrix} I_q & 0_{q \times nq} \\ -\Delta_{n-1} y_{1,n} s_0^{\dagger} & \Delta_{n-1} \end{bmatrix} = \text{diag}(I_q, \Delta_{n-1}) \begin{bmatrix} I_q & 0_{q \times nq} \\ -y_{1,n} s_0^{\dagger} & I_{nq} \end{bmatrix}.$$

Using (9.10), (7.2), and (9.9) we obtain

$$\mathbf{V}_{n,n}H_n = \mathrm{diag}(I_q, \Delta_{n-1}) \cdot \begin{bmatrix} I_q & 0_{q \times nq} \\ -y_{1,n}s_0^\dagger & I_{nq} \end{bmatrix} \begin{bmatrix} s_0 & z_{1,n} \\ y_{1,n} & G_{n-1} \end{bmatrix}$$

$$= \mathrm{diag}(I_q, \Delta_{n-1}) \cdot \begin{bmatrix} s_0 & z_{1,n} \\ -y_{1,n}s_0^\dagger s_0 + y_{1,n} & -y_{1,n}s_0^\dagger z_{1,n} + G_{n-1} \end{bmatrix} \quad (9.11)$$

$$= \begin{bmatrix} s_0 & z_{1,n} \\ 0_{nq \times q} & \Delta_{n-1}(G_{n-1} - y_{1,n}s_0^\dagger z_{1,n}) \end{bmatrix}.$$

Because of $(s_j)_{j=0}^{2n} \in \mathcal{H}_{q,2n}^\geq$, part (b) of Lemma 3.2 implies $s_j^* = s_j$ for every $j \in \mathbb{Z}_{0,n}$. From part (a) of Lemma A.1 and (2.2) we see that

$$(-y_{1,n}s_0^\dagger)^* = -(s_0^\dagger)^* y_{1,n}^* = -(s_0^*)^\dagger y_{1,n}^* = -s_0^\dagger z_{1,n}.$$

Hence, in view of (9.10), we get

$$\mathbf{V}_{n,n}^* = \begin{bmatrix} I_q & -s_0^\dagger z_{1,n} \\ 0_{nq \times q} & I_{nq} \end{bmatrix} \cdot \mathrm{diag}(I_q, \Delta_{n-1}^*). \quad (9.12)$$

Since $(s_j)_{j=0}^{2n}$ belongs to $\mathcal{H}_{q,2n}^\geq$, Corollary 8.21 yields (8.10). From (9.3) we have $(s_j)_{j=0}^{2n-1} \in \mathcal{D}_{q,2n-1}$. Thus, using Lemma 8.18, we see that

$$\Delta_{n-1} \in \mathfrak{L}_{q,n-1} \quad (9.13)$$

which in view of part (c) of Remark B.2 implies $\Delta_{n-1}^* \in \mathfrak{U}_{q,n-1}$. Let $\Gamma_{n-1;q,q}$ be defined via (6.61). Then taking additionally into account (9.13), and Definition B.1 we obtain $\Delta_{n-1}\Gamma_{n-1;q,q}\Delta_{n-1}^* = \Gamma_{n-1;q,q}$. Hence, in view of (8.10), we conclude that

$$H_{n-1}^{(1)} - \Gamma_{n-1;q,q} = \Delta_{n-1}(G_{n-1} - y_{1,n}s_0^\dagger z_{1,n})\Delta_{n-1}^*. \quad (9.14)$$

Using (9.11), (9.12), the second equation in (9.9), (9.14) and (6.61), we see that

$$\mathbf{V}_{n,n}H_n\mathbf{V}_{n,n}^*$$

$$= \begin{bmatrix} s_0 & z_{1,n} \\ 0_{nq \times q} & \Delta_{n-1}(G_{n-1} - y_{1,n}s_0^\dagger z_{1,n}) \end{bmatrix} \begin{bmatrix} I_q & -s_0^\dagger z_{1,n} \\ 0_{q \times nq} & I_{nq} \end{bmatrix} \cdot \mathrm{diag}(I_q, \Delta_{n-1}^*)$$

$$= \begin{bmatrix} s_0 & -s_0 s_0^\dagger z_{1,n} + z_{1,n} \\ 0_{nq \times q} & \Delta_{n-1}(G_{n-1} - y_{1,n}s_0^\dagger z_{1,n}) \end{bmatrix} \cdot \mathrm{diag}(I_q, \Delta_{n-1}^*) \quad (9.15)$$

$$= \mathrm{diag}\left(s_0, \Delta_{n-1}(G_{n-1} - y_{1,n}s_0^\dagger z_{1,n})\Delta_{n-1}^*\right) = \mathrm{diag}(s_0, H_{n-1}^{(1)} - \Gamma_{n-1;q,q})$$

$$= \mathrm{diag}(s_0^{(0)}, H_{n-1}^{(1)}) + \mathrm{diag}\left(0_{nq \times nq}, s_{2n}^{(0)} - s_0^{(0)}(s_0^{(0)})^\dagger s_{2n}^{(0)}(s_0^{(0)})^\dagger s_0^{(0)}\right).$$

Because of (8.1), in the case $n = 1$ formula (9.15) shows that part (c) is proved. Now we consider the case $n > 1$. Then, in view of our considerations above, there is some $m \in \mathbb{Z}_{2,n}$ such that the assertion stated in part (c) is true for $n = m - 1$. We argue inductively and check that part (c) is valid for $n = m$ as well. Clearly, $(s_j)_{j=0}^{2n} \in \mathcal{H}_{q,2n}^\geq$ implies

$$(s_j)_{j=0}^{2m} \in \mathcal{H}_{q,2m}^\geq. \quad (9.16)$$

Let $\tilde{s}_j := s_j^{(1)}$ for every $j \in \mathbb{Z}_{0,2(m-1)}$. In view of (9.16) and Proposition 8.12, we obtain that

$$(\tilde{s}_j)_{j=0}^{2(m-1)} \in \mathcal{H}_{q,2(m-1)}^{\geq}. \tag{9.17}$$

Keeping Remark 9.2 in mind and using obvious notation, we see that $\tilde{V}_{k,m-1} = V_{k,m}$ holds for every $k \in \mathbb{Z}_{1,m-1}$. Hence, it is readily proved that with obvious notation

$$\mathbb{V}_m = \mathrm{diag}(I_q, \tilde{\mathbb{V}}_{m-1}) \cdot \mathbf{V}_{m,m} \tag{9.18}$$

holds. In view of (9.16) and (9.15), the same reasoning as above leads us to

$$\mathbf{V}_{m,m} H_m \mathbf{V}_{m,m}^*$$
$$= \mathrm{diag}(s_0^{(0)}, H_{m-1}^{(1)}) + \mathrm{diag}\left(0_{mq \times mq}, s_{2m}^{(0)} - s_0^{(0)}(s_0^{(0)})^\dagger s_{2m}^{(0)}(s_0^{(0)})^\dagger s_0^{(0)}\right). \tag{9.19}$$

Furthermore, because of (9.17) and our assumption that part (c) is valid for $n = m - 1$, with obvious notation we have

$$\tilde{\mathbb{V}}_{m-1}\tilde{H}_{m-1}\tilde{\mathbb{V}}_{m-1}^* = \mathrm{diag}\left(\tilde{s}_0^{(0)}, \tilde{s}_0^{(1)}, \ldots, \tilde{s}_0^{(m-2)},\right.$$

$$\left.\tilde{s}_0^{(m-1)} + \sum_{k=0}^{m-2}\left[\tilde{s}_{2(m-1-k)}^{(k)} - \tilde{s}_0^{(k)}(\tilde{s}_0^{(k)})^\dagger \tilde{s}_{2(m-1-k)}^{(k)}(\tilde{s}_0^{(k)})^\dagger \tilde{s}_0^{(k)}\right]\right)$$

which, recalling Remark 9.2, implies

$$\tilde{\mathbb{V}}_{m-1}H_{m-1}^{(1)}\tilde{\mathbb{V}}_{m-1}^* = \tilde{\mathbb{V}}_{m-1}\tilde{H}_{m-1}\tilde{\mathbb{V}}_{m-1}^*$$
$$= \mathrm{diag}\left(s_0^{(1)}, s_0^{(2)}, \ldots s_0^{(m-1)},\right.$$

$$\left.s_0^{(m)} + \sum_{k=0}^{m-2}\left[s_{2(m-1-k)}^{(k+1)} - s_0^{(k+1)}(s_0^{(k+1)})^\dagger s_{2(m-1-k)}^{(k+1)}(s_0^{(k+1)})^\dagger s_0^{(k+1)}\right]\right)$$

$$= \mathrm{diag}\left(s_0^{(1)}, s_0^{(2)}, \ldots, s_0^{(m)} + \sum_{j=1}^{m-1}\left[s_{2(m-j)}^{(j)} - s_0^{(j)}(s_0^{(j)})^\dagger s_{2(m-j)}^{(j)}(s_0^{(j)})^\dagger s_0^{(j)}\right]\right). \tag{9.20}$$

Because of (9.17) and part (a), we have $\tilde{\mathbb{V}}_{m-1} \in \mathcal{L}_{q,m-1}$. Consequently, the matrix $\mathrm{diag}(I_q, \tilde{\mathbb{V}}_{m-1})$ belongs to $\mathcal{L}_{q,m}$, which in view of part (c) of Remark B.2, shows that $\mathrm{diag}(I_q, \tilde{\mathbb{V}}_{m-1}^*) \in \mathfrak{U}_{q,m}$. Thus, we obtain that

$$\mathrm{diag}(I_q, \tilde{\mathbb{V}}_{m-1}) \cdot \mathrm{diag}\left(0_{mq \times mq}, s_{2m}^{(0)} - s_0^{(0)}(s_0^{(0)})^\dagger s_{2m}^{(0)}(s_0^{(0)})^\dagger s_0^{(0)}\right) \cdot \mathrm{diag}(I_q, \tilde{\mathbb{V}}_{m-1*}^*)$$

$$= \mathrm{diag}\left(0_{mq \times mq}, s_{2m}^{(0)} - s_0^{(0)}(s_0^{(0)})^\dagger s_{2m}^{(0)}(s_0^{(0)})^\dagger s_0^{(0)}\right). \tag{9.21}$$

Hence, using (9.18), (9.19), (9.21), and (9.20), we conclude that

$$\mathbb{V}_m H_m \mathbb{V}_m^* = \mathrm{diag}(I_q, \tilde{\mathbb{V}}_{m-1}) \mathbf{V}_{m,m} H_m \mathbf{V}_{m,m}^* \, \mathrm{diag}(I_q, \tilde{\mathbb{V}}_{m-1}^*)$$

$$= \mathrm{diag}(I_q, \tilde{\mathbb{V}}_{m-1})$$

$$\times \left[\mathrm{diag}(s_0^{(0)}, H_{m-1}^{(1)}) + \mathrm{diag}\left(0_{mq \times mq}, s_{2m}^{(0)} - s_0^{(0)}(s_0^{(0)})^\dagger s_{2m}^{(0)}(s_0^{(0)})^\dagger s_0^{(0)} \right) \right]$$

$$\times \mathrm{diag}(I_q, \tilde{\mathbb{V}}_{m-1}^*)$$

$$= \mathrm{diag}(s_0^{(0)}, \tilde{\mathbb{V}}_{m-1} H_{m-1}^{(1)} \tilde{\mathbb{V}}_{m-1}^*) + \mathrm{diag}\left(0_{mq \times mq}, s_{2m}^{(0)} - s_0^{(0)}(s_0^{(0)})^\dagger s_{2m}^{(0)}(s_0^{(0)})^\dagger s_0^{(0)} \right)$$

$$= \mathrm{diag}\left(s_0^{(0)}, s_0^{(1)}, \ldots, s_0^{(m)} + \sum_{j=0}^{m-1} \left[s_{2(m-j)}^{(j)} - s_0^{(j)}(s_0^{(j)})^\dagger s_{2(m-j)}^{(j)}(s_0^{(j)})^\dagger s_0^{(j)} \right] \right).$$

Hence part (c) holds true in the case $n = m$ as well.

(d) Let $j \in \mathbb{Z}_{0,n-1}$. In view of $(s_k)_{k=0}^{2n} \in \mathcal{H}_{q,2n}^{\geq,e}$, the application of Proposition 9.4 yields $(s_k^{(j)})_{k=0}^{2(n-j)} \in \mathcal{H}_{q,2(n-j)}^{\geq,e}$. Then, Proposition 4.24 implies that $(s_k^{(j)})_{k=0}^{2(n-j)}$ belongs to $\mathcal{D}_{q \times q, 2(n-j)}$. Hence, from Proposition 4.8 we infer

$$s_{2(n-j)}^{(j)} = s_0^{(j)}(s_0^{(j)})^\dagger s_{2(n-j)}^{(j)}(s_0^{(j)})^\dagger s_0^{(j)}.$$

Thus, the application of part (c) completes the proof. □

Since the sequence $(s_0^{(k)})_{k=0}^n$ coincides with the sequence defined by Chen and Hu [9, formula (4.9) at p. 206–207], we see that in the special case $(s_j)_{j=0}^{2n} \in \mathcal{H}_{q,2n}^{\geq,e}$ the block diagonalization of H_n which is contained in part (d) of Proposition 9.9 corresponds to the block diagonalization of H_n, which was attained in Chen/Hu [9, Theorem 3.9].

Proposition 9.10. *Let* $n \in \mathbb{N}$ *and let* $(s_j)_{j=0}^{2n} \in \mathcal{H}_{q,2n}^{\geq}$. *Then* $L_k = s_0^{(k)}$ *for each* $k \in \mathbb{Z}_{0,n-1}$ *and*

$$L_n = s_0^{(n)} + \sum_{j=0}^{n-1} \left[s_{2(n-j)}^{(j)} - s_0^{(j)}(s_0^{(j)})^\dagger s_{2(n-j)}^{(j)}(s_0^{(j)})^\dagger s_0^{(j)} \right].$$

Proof. According to parts (a) and (c) of Proposition 9.8, we have

$$\mathbb{F}_n \in \mathfrak{U}_{q,n} \tag{9.22}$$

and

$$\mathbb{F}_n^* H_n \mathbb{F}_n = \mathrm{diag}(L_0, L_1, \ldots, L_n). \tag{9.23}$$

From part (a) of Proposition 9.9 and part (c) of Remark B.2 we infer that

$$\mathbb{V}_n^* \in \mathfrak{U}_{q,n}, \tag{9.24}$$

whereas part (c) of Proposition 9.9 yields

$(\mathbb{V}_n^*)^* H_n \mathbb{V}_n^*$

$$= \operatorname{diag}\left(s_0^{(0)}, s_0^{(1)}, \ldots, s_0^{(n)} + \sum_{j=0}^{n-1}\left[s_{2(n-j)}^{(j)} - s_0^{(j)}(s_0^{(j)})^\dagger s_{2(n-j)}^{(j)}(s_0^{(j)})^\dagger s_0^{(j)}\right]\right).$$

$$(9.25)$$

Hence, application of part (a) of Lemma B.5 completes the proof.   □

Corollary 9.11. *Let* $m \in \mathbb{N}_0$ *and let* $(s_j)_{j=0}^m \in \mathcal{H}_{q,m}^{\geq,e}$. *Then* $L_k = s_0^{(k)}$ *for each* $k \in \mathbb{N}_0$ *with* $2k \leq m$.

Proof. If $m = 2n$ with $n \in \mathbb{N}_0$, then we choose matrices s_{2n+1} and s_{2n+2} from $\mathbb{C}^{q \times q}$ such that $(s_j)_{j=0}^{2n+2} \in \mathcal{H}_{q,2n+2}^{\geq}$. If $m = 2n + 1$ with some $n \in \mathbb{N}_0$, then we choose a matrix s_{2n+2} from $\mathbb{C}^{q \times q}$ such that $(s_j)_{j=0}^{2n+2} \in \mathcal{H}_{q,2n+2}^{\geq}$. Consequently, the assertion follows by applying Proposition 9.10 to the sequence $(s_j)_{j=0}^{2n+2}$ and taking Remark 8.2 into account.   □

Proposition 9.12. *Let* $n \in \mathbb{N}_0$ *and let* $(s_j)_{j=0}^{2n} \in \mathcal{H}_{q,2n}^{\geq,e}$. *For each* $k \in \mathbb{Z}_{0,n}$ *let* $(s_j^{(k)})_{j=0}^{2(n-k)}$ *be the kth Schur transform of* $(s_j)_{j=0}^{2n}$. *Then*

$$\operatorname{rank} H_n = \sum_{k=0}^n \operatorname{rank} s_0^{(k)}.$$

Proof. The case $n = 0$ is trivial. We now consider the case $n > 0$. Because of $(s_j)_{j=0}^{2n} \in \mathcal{H}_{q,2n}^{\geq,e}$ we have $(s_j)_{j=0}^{2n} \in \mathcal{H}_{q,2n}^{\geq}$. From [12, Remark 2.1 (a)] we get that $\operatorname{rank} H_n = \sum_{k=0}^n \operatorname{rank} L_k$. Hence, Corollary 9.11 yields the asserted equation.   □

An alternate approach to Corollary 9.11 which is based on orthogonal matrix polynomials can be found in Chen/Hu [9, Lemma 4.6, formula (4.22)].

A closer look at Proposition 9.10 shows that the sequence $(s_0^{(k)})_{k=0}^n$ is the part $(D_k)_{k=0}^n$ of the canonical Hankel parametrization $[(C_k)_{k=1}^n, (D_k)_{k=0}^n]$ of the sequence $(s_j)_{j=0}^{2n} \in \mathcal{H}_{q,2n}^{\geq}$. Now we want to show that the remaining part $(C_k)_{k=1}^n$ can also be described in terms of the sequences of Schur transforms of $(s_j)_{j=0}^{2n}$. Our strategy to realize this aim is as follows. We take into account that the part $(C_k)_{k=1}^n$ of the canonical Hankel parametrization of $(s_j)_{j=0}^{2n}$ is described in terms of the matrices Λ_k defined in (2.5). Then we will apply [12, Lemma 3.14] to get the identity

$$\Lambda_k = M_k + L_k L_{k-1}^\dagger (s_{2k} - M_{k-1}),$$

where the matrices M_{k-1} and M_k are defined via (2.4) and (2.6). To generate the desired connection with the sequences of Schur transforms of $(s_j)_{j=0}^{2n}$, we are looking for an expression of the difference $s_{2k} - M_{k-1}$ in terms of the sequences of Schur transforms of $(s_j)_{j=0}^{2n}$. Following this plan, we are led to the following result which is, by itself, of interest.

Proposition 9.13. *Let* $n \in \mathbb{N}$ *and let* $(s_j)_{j=0}^{2n-1} \in \mathcal{H}_{q,2n-1}^{\geq,e}$. *Then*

$$M_{n-1} - s_{2n-1} = L_{n-1} \sum_{j=0}^{n-1} (s_1^{(j)})^\sharp s_0^{(j)}.$$

Proof. The starting point of our proof consists of the block diagonalizations of H_n obtained in part (c) of Proposition 9.8 and part (c) of Proposition 9.9. The crucial step of the proof is then the application of part (c) of Lemma B.5. Using Proposition 9.8, Proposition 9.9, Remark B.2, and the notation given there, we get that (9.22), (9.23), (9.24), and (9.25) are satisfied. Let

$$\mathbb{F}_n = (\mathbb{F}_{lm}^{(n)})_{l,m=0}^n, \qquad \mathbb{V}_n = (\mathbb{V}_{lm}^{(n)})_{l,m=0}^n \qquad \text{and} \qquad \mathbb{V}_n^* = (\mathbb{W}_{lm}^{(n)})_{l,m=0}^n$$

be the $q \times q$ block representations of $\mathbb{F}_n$, $\mathbb{V}_n$, and $\mathbb{V}_n^*$, respectively. From (9.22), (9.24), (9.23), (9.25), and part (c) of Lemma B.5, we see that

$$L_{n-1} \mathbb{F}_{n-1,n}^{(n)} = L_{n-1} \mathbb{W}_{n-1,n}^{(n)}. \tag{9.26}$$

Now we will study both sides of (9.26) separately. Since $(s_j)_{j=0}^{2n}$ belongs to $\mathcal{H}_{q,2n}^{\geq}$, application of part (b) of Proposition 9.8 provides us with

$$L_{n-1} \mathbb{F}_{n-1,n}^{(n)} = M_{n-1} - s_{2n-1} \tag{9.27}$$

whereas part (b) of Proposition 9.9 yields

$$\mathbb{W}_{n-1,n}^{(n)} = (\mathbb{V}_{n,n-1}^{(n)})^* = \left[\sum_{j=0}^{n-1} s_0^{(j)} (s_1^{(j)})^\sharp \right]^* = \sum_{j=0}^{n-1} \left[(s_1^{(j)})^\sharp \right]^* (s_0^{(j)})^*. \tag{9.28}$$

Because of Proposition 9.4, for each $k \in \mathbb{Z}_{0,n-1}$ the sequence $(s_j^{(k)})_{j=0}^{2(n-k)-1}$ belongs to $\mathcal{H}_{q,2(n-k)-1}^{\geq,e}$. Consequently, from part (b) of Lemma 3.1 and [16, Proposition 5.16] we conclude that $(s_0^{(k)})^* = s_0^{(k)}$ and $[(s_1^{(k)})^\sharp]^* = (s_1^{(k)})^\sharp$ are fulfilled for each $k \in \mathbb{Z}_{0,n-1}$, respectively. Hence (9.28) provides us with

$$\mathbb{W}_{n-1,n}^{(n)} = \sum_{j=0}^{n-1} (s_1^{(j)})^\sharp s_0^{(j)} \tag{9.29}$$

and, by virtue of (9.27), (9.26), and (9.29), the proof is complete. $\qquad \square$

Now we are able to prove the first main result of this section.

Theorem 9.14. *Let* $n \in \mathbb{N}$, *let* $(s_j)_{j=0}^{2n} \in \mathcal{H}_{q,2n}^{\geq}$, *and let* $[(C_k)_{k=1}^n, (D_k)_{k=0}^n]$ *be the canonical Hankel parametrization of* $(s_j)_{j=0}^{2n}$. *Then* $C_k = s_1^{(k-1)}$ *for every* $k \in \mathbb{Z}_{1,n}$, $D_k = s_0^{(k)}$ *for every* $k \in \mathbb{Z}_{0,n-1}$, *and*

$$D_n = s_0^{(n)} + \sum_{j=0}^{n-1} \left[s_{2(n-j)}^{(j)} - s_0^{(j)} (s_0^{(j)})^\dagger s_{2(n-j)}^{(j)} (s_0^{(j)})^\dagger s_0^{(j)} \right]. \tag{9.30}$$

Proof. From Proposition 9.10 we get

$$D_k = L_k = s_0^{(k)} \qquad \text{for each } k \in \mathbb{Z}_{0,n-1} \qquad (9.31)$$

and

$$D_n = L_n = s_0^{(n)} + \sum_{j=0}^{n-1} \left[s_{2(n-j)}^{(j)} - s_0^{(j)}(s_0^{(j)})^\dagger s_{2(n-j)}^{(j)}(s_0^{(j)})^\dagger s_0^{(j)} \right].$$

It remains to prove $C_{k+1} = s_1^{(k)}$ for each $k \in \mathbb{Z}_{0,n-1}$. In the case $k = 0$ we have

$$C_{k+1} = C_1 = s_1 - \Lambda_0 = s_1 = s_1^{(0)} = s_1^{(k)}.$$

If $n = 1$, then the proof is complete. Now we suppose $n > 1$. We consider an arbitrary $k \in \mathbb{Z}_{1,n-1}$. In view of part (a) of Proposition 9.3, the sequence $(s_j^{(k)})_{j=0}^{2(n-k)}$ belongs to $\mathcal{H}_{q,2(n-k)}^{\geq}$. Consequently, from part (c) of Lemma 3.2 and parts (b) and (c) of Lemma A.1 we get

$$s_1^{(k)}(s_0^{(k)})^\dagger s_0^{(k)} = s_1^{(k)} \qquad \text{and} \qquad s_0^{(k)}(s_0^{(k)})^\dagger s_1^{(k)} = s_1^{(k)}. \qquad (9.32)$$

Because of $(s_j)_{j=0}^{2n} \in \mathcal{H}_{q,2n}^{\geq}$ and $k \in \mathbb{Z}_{1,n-1}$, we have

$$(s_j)_{j=0}^{2k} \in \mathcal{H}_{q,2k}^{\geq,e}. \qquad (9.33)$$

Hence, [12, Proposition 2.13] yields $\mathcal{N}(L_{k-1}) \subseteq \mathcal{N}(L_k)$. Taking into account part (b) of Lemma A.1, we conclude that

$$L_k L_{k-1}^\dagger L_{k-1} = L_k. \qquad (9.34)$$

Because of $(s_j)_{j=0}^{2n} \in \mathcal{H}_{q,2n}^{\geq}$ and $k \in \mathbb{Z}_{1,n-1}$, we get $(s_j)_{j=0}^{2m-1} \in \mathcal{H}_{q,2m-1}^{\geq,e}$ for each $m \in \{k, k+1\}$. Consequently, Remark 4.2 and Proposition 9.13 yield

$$M_{m-1} - s_{2m-1} = L_{m-1} \sum_{l=0}^{m-1} (s_1^{(l)})^\sharp s_0^{(l)} \qquad \text{for each } m \in \{k, k+1\}. \qquad (9.35)$$

In view of (9.33) and [12, Lemma 3.14], we obtain

$$\Lambda_k = M_k + L_k L_{k-1}^\dagger (s_{2k-1} - M_{k-1}), \qquad (9.36)$$

where Λ_k is defined in (2.5). From (9.32), Definition 4.1, (9.31), (9.34), (9.35) and (9.36) we get that

$$s_1^{(k)} = s_0^{(k)}(s_0^{(k)})^\dagger s_1^{(k)}(s_0^{(k)})^\dagger s_0^{(k)} = -s_0^{(k)} \left[-(s_0^{(k)})^\dagger s_1^{(k)}(s_0^{(k)})^\sharp \right] s_0^{(k)}$$

$$= -s_0^{(k)}(s_1^{(k)})^\sharp s_0^{(k)} = -L_k(s_1^{(k)})^\sharp s_0^{(k)} = L_k \sum_{l=0}^{k-1}(s_1^{(l)})^\sharp s_0^{(l)} - L_k \sum_{l=0}^{k}(s_1^{(l)})^\sharp s_0^{(l)}$$

$$= L_k L_{k-1}^\dagger L_{k-1} \sum_{l=0}^{k-1}(s_1^{(l)})^\sharp s_0^{(l)} - L_k \sum_{l=0}^{k}(s_1^{(l)})^\sharp s_0^{(l)}$$

$$= L_k L_{k-1}^\dagger (M_{k-1} - s_{2k-1}) - (M_k - s_{2k+1}) = s_{2k+1} - \Lambda_k = C_{k+1}.$$

The proof is complete. $\qquad \square$

For the subclass $\mathcal{H}_{q,2\kappa}^{\geq,\mathrm{e}}$ of $\mathcal{H}_{q,2\kappa}^{\geq}$, the interrelation between the canonical Hankel parametrization and Schur transform can be simplified:

Theorem 9.15. *Let $\kappa \in \mathbb{N} \cup \{+\infty\}$ and let $(s_j)_{j=0}^{2\kappa} \in \mathcal{H}_{q,2\kappa}^{\geq,\mathrm{e}}$. Then*

$$[(s_1^{(k-1)})_{k=1}^{\kappa}, (s_0^{(k)})_{k=0}^{\kappa}]$$

is the canonical Hankel parametrization of $(s_j)_{j=0}^{2\kappa}$.

Proof. We choose matrices s_{2n+1} and s_{2n+2} such that $(s_j)_{j=0}^{2n+2} \in \mathcal{H}_{q,2n+2}^{\geq}$. Then the assertion follows by applying Theorem 9.14 to the sequence $(s_j)_{j=0}^{2n+2}$ and taking Remark 8.2 into account. $\qquad\square$

Now we characterize the membership of a finite sequence $(s_j)_{j=0}^{2n} \in \mathcal{H}_{q,2n}^{\geq}$ to the class $\mathcal{H}_{q,2n}^{>}$ in terms of the sequence of its Schur transforms.

Proposition 9.16. *Let $n \in \mathbb{N}_0$ and let $(s_j)_{j=0}^{2n} \in \mathcal{H}_{q,2n}^{\geq}$. For each $k \in \mathbb{Z}_{0,n}$ let $(s_j^{(k)})_{j=0}^{2(n-k)}$ be the kth Schur transform of $(s_j)_{j=0}^{2n}$. Then:*

(a) *If $n = 0$, then $\operatorname{rank} H_n = \operatorname{rank} s_0^{(n)}$. If $n \geq 1$, then*

$$\operatorname{rank} H_n = \sum_{k=0}^{n-1} \operatorname{rank} s_0^{(k)}$$

$$+ \operatorname{rank}\left(s_0^{(n)} + \sum_{j=0}^{n-1}[s_{2(n-j)}^{(j)} - s_0^{(j)}(s_0^{(j)})^\dagger s_{2(n-j)}^{(j)}(s_0^{(j)})^\dagger s_0^{(j)}]\right).$$

(b) *For all $k \in \mathbb{Z}_{0,n}$, the matrix $s_0^{(k)}$ is non-negative Hermitian.*
(c) *The following statements are equivalent:*
   (i) $(s_j)_{j=0}^{2n} \in \mathcal{H}_{q,2n}^{>}$.
   (ii) *For all $k \in \mathbb{Z}_{0,n}$, the matrix $s_0^{(k)}$ is positive Hermitian.*
   (iii) *The matrix $s_0^{(n)}$ is non-singular.*

Proof. (a) In view of (1.2), the case $n = 0$ is trivial. We now consider the case $n > 0$. Because of $(s_j)_{j=0}^{2n} \in \mathcal{H}_{q,2n}^{\geq}$ and [12, Remark 2.1 (a)], we have then $\operatorname{rank} H_n = \sum_{k=0}^{n} \operatorname{rank} L_k$. Using Proposition 9.10 we obtain the asserted equation.

(b) Follows from part (a) of Proposition 9.3 and part (a) of Lemma 3.2.

(c) "(i)⇒(ii)": In view of (1.2), the case $n = 0$ is trivial. Now we consider the case $n > 0$. In view of (i) part (d) of [12, Proposition 2.30] yields $D_k \in \mathbb{C}_{>}^{q \times q}$ for every $k \in \mathbb{Z}_{0,n}$. Because of (i), we have $(s_j)_{j=0}^{2n} \in \mathcal{H}_{q,2n}^{\geq,\mathrm{e}}$. Thus, Theorem 9.15 implies $(s_0^{(k)})_{k=0}^{n} = (D_k)_{k=0}^{n}$. Hence, (ii) holds true.

"(ii)⇒(i)": Keeping part (a) in mind, from (ii) we obtain

$$\operatorname{rank} H_n = \sum_{k=0}^{n} \operatorname{rank} s_0^{(k)} = (n+1)q$$

which, because of $(s_j)_{j=0}^{2n} \in \mathcal{H}_{q,2n}^{\geq}$, is equivalent to (i).

"(ii)⇒(iii)": This implication is obvious.

"(iii)⇒(ii)": This follows from Proposition 9.7 and part (b). □

Now we characterize the membership of sequences $(s_j)_{j=0}^\infty \in \mathcal{H}_{q,\infty}^\geq$ to the class $\mathcal{H}_{q,\infty}^{\geq,\mathrm{cd},n}$ in terms of the sequence of its Schur transforms.

Proposition 9.17. *Let* $(s_j)_{j=0}^\infty \in \mathcal{H}_{q,\infty}^\geq$ *and let* $n \in \mathbb{N}_0$. *Then the following statements are equivalent:*

(i) $(s_j)_{j=0}^\infty \in \mathcal{H}_{q,\infty}^{\geq,\mathrm{cd},n}$.

(ii) $s_0^{(n)} = 0_{q \times q}$.

(iii) $s_j^{(k)} = 0_{q \times q}$ *for each integer* $k \geq n$ *and each* $j \in \mathbb{N}_0$.

Proof. "(i)⇔(ii)": Since $(s_j)_{j=0}^\infty$ belongs to $\mathcal{H}_{q,\infty}^\geq$, we have $(s_j)_{j=0}^{2n} \in \mathcal{H}_{q,2n}^{\geq,\mathrm{e}}$. Hence, the equivalence of (i) and (ii) is an immediate consequence of Corollary 9.11.

"(ii)⇒(iii)": Because of $(s_j)_{j=0}^\infty \in \mathcal{H}_{q,\infty}^\geq$ and part (a) of Proposition 9.5 we have $(s_j^{(n)})_{j=0}^\infty \in \mathcal{H}_{q,\infty}^\geq$. Hence, part (d) of Lemma 3.1 yields $\mathcal{N}(s_0^{(n)}) \subseteq \mathcal{N}(s_j^{(n)})$ for every $j \in \mathbb{N}_0$. Furthermore, Proposition 9.6 provides us with $\mathcal{N}(s_0^{(n)}) \subseteq \mathcal{N}(s_j^{(k)})$ for every $k \in \mathbb{Z}_{n+1,\infty}$ and every $j \in \mathbb{N}_0$. Thus, in view of (ii), then (iii) holds.

"(iii)⇒(ii)": This implication is obvious. □

The following result shows how the canonical Hankel parametrization of the lth Schur transform $(s_j^{(l)})_{j=0}^{2(n-l)}$ of a sequence $(s_j)_{j=0}^{2n} \in \mathcal{H}_{q,2n}^\geq$ is related to the canonical Hankel parametrization of $(s_j)_{j=0}^{2n}$.

Theorem 9.18. *Let* $n \in \mathbb{Z}_{2,\infty}$, *let* $(s_j)_{j=0}^{2n} \in \mathcal{H}_{q,2n}^\geq$, *and let* $l \in \mathbb{Z}_{1,n-1}$. *Let* $[(C_k)_{k=1}^n, (D_k)_{k=0}^n]$ *and* $[(C_k^{(l)})_{k=1}^{n-l}, (D_k^{(l)})_{k=0}^{n-l}]$ *be the canonical Hankel parametrizations of* $(s_j)_{j=0}^{2n}$ *and* $(s_j^{(l)})_{j=0}^{2(n-l)}$, *respectively. Then*

$$C_k^{(l)} = C_{l+k} \qquad \text{for every } k \in \mathbb{Z}_{1,n-l},$$

$$D_k^{(l)} = D_{l+k} \qquad \text{for every } k \in \mathbb{Z}_{0,n-l-1}$$

and

$$D_{n-l}^{(l)} = D_n - \sum_{j=0}^{l-1} \left[s_{2(n-j)}^{(j)} - s_0^{(j)} (s_0^{(j)})^\dagger s_{2(n-j)}^{(j)} (s_0^{(j)})^\dagger s_0^{(j)} \right].$$

Proof. Because of part (a) of Proposition 9.3, we have $(s_j^{(l)})_{j=0}^{2(n-l)} \in \mathcal{H}_{q,2(n-l)}^\geq$. Hence, the application of Theorem 9.14 to the sequence $(s_j^{(l)})_{j=0}^{2(n-l)}$ yields

$$C_k^{(l)} = (s_1^{(l)})^{(k-1)} \qquad \text{for every } k \in \mathbb{Z}_{1,n-l}, \tag{9.37}$$

$$D_k^{(l)} = (s_0^{(l)})^{(k)} \qquad \text{for every } k \in \mathbb{Z}_{0,n-l-1}, \tag{9.38}$$

and

$$D_{n-l}^{(l)}$$
$$= (s_0^{(l)})^{(n-l)}$$
$$+ \sum_{r=0}^{n-l-1} \left((s_{2(n-l-r)}^{(l)})^{(r)} - (s_0^{(l)})^{(r)} \left[(s_0^{(l)})^{(r)} \right]^\dagger (s_{2(n-l-r)}^{(l)})^{(r)} \left[(s_0^{(l)})^{(r)} \right]^\dagger (s_0^{(l)})^{(r)} \right).$$

$$(9.39)$$

Since $(s_j)_{j=0}^{2n}$ belongs to $\mathcal{H}_{q,2n}^\geq$ and from Theorem 9.14, we see that

$$C_k = s_1^{(k-1)} \qquad \text{for every } k \in \mathbb{Z}_{1,n}, \qquad (9.40)$$

$$D_k = s_0^{(k)} \qquad \text{for every } k \in \mathbb{Z}_{0,n-1}, \qquad (9.41)$$

and

$$D_n = s_0^{(n)} + \sum_{j=0}^{n-1} \left[s_{2(n-j)}^{(j)} - s_0^{(j)}(s_0^{(j)})^\dagger s_{2(n-j)}^{(j)}(s_0^{(j)})^\dagger s_0^{(j)} \right]. \qquad (9.42)$$

From Remark 9.2, (9.37), (9.40), (9.38), (9.41), (9.39), and (9.42) we conclude that

$$C_k^{(l)} = (s_1^{(l)})^{(k-1)} = s_1^{(l+k-1)} = C_{l+k} \qquad \text{for every } k \in \mathbb{Z}_{1,n-l},$$

$$D_k^{(l)} = (s_0^{(l)})^{(k)} = s_0^{(l+k)} = D_{l+k} \qquad \text{for every } k \in \mathbb{Z}_{0,n-l-1},$$

and

$$D_{n-l}^{(l)} = s_0^{(n)} + \sum_{r=0}^{n-l-1} \left[s_{2(n-l-r)}^{(l+r)} - s_0^{(l+r)}(s_0^{(l+r)})^\dagger s_{2(n-l-r)}^{(l+r)}(s_0^{(l+r)})^\dagger s_0^{(l+r)} \right]$$

$$= s_0^{(n)} + \sum_{j=l}^{n-1} \left[s_{2(n-j)}^{(j)} - s_0^{(j)}(s_0^{(j)})^\dagger s_{2(n-j)}^{(j)}(s_0^{(j)})^\dagger s_0^{(j)} \right]$$

$$= D_n - \sum_{j=0}^{l-1} \left[s_{2(n-j)}^{(j)} - s_0^{(j)}(s_0^{(j)})^\dagger s_{2(n-j)}^{(j)}(s_0^{(j)})^\dagger s_0^{(j)} \right].$$

The proof is complete. $\qquad\qquad\qquad\qquad\qquad\qquad\qquad\qquad\qquad\quad \square$

The following two theorems indicate that, in the case of sequences $(s_j)_{j=0}^{2n} \in \mathcal{H}_{q,2n}^{\geq,e}$ or $(s_j)_{j=0}^\infty \in \mathcal{H}_{q,\infty}^\geq$, the canonical Hankel parametrization can be considered as a Schur parametrization with respect to the Schur algorithm constructed in this section.

Theorem 9.19. *Let $n \in \mathbb{N}$, let $(s_j)_{j=0}^{2n} \in \mathcal{H}_{q,2n}^{\geq,e}$, and let $[(C_k)_{k=1}^n, (D_k)_{k=0}^n]$ be the canonical Hankel parametrization of $(s_j)_{j=0}^{2n}$. For each $l \in \mathbb{Z}_{1,n-1}$, then the pair $[(C_{l+k})_{k=1}^{n-l}, (D_{l+k})_{k=0}^{n-l}]$ is the canonical Hankel parametrization of $(s_j^{(l)})_{j=0}^{2(n-l)}$.*

Proof. Let the matrices s_{2n+1} and s_{2n+2} from $\mathbb{C}^{q \times q}$ be chosen such that $(s_j)_{j=0}^{2n+2} \in \mathcal{H}_{q,2n+2}^{\geq}$. Then applying Theorem 9.18 to the sequence $(s_j)_{j=0}^{2n+2}$ and taking Remark 8.2 into account we obtain the assertion. $\qquad\square$

Theorem 9.20. *Let $(s_j)_{j=0}^{\infty} \in \mathcal{H}_{q,\infty}^{\geq}$. For each $k \in \mathbb{N}_0$ denote by $(s_j^{(k)})_{j=0}^{\infty}$ the kth Schur transform of $(s_j)_{j=0}^{\infty}$. Then:*

(a) *Let $[(C_k)_{k=1}^{\infty}, (D_k)_{k=0}^{\infty}]$ be the canonical Hankel parametrization of $(s_j)_{j=0}^{\infty}$. For each $l \in \mathbb{N}$ then $[(C_{l+k})_{k=1}^{\infty}, (D_{l+k})_{k=0}^{\infty}]$ is the canonical Hankel parametrization of $(s_j^{(l)})_{j=0}^{\infty}$.*

(b) *For all $k \in \mathbb{N}_0$, the matrix $s_0^{(k)}$ is non-negative Hermitian.*

(c) *The following statements are equivalent:*

  (i) *$(s_j)_{j=0}^{\infty} \in \mathcal{H}_{q,\infty}^{>}$.*

  (ii) *For all $k \in \mathbb{N}_0$, the matrix $s_0^{(k)}$ is positive Hermitian.*

  (iii) *There is some $n \in \mathbb{N}_0$ such that for all $k \in \mathbb{N}_0$ with $k \geq n$ the matrix $s_0^{(k)}$ is non-singular.*

Proof. The combination of Theorem 9.19 and Remark 8.2 implies (a), whereas the assertions of (b) and (c) follow from part (b) and part (c) of Proposition 9.16, respectively. $\qquad\square$

Proposition 9.21. *Let $\kappa \in \mathbb{N} \cup \{+\infty\}$, let $(s_j)_{j=0}^{2\kappa} \in \mathcal{H}_{q,2\kappa}^{\geq}$, let $l \in \mathbb{Z}_{0,\kappa-1}$, let $(s_j^{(l)})_{j=0}^{2(\kappa-l)}$ be the lth Schur transform of $(s_j)_{j=0}^{\infty}$ and let $[(C_k)_{k=1}^{\kappa}, (D_k)_{k=0}^{\kappa}]$ be the canonical Hankel parametrization of $(s_j)_{j=0}^{2\kappa}$. Then the following statements are equivalent:*

(i) *$s_{2k-1}^{(l)} = 0_{q \times q}$ for each $k \in \mathbb{Z}_{1,\kappa-l}$.*

(ii) *$C_k = 0_{q \times q}$ for each $k \in \mathbb{Z}_{l+1,\kappa}$.*

Proof. In view of part (a) of Proposition 9.3 and part (a) of Proposition 9.5 we have $(s_j^{(l)})_{j=0}^{2(\kappa-l)} \in \mathcal{H}_{q,2(\kappa-l)}^{\geq}$. Now the assertion follows from [15, Proposition 2.14, Proposition 2.17] and Theorem 9.18. $\qquad\square$

Now we compute the Schur transforms of two classical Hankel positive definite number sequences.

Example 9.22. Let $(\gamma_n)_{n=0}^{\infty}$ be the sequence of Catalan numbers, i.e., $\gamma_n := \frac{1}{n+1}\binom{2n}{n}$ for each $n \in \mathbb{N}_0$. Then the sequence $(s_j)_{j=0}^{\infty}$ given for every $j \in \mathbb{N}_0$ by $s_j := \gamma_{j+1}$ belongs to $\mathcal{H}_{1,\infty}^{\geq}$. Indeed, $s_j = \int_{\mathbb{R}} t^j \mu(dt)$ where $\mu \colon \mathfrak{B}_{\mathbb{R}} \to [0,+\infty)$ is defined by $\mu(B) := \int_B h\, d\lambda$, where λ is the Lebesgue-Borel measure defined on $\mathfrak{B}_{\mathbb{R}}$ and where $h \colon \mathbb{R} \to [0,+\infty)$ is given by

$$h(t) := \begin{cases} \frac{t}{2\pi}\sqrt{\frac{4-t}{t}} & \text{if } t \in (0,4] \\ 0 & \text{if } t \in \mathbb{R} \setminus (0,4] \end{cases}$$

(see [26, Formulas (11) and (12)]). Furthermore, in view of Definition 4.1, the recurrence formula $\gamma_n = \sum_{l=0}^{n-1} \gamma_l \gamma_{n-1-l}$, which holds for each $n \in \mathbb{N}$, Definition 8.1, Definition 9.1, and Theorem 9.15, one can easily check that the following statements hold:

(a) $s_0^\sharp = 1$, $s_1^\sharp = -2$, and $s_j^\sharp = -\gamma_{j-1}$ for each integer j with $j \geq 2$.

(b) $(s_j^{(r)})_{j=0}^\infty = (\gamma_{j+1})_{j=0}^\infty$ and in particular $s_0^{(r)} = 1$ and $s_1^{(r)} = 2$ for each $r \in \mathbb{N}_0$.

(c) $[(2)_{k=1}^\infty, (1)_{k=0}^\infty]$ is the canonical Hankel parametrization of $(s_j)_{j=0}^\infty$.

In view of (c), we see that [15, Proposition 2.15, part (c)] implies that $(s_j)_{j=0}^\infty$ belongs to $\mathcal{H}_{1,\infty}^{\geq}$.

Example 9.23. Let $(m_n)_{n=0}^\infty$ be the sequence of Motzkin numbers, i.e.,

$$m_n := \frac{(-1)^{n+1}}{2^{2n+5}} \sum_{j=0}^{n+2} \frac{(-3)^j}{(2j-1)[2(n+2-j)-1]} \binom{2j}{j} \binom{2(n+2-j)}{n+2-j}$$

for each $n \in \mathbb{N}_0$. Then the sequence $(s_j)_{j=0}^\infty$ given for every $j \in \mathbb{N}_0$ by $s_j := m_j$ belongs to $\mathcal{H}_{1,\infty}^{\geq}$. Indeed $s_j = \int_\mathbb{R} t^j \mu(dt)$ where $\mu \colon \mathfrak{B}_\mathbb{R} \to [0,+\infty)$ is defined by $\mu(B) := \int_B h d\lambda$, where λ is the Lebesgue-Borel measure defined on $\mathfrak{B}_\mathbb{R}$ and where $h \colon \mathbb{R} \to [0,+\infty)$ is given by

$$h(t) := \begin{cases} \frac{1}{2\pi}\sqrt{(3-t)(1+t)} & \text{if } t \in [-1,3] \\ 0 & \text{if } t \in \mathbb{R} \setminus [-1,3] \end{cases}$$

(see [31, A001006]). Moreover, in view of Definition 4.1, the recurrence formula $m_n = m_{n-1} + \sum_{l=0}^{n-2} m_l m_{n-2-l}$ which holds for each $n \in \mathbb{Z}_{2,\infty}$, Definition 8.1, Definition 9.1, and Theorem 9.15, one can easily check that the following statements hold:

(a) $s_0^\sharp = 1$, $s_1^\sharp = -1$, and $s_j^\sharp = -m_{j-2}$ for each integer j with $j \geq 2$.

(b) $(s_j^{(r)})_{j=0}^\infty = (m_j)_{j=0}^\infty$ and in particular $s_0^{(r)} = 1$ and $s_1^{(r)} = 1$ for each $r \in \mathbb{N}_0$.

(c) $[(1)_{k=1}^\infty, (1)_{k=0}^\infty]$ is the canonical Hankel parametrization of $(s_j)_{j=0}^\infty$.

In view of (c), we see that [15, Proposition 2.15, part (c)] implies that $(s_j)_{j=0}^\infty$ belongs to $\mathcal{H}_{1,\infty}^{\geq}$.

10. Recovering the original sequence from the first Schur transform and first two matrices

The considerations in Section 8 suggest the study of a natural inverse question associated with the first Schur transform of a (finite or infinite) sequence of complex $p \times q$-matrices. The main theme of this section is the treatment of this inverse problem, which will be explained below in more detail.

It should be mentioned that a similar task was treated in the papers [5] and [6] where the inverse problem associated with the Schur-Potapov algorithm for strict $p \times q$-Schur sequences was handled.

Let $\kappa \in \mathbb{N} \cup \{+\infty\}$. For each sequence $(s_j)_{j=0}^{\kappa}$ of complex $p \times q$-matrices, the first Schur transform $(s_j^{(1)})_{j=0}^{\kappa-2}$ is given by Definition 8.1. Conversely, we consider the question: If the first Schur transform $(s_j^{(1)})_{j=0}^{\kappa-2}$ and the matrices s_0 and s_1 are known, how one can recover the original sequence $(s_j)_{j=0}^{\kappa}$? If $n \in \mathbb{N}$ and if $(s_j)_{j=0}^{2n}$ belongs to $\mathcal{H}_{q,2n}^{\geq}$, then Theorem 9.14, Definition 9.1, and Remark 2.4 yield such a possibility to express the sequence $(s_j)_{j=0}^{2n}$ by $(s_j^{(1)})_{j=0}^{2(n-1)}$ and the matrices s_0 and s_1. In view of the definition of the canonical Hankel parametrization of a finite sequence of complex $q \times q$-matrices and the formulas (2.3), (2.4), and (2.5), this way of computation is not very comfortable. That's why it seems to be more advantageous to construct a recursive procedure to recover the original sequence $(s_j)_{j=0}^{\kappa}$ from its first Schur transform and the matrices s_0 and s_1. To realize this aim, first we introduce the central construction of this section.

Definition 10.1. Let $\kappa \in \mathbb{N}_0 \cup \{+\infty\}$, let $(t_j)_{j=0}^{\kappa}$ be a sequence of complex $p \times q$-matrices, and let A and B be complex $p \times q$-matrices. Define

$$t_0^{(-1,A,B)} := A, \qquad\qquad t_1^{(-1,A,B)} := AA^{\dagger}BA^{\dagger}A \qquad (10.1)$$

and, recursively for each $m \in \mathbb{Z}_{2,\kappa+2}$, moreover

$$t_m^{(-1,A,B)} := \sum_{j=0}^{m-2} AA^{\dagger}t_{m-j-2}A^{\dagger}t_j^{(-1,A,B)} + AA^{\dagger}BA^{\dagger}t_{m-1}^{(-1,A,B)}. \qquad (10.2)$$

Then the sequence $(t_j^{(-1,A,B)})_{j=0}^{\kappa+2}$ is called the *first inverse Schur transform corresponding to* $[(t_j)_{j=0}^{\kappa}, A, B]$.

Remark 10.2. Let $\kappa \in \mathbb{N}_0 \cup \{+\infty\}$, let $(t_j)_{j=0}^{\kappa}$ be a sequence from $\mathbb{C}^{p \times q}$, and let $A, B \in \mathbb{C}^{p \times q}$. Denote by $(t_j^{(-1,A,B)})_{j=0}^{\kappa+2}$ the first inverse Schur transform corresponding to $[(t_j)_{j=0}^{\kappa}, A, B]$. In view of Definition 10.1, for each $l \in \mathbb{Z}_{0,\kappa}$ then $(t_j^{(-1,A,B)})_{j=0}^{l+2}$ is the first inverse Schur transform corresponding to $[(t_j)_{j=0}^{l}, A, B]$.

The following observation expresses an essential feature of our construction.

Remark 10.3. Let $\kappa \in \mathbb{N}_0 \cup \{+\infty\}$, let $(t_j)_{j=0}^{\kappa}$ be a sequence of complex $p \times q$-matrices, and let A and B be complex $p \times q$-matrices. From (10.1) and (10.2) one can easily see then that $(t_j^{(-1,A,B)})_{j=0}^{\kappa+2}$ belongs to $\mathcal{D}_{p \times q,\kappa+2}$.

Given $\kappa \in \mathbb{N}_0 \cup \{+\infty\}$ and a sequence $(t_j)_{j=0}^{\kappa}$ of complex $p \times q$-matrices we want to determine complex $p \times q$-matrices A and B such that the sequence $(t_j)_{j=0}^{\kappa}$ turns out to be exactly the first Schur transform of the first inverse Schur transform corresponding to $[(t_j)_{j=0}^{\kappa}, A, B]$. To realize this goal, we still need a little preparation.

Lemma 10.4. *Let* $\kappa \in \mathbb{N}_0 \cup \{+\infty\}$, *let* $(t_j)_{j=0}^{\kappa}$ *be a sequence of complex* $p \times q$-*matrices, and let* A *and* B *be complex* $p \times q$-*matrices. Then the elements of the* reciprocal sequence $((t_j^{(-1,A,B)})^\sharp)_{j=0}^{\kappa+2}$ corresponding to $(t_j^{(-1,A,B)})_{j=0}^{\kappa+2}$ can be represented by $(t_0^{(-1,A,B)})^\sharp = A^\dagger$, $(t_1^{(-1,A,B)})^\sharp = -A^\dagger B A^\dagger$, and, for each $j \in \mathbb{Z}_{2,\kappa+2}$, by $(t_j^{(-1,A,B)})^\sharp = -A^\dagger t_{j-2} A^\dagger$. Furthermore, $((t_j^{(-1,A,B)})^\sharp)^\sharp = t_j^{(-1,A,B)}$ for each $j \in \mathbb{Z}_{0,\kappa+2}$.

Proof. Let $r_0 := A^\dagger$, let $r_1 := -A^\dagger B A^\dagger$, and, for each $j \in \mathbb{Z}_{2,\kappa+2}$, let $r_j := -A^\dagger t_{j-2} A^\dagger$. We will show by induction that

$$(r_j^\sharp)_{j=0}^{\kappa+2} = (t_j^{(-1,A,B)})_{j=0}^{\kappa+2}. \tag{10.3}$$

First we get $r_0^\sharp = (A^\dagger)^\dagger = t_0^{(-1,A,B)}$ and $r_1^\sharp = -r_0^\dagger r_1 r_0^\sharp = t_1^{(-1,A,B)}$. Furthermore, we have that

$$r_2^\sharp = -r_0^\dagger r_2 r_0^\sharp - r_0^\dagger r_1 r_1^\sharp = AA^\dagger t_0 A^\dagger t_0^{(-1,A,B)} + AA^\dagger B A^\dagger t_1^{(-1,A,B)} = t_2^{(-1,A,B)}.$$

If $\kappa = 0$, then

$$r_j^\sharp = t_j^{(-1,A,B)} \tag{10.4}$$

is verified for each $j \in \mathbb{Z}_{0,\kappa+2}$. Let us consider the case $\kappa \geq 1$. Thus, we have already proved that there is an $m \in \mathbb{Z}_{0,\kappa-1}$ such that (10.4) is satisfied for each $j \in \mathbb{Z}_{0,m+2}$. In view of (10.2), we conclude that

$$r_{m+3}^\sharp = -\sum_{j=0}^{m+1} r_0^\dagger r_{m+3-j} r_j^\sharp - r_0^\dagger r_1 r_{m+2}^\sharp$$

$$= \left(\sum_{j=0}^{m+1} AA^\dagger t_{m+1-j}^{(-1,A,B)} A^\dagger t_j^{(-1,A,B)} \right) + AA^\dagger B A t_{m+2}^{(-1,A,B)} = t_{m+3}^{(-1,A,B)}.$$

Consequently, (10.4) is proved inductively for each $j \in \mathbb{Z}_{0,\kappa+2}$. Obviously, $(r_j)_{j=0}^{\kappa+2}$ belongs to $\mathcal{D}_{q\times p,\kappa+2}$. Thus part (d) of Proposition 4.10 yields $(r_j^\sharp)^\sharp = r_j$ for each $j \in \mathbb{Z}_{0,\kappa+2}$. Consequently, from (10.3) then $(t_j^{(-1,A,B)})^\sharp = r_j$ follows for each $j \in \mathbb{Z}_{0,\kappa+2}$. $\qquad\square$

Lemma 10.5. *Let $\kappa \in \mathbb{N}_0 \cup \{+\infty\}$, let $(t_j)_{j=0}^\kappa$ be a sequence of Hermitian $q \times q$-matrices, and let $A, B \in \mathbb{C}_{\mathrm{H}}^{q\times q}$. Then $(t_j^{(-1,A,B)})_{j=0}^{\kappa+2}$ is a sequence of Hermitian $q \times q$-matrices.*

Proof. Let $(s_j)_{j=0}^{\kappa+2}$ be the reciprocal sequence corresponding to $(t_j^{(-1,A,B)})_{j=0}^{\kappa+2}$. Because of Lemma 10.4, we have that $s_j^\sharp = t_j^{(-1,A,B)}$ for every $j \in \mathbb{Z}_{0,\kappa+2}$ and moreover $s_0 = A^\dagger$, $s_1 = -A^\dagger B A^\dagger$, and $s_j = -A^\dagger t_{j-2} A^\dagger$ for every $j \in \mathbb{Z}_{2,\kappa+2}$. Taking into account the particular choice of A, B and $(t_j)_{j=0}^\kappa$ and part (a) of Lemma A.1, we see that $(s_j)_{j=0}^{\kappa+2}$ is a sequence of Hermitian $q \times q$-matrices. Part (a) of Proposition 4.9 yields $(s_j)_{j=0}^{\kappa+2} \in \mathcal{D}_{q\times q,\kappa+2}$. Hence, [16, Corollary 5.17] provides that $(s_j^\sharp)_{j=0}^{\kappa+2}$ is a sequence of Hermitian $q \times q$-matrices, which completes the proof. $\qquad\square$

Lemma 10.6. *Let $\kappa \in \mathbb{N}_0 \cup \{+\infty\}$, let $(t_j)_{j=0}^{\kappa}$ be a sequence of complex $p \times q$-matrices, and let A and B be complex $p \times q$-matrices. Then:*

(a) *The first Schur transform $((t_j^{(-1,A,B)})^{(1)})_{j=0}^{\kappa}$ of $(t_j^{(-1,A,B)})_{j=0}^{\kappa+2}$ is given by*
$$(t_j^{(-1,A,B)})^{(1)} = AA^{\dagger}t_j A^{\dagger} A \text{ for each } j \in \mathbb{Z}_{0,\kappa}.$$

(b) *Let $\bigcup_{j=0}^{\kappa} \mathcal{R}(t_j) \subseteq \mathcal{R}(A)$ and $\mathcal{N}(A) \subseteq \bigcap_{j=0}^{\kappa} \mathcal{N}(t_j)$. Then $(t_j^{(-1,A,B)})^{(1)} = t_j$ for each $j \in \mathbb{Z}_{0,\kappa}$.*

Proof. (a) Because of Definition 8.1 and Lemma 10.4, for each $j \in \mathbb{Z}_{0,\kappa}$, we have
$$(t_j^{(-1,A,B)})^{(1)} = -t_0^{(-1,A,B)}(t_{j+2}^{(-1,A,B)})^{\sharp}t_0^{(-1,A,B)}$$
$$= -t_0^{(-1,A,B)}(-A^{\dagger}t_j A^{\dagger})t_0^{(-1,A,B)} = AA^{\dagger}t_j A^{\dagger} A.$$

(b) Combine parts (b) and (c) of Lemma A.1 and part (a). □

Roughly speaking, the content of our following considerations can be described as follows. Let $n \in \mathbb{N}_0$ and let $(t_j)_{j=0}^{2n}$ be a sequence which belongs to $\mathcal{H}_{q,2n}^{\geq}$ or to one of its distinguished subclasses $\mathcal{H}_{q,2n}^{\geq,\mathrm{e}}$, $\mathcal{H}_{q,2n}^{>}$, and $\mathcal{H}_{q,2n}^{\geq,\mathrm{cd}}$. Then we are looking for complex $q \times q$-matrices A and B such that the first inverse Schur transform corresponding to $[(t_j)_{j=0}^{2n}, A, B]$ belongs to $\mathcal{H}_{q,2(n+1)}^{\geq}$, $\mathcal{H}_{q,2(n+1)}^{\geq,\mathrm{e}}$, $\mathcal{H}_{q,2(n+1)}^{>}$, and $\mathcal{H}_{q,2(n+1)}^{\geq,\mathrm{cd}}$, respectively.

Proposition 10.7. *Let $n \in \mathbb{N}_0$, let $(t_j)_{j=0}^{2n} \in \mathcal{H}_{q,2n}^{\geq}$, let $A \in \mathbb{C}_{\geq}^{q \times q}$, and let $B \in \mathbb{C}_{\mathrm{H}}^{q \times q}$. Then*
$$(t_j^{(-1,A,B)})_{j=0}^{2(n+1)} \in \mathcal{H}_{q,2(n+1)}^{\geq} \tag{10.5}$$
and
$$\mathrm{rank}\left([t_{j+k}^{(-1,A,B)}]_{j,k=0}^{n+1}\right)$$
$$= \mathrm{rank}\, A + \mathrm{rank}\left([I_{n+1} \otimes (AA^{\dagger})][t_{j+k}]_{j,k=0}^{n}][I_{n+1} \otimes (AA^{\dagger})]^*\right). \tag{10.6}$$

Proof. Let $s_j := t_j^{(-1,A,B)}$ for each $j \in \mathbb{Z}_{0,2(n+1)}$. We want to apply a well-known characterization of non-negative Hermitian block matrices (see, e.g., [10, Lemma 1.1.9 (a), Lemma 1.1.7 (a)]) to the matrix H_{n+1}. Part (b) of Lemma 3.2 yields that $(t_j)_{j=0}^{2n}$ is a sequence of Hermitian $q \times q$-matrices. Since the matrices A and B are both Hermitian, Lemma 10.5 provides that $(t_j^{(-1,A,B)})_{j=0}^{2(n+1)}$ is a sequence of Hermitian $q \times q$-matrices. Hence,
$$s_j^* = s_j \qquad \text{for every } j \in \mathbb{Z}_{0,n+1}, \tag{10.7}$$
which, in view of (2.2), in particular implies
$$z_{1,n+1}^* = y_{1,n+1}. \tag{10.8}$$
Because of (7.2), we get the block representation
$$H_{n+1} = \begin{bmatrix} s_0 & z_{1,n+1} \\ z_{1,n+1}^* & G_n \end{bmatrix}. \tag{10.9}$$

From Definition 10.1 we see that

$$s_0 = t_0^{(-1,A,B)} = A \tag{10.10}$$

and, in particular, $s_0 \in \mathbb{C}^{q \times q}_{\geq}$. From Remark 10.3 we obtain

$$(s_j)_{j=0}^{2(n+1)} \in \mathcal{D}_{q \times q, 2(n+1)} \tag{10.11}$$

and, in view of Definition 4.5, consequently

$$\mathcal{R}(z_{1,n+1}) \subseteq \mathcal{R}(s_0). \tag{10.12}$$

Because of (10.11) and Lemma 8.18, the matrices Δ_n and ∇_n are both non-singular. In view of (10.7), we see from (4.6) that $\mathbb{S}_n = \mathbf{S}_n^*$. Hence, using part (a) of Lemma A.1 we get

$$(\mathbf{S}_n^\dagger)^* = (\mathbf{S}_n^*)^\dagger = \mathbb{S}_n^\dagger, \tag{10.13}$$

whereas (10.7) and part (a) of Lemma A.1 yield $(s_0 s_0^\dagger)^* = s_0^\dagger s_0$. Taking additionally into account (8.2), (10.13), (10.7), and (8.3), we conclude that $\Delta_n^* = \nabla_n$. Thus, in view of (10.11) and Corollary 8.20 we obtain

$$G_n - y_{1,n+1} s_0^\dagger z_{1,n+1} = \Delta_n^{-1} H_n^{(1)} \nabla_n^{-1} = \Delta_n^{-1} H_n^{(1)} \Delta_n^{-*}. \tag{10.14}$$

Since the matrix A is Hermitian, from part (a) of Lemma A.1 we get $(AA^\dagger)^* = A^\dagger A$. Using part (a) of Lemma 10.6 we have that $s_j = (t_j^{(-1,A,B)})^{(1)} = (AA^\dagger)t_j(AA^\dagger)^*$ for every $j \in \mathbb{Z}_{0,2n}$ and thus in particular

$$H_n^{(1)} = [s_{j+k}^{(1)}]_{j,k=0}^n = [I_{n+1} \otimes (AA^\dagger)] [[t_{j+k}]_{j,k=0}^n] [I_{n+1} \otimes (AA^\dagger)]^*. \tag{10.15}$$

Because of $(t_j)_{j=0}^{2n} \in \mathcal{H}^{\geq}_{q,2n}$, the matrix $[t_{j+k}]_{j,k=0}^n$ is non-negative Hermitian. Consequently, the matrix on the right-hand side of (10.15) is non-negative Hermitian. From (10.15) we get $H_n^{(1)} \in \mathbb{C}^{(n+1)q \times (n+1)q}_{\geq}$. By virtue of (10.8) and (10.14), then

$$G_n - z_{1,n+1}^* s_0^\dagger z_{1,n+1} \in \mathbb{C}^{(n+1)q \times (n+1)q}_{\geq} \tag{10.16}$$

follows. In view of (10.9), $s_0 \in \mathbb{C}^{q \times q}_{\geq}$, (10.12), and (10.16), the application of [10, Lemma 1.1.9 (a)] yields $H_{n+1} \in \mathbb{C}^{(n+2)q \times (n+2)q}_{\geq}$, i.e., $(t_j^{(-1,A,B)})_{j=0}^{2(n+1)} \in \mathcal{H}^{\geq}_{q,2(n+1)}$. Moreover, because of (10.9), (10.12), (10.7) and [10, Lemma 1.1.7 (a)], we have

$$\operatorname{rank} H_{n+1} = \operatorname{rank} s_0 + \operatorname{rank}(G_n - z_{1,n+1}^* s_0^\dagger z_{1,n+1}).$$

Taking additionally into account (10.10), (10.14) and (10.15), we obtain the asserted equation (10.6). □

Corollary 10.8. *Let* $\kappa \in \mathbb{N}_0 \cup \{+\infty\}$, *let* $(t_j)_{j=0}^\kappa \in \mathcal{H}^{\geq,e}_{q,\kappa}$, *let* $A \in \mathbb{C}^{q \times q}_{\geq}$, *and let* $B \in \mathbb{C}^{q \times q}_{\mathrm{H}}$. *Then* $(t_j^{(-1,A,B)})_{j=0}^{\kappa+2} \in \mathcal{H}^{\geq,e}_{q,\kappa+2}$.

Proof. This follows from Definition 10.1, Remark 10.2 and Proposition 10.7. □

Corollary 10.9. *Let* $\kappa \in \mathbb{N}_0 \cup \{+\infty\}$, *let* $(t_j)_{j=0}^{2\kappa} \in \mathcal{H}^{>}_{q,2\kappa}$, *let* $A \in \mathbb{C}^{q \times q}_{>}$, *and let* $B \in \mathbb{C}^{q \times q}_{\mathrm{H}}$. *Then* $(t_j^{(-1,A,B)})_{j=0}^{2(\kappa+1)} \in \mathcal{H}^{>}_{q,2(\kappa+1)}$.

Proof. Let $n \in \mathbb{Z}_{0,\kappa}$. Because of $(t_j)_{j=0}^{2\kappa} \in \mathcal{H}_{q,2\kappa}^>$ we have $(t_j)_{j=0}^{2n} \in \mathcal{H}_{q,2n}^{\geq}$ and $\text{rank}[[t_{j+k}]_{j,k=0}^n] = (n+1)q$. Because of $A \in \mathbb{C}_>^{q \times q}$ we see that $A \in \mathbb{C}_{\geq}^{q \times q}$ and $\text{rank } A = q$ hold. Consequently, $I_{n+1} \otimes (AA^\dagger) = I_{n+1}$. Hence, the application of Proposition 10.7 yields (10.5) and

$$\text{rank}\left[[t_{j+k}^{(-1,A,B)}]_{j,k=0}^{n+1}\right] = \text{rank } A + \text{rank}\left[[t_{j+k}]_{j,k=0}^n\right] = (n+2)q.$$

Thus, $[t_{j+k}^{(-1,A,B)}]_{j,k=0}^{n+1} \in \mathbb{C}_>^{(n+2)q \times (n+2)q}$ for each $n \in \mathbb{Z}_{0,\kappa}$. $\qquad\square$

Proposition 10.10. *Let* $n \in \mathbb{N}_0$, *let* $(t_j)_{j=0}^{2n} \in \mathcal{H}_{q,2n}^{\geq,\text{cd}}$, *let* $A \in \mathbb{C}_{\geq}^{q \times q}$ *be such that in the case* $n > 0$ *the relation* $\mathcal{N}(A) \subseteq \mathcal{N}(t_0)$ *holds, and let* $B \in \mathbb{C}_{\text{H}}^{q \times q}$. *Then* $(t_j^{(-1,A,B)})_{j=0}^{2(n+1)} \in \mathcal{H}_{q,2(n+1)}^{\geq,\text{cd}}$.

Proof. Let $s_j := t_j^{(-1,A,B)}$ for each $j \in \mathbb{Z}_{0,2(n+1)}$. Since $(t_j)_{j=0}^{2n}$ belongs to $\mathcal{H}_{q,2n}^{\geq,\text{cd}}$, we have

$$(t_j)_{j=0}^{2n} \in \mathcal{H}_{q,2n}^{\geq}. \tag{10.17}$$

From (10.17) and Proposition 10.7 we conclude that (10.5) and

$$\text{rank } H_{n+1} = \text{rank } A + \text{rank}\left([I_{n+1} \otimes (AA^\dagger)] [[t_{j+k}]_{j,k=0}^n] [I_{n+1} \otimes (AA^\dagger)]^*\right). \tag{10.18}$$

We first consider the case $n = 0$. Because of $(t_j)_{j=0}^0 \in \mathcal{H}_{q,0}^{\geq,\text{cd}}$ and (2.1) then $t_0 = 0_{q \times q}$ holds. Using Definition 10.1, we obtain $s_0 = A$, $s_1 = AA^\dagger B A^\dagger A$ and $s_2 = AA^\dagger B A^\dagger B A^\dagger A$. From (2.1), (2.2) and (1.2), we get that $L_1 = 0_{q \times q}$ which, in combination with (10.5), implies $(t_j^{(-1,A,B)})_{j=0}^2 \in \mathcal{H}_{q,2}^{\geq,\text{cd}}$.

We now consider the case $n > 0$. Because of $A \in \mathbb{C}_{\geq}^{q \times q}$, we have $A^* = A$ and hence

$$\mathcal{N}(AA^\dagger) = \mathcal{R}\left((AA^\dagger)^*\right)^\perp = \mathcal{R}(AA^\dagger)^\perp = \mathcal{R}(A)^\perp = \mathcal{N}(A^*) = \mathcal{N}(A).$$

Recalling (10.17), part (c) of Lemma 3.2 yields $\mathcal{N}(t_0) \subseteq \mathcal{N}(t_j)$ for each $j \in \mathbb{Z}_{0,2n-1}$, which, in view of $\mathcal{N}(A) \subseteq \mathcal{N}(t_0)$, implies

$$\mathcal{N}(AA^\dagger) = \mathcal{N}(A) \subseteq \mathcal{N}(t_0) \subseteq \bigcap_{j=0}^{2n-1} \mathcal{N}(t_j).$$

Since $(t_j)_{j=0}^{2n}$ belongs to $\mathcal{H}_{q,2n}^{\geq,\text{cd}}$, Proposition 3.12 provides us with

$$\left((AA^\dagger)t_j(AA^\dagger)^*\right)_{j=0}^{2n} \in \mathcal{H}_{q,2n}^{\geq,\text{cd}}.$$

Consequently, the application of [12, Remark 2.1 (a)] yields

$$\text{rank}\left[(AA^\dagger)t_{j+k}(AA^\dagger)^*\right]_{j,k=0}^n = \text{rank}\left[(AA^\dagger)t_{j+k}(AA^\dagger)^*\right]_{j,k=0}^{n-1}. \tag{10.19}$$

Because of (10.17), we have $(t_j)_{j=0}^{2(n-1)} \in \mathcal{H}_{q,2(n-1)}^{\geq}$. Hence, Proposition 10.7 yields

$$\text{rank } H_n = \text{rank } A + \text{rank}\left([I_n \otimes (AA^\dagger)] [[t_{j+k}]_{j,k=0}^{n-1}] [I_n \otimes (AA^\dagger)]^*\right). \tag{10.20}$$

From (10.18), (10.19), and (10.20) we obtain

$$
\begin{aligned}
\operatorname{rank} H_{n+1} &= \operatorname{rank} A + \operatorname{rank} \left(\left[I_{n+1} \otimes (AA^\dagger) \right] \left[t_{j+k} \right]_{j,k=0}^{n} \left[I_{n+1} \otimes (AA^\dagger) \right]^* \right) \\
&= \operatorname{rank} A + \operatorname{rank} \left[(AA^\dagger) t_{j+k} (AA^\dagger)^* \right]_{j,k=0}^{n} \\
&= \operatorname{rank} A + \operatorname{rank} \left[(AA^\dagger) t_{j+k} (AA^\dagger)^* \right]_{j,k=0}^{n-1} \\
&= \operatorname{rank} A + \operatorname{rank} \left(\left[I_{n} \otimes (AA^\dagger) \right] \left[t_{j+k} \right]_{j,k=0}^{n-1} \left[I_{n} \otimes (AA^\dagger) \right]^* \right) \\
&= \operatorname{rank} H_{n},
\end{aligned}
$$

which in view of (10.5) and [12, Remark 2.1 (a)] implies the assertion.   □

Corollary 10.11. *Let $n \in \mathbb{N}_0$, let $(t_j)_{j=0}^{\infty} \in \mathcal{H}_{q,\infty}^{\geq,\mathrm{cd},n}$, let $A \in \mathbb{C}_{\geq}^{q \times q}$ be such that in the case $n > 0$ the relation $\mathcal{N}(A) \subseteq \mathcal{N}(t_0)$ holds, and let $B \in \mathbb{C}_{\mathrm{H}}^{q \times q}$. Then $(t_j^{(-1,A,B)})_{j=0}^{\infty} \in \mathcal{H}_{q,\infty}^{\geq,\mathrm{cd},n+1}$.*

Proof. Since $(t_j)_{j=0}^{\infty}$ belongs to $\mathcal{H}_{q,\infty}^{\geq}$, Corollary 10.8 implies $(t_j^{(-1,A,B)})_{j=0}^{\infty} \in \mathcal{H}_{q,\infty}^{\geq}$. In view of $(t_j)_{j=0}^{2n} \in \mathcal{H}_{q,2n}^{\geq,\mathrm{cd}}$ we see from Proposition 10.10 and Remark 10.2 that $(t_j^{(-1,A,B)})_{j=0}^{2(n+1)} \in \mathcal{H}_{q,2(n+1)}^{\geq,\mathrm{cd}}$. This completes the proof.   □

Lemma 10.12. *Let $\kappa \in \mathbb{N}_0 \cup \{+\infty\}$ with $\kappa \geq 2$ and let $(s_j)_{j=0}^{\kappa}$ be a sequence of complex $p \times q$-matrices. For each $j \in \mathbb{Z}_{0,\kappa}$, then*

$$
(s_j^{(1)})^{(-1,s_0,s_1)} = (s^\sharp)_j^\sharp.
$$

Proof. Let $r_0 := s_0^\dagger$, let $r_1 := -s_0^\dagger s_1 s_0^\dagger$, and for each $j \in \mathbb{Z}_{2,\kappa+2}$, let $r_j := -s_0^\dagger s_{j-2}^{(1)} s_0^\dagger$. From the proof of Lemma 10.4 we get that $r_j^\sharp = (s_j^{(1)})^{(-1,s_0,s_1)}$ for each $j \in \mathbb{Z}_{0,\kappa}$. Since Definition 4.1 and part (b) of Remark 8.5 imply that $r_j = s_j^\sharp$ holds for each $j \in \mathbb{Z}_{0,\kappa}$, the assertion follows.   □

Given the first Schur transform of a sequence $(s_j)_{j=0}^{\kappa} \in \mathcal{D}_{p \times q,\kappa}$ and the first two matrices s_0 and s_1, now we want to recover the original sequence $(s_j)_{j=0}^{\kappa}$ via its first inverse Schur transform corresponding to $[(s_j^{(1)})_{j=0}^{\kappa-2}, s_0, s_1]$.

Theorem 10.13. *Let $\kappa \in \mathbb{Z}_{2,\infty} \cup \{+\infty\}$ and let $(s_j)_{j=0}^{\kappa} \in \mathcal{D}_{p \times q,\kappa}$. For each $j \in \mathbb{Z}_{0,\kappa}$, then*

$$
(s_j^{(1)})^{(-1,s_0,s_1)} = s_j. \tag{10.21}
$$

Proof. From part (d) of Proposition 4.10 we know that $(s_j^\sharp)^\sharp = s_j$ holds for each $j \in \mathbb{Z}_{0,\kappa}$. Thus, the application of Lemma 10.12 completes the proof.   □

Corollary 10.14. *Let $\kappa \in \mathbb{Z}_{2,\infty} \cup \{+\infty\}$ and let $(s_j)_{j=0}^{\kappa} \in \mathcal{H}_{q,\kappa}^{\geq,\mathrm{e}}$. Then (10.21) is fulfilled for each $j \in \mathbb{Z}_{0,\kappa}$.*

Proof. Use Proposition 4.24 and Theorem 10.13.   □

Appendix A. The Moore-Penrose inverse of a complex matrix

In this appendix, we state some facts on the Moore-Penrose inverse of a matrix. For a comprehensive exposition of the theory of generalized inverses we refer, e.g., to the monograph Rao/Mitra [27]. Let $p, q, r, s \in \mathbb{N}$ and let $A \in \mathbb{C}^{p \times q}$. Then by definition the *Moore-Penrose inverse* $A^\dagger$ of A is the unique matrix $A^\dagger \in \mathbb{C}^{q \times p}$ which satisfies the conditions $AA^\dagger A = A$, $A^\dagger AA^\dagger = A^\dagger$, $(AA^\dagger)^* = AA^\dagger$, and $(A^\dagger A)^* = A^\dagger A$. For the convenience of the reader, we state some well-known results on Moore-Penrose inverses of matrices, the proof of which is elementary.

Lemma A.1. *Let $A \in \mathbb{C}^{p \times q}$.*

(a) $(A^\dagger)^\dagger = A$, $(A^\dagger)^* = (A^*)^\dagger$, $\mathcal{R}(A^\dagger) = \mathcal{R}(A^*)$ *and* $\mathcal{N}(A^\dagger) = \mathcal{N}(A^*)$.

(b) *Let $r \in \mathbb{N}$ and $B \in \mathbb{C}^{r \times q}$. Then $\mathcal{N}(A) \subseteq \mathcal{N}(B)$ if and only if $BA^\dagger A = B$.*

(c) *Let $s \in \mathbb{N}$ and $C \in \mathbb{C}^{p \times s}$. Then $\mathcal{R}(C) \subseteq \mathcal{R}(A)$ if and only if $AA^\dagger C = C$.*

(d) *For each $p \times p$ unitary complex matrix U and each $q \times q$ unitary complex matrix V the equation $(UAV)^\dagger = V^* A^\dagger U^*$ holds.*

Now we want to pay some attention to a further generalized inverse of a square matrix. Let $A \in \mathbb{C}^{q \times q}$. For each $k \in \mathbb{N}$, then $\mathcal{R}(A^{k+1}) \subseteq \mathcal{R}(A^k)$ and $0 \le \dim \mathcal{R}(A^{k+1}) \le \dim \mathcal{R}(A^k) \le q$. Thus, there exists a minimal $m \in \mathbb{Z}_{0,q}$ such that $\dim \mathcal{R}(A^{j+1}) = \dim \mathcal{R}(A^j)$ for all $j \in \mathbb{Z}_{m,+\infty}$. This number $m \in \mathbb{Z}_{0,q}$ is called the *index of A*. Furthermore, it is well known and easily checked that there is a unique matrix $A^D \in \mathbb{C}^{q \times q}$ satisfying the equations $A^D AA^D = A^D$, $A^D A = AA^D$ and $A^D A^{m+1} = A^m$ (see, e.g., [4, Theorem 7, p. 164]). This unique matrix $A^D \in \mathbb{C}^{q \times q}$ is called the *Drazin inverse* of A. The following result (see, e.g., Campbell/Meyer [8, Theorem 7.3.4, p. 129] indicates an important relation between the Moore-Penrose inverse and the Drazin inverse.

Proposition A.2. *Let $A \in \mathbb{C}^{q \times q}$. Then $A^\dagger = A^D$ if and only if $\mathcal{R}(A) = \mathcal{R}(A^*)$.*

Appendix B. On two particular multiplicative groups of triangular block matrices

In this appendix we study two multiplicative groups of block triangular matrices.

Definition B.1. Let $n \in \mathbb{N}_0$. Then let $\mathfrak{L}_{q,n}$ (respectively, $\mathfrak{U}_{q,n}$) be the set of all $A \in \mathbb{C}^{(n+1)q \times (n+1)q}$ which satisfy the following condition: If $A = [A_{jk}]_{j,k=0}^n$ is the $q \times q$ block representation of A, then $A_{jj} = I_q$ for each $j \in \mathbb{Z}_{0,n}$ and $A_{jk} = 0_{q \times q}$ for every choice of integers j and k with $0 \le j < k \le n$ (respectively, $A_{jj} = I_q$ for each $j \in \mathbb{Z}_{0,n}$ and $A_{jk} = 0_{q \times q}$ for every choice of integers j and k with $0 \le k < j \le n$).

Remark B.2. Let $n \in \mathbb{N}_0$. Then:

(a) The set $\mathfrak{L}_{q,n}$ is a subgroup of the general linear group $\mathrm{GL}((n+1)q, \mathbb{C})$ of all non-singular complex $(n+1)q \times (n+1)q$ matrices. If $A \in \mathfrak{L}_{q,n}$ and if $A^{-1} = (\tilde{A}_{jk})_{j,k=0}^n$ denotes the $q \times q$ block partition of A^{-1}, then $\tilde{A}_{r+1,r} = -A_{r+1,r}$ for each $r \in \mathbb{Z}_{0,n-1}$.

(b) The set $\mathfrak{U}_{q,n}$ is a subgroup of $\mathrm{GL}((n+1)q,\mathbb{C})$. If $B \in \mathfrak{U}_{q,n}$ is non-singular and if $B^{-1} = (\tilde{B}_{jk})_{j,k=0}^{n}$ denotes the $q \times q$ block partition of B^{-1}, then
$$\tilde{B}_{r,r+1} = -B_{r,r+1} \text{ for each } r \in \mathbb{Z}_{0,n-1}.$$
(c) Let $A \in \mathbb{C}^{(n+1)q\times(n+1)q}$. Then $A \in \mathfrak{L}_{q,n}$ if and only if $A^* \in \mathfrak{U}_{q,n}$.

Remark B.3. Let $m, n \in \mathbb{N}$, let $(E_l)_{l=1}^{m}$ be a sequence from $\mathbb{C}^{(n+1)q\times(n+1)q}$, and let $E_0 := E_1 E_2 \cdots E_m$. For each $l \in \mathbb{Z}_{0,m}$, let $E_l = (E_{jk}^{(l)})_{j,k=0}^{n}$ be the $q \times q$ block partition of E_l. Then one can easily prove by induction that the following statements hold true:

(a) If $E_l \in \mathfrak{L}_{q,n}$ for each $l \in \mathbb{Z}_{1,m}$, then $E_{r+1,r}^{(0)} = \sum_{l=1}^{m} E_{r+1,r}^{(l)}$ for each $r \in \mathbb{Z}_{0,n-1}$.
(b) If $E_l \in \mathfrak{U}_{q,n}$ for each $l \in \mathbb{Z}_{1,m}$, then $E_{r,r+1}^{(0)} = \sum_{l=1}^{m} E_{r,r+1}^{(l)}$ for each $r \in \mathbb{Z}_{0,n-1}$.

The following two lemmas were proved in [15, Lemma A.3, Lemma A.4].

Lemma B.4. *Let $n \in \mathbb{N}_0$ and let $(A_j)_{j=0}^{n}$ and $(B_j)_{j=0}^{n}$ be sequences of complex $p \times q$ matrices. Further, let $E, V \in \mathfrak{L}_{p,n}$ and $F, W \in \mathfrak{U}_{q,n}$ be such that*
$$E \cdot \mathrm{diag}(A_0, A_1, \ldots, A_n) \cdot F = V \cdot \mathrm{diag}(B_0, B_1, \ldots, B_n) \cdot W. \qquad (\text{B.1})$$
Then $A_j = B_j$ for each $j \in \mathbb{Z}_{0,n}$ and, moreover, $(E - V) \cdot \mathrm{diag}(A_0, A_1, \ldots, A_n) = 0$ and $\mathrm{diag}(B_0, B_1, \ldots, B_n) \cdot (F - W) = 0$.

Lemma B.5. *Let $n \in \mathbb{N}_0$ and let $H \in \mathbb{C}^{(n+1)q\times(n+1)q}$. Suppose that $(A_j)_{j=0}^{n}$ and $(B_j)_{j=0}^{n}$ are sequences of complex $q \times q$-matrices and that $F, W \in \mathfrak{U}_{q,n}$ are such that $F^*HF = \mathrm{diag}(A_0, A_1, \ldots, A_n)$ and $W^*HW = \mathrm{diag}(B_0, B_1, \ldots, B_n)$ are fulfilled. Then the following statements hold:*

(a) *$A_j = B_j$ for each $j \in \mathbb{Z}_{0,n}$.*
(b) *The matrices F and W are both non-singular. Moreover, $\mathrm{diag}(A_0, A_1, \ldots, A_n) \cdot (F^{-1} - W^{-1}) = 0$ and $\mathrm{diag}(A_0^*, A_1^*, \ldots, A_n^*) \cdot (F^{-1} - W^{-1}) = 0$ hold.*
(c) *If $n \geq 1$, then $A_r F_{r,r+1} = A_r W_{r,r+1}$ and $A_r^* F_{r,r+1} = A_r^* W_{r,r+1}$ for each $r \in \mathbb{Z}_{0,n-1}$, where $F = (F_{jk})_{j,k=0}^{n}$ and $W = (W_{jk})_{j,k=0}^{n}$ are the $q \times q$ block representations of F and W, respectively.*

References

[1] D. Alpay, A. Dijksma, and H. Langer, *J_l-unitary factorization and the Schur algorithm for Nevanlinna functions in an indefinite setting*, Linear Algebra Appl. **419** (2006), no. 2-3, 675–709. MR2277998 (2007j:47024).

[2] D. Alpay, A. Dijksma, H. Langer, and Y. Shondin, *The Schur transformation for generalized Nevanlinna functions: interpolation and self-adjoint operator realizations*, Complex Anal. Oper. Theory **1** (2007), no. 2, 169–210. MR2302053 (2008b:47023).

[3] N.I. Akhiezer, *The classical moment problem and some related questions in analysis*, Translated by N. Kemmer, Hafner Publishing Co., New York, 1965. MR0184042 (32 #1518).

[4] A. Ben-Israel and T.N.E. Greville, *Generalized inverses – Theory and applications*, 2nd ed., CMS Books in Mathematics/Ouvrages de Mathématiques de la SMC, 15, Springer-Verlag, New York, 2003. MR1987382 (2004b:15008).

[5] S. Bogner, B. Fritzsche, and B. Kirstein, *The Schur-Potapov algorithm for sequences of complex $p \times q$ matrices. I*, Complex Anal. Oper. Theory **1** (2007), no. 1, 55–95. MR2276733 (2007j:47027).

[6] ———, *The Schur-Potapov algorithm for sequences of complex $p \times q$ matrices. II*, Complex Anal. Oper. Theory **1** (2007), no. 2, 235–278. MR2302055 (2008b:47028).

[7] V.A. Bolotnikov, *On degenerate Hamburger moment problem and extensions of nonnegative Hankel block matrices*, Integral Equations Operator Theory **25** (1996), no. 3, 253–276. MR1395706 (97k:44011).

[8] S.L. Campbell and C.D. Meyer Jr., *Generalized inverses of linear transformations*, Dover Publications Inc., New York, 1991. Corrected reprint of the 1979 original. MR1105324 (92a:15003).

[9] G.-N. Chen and Y.-J. Hu, *The truncated Hamburger matrix moment problems in the nondegenerate and degenerate cases, and matrix continued fractions*, Linear Algebra Appl. **277** (1998), no. 1-3, 199–236. MR1624548 (99j:44015).

[10] V.K. Dubovoj, B. Fritzsche, and B. Kirstein, *Matricial version of the classical Schur problem*, Teubner-Texte zur Mathematik [Teubner Texts in Mathematics], vol. 129, B.G. Teubner Verlagsgesellschaft mbH, Stuttgart, 1992. With German, French and Russian summaries. MR1152328 (93e:47021).

[11] Yu.M. Dyukarev, *Multiplicative and additive Stieltjes classes of analytic matrix-valued functions and interpolation problems connected with them. II*, Teor. Funktsiĭ Funktsional. Anal. i Prilozhen. **38** (1982), 40–48, 127 (Russian). MR686076 (84i:30042).

[12] Y.M. Dyukarev, B. Fritzsche, B. Kirstein, C. Mädler, and H.C. Thiele, *On distinguished solutions of truncated matricial Hamburger moment problems*, Complex Anal. Oper. Theory **3** (2009), no. 4, 759–834, DOI 10.1007/s11785-008-0061-2. MR2570113

[13] Yu.M. Dyukarev and V.È. Katsnel'son, *Multiplicative and additive Stieltjes classes of analytic matrix-valued functions and interpolation problems connected with them. I*, Teor. Funktsiĭ Funktsional. Anal. i Prilozhen. **36** (1981), 13–27, 126 (Russian). MR645305 (83i:30050).

[14] ———, *Multiplicative and additive Stieltjes classes of analytic matrix-valued functions, and interpolation problems connected with them. III*, Teor. Funktsiĭ Funktsional. Anal. i Prilozhen. **41** (1984), 64–70 (Russian). MR752057 (86a:30056).

[15] B. Fritzsche, B. Kirstein, and C. Mädler, *on Hankel nonnegative definite sequences, the canonical Hankel parametrization, and orthogonal matrix polynomials*, Complex Anal. Oper. Theory **5** (2011), 447–511.

[16] B. Fritzsche, B. Kirstein, C. Mädler, and T. Schwarz, *On the Concept of Invertibility for Sequences of Complex $p \times q$-Matrices and its Application to Holomorphic $p \times q$-Matrix-Valued Functions* **This volume**.

[17] M.I. Gekhtman and M.E. Shmoĭsh, *On invertibility of nonsquare generalized Bezoutians*, Linear Algebra Appl. **223/224** (1995), 205–241. Special issue honoring Miroslav Fiedler and Vlastimil Pták. MR1340693 (96g:15024).

[18] I.S. Kats, *On Hilbert spaces generated by monotone Hermitian matrix-functions*, Har'kov Gos. Univ. Uč. Zap. 34 = Zap. Mat. Otd. Fiz.-Mat. Fak. i Har'kov. Mat. Obšč. (4) **22** (1950), 95–113 (1951) (Russian). MR0080280 (18,222b).

[19] V.È. Katsnel'son, *Continual analogues of the Hamburger-Nevanlinna theorem and fundamental matrix inequalities of classical problems. I*, Teor. Funktsiĭ Funktsional. Anal. i Prilozhen. **36** (1981), 31–48, 127 (Russian). MR645308 (84k:44016a).

[20] ──────, *Continual analogues of the Hamburger-Nevanlinna theorem and fundamental matrix inequalities of classical problems. II*, Teor. Funktsiĭ Funktsional. Anal. i Prilozhen. **37** (1982), 31–48 (Russian). MR701996 (84k:44016b).

[21] ──────, *Continual analogues of the Hamburger-Nevanlinna theorem and fundamental matrix inequalities of classical problems. III*, Teor. Funktsiĭ Funktsional. Anal. i Prilozhen. **39** (1983), 61–73 (Russian). MR734686 (84k:44016c).

[22] ──────, *Continual analogues of the Hamburger-Nevanlinna theorem, and fundamental matrix inequalities of classical problems. IV*, Teor. Funktsiĭ Funktsional. Anal. i Prilozhen. **40** (1983), 79–90 (Russian). MR738449 (86b:30058).

[23] ──────, *Methods of J-theory in continuous interpolation problems of analysis. Part I*, T. Ando Hokkaido University, Sapporo, 1985. Translated from the Russian and with a foreword by T. Ando. MR777324 (86i:47048).

[24] I.V. Kovalishina, *Analytic theory of a class of interpolation problems*, Izv. Akad. Nauk SSSR Ser. Mat. **47** (1983), no. 3, 455–497 (Russian). MR703593 (84i:30043).

[25] R. Nevanlinna, *Asymptotische Entwicklungen beschränkter Funktionen und das Stieltjessche Momentenproblem*, Ann. Acad. Sci. Fenn. **A18** (1922), no. 5, 1–53.

[26] K.A. Penson and J.-M. Sixdeniers, *Integral representations of Catalan and related numbers*, J. Integer Seq. **4** (2001), no. 2, Article 01.2.5, 6 pp. (electronic). MR1892306 (2002m:05007).

[27] C.R. Rao and S.K. Mitra, *Generalized inverse of matrices and its applications*, John Wiley & Sons, Inc., New York-London-Sydney, 1971. MR0338013 (49 #2780).

[28] M. Rosenberg, *The square-integrability of matrix-valued functions with respect to a non-negative Hermitian measure*, Duke Math. J. **31** (1964), 291–298. MR0163346 (29 #649).

[29] ──────, *Operators as spectral integrals of operator-valued functions from the study of multivariate stationary stochastic processes*, J. Multivariate Anal. **4** (1974), 166–209. MR0378068 (51 #14237).

[30] ──────, *Spectral integrals of operator-valued functions. II. From the study of stationary processes*, J. Multivariate Anal. **6** (1976), no. 4, 538–571. MR0436302 (55 #9249).

[31] N.J.A. Sloane, *The On-Line Encyclopedia of Integer Sequences*, Dec. 2008. Published electronically at http://www.research.att.com/~njas/sequences/.

Bernd Fritzsche, Bernd Kirstein, Conrad Mädler and Tilo Schwarz
Mathematisches Institut
Universität Leipzig
Augustusplatz 10/11
D-04109 Leipzig, Germany
e-mail: {fritzsche,kirstein,maedler,tilo.schwarz}@math.uni-leipzig.de

Operator Theory:
Advances and Applications, Vol. 226, 193–210
© 2012 Springer Basel

Multiplicative Structure of the Resolvent Matrix for the Truncated Hausdorff Matrix Moment Problem

Abdon Eddy Choque Rivero

Abstract. The multiplicative structure of the resolvent matrix of the Hausdorff Matrix Moment (HMM) problem is described in the case of an odd number of moments. We use the Fundamental Matrix Inequality approach, previously used in obtaining the Blaschke–Potapov product of the resolvent matrix for the Hamburger and Stieltjes matrix moment problem studied in [10] and [7], respectively. The case of an even number of moments for the HMM problem was considered in [12].

Mathematics Subject Classification (2010). Primary 44A60, 47A57.

Keywords. Resolvent matrix, multiplicative structure, Blaschke–Potapov product.

1. Introduction

Throughout this paper, let q be a positive integer. We will use $\mathbb{C}$, $\mathbb{R}$, $\mathbb{N}_0$ and $\mathbb{N}$ to denote the set of all complex numbers, the set of all real numbers, the set of all non-negative integers, and the set of all positive integers, respectively. The set of all complex $q \times q$ matrices will be denoted by $\mathbb{C}^{q \times q}$.

Our goal will be to obtain the multiplicative structure of the resolvent matrix of the Hausdorff matrix moment (HMM) problem in the case of an odd number of moments. The resolvent matrix (RM) of the HMM problem in the even and odd cases of moments, first determined in [1] and [2], is a $2q \times 2q$ matrix polynomial.

We use the fundamental matrix inequality (FMI) approach, previously used in obtaining the Blaschke–Potapov product of the RM for the Hamburger matrix moment problem and discussed in [10]. The multiplicative structure of the Stieltjes matrix moment problem was considered in [7]. The case of an even number of

This work was supported by CIC–UMSNH, by PROMEP Red de CAs and by CONACyT grant 153184, México.

moments for the HMM problem was considered in [12]. The FMI method was successfully used to solve interpolation problems on a finite interval $[a, b]$, see [3], [4], as well as other papers dealing with interpolation problems on the real axis [9], [8], [13] and on the complex unit circle [6].

In this paper we use the same form of the RM of the HMM problem in the odd case of moments as was given in [5]. This paper focused on the construction of orthogonal matrix polynomials on the finite interval $[a, b]$ of the real axis (including the case of an odd number of moments).

Let

$$\widetilde{J}_q := \begin{pmatrix} 0_{q \times q} & -iI_q \\ iI_q & 0_{q \times q} \end{pmatrix}, \tag{1.1}$$

where $0_{q \times q}$ and I_q denote the $q \times q$ null and identity matrices, respectively, and let $\Pi_+ := \{w \in \mathbb{C} : \operatorname{Im} w \in (0, +\infty)\}$.

Definition 1.1. A matrix function $W : \mathbb{C} \to \mathbb{C}^{2q \times 2q}$ is said to belong to the Potapov class $\mathfrak{P}_{\widetilde{J}_q}$ in Π_+ if W satisfies the following two conditions:

1) $\widetilde{J}_q - W^*(z)\widetilde{J}_q W(z) \geq 0_{2q \times 2q}$, $z \in \Pi_+$.
2) $\widetilde{J}_q - W^*(x)\widetilde{J}_q W(x) = 0_{2q \times 2q}$, $x \in \mathbb{R}$.

The resolvent matrix U_j (2.16) which we study in this paper belongs to the Potapov class of functions.

The main result of this paper is Theorem 4.2 and its corollary which gives us a multiplicative representation of the RM (2.16),

$$U_n(z) = b_1(z) \times \cdots \times b_n(z).$$

Compared with previous papers on multiplicative representations of the RM solved with the help of the Potapov method (see [12], [7], [10]) the Blaschke–Potapov product representation of the RM of the HMM problem in the case of an odd number of moments turned out to be much more difficult. This difficulty, in essence, arises from the fact that the two block Hankel matrices appearing in the RM (see (2.16)), have different sizes. A number of nontrivial coupling identities (see Remark 3.3 and Proposition 3.4) were required for the construction of the multiplicative representation of the corresponding RM.

For the sake of completeness, in the Appendix we include the Blaschke–Potapov product for the case of an even number of moments [12]. In both cases, i.e., for an even as well as an odd number of moments), it is assumed that the information blocks, $H_{1,j}$ and $H_{2,j}$ (see (2.2) and (2.3)), are positive definite.

2. Notation and preliminaries

2.1. Hausdorff matrix moment problem and the resolvent matrix

Let p and m be positive integers. We will use $0_{p \times m}$ to denote the null matrix belonging to $\mathbb{C}^{p \times m}$ and I_p for the identity matrix belonging to $\mathbb{C}^{p \times p}$. In cases where the size of the null and the identity matrix are clear, we will omit the indices.

The HMM problem for an interval $[a, b]$ can be formulated as follows: Let a finite sequence of complex $q \times q$ matrices $(s_j)_{j=0}^{m}$ be given. Find the set $M_{\geq}^{q}[[a, b], \mathfrak{B} \cap [a, b]; (s_j)_{j=0}^{m}]$ of all non-negative Hermitian $q \times q$ measures σ defined on the Borel σ-algebra $\mathfrak{B} \cap [a, b]$ such that

$$s_j = \int_{[a,b]} t^j d\sigma(t) \tag{2.1}$$

for each integer j with $0 \leq j \leq m$.

It was proved in [2], for $m = 2n$ and $n \geq 1$, that the HMM problem is solvable if and only if the block matrices $H_{1,n}$ and $H_{2,n}$ are positive semidefinite, where

$$H_{1,n} := \widetilde{H}_{0,n} \tag{2.2}$$

and

$$H_{2,n} := -ab\widetilde{H}_{0,n-1} + (a + b)\widetilde{H}_{1,n-1} - \widetilde{H}_{2,n-1} \tag{2.3}$$

are defined with help of the Hankel matrices,

$$\widetilde{H}_{0,n} := \begin{pmatrix} s_0 & s_1 & \cdots & s_n \\ s_1 & s_2 & \cdots & s_{n+1} \\ \vdots & \vdots & \vdots & \vdots \\ s_n & s_{n+1} & \cdots & s_{2n} \end{pmatrix}, \quad \widetilde{H}_{1,n-1} := \begin{pmatrix} s_1 & s_2 & \cdots & s_n \\ s_2 & s_3 & \cdots & s_{n+1} \\ \vdots & \vdots & \vdots & \vdots \\ s_n & s_{n+1} & \cdots & s_{2n-1} \end{pmatrix}$$

and

$$\widetilde{H}_{2,n-1} := \begin{pmatrix} s_2 & s_3 & \cdots & s_{n+1} \\ s_3 & s_4 & \cdots & s_{n+2} \\ \vdots & \vdots & \vdots & \vdots \\ s_{n+1} & s_{n+2} & \cdots & s_{2n} \end{pmatrix}.$$

The set of solutions to the HMM problem is given in terms of the resolvent matrix of the HMM problem (see [2, Theorem 6.14]).

Throughout this paper, we assume that $H_{1,n}$ and $H_{2,n}$ are positive definite.

2.2. The resolvent matrix of the HMM problem

Let $R_{1,j} : \mathbb{C} \to \mathbb{C}^{(j+1)q \times (j+1)q}$ and $R_{2,j} : \mathbb{C} \to \mathbb{C}^{jq \times jq}$ be given by

$$R_{1,j}(z) := (I_{(j+1)q} - zT_j)^{-1}, \quad j \geq 0, \tag{2.4}$$

$$R_{2,j}(z) := \begin{cases} 0_{q \times q}; & j = 0, \\ R_{1,j}(z); & j \geq 1, \end{cases} \tag{2.5}$$

with

$$T_0 := 0_{q \times q}, \tag{2.6}$$

$$T_j := \begin{pmatrix} 0_{q \times jq} & 0_{q \times q} \\ I_{jq} & 0_{jq \times q} \end{pmatrix}, \quad j \geq 0. \tag{2.7}$$

Observe that, for each $j \in \mathbb{N}_0$, the matrix-valued function $R_{1,j}$ can be represented via

$$R_{1,j}(z) = \begin{pmatrix} I_q & 0_{q\times q} & 0_{q\times q} & \cdots & 0_{q\times q} & 0_{q\times q} \\ zI_q & I_q & 0_{q\times q} & \cdots & 0_{q\times q} & 0_{q\times q} \\ z^2 I_q & zI_q & I_q & \cdots & 0_{q\times q} & 0_{q\times q} \\ \vdots & \vdots & \vdots & \ddots & \vdots & \vdots \\ z^j I_q & z^{j-1} I_q & z^{j-2} I_q & \cdots & zI_q & I_q \end{pmatrix} \tag{2.8}$$

and

$$R_{1,j}(z) = \sum_{l=0}^{j} z^l T_j^l. \tag{2.9}$$

Let

$$v_{1,0} := I_q, \, v_{1,\ell} := 0_{q\times q}, \quad \text{if} \quad \ell < 0, \tag{2.10}$$

$$v_{1,j} := \begin{pmatrix} I_q \\ 0_{jq\times q} \end{pmatrix} = \begin{pmatrix} v_{1,j-1} \\ 0_{q\times q} \end{pmatrix}, \quad \forall j \in \mathbb{N} \tag{2.11}$$

and

$$v_{2,j} := v_{1,j-1}, \quad \forall j \in \mathbb{N}. \tag{2.12}$$

Furthermore, let

$$y_{[j,k]} := \begin{pmatrix} s_j \\ s_{j+1} \\ \vdots \\ s_k \end{pmatrix}, \, 0 \le j \le k \le 2n \tag{2.13}$$

with $y_{[j,k]} = 0_{q\times q}$, if $j > k$,

$$u_{1,0} := 0_{q\times q}, \, u_{1,j} := \begin{pmatrix} 0_{q\times q} \\ -y_{[0,j-1]} \end{pmatrix}, \quad 1 \le j \le n, \tag{2.14}$$

and

$$u_{2,0} = 0_{q\times q}, \, u_{2,j} := -ab u_{1,j-1} - (a+b)y_{[0,j-1]} + y_{[1,j]}, \, 1 \le j \le n. \tag{2.15}$$

We will consider a sequence of resolvent matrices $(U_j)_{j=1}^n$ of the HMM problem for an odd number of given data matrices $s_0, s_1, \ldots, s_{2n}$,

$$U_j(z) := \begin{pmatrix} U_{11,j}(z) & U_{12,j}(z) \\ U_{21,j}(z) & U_{22,j}(z) \end{pmatrix}, \, z \in \mathbb{C}, \quad 1 \le j \le n, \tag{2.16}$$

with

$$U_{11,j}(z) := I_q - (z-a)u_{1,j}^*(R_{1,j}(\bar{z}))^* H_{1,j}^{-1} R_{1,j}(a)v_{1,j}, \tag{2.17}$$

$$U_{12,j}(z) := (b-a)^{-1}(s_0 + (u_{2,j}^* + z s_0 v_{2,j}^*)(R_{2,j}(\bar{z}))^* H_{2,j}^{-1} R_{2,j}(a)(u_{2,j} + a v_{2,j} s_0)), \tag{2.18}$$

$$U_{21,j}(z) := -(z-a)v_{1,j}^*(R_{1,j}(\bar{z}))^* H_{1,j}^{-1} R_{1,j}(a)v_{1,j} \tag{2.19}$$

and

$$U_{22,j}(z) := (b-z)(b-a)^{-1}(I_q + (z-a)v_{2,j}^*(R_{2,j}(\bar{z}))^* H_{2,j}^{-1} R_{2,j}(a)(u_{2,j} + av_{2,j}s_0)).$$

Observe that U_j is a $2q \times 2q$ matrix polynomial.

The matrix polynomials U_j, $1 \leq j \leq n$ have interesting and constructive properties with respect to the signature matrix (1.1). In particular, each U_j belongs to the Potapov class (see Definition 1.1 as well as [6, Theorem 1.3.3] and [1, Proposition 6.3]).

Let

$$\widehat{s}_j := -abs_j + (a+b)s_{j+1} - s_{j+2},\ 0 \leq j \leq 2n-2, \tag{2.20}$$

$$\widehat{u}_{2,j}(z) := u_{2,j} + zv_{2,j}s_0,\ 1 \leq j \leq 2n \tag{2.21}$$

and

$$\lambda_1 = I_q, \quad \lambda_j := \begin{pmatrix} 0_{(j-1)q \times q} \\ I_q \end{pmatrix},\ j \in \mathbb{N}, \tag{2.22}$$

where λ_j is a $jq \times q$ matrix.

We furthermore define the following matrices:

$$Y_{1,j} := \begin{pmatrix} s_j \\ s_{j+1} \\ \vdots \\ s_{2j-1} \end{pmatrix},\ 1 \leq j \leq n, \qquad Y_{2,j} := \begin{pmatrix} \widehat{s}_{j-1} \\ \widehat{s}_j \\ \vdots \\ \widehat{s}_{2j-3} \end{pmatrix},\ 2 \leq j \leq n, \tag{2.23}$$

$$\widehat{H}_{1,j} := s_{2j} - Y_{1,j}^* H_{1,j-1}^{-1} Y_{1,j},\ 1 \leq j \leq n, \tag{2.24}$$

$$\widehat{H}_{2,j} := \widehat{s}_{2j-2} - Y_{2,j}^* H_{2,j-1}^{-1} Y_{2,j},\ 2 \leq j \leq n, \tag{2.25}$$

$$\widehat{v}_{1,j}(z) := z^j I_q - Y_{1,j}^* H_{1,j-1}^{-1} R_{1,j-1}(z)v_{1,j-1},\ 1 \leq j \leq n, \tag{2.26}$$

$$\widehat{v}_{2,j}(z) := z^{j-1} I_q - Y_{2,j}^* H_{2,j-1}^{-1} R_{2,j-1}(z)v_{2,j-1},\ 2 \leq j \leq n, \tag{2.27}$$

$$\widehat{w}_{1,j}(z) := (-Y_{1,j}^* + z\lambda_j^* H_{1,j-1}) H_{1,j-1}^{-1} R_{1,j-1}(z)u_{1,j-1} - s_{j-1},\ 1 \leq j \leq n, \tag{2.28}$$

and

$$\widehat{w}_{2,j}(z) := (-Y_{2,j}^* + z\lambda_{j-1}^* H_{2,j-1}) H_{2,j-1}^{-1} R_{2,j-1}(z)\widehat{u}_{2,j-1}(z) - \widehat{s}_{j-2},\ 2 \leq j \leq n. \tag{2.29}$$

The inverse of $H_{1,j}$, for $1 \leq j \leq n$ and the inverse of $H_{2,j}$, for $2 \leq j \leq n$ can be written in the form

$$H_{k,j}^{-1} = \begin{pmatrix} H_{k,j-1}^{-1} & 0_{jq \times q} \\ 0_{q \times jq} & 0_{q \times q} \end{pmatrix} + \begin{pmatrix} -H_{k,j-1}^{-1} Y_{k,j} \\ I_q \end{pmatrix} \widehat{H}_{k,j}^{-1} (-Y_{k,j}^* H_{k,j-1}^{-1}, I_q). \tag{2.30}$$

The following proposition describes a form of the RM which is suitable for our purposes.

Proposition 2.1. *The resolvent matrix $U_j(z)$ defined by (2.16) can be written in the form,*

$$U_{11,j}(z) = U_{11,j-1}(z) - (z-a)\widehat{w}_{1,j}^*(z)\widehat{H}_{1,j}^{-1}\widehat{v}_{1,j}(a), \tag{2.31}$$

$$U_{12,j}(z) = U_{12,j-1}(z) + (b-a)^{-1}\widehat{w}_{2,j}^*(z)\widehat{H}_{2,j}^{-1}\widehat{w}_{2,j}(a), \tag{2.32}$$

$$U_{21,j}(z) = U_{21,j-1}(z) - (z-a)\widehat{v}_{1,j}^*(z)\widehat{H}_{1,j}^{-1}\widehat{v}_{1,j}(a) \tag{2.33}$$

and

$$U_{22,j}(z) = U_{22,j-1}(z) + \frac{(z-a)(b-z)}{b-a}\widehat{v}_{2,j}^*(z)\widehat{H}_{2,j}^{-1}\widehat{w}_{2,j}(a) \tag{2.34}$$

for $2 \le j \le n$.

Proof. The proof is by direct calculation. Use (2.30) and the identities

$$R_{1,j}(z) = \left(\begin{array}{c|c} R_{1,j-1}(z) & 0_{jq \times q} \\ (z^j I_q, z^{j-1} I_q, \ldots, z I_q) & I_q \end{array} \right), \quad u_{1,j} = \left(\begin{array}{c} u_{1,j-1} \\ -s_{j-1} \end{array} \right)$$

and

$$u_{2,j} = \left(\begin{array}{c} u_{2,j-1} \\ -\widehat{s}_{j-2} \end{array} \right). \tag*{$\square$}$$

3. Main algebraic identities

In this section we obtain essential identities concerning the block matrices introduced in Section 2. We derive coupling identities between the block Hankel matrices $H_{1,j}$ and $H_{2,j}$, $1 \le j \le n$. As a byproduct, we obtain a number of auxiliary identities, each of interest on their own.

For each positive integer n, let

$$L_{1,n} := (\delta_{j,k+1} I_q)_{\substack{j=0,\ldots,n \\ k=0,\ldots,n-1}}, \quad \text{and} \quad L_{2,n} := (\delta_{j,k} I_q)_{\substack{j=0,\ldots,n \\ k=0,\ldots,n-1}}, \tag{3.1}$$

where $\delta_{j,k}$ is the Kronecker symbol: $\delta_{j,k} := 1$ if $j = k$ and $\delta_{j,k} := 0$ if $j \neq k$.

Remark 3.1. Let $j \in \mathbb{N}$. Then the following identities hold:

$$\lambda_j - L_{1,j}^* \lambda_{j+1} = 0, \tag{3.2}$$

$$v_{1,j}^* L_{1,j} = 0, \tag{3.3}$$

$$v_{2,j}^* - v_{1,j}^* L_{2,j} = 0, \tag{3.4}$$

$$L_{1,j} L_{1,j}^* - T_j T_j^* = 0, \tag{3.5}$$

$$L_{2,j} L_{1,j}^* - T_j^* = 0, \tag{3.6}$$

$$L_{2,j} T_{j-1}^* - T_j^* L_{2,j} = 0, \tag{3.7}$$

and

$$v_{1,j} v_{1,j}^* - I_{(j+1)q} + L_{1,j} L_{1,j}^* = 0. \tag{3.8}$$

Remark 3.2. Let $j \in \mathbb{N}$. Then, for all complex numbers z and w, the identities

$$v_{1,j}^*(R_{1,j}(\bar{z}))^* \lambda_{j+1} - z^j I_q = 0, \tag{3.9}$$

$$R_{1,j}(z) - R_{1,j}(w) = (z-w)R_{1,j}(z)T_j R_{1,j}(w), \tag{3.10}$$

$$zR_{1,j}(z) - wR_{1,j}(w) = (z-w)R_{1,j}(z)R_{1,j}(w), \tag{3.11}$$

$$(I_{(j+1)q} - zT_j^*)L_{2,j}(R_{2,j}(\bar{z}))^* - L_{2,j} = 0, \tag{3.12}$$

$$(z-w)v_{1,j}^*(R_{1,j}(\bar{z}))^*(R_{1,j}(\bar{w}))^*\lambda_{j+1} - z^{j+1}I_q + w^{j+1}I_q = 0, \tag{3.13}$$

$$(z-w)v_{1,j}^*(R_{1,j}(\bar{z}))^*T_j^*(R_{1,j}(\bar{w}))^*\lambda_{j+1} - z^j I_q + w^j I_q = 0, \tag{3.14}$$

$$(z-w)v_{2,j}^*(R_{2,j}(\bar{z}))^*L_{1,j}^*(R_{1,j}(\bar{z}))^*\lambda_{j+1} - z^j I_q + w^j I_q = 0, \tag{3.15}$$

$$(R_{1,j}(\bar{z}))^* \left(L_{1,j} - zL_{1,j}T_{j-1}^* - zL_{2,j} + z^2 T_j^* L_{2,j} \right) (R_{2,j}(\bar{z}))^* - L_{1,j} = 0, \tag{3.16}$$

and

$$(R_{1,j}(\bar{z}))^*T_j^* - L_{2,j}(R_{2,j}(\bar{z}))^*L_{1,j}^* = 0, \tag{3.17}$$

hold.

Remark 3.3. Let $(s_j)_{j=0}^{2n}$ be a sequence of $q \times q$ matrices. Then the following identities hold:

$$\widetilde{H}_{0,j-1}L_{1,j}^*T_j^{*^{j-1}}\lambda_{j+1} - L_{2,j}^*H_{1,j}v_{1,j} = 0, \quad 1 \leq j \leq n, \tag{3.18}$$

$$\widetilde{H}_{1,j-1}L_{1,j}^*T_j^{*^{j-1}}\lambda_{j+1} - L_{1,j}^*H_{1,j}v_{1,j} = 0, \quad 1 \leq j \leq n, \tag{3.19}$$

$$\widetilde{H}_{1,j-1}L_{1,j}^*\lambda_{j+1} - y_{[j,2j-1]} = 0, \quad 1 \leq j \leq n, \tag{3.20}$$

$$\widetilde{H}_{2,j-1}L_{1,j}^*\lambda_{j+1} - y_{[j+1,2j]} = 0, \quad 1 \leq j \leq n, \tag{3.21}$$

$$L_{2,j}^*Y_{1,j+1} - y_{[j+1,2j]} = 0, \quad 1 \leq j \leq n, \tag{3.22}$$

$$L_{1,j}^*Y_{1,j+1} - y_{[j+2,2j+1]} = 0, \quad 1 \leq j \leq n-1, \tag{3.23}$$

$$L_{1,j}y_{[j+2,2j+1]} + v_{1,j}s_{j+1} - Y_{1,j+1} = 0, \quad 1 \leq j \leq n-1, \tag{3.24}$$

$$-L_{1,j}y_{[j+1,2j]} + H_{1,j}L_{1,j}\lambda_j - v_{1,j}s_j = 0, \quad 1 \leq j \leq n, \tag{3.25}$$

$$L_{1,j}y_{[j,2j-1]} + v_{1,j}s_{j-1} - H_{1,j}L_{2,j}\lambda_j = 0, \quad 1 \leq j \leq n, \tag{3.26}$$

$$L_{1,j}\widetilde{H}_{0,j-1} - T_j H_{1,j}L_{2,j} = 0, \quad 1 \leq j \leq n, \tag{3.27}$$

$$L_{1,j}\widetilde{H}_{1,j-1} - T_j H_{1,j}L_{1,j} = 0, \quad 1 \leq j \leq n, \tag{3.28}$$

$$L_{1,j}\widetilde{H}_{2,j-1}\lambda_j - T_j Y_{1,j+1} = 0, \quad 1 \leq j \leq n-1, \tag{3.29}$$

$$(L_{1,j}^* - bL_{2,j}^*)H_{1,j}(-v_{1,j}v_{1,j}^* + I_{(j+1)q} - aT_j^*) + H_{2,j}L_{1,j}^* = 0, \quad 1 \leq j \leq n, \tag{3.30}$$

$$u_{1,j}^* + v_{1,j}^*H_{1,j}T_j^* = 0, \quad 1 \leq j \leq n, \tag{3.31}$$

$$s_0 v_{1,j}^* + v_{1,j}^*H_{1,j}(T_jT_j^* - I_{(j+1)q}) = 0, \quad 1 \leq j \leq n, \tag{3.32}$$

$$v_{1,j}^*Y_{1,j+1} = s_{j+1}, \quad 1 \leq j \leq n-1, \tag{3.33}$$

$$v_{1,j}^*H_{1,j}\lambda_{j+1} = s_j, \quad 1 \leq j \leq n, \tag{3.34}$$

$$v_{1,j}^*H_{1,j}L_{2,j}\lambda_j = s_{j-1}, \quad 1 \leq j \leq n, \tag{3.35}$$

$$H_{2,j} + abL_{2,j}^* H_{1,j} L_{1,j} - (a+b)L_{2,j}^* H_{1,j} L_{1,j} + L_{1,j}^* H_{1,j} L_{1,j} = 0, \quad 1 \le j \le n, \tag{3.36}$$

$$L_{1,j} H_{1,j} T_j^* - L_{2,j}^* H_{1,j} L_{1,j} L_{1,j}^* = 0, \quad 1 \le j \le n, \tag{3.37}$$

$$u_{2,j}^* T_{j-1}^* + s_0 v_{2,j}^* + u_{1,j}^* (L_{1,j}(I_{jq} - aT_{j-1}^*)$$
$$- b(I_{(j+1)q} - aT_j^*) L_{2,j}) = 0, \quad 1 \le j \le n, \tag{3.38}$$

$$u_{1,j}(R_{1,j}(\bar{z}))^* [L_{1,j} T_{j-1}^* - T_j^* L_{1,j}] = 0, \ z \in \mathbb{C}, \ 1 \le j \le 2n+1 \tag{3.39}$$

and

$$u_{1,j}(R_{1,j}(\bar{z}))^* (L_{1,j} - bL_{2,j}) L_{1,j}^* - u_{1,j}(R_{1,j}(\bar{z}))^* (I_{(j+1)q} - bT_j^*) = 0,$$
$$z \in \mathbb{C}, \ 1 \le j \le 2n+1. \tag{3.40}$$

Moreover, for $0 \le k \le j$ and $1 \le j \le n$, the following identities are valid:

$$\widetilde{H}_{0,j-1} L_{1,j}^* T_j^{*^{j-k-2}} \lambda_{j+1} - \widetilde{H}_{1,j-1} L_{1,j}^* T_j^{*^{j-k-1}} \lambda_{j+1} = 0, \tag{3.41}$$

$$\widetilde{H}_{1,j-1} L_{1,j}^* T_j^{*^{j-k-2}} \lambda_{j+1} - \widetilde{H}_{2,j-1} L_{1,j}^* T_j^{*^{j-k-1}} \lambda_{j+1} = 0. \tag{3.42}$$

Proposition 3.4 (Coupling Identities). *Let* $(s_j)_{j=0}^{2n}$ *be a sequence of complex* $q \times q$ *matrices. Then, for* $1 \le j \le n$, *the following identities hold:*

$$(u_{2,j}^* + as_0 v_{2,j}^*)(R_{2,j}(a))^* - v_{1,j}^* H_{1,j}(L_{1,j} - bL_{2,j}) = 0, \tag{3.43}$$

$$R_{1,j}(a) v_{1,j}(u_{2,j}^* + as_0 v_{2,j}^*) R_{2,j}(a) - R_{1,j}(a) L_{1,j} H_{2,j} - H_{1,j}(L_{1,j} - bL_{2,j}) = 0, \tag{3.44}$$

and

$$(u_{2,j}^* + as_0 v_{2,j}^*)(R_{2,j}(a))^* - (u_{2,j}^* + zs_0 v_{2,j}^*)(R_{2,j}(\bar{z}))^*$$
$$- (z-a) u_{1,j}^* (R_{1,j}(\bar{z}))^* (L_{1,j} - bL_{2,j}) = 0. \tag{3.45}$$

Moreover, for $1 \le j \le n-1$, *the following the identities are valid:*

$$\left(L_{1,j}^* - bL_{2,j}^* \right) H_{1,j} v_{1,j} a^{j+1} - \left(L_{1,j}^* - bL_{2,j}^* \right) Y_{1,j+1}$$
$$- aH_{2,j} L_{1,j}^* (R_{1,j}(a))^* \lambda_{j+1} - Y_{2,j+1} = 0, \tag{3.46}$$

$$aH_{1,j}(L_{1,j} - bL_{2,j})\lambda_j + bH_{1,j}\lambda_{j+1} - Y_{1,j+1} - L_{1,j} Y_{2,j+1} - v_{1,j} \hat{s}_{j-1} = 0, \tag{3.47}$$

$$aL_{1,j} H_{2,j} \lambda_j - a^2 T_j H_{1,j}(L_{1;j} - bL_{2,j})\lambda_j + aT_j Y_{1,j+1} - abT_j H_{1,j}\lambda_{j+1} = 0, \tag{3.48}$$

$$R_{2,j}(a)(u_{2,j} + av_{2,j} s_0)\hat{v}_{1,j+1}^*(a)$$
$$- H_{2,j} L_{1,j}^* (R_{1,j}(a))^* (-H_{1,j}^{-1} Y_{1,j+1} + a\lambda_{j+1}) + Y_{2,j+1} = 0, \tag{3.49}$$

$$R_{1,j}(a) v_{1,j} \hat{w}_{2,j+1}(a) + H_{1,j}(L_{1,j} - bL_{2,j}) H_{2,j}^{-1} Y_{2,j+1}$$
$$+ bH_{1,j}\lambda_{j+1} - Y_{1,j+1} = 0. \tag{3.50}$$

Proof. Identity (3.43) is equivalent to the equality

$$u_{2,j}^* + as_0 v_{2,j}^* - v_{1,j}^* H_{1,j}(L_{1,j} - bL_{2,j})(I_{q(j-1)} - aT_{j-1}^*) = 0.$$

The latter readily follows from (2.7), (2.11), (2.2), (2.15) and (3.1).
    Identity (3.44) follows directly from identity (2.3) [2, Proposition 2.5].

Let $\Delta_{(3.46)}$ denote the left-hand side of (3.46). Then

$$
\begin{aligned}
\Delta_{(3.46)} &= \left(L_{1,j}^* - bL_{2,j}^*\right)H_{1,j}v_{1,j}a^{j+1} - \left(L_{1,j}^* - bL_{2,j}^*\right)Y_{1,j+1} + (a^2b\widetilde{H}_{0,j-1} \\
&\quad - a\widetilde{H}_{1,j-1} + \widetilde{H}_{2,j-1})(L_{1,j}^*\lambda_{j+1} + aL_{1,j}^*T_j^*\lambda_{j+1} + \cdots + a^j L_{1,j}^*T_j^{*j}\lambda_{j+1}) \\
&\quad + aby_{[j,2j-1]} - (a+b)y_{[j+1,2j]} + y_{[j+2,2j+1]} \\
&= a^{j+1}b\widetilde{H}_{0,j-1}L_{1,j}^*T_j^{*j-1}\lambda_{j+1} + \cdots + a^3 b\widetilde{H}_{0,j-1}L_{1,j}^*T_j^*\lambda_{j+1} \\
&\quad + a^2 b\widetilde{H}_{0,j-1}L_{1,j}^*\lambda_{j+1} - a^{j+1}\widetilde{H}_{1,j-1}L_{1,j}^*T_j^{*j-1}\lambda_{j+1} - \cdots \\
&\quad - a^3 b\widetilde{H}_{1,j-1}L_{1,j}^*T_j^*\lambda_{j+1} - a^2\widetilde{H}_{1,j-1}L_{1,j}^*\lambda_{j+1} - a^j b\widetilde{H}_{1,j-1}L_{1,j}^*T_j^{*j-1}\lambda_{j+1} \\
&\quad - \cdots - a^2 b\widetilde{H}_{1,j-1}L_{1,j}^*T_j^*\lambda_{j+1} - ab\widetilde{H}_{1,j-1}L_{1,j}^*\lambda_{j+1} \\
&\quad - a^j\widetilde{H}_{2,j-1}L_{1,j}^*T_j^{*j-1}\lambda_{j+1} - \cdots - a^2\widetilde{H}_{2,j-1}L_{1,j}^*T_j^*\lambda_{j+1} \\
&\quad - a\widetilde{H}_{2,j-1}L_{1,j}^*\lambda_{j+1} + a^{j+1}L_{1,j}^*H_{1,j-1}v_{1,j} - a^{j+1}bL_{1,j}^*H_{1,j-1}v_{1,j} \\
&\quad + aby_{[j,2j-1]} - bL_{1,j}^*y_{[j+1,2j+1]} + by_{[j+1,2j]} - ay_{[j+1,2j]} \\
&\quad - L_{1,j}y_{[j+1,2j+1]} + y_{[j+2,2j+1]} \\
&= \sum_{l=1}^{j+1} A_{l,1}a^l b + \sum_{l=1}^{j+1} A_{l,0}a^l + A_{0,1}b + A_{0,0}
\end{aligned}
$$

where

$$
\begin{aligned}
A_{j+1,1} &:= -L_{2,j}^* H_{1,j}v_{1,j} + \widetilde{H}_{0,j-1}L_{1,j}^*T_j^{*j-1}\lambda_j, \\
A_{l,1} &:= \widetilde{H}_{0,j-1}L_{1,j}T_j^{*j-2}\lambda_j - H_{1,j-1}L_{1,j}T_j^{*j-1}\lambda_j, 2 \le l \le j, \\
A_{1,1} &:= H_{1,j-1}L_{1,j}^*\lambda_j + y_{[j,2j-1]}, \\
A_{j+1,0} &:= L_{1,j}^* H_{1,j}v_{1,j} - \widetilde{H}_{1,j-1}L_{1,j}^*T_j^{*j-1}\lambda_j, \\
A_{l,0} &:= \widetilde{H}_{1,j-1}L_{1,j}T_j^{*j-2}\lambda_j - H_{2,j-1}L_{1,j}T_j^{*j-1}\lambda_j, 2 \le l \le j, \\
A_{1,0} &:= \widetilde{H}_{2,j-1}L_{1,j}^*\lambda_j + y_{[j+2,2j+1]}, \\
A_{0,1} &:= L_{2,j}^* y_{[j+1,2j+1]} + y_{[j+1,2j]}, \\
A_{0,0} &:= L_{1,j}^* y_{[j+1,2j+1]} - y_{[j+2,2j-1]}.
\end{aligned}
$$

Taking (3.18)–(3.23) into account, it follows that $A_{l,k} = 0$ for $j,k \in \{0,\ldots,j+1\}$. Thus we get $\Delta_{(3.46)} = 0$. This proves the identity (3.46).

We next prove (3.47). Let $\Delta_{(3.47)}$ denote the left-hand side of (3.47). Then

$$
\begin{aligned}
\Delta_{(3.47)} &= ab(L_{1,j}y_{[j,2j-1]} + v_{1,j}s_{j-1} - H_{1,j}L_{2,j}\lambda_j) \\
&\quad + b(-L_{1,j}y_{[j+1,2j]} - v_{1,j}s_j + H_{1,j}\lambda_{j+1}) \\
&\quad \times a(-L_{1,j}y_{[j+1,2j]} + H_{1,j}L_{1,j}\lambda_j - v_{1,j}s_j) \\
&\quad + L_{1,j}y_{[j+2,2j+1]} + v_{1,j}s_{j+1} - Y_{1,j+1}.
\end{aligned}
$$

By (3.24), (3.25) and (3.26), we have $\Delta_{(3.47)} = 0$, which proves (3.47).

We now prove (3.48). Let $\Delta_{(3.48)}$ denote the left-hand side of (3.48). We have

$$\Delta_{(3.48)} = a^2 b(-L_{1,j}\widetilde{H}_{0,j-1} + T_j H_{1,j} L_{2,j})\lambda_j + a^2 (L_{1,j}\widetilde{H}_{1,j-1} - T_j H_{1,j} L_{2,j})\lambda_j$$
$$ab(L_{1,j}\widetilde{H}_{1,j-1}\lambda_j - T_j H_{1,j} L_{2,j}\lambda_{j+1}) + a(-L_{1,j}\widetilde{H}_{2,j-1}\lambda_j + T_j Y_{1,j+1}).$$

According to (3.27), (3.28) and (3.29), we have $\Delta_{(3.48)} = 0$, which proves (3.48).

Now we prove (3.49). Let the left-hand side of this identity be denoted by $\Delta_{(3.49)}$. Using (3.43) and (2.26), we have

$$\begin{aligned}
\Delta_{(3.49)} &= (L_{1,j}^* - bL_{2,j}^*)H_{1,j}v_{1,j}a^{j+1} + [-(L_{1,j}^* - bL_{2,j}^*)H_{1,j}v_{1,j}v_{1,j}^* + H_{2,j}L_{1,j}^*] \\
&\quad \cdot (R_{1,j}(a))^* H_{1,j}^{-1} Y_{1,j+1} + aH_{2,j}L_{1,j}^*(R_{1,j}(a))^*\lambda_{j+1} + Y_{2,j+1} \\
&= [(L_{1,j}^* - bL_{2,j}^*)H_{1,j}(-v_{1,j}v_{1,j}^* + I_{(j+1)q} - aT_j^*) + H_{2,j}L_{1,j}^*] \\
&\quad \cdot (R_{1,j}(a))^* H_{1,j}^{-1} Y_{1,j+1} \\
&= 0.
\end{aligned}$$

In the second equality we used (3.46) and in the last equality we used (3.30). This proves the identity (3.49).

We next prove identity (3.50). Let $\Delta_{(3.50)}$ denote the left-hand side of (3.50). Using (3.44) we have

$$\begin{aligned}
\Delta_{(3.50)} &= R_{1,j}(a)L_{1,j}(-Y_{2,j+1} + aH_{2,j}\lambda_j) + aH_{1,j}(L_{1,j} - bL_{2,j})\lambda_j \\
&\quad - R_{1,j}(a)v_{1,j}\widehat{s}_{j-1} + bH_{1,j}\lambda_{j+1} - Y_{1,j+1} \\
&= R_{1,j}(a)\,[aH_{1,j}L_{1,j}\lambda_j + bH_{1,j}\lambda_{j+1} - abH_{1,j}L_{2,j}\lambda_j - L_{1,j}Y_{2,j+1} - Y_{1,j+1} \\
&\quad - v_{1,j}\widehat{s}_{j-1} + aL_{1,j}H_{2,j}\lambda_j - a^2 T_j H_{1,j}(L_{1,j} - bL_{2,j})\lambda_j + aT_j Y_{1,j+1} \\
&\quad - abT_j H_{1,j}\lambda_{j+1}] \\
&= 0.
\end{aligned}$$

In the last equality we used (3.47) and (3.48). This proves the identity (3.50).

Let $\Delta_{(3.45)}$ be the left-hand side of (3.45). Using (3.10) and (3.11) we have, for all $z \in \mathbb{C}$:

$$\begin{aligned}
\Delta_{(3.45)} &= u_{2,j}^*((R_{2,j}(a))^* - (R_{2,j}(\bar{z}))^*) + s_0 v_{2,j}(a(R_{2,j}(a))^* - z(R_{2,j}(\bar{z}))^*) \\
&\quad - (z-a)u_{1,j}^*(R_{1,j}(\bar{z}))^*(L_{1,j} - bL_{2,j}) \\
&= -(z-a)[(u_{2,j}^*T_{j-1}^* + s_0 v_{2,j})(R_{2,j}(a))^*(R_{2,j}(\bar{z}))^* \\
&\quad + u_{1,j}^*(R_{1,j}(\bar{z}))^*(L_{1,j} - bL_{2,j})] \\
&= (z-a)(u_{1,j}^*[L_{1,j}(I_{jq} - aT_{j-1}^*) - b(I_{(j+1)q} - aT_j^*)L_{2,j}](R_{2,j}(a))^*(R_{2,j}(\bar{z}))^* \\
&\quad - u_{1,j}^*(R_{1,j}(\bar{z}))^*(L_{1,j} - bL_{2,j})) \\
&= (z-a)u_{1,j}^*[(L_{1,j} - bL_{2,j})(R_{2,j}(\bar{z}))^* - (R_{1,j}(\bar{z}))^*(L_{1,j} - bL_{2,j})] \\
&= -z(z-a)u_{1,j}^*(R_{1,j}(\bar{z}))^*[(L_{1,j} - bL_{2,j})T_{j-1}^* - T_j^*(L_{1,j} - bL_{2,j})](R_{2,j}(\bar{z}))^* \\
&= -z(z-a)u_{1,j}^*(R_{1,j}(\bar{z}))^*(L_{1,j}T_{j-1}^* - T_j^*L_{2,j})(R_{2,j}(\bar{z}))^* \\
&= 0.
\end{aligned}$$

In the third equality we used (3.38) while (3.12) was used in the fourth. In the last equality we used (3.39). This proves identity (3.45). Therefore, the proposition is proved. $\qquad\square$

4. The Blaschke–Potapov factors

Using the identities obtained in Section 3, we show that the functions defined in Definition 4.1 are the Blaschke–Potapov factors corresponding to the HMM problem in the case of an odd number of moments.

Definition 4.1. Let

$$b_1(z) := U_1(z), \tag{4.1}$$

and

$$b_j(z) := \begin{pmatrix} I_q - \frac{z-a}{b-a}\widehat{w}^*_{2,j}(a)\widehat{H}^{-1}_{1,j}\widehat{v}_{1,j}(a) & \frac{b-z}{(b-a)^2}\widehat{w}^*_{2,j}(a)\widehat{H}^{-1}_{2,j}\widehat{w}_{2,j}(a) \\ -(z-a)\widehat{v}^*_{1,j}(a)\widehat{H}^{-1}_{1,j}\widehat{v}_{1,j}(a) & I_q - \frac{z-a}{b-a}\widehat{v}^*_{1,j}(a)\widehat{H}^{-1}_{2,j}\widehat{w}_{2,j}(a) \end{pmatrix}, \tag{4.2}$$

for $2 \le j \le n$.

The matrix function b_j is called the Blaschke–Potapov factor of the HMM problem in the case of an odd number of moments.

We write

$$B_{11,j}(a) := \widehat{w}^*_{2,j}(a)\widehat{H}^{-1}_{1,j}\widehat{v}_{1,j}(a), \tag{4.3}$$

$$B_{12,j}(a) := \widehat{w}^*_{2,j}(a)\widehat{H}^{-1}_{2,j}\widehat{w}_{2,j}(a), \tag{4.4}$$

$$B_{21,j}(a) := \widehat{v}^*_{1,j}(a)\widehat{H}^{-1}_{1,j}\widehat{v}_{1,j}(a) \tag{4.5}$$

and

$$B_{22,j}(a) := \widehat{v}^*_{1,j}(a)\widehat{H}^{-1}_{2,j}\widehat{w}_{2,j}(a) \tag{4.6}$$

for $2 \le j \le n$.

Theorem 4.2. *Let the resolvent matrix U_j, $1 \le j \le n$, of the Hausdorff matrix moment problem with an odd number of moments be defined as in (2.16). Let b_j be defined as in (4.1) and (4.2), then*

$$U_{j+1}(z) = U_j(z)b_{j+1}(z), \quad z \in \mathbb{C}, \quad 1 \le j \le n-1. \tag{4.7}$$

Proof. Using (2.16), (4.3), (4.4), (4.5) and (4.6), equation (4.7) can be written in the equivalent form:

$$\begin{pmatrix} U_{11,j+1}(z) & U_{12,j+1}(z) \\ U_{21,j+1}(z) & U_{22,j+1}(z) \end{pmatrix} \tag{4.8}$$

$$- \begin{pmatrix} U_{11,j}(z) & U_{12,j}(z) \\ U_{21,j}(z) & U_{22,j}(z) \end{pmatrix} \begin{pmatrix} I_q - \frac{z-a}{b-a}B_{11,j+1}(a) & \frac{b-z}{(b-a)^2}B_{12,j+1}(a) \\ -(z-a)B_{21,j+1}(a) & I_q - \frac{z-a}{b-a}B_{22,j+1}(a) \end{pmatrix} = 0.$$

The left-hand side of (4.8) is equivalent to the following four equalities:

$$\Upsilon_{11,j} := U_{11,j+1}(z) - U_{11,j}(z) \left[I_q - \frac{z-a}{b-a} B_{11,j+1}(a) \right]$$
$$+ (z-a)U_{12,j}(z)B_{21,j+1}(a),$$

$$\Upsilon_{12,j} := U_{12,j+1}(z) - \frac{b-z}{(b-a)^2} U_{11,j}(z)B_{12,j+1}(a)$$
$$- U_{12,j}(z) \left[I_q - \frac{z-a}{b-a} B_{22,j+1}(a) \right],$$

$$\Upsilon_{21,j} := U_{21,j+1}(z) - U_{21,j}(z) \left[I_q - \frac{z-a}{b-a} B_{11,j+1}(a) \right]$$
$$+ (z-a)U_{22,j}(z)B_{21,j+1}(a)$$

and

$$\Upsilon_{22,j} := U_{22,j+1}(z) - \frac{b-z}{(b-a)^2} U_{21,j}(z)B_{12,j+1}(a) - U_{22,j}(z)$$
$$\cdot \left[I_q - \frac{z-a}{b-a} B_{22,j+1}(a) \right].$$

Taking into account (4.3) and (4.5), we have

$$\Upsilon_{11,j} = - \left[(b-a)\widehat{w}^*_{1,j+1}(z) - U_{11,j}(z)\widehat{w}^*_{2,j+1}(a) \right.$$
$$\left. - (b-a)U_{12,j}(z)\widehat{v}^*_{2,j+1}(a) \right] \frac{z-a}{b-a} \widehat{H}^{-1}_{1,j+1}\widehat{v}_{1,j+1}(a),$$

$$\Upsilon_{12,j} = \left[(b-a)\widehat{w}^*_{2,j+1}(z) - (b-z)U_{11,j}(z)\widehat{w}^*_{2,j+1}(a) \right.$$
$$\left. + (b-a)(z-a)U_{12,j}(z)\widehat{v}^*_{1,j+1}(a) \right] (b-a)^{-2}\widehat{H}^{-1}_{1,j+1}\widehat{v}_{1,j+1}(a),$$

$$\Upsilon_{21,j} = \left[(b-a)\widehat{v}^*_{1,j+1}(z) - U_{21,j}(z)\widehat{w}^*_{2,j+1}(a) \right.$$
$$\left. - (b-a)U_{22,j}(z)\widehat{v}^*_{1,j+1}(a) \right] \frac{z-a}{b-a} \widehat{H}^{-1}_{1,j+1}\widehat{v}_{1,j+1}(a)$$

and

$$\Upsilon_{22,j} = \left[(b-a)(b-z)(z-a)\widehat{v}^*_{2,j+1}(z) - (b-z)U_{21,j}(z)\widehat{w}^*_{2,j+1}(a) \right.$$
$$\left. + (b-a)(z-a)U_{22,j}(z)\widehat{v}^*_{1,j+1}(a) \right] (b-a)^{-2}\widehat{H}^{-1}_{1,j+1}\widehat{v}_{1,j+1}(a).$$

We will demonstrate that the factors in the square brackets, denoted by

$$\widetilde{\Upsilon}_{11,j} := (b-a)\widehat{w}^*_{1,j+1}(z) - U_{11,j}(z)\widehat{w}^*_{2,j+1}(a) - (b-a)U_{12,j}(z)\widehat{v}^*_{2,j+1}(z),$$

$$\widetilde{\Upsilon}_{12,j} := (b-a)\widehat{w}^*_{2,j+1}(z) - (b-z)U_{11,j}(z)\widehat{w}^*_{2,j+1}(a)$$
$$+ (b-a)(z-a)U_{12,j}(z)\widehat{v}^*_{1,j+1}(a),$$

$$\widetilde{\Upsilon}_{21,j} := (b-a)\widehat{v}^*_{1,j+1}(z) - U_{21,j}(z)\widehat{w}^*_{2,j+1}(a) - (b-a)U_{22,j}(z)\widehat{v}^*_{1,j+1}(a)$$

and

$$\widetilde{\Upsilon}_{22,j} := (b-a)(b-z)(z-a)\widehat{v}_{2,j+1}^*(z) - (b-z)U_{21,j}(z)\widehat{w}_{2,j+1}^*(a)$$
$$+ (b-a)(z-a)U_{22,j}(z)\widehat{v}_{1,j+1}^*(a)$$

are equal to zero, i.e.,

$$\widetilde{\Upsilon}_{\ell k,j} = 0, \quad \ell, k \in \{1,2\}, \quad 1 \le j \le n-1. \tag{4.9}$$

Using (2.17), (2.18), (2.28), (2.29) and (2.26), we have

$$\widetilde{\Upsilon}_{11,j} := -(b-a)s_j - \widehat{s}_{j-1} - a^{j+1}s_0 - (b-a)u_{1,j}^*(R_{1,j}(\bar{z}))^* H_{1,j}^{-1} Y_{1,j+1}$$
$$+ (z-a)u_{1,j}^*(R_{1,j}(\bar{z}))^* H_{1,j}^{-1} Y_{1,j+1} + z(b-a)u_{1,j}^*(R_{1,j}(\bar{z}))^* \lambda_{j+1}$$
$$- b(z-a)u_{1,j}^*(R_{1,j}(\bar{z}))^* \lambda_{j+1} + (u_{2,j}^* + as_0 v_{2,j}^*)(R_{2,j}(a))^* H_{2,j}^{-1} Y_{2,j+1}$$
$$- a(u_{2,j}^* + as_0 v_{2,j}^*)(R_{2,j}(a))^* \lambda_j + s_0 v_{1,j}^*(R_{1,j}(a))^* H_{1,j}^{-1} Y_{1,j+1}$$
$$- (z-a)u_{1,j}^*(R_{1,j}(\bar{z}))^*(L_{1,j} - bL_{2,j}) H_{2,j}^{-1} Y_{2,j+1}$$
$$+ (u_{2,j}^* + zs_0 v_{2,j}^*)(R_{2,j}(\bar{z}))^* L_{1,j}^*(R_{1,j}(a))^* H_{1,j}^{-1} Y_{1,j+1}$$
$$- (u_{2,j}^* + zs_0 v_{2,j}^*)(R_{2,j}(\bar{z}))^* H_{2,j}^{-1} Y_{2,j+1}$$
$$- a(u_{2,j}^* + zs_0 v_{2,j}^*)(R_{2,j}(\bar{z}))^* L_{1,j}^*(R_{1,j}(a))^* \lambda_{j+1}$$
$$= -(b-a)s_j - \widehat{s}_{j-1} - a^{j+1}s_0 - u_{1,j}^*(R_{1,j}(\bar{z}))^* \left[(b-z)I_{(j+1)q}\right.$$
$$+ (z-a)(L_{1,j} - bL_{2,j})L_{1,j}^*(R_{1,j}(a))^* \big] H_{1,j}^{-1} Y_{1,j+1}$$
$$+ a(b-z)u_{1,j}^*(R_{1,j}(\bar{z}))^* \lambda_{j+1}$$
$$- av_{1,j}^* H_{1,j}(L_{1,j} - bL_{2,j})L_{1,j}^* \left[I_{(j+1)q} + (R_{1,j}(a))^*\right] \lambda_{j+1}$$
$$+ v_{1,j}^* H_{1,j}(L_{1,j} - bL_{2,j})L_{1,j}^*(R_{1,j}(a))^* H_{1,j}^{-1} Y_{1,j+1}$$
$$+ s_0 v_{1,j}^*(R_{1,j}(a))^* H_{1,j}^{-1} Y_{1,j+1}$$
$$+ a(z-a)u_{1,j}^*(R_{1,j}(\bar{z}))^*(L_{1,j} - bL_{2,j})L_{1,j}^*(R_{1,j}(a))^* \lambda_{j+1}$$
$$= -(b-a)s_j - \widehat{s}_{j-1} - a^{j+1}s_0 + v_{1,j}^* Y_{1,j+1} - av_{1,j}^* H_{1,j} \lambda_{j+1}$$
$$+ as_0 v_{1,j}^*(R_{1,j}(a))^* \lambda_{j+1} - av_{1,j}^* H_{1,j} L_{1,j} \lambda_j + abv_{1,j}^* H_{1,j} L_{2,j} \lambda_j$$
$$= 0. \tag{4.10}$$

In the second equality we used (3.2), (3.40), (3.43) and (3.45). In the third equality we used (3.31), (3.32), (3.5) and (3.6). The last equality follows from (3.33)–(3.35) and (3.9). This proves equation (4.9) for $\ell = k = 1$.

Using (2.17), (2.18), (2.29) and (2.26), we have

$$\widetilde{\Upsilon}_{12,j} = (z-a)s_0\widehat{v}_{1,j+1}^*(a) - (b-a)\widehat{s}_{j-1} + (b-z)\widehat{s}_{j-1}$$
$$+ (b-z)(z-a)u_{1,j}^*(R_{1,j}(\bar{z}))^* H_{1,j}^{-1} Y_{1,j+1}$$
$$+ (z-b)(u_{2,j}^* + zs_0 v_{2,j}^*)(R_{2,j}(\bar{z}))^* H_{2,j}^{-1} Y_{2,j+1}$$
$$+ z(b-a)(u_{2,j}^* + zs_0 v_{2,j}^*)(R_{2,j}(\bar{z}))^* \lambda_j$$

$$
\begin{aligned}
&+ (z-a)\left(u_{2,j}^* + z s_0 v_{2,j}^*\right)(R_{2,j}(\bar{z}))^* L_{1,j}^*(R_{1,j}(a))^* \left(-H_{1,j}^{-1} Y_{1,j+1} + a\lambda_{j+1}\right)\\
&+ (b-z)\left(u_{2,j}^* + a s_0 v_{2,j}^*\right)(R_{2,j}(a))^* H_{2,j}^{-1} Y_{2,j+1}\\
&- a(b-z)\left(u_{2,j}^* + a s_0 v_{2,j}^*\right)(R_{2,j}(a))^* \lambda_j\\
&+ (b-z)(z-a)u_{1,j}^*(R_{1,j}(\bar{z}))^* \left[-(L_{1,j} - bL_{2,j}) H_{2,j}^{-1} Y_{2,j+1} - b\lambda_{j+1}\right]\\
&= (z-a)\left[s_0 \widehat{v}_{1,j+1}^*(a) - \widehat{s}_{j-1} + (b-z)u_{1,j}^*(R_{1,j}(\bar{z}))^* H_{1,j}^{-1} Y_{1,j+1}\right.\\
&\quad + (z-a)u_{1,j}^*(R_{1,j}(\bar{z}))^* \left(I_{(j+1)q} - bT_j^*\right)(R_{1,j}(a))^* H_{1,j}^{-1} Y_{1,j+1}\\
&\quad - a(z-a)u_{1,j}^*(R_{1,j}(\bar{z}))^* \left(I_{(j+1)q} - bT_j^*\right)(R_{1,j}(a))^* \lambda_{j+1}\\
&\quad - z(b-a)u_{1,j}^*(R_{1,j}(\bar{z}))^* \left(I_{(j+1)q} - bT_j^*\right)\lambda_{j+1}\\
&\quad - b(b-z)u_{1,j}^*(R_{1,j}(\bar{z}))^* \lambda_{j+1} + b\left(u_{2,j}^* + a s_0 v_{2,j}^*\right)(R_{2,j}(a))^* \lambda_j\\
&\quad + a\left(u_{2,j}^* + a s_0 v_{2,j}^*\right)(R_{2,j}(a))^* L_{1,j}^*(R_{1,j}(a))^* \lambda_{j+1}\\
&\quad \left. - \left(u_{2,j}^* + a s_0 v_{2,j}^*\right)(R_{2,j}(a))^* L_{1,j}^*(R_{1,j}(a))^* H_{1,j}^{-1} Y_{1,j+1}\right]\\
&= (z-a)\left[a^{j+1} s_0 - \widehat{s}_{j-1} - v_{1,j}^* H_{1,j}\left(I_{(j+1)q} - T_j T_j^*\right)(R_{1,j}(a))^* H_{1,j}^{-1} Y_{1,j+1}\right.\\
&\quad - (b-a)v_{1,j}^* H_{1,j} T_j^*(R_{1,j}(a))^* H_{1,j}^{-1} Y_{1,j+1}\\
&\quad - v_{1,j}^* H_{1,j}(L_{1,j} - bL_{2,j}) L_{1,j}^*(R_{1,j}(a))^* H_{1,j}^{-1} Y_{1,j+1}\\
&\quad - a^2 v_{1,j}^* H_{1,j} T_j^*\left(I_{(j+1)q} - bT_j^*\right)(R_{1,j}(a))^* \lambda_{j+1}\\
&\quad + b^2 v_{1,j}^* H_{1,j} T_j^* \lambda_{j+1} + b v_{1,j}^* H_{1,j}(L_{1,j} - bL_{2,j}) L_{1,j}^* \lambda_{j+1}\\
&\quad \left. + a v_{1,j}^* H_{1,j}(L_{1,j} - bL_{2,j}) L_{1,j}^*(R_{1,j}(a))^* \lambda_{j+1}\right]\\
&= (z-a)\left[a^{j+1} s_0 - \widehat{s}_{j-1} - v_{1,j}^* Y_{1,j+1} + b v_{1,j}^* H_{1,j} L_{1,j} L_{1,j}^* \lambda_{j+1}\right.\\
&\quad + a v_{1,j}^* H_{1,j}\left(-I_{(j+1)q} + T_j T_j^*\right)(R_{1,j}(a))^* \lambda_{j+1} + a v_{1,j}^* H_{1,j} \lambda_{j+1}\\
&\quad \left. - ab v_{1,j}^* H_{1,j} T_j^* \lambda_{j+1}\right]\\
&= 0.
\end{aligned}
\tag{4.11}
$$

In the second equality we used (3.45) (3.40) and (3.2). In the third equality we used (2.26), (3.43) and (3.31). In the fourth equality we used (3.32), (3.5) and (3.6). The last equality follows from (3.33)–(3.35) and (3.32).

Using (2.26), (2.19), (2.29), (2.20) and (2.26), we have

$$
\begin{aligned}
\widetilde{\Upsilon}_{21,j} =\ & (b-a)z^{j+1} I_q - (b-z)a^{j+1} I_q + (b-z)v_{1,j}^*(R_{1,j}(a))^* H_{1,j}^{-1} Y_{1,j+1}\\
&- (b-a)v_{1,j}^*(R_{1,j}(\bar{z}))^* H_{1,j}^{-1} Y_{1,j+1} + (z-a)v_{1,j}^*(R_{1,j}(\bar{z}))^* H_{1,j}^{-1} Y_{1,j+1}\\
&- (z-a)v_{1,j}^*(R_{1,j}(\bar{z}))^* (L_{1,j} - bL_{2,j}) H_{2,j}^{-1} Y_{2,j+1}\\
&- b(z-a)v_{1,j}^*(R_{1,j}(\bar{z}))^* \lambda_{j+1}\\
&+ (z-a)(b-z)v_{2,j}^*(R_{2,j}(\bar{z}))^* L_{1,j}^*(R_{1,j}(a))^* H_{1,j}^{-1} Y_{1,j+1}\\
&- a(z-a)(b-z)v_{2,j}^*(R_{2,j}(\bar{z}))^* L_{1,j}^*(R_{1,j}(a))^* \lambda_{j+1}\\
&- (z-a)(b-z)v_{2,j}^*(R_{2,j}(\bar{z}))^* H_{2,j}^{-1} Y_{2,j+1}
\end{aligned}
$$

$$= (b-a)z^{j+1}I_q - (b-z)a^{j+1}I_q$$
$$- (z-a)v_{1,j}^*(R_{1,j}(\bar{z}))^* \left(L_{1,j} - zL_{1,j}T_{j-1}^* - zL_{2,j} + z^j T_j^* L_{2,j}\right)$$
$$\times (R_{2,j}(\bar{z}))^* H_{2,j}^{-1} Y_{2,j+1}$$
$$- (z-a)v_{1,j}^*(R_{1,j}(\bar{z}))^* \left(bI_{(j+1)q} - azT_j^*\right)(R_{1,j}(a))^*\lambda_{j+1}$$
$$= 0. \tag{4.12}$$

In the second equality we used (3.10), (3.4), (3.17) and (3.7). In the third equality we used (3.16), (3.3), (3.13) and (3.14).

Now we consider $\widetilde{\Upsilon}_{22,j}$. Using (2.19) and (2.20), we have

$$\widetilde{\Upsilon}_{22,j} = (b-z)(z-a)\big[(b-a)\widehat{v}_{2,j+1}^*(z) + v_{1,j}^*(R_{1,j}(\bar{z}))^* H_{1,j}^{-1} R_{1,j}(a)v_{1,j}\widehat{w}_{2,j+1}^*(a)$$
$$+\widehat{v}_{1,j+1}^*(a) + (z-a)v_{2,j}^*(R_{2,j}(\bar{z}))^* H_{2,j}^{-1} R_{2,j}(a)(u_{2,j} + as_0 v_{2,j})\widehat{v}_{1,j+1}^*(a)\big]$$
$$= (b-z)(z-a)\big[(b-a)z^j - (b-a)v_{2,j}^*(R_{2,j}(\bar{z}))^* H_{2,j}^{-1} Y_{2,j+1}$$
$$+v_{1,j}^*(R_{1,j}(\bar{z}))^* H_{1,j}^{-1}\left(H_{1,j}\left[-(L_{1,j} - bL_{2,j})H_{2,j}^{-1} Y_{2,j+1} - b\lambda_{j+1}\right] + Y_{1,j+1}\right)$$
$$+a^{j+1} - v_{1,j}^*(R_{1,j}(a))^* H_{1,j}^{-1} Y_{1,j+1} + (z-a)v_{2,j}^*(R_{2,j}(\bar{z}))^* H_{2,j}^{-1}$$
$$\cdot\left(H_{2,j}L_{1,j}^*(R_{1,j}(a))^*\left(-H_{1,j}^{-1} Y_{1,j+1} + a\lambda_{j+1}\right) + Y_{2,j+1}\right)\big]$$
$$= (b-z)(z-a)\big((b-a)z^j + a^{j+1} - bv_{1,j}^*(R_{1,j}(\bar{z}))^*\lambda_{j+1}$$
$$+v_{1,j}^*\left[-(R_{1,j}(\bar{z}))^*(L_{1,j} - bL_{2,j}) + (z-b)L_{2,j}(R_{2,j}(\bar{z}))^*\right]H_{2,j}^{-1} Y_{2,j+1}$$
$$+a(z-a)v_{2,j}^*(R_{2,j}(\bar{z}))^* L_{1,j}^*(R_{1,j}(a))^*\lambda_{j+1}\big)$$
$$= (b-z)(z-a)\big[(b-a)z^j + a^{j+1} + a(z-a)v_{2,j}^*(R_{2,j}(\bar{z}))^* L_{1,j}^*(R_{1,j}(a))^*\lambda_{j+1}$$
$$-bv_{1,j}^*(R_{1,j}(\bar{z}))^*\lambda_{j+1}\big]$$
$$= 0. \tag{4.13}$$

In the second equality we used (2.27), (2.29), (3.50) and (3.49). In the third equality we used (3.4) and (3.17). In the fourth equality we used (3.7) and (3.16). In the last equality we used (3.9) and (3.15). This proves the equality (4.9) for $\ell = k = 2$.

By (4.10), (4.11), (4.12) and (4.13) the equality (4.8) and consequently (4.7) is proved. The theorem is proved. $\square$

Corollary 4.3. *Let the elementary factors of the Blaschke–Potapov product be given by (4.1) and (4.2). Then the resolvent matrix of the HMM problem (2.16) admits a Blaschke–Potapov multiplicative representation of the form*

$$U_n = b_1 \times \cdots \times b_n. \tag{4.14}$$

The proof follows immediately from Theorem 4.2.

Appendix A. Blaschke–Potapov representation of the resolvent matrix, the case of an even number of moments

In this appendix we reproduce the Blaschke–Potapov representation of the RM for the case of an even number of moments $(s_j)_{j=0}^{2n+1}$, given in [12].

We assume, in this section, that $0 \leq j \leq n$. Let

$$
\begin{aligned}
H_j &:= \{s_{k+l}\}_{k,l=0}^{j}, \\
K_j &:= \{s_{k+l+1}\}_{k,l=0}^{j}, \\
H_{1,j} &:= -aH_j + K_j, \\
H_{2,j} &:= bH_j - K_j,
\end{aligned}
\tag{A.1}
$$

and set

$$
u_{1,0} := -s_0, \; u_{2,0} = s_0,
\tag{A.2}
$$

$$
v_0 := I_q.
\tag{A.3}
$$

For every $1 \leq j \leq n-1$, let

$$
u_{1,j} := -\begin{pmatrix} s_0 \\ s_1 \\ \vdots \\ s_j \end{pmatrix} + a \begin{pmatrix} 0_{q \times q} \\ s_0 \\ \vdots \\ s_{j-1} \end{pmatrix}, \quad
u_{2,j} := \begin{pmatrix} s_0 \\ s_1 \\ \vdots \\ s_j \end{pmatrix} - b \begin{pmatrix} 0_{q \times q} \\ s_0 \\ \vdots \\ s_{j-1} \end{pmatrix},
\tag{A.4}
$$

$$
v_j := v_{1,j}, \; R_j(z) := R_{1,j}(z).
\tag{A.5}
$$

Assume that the block matrices $H_{1,j}$ and $H_{2,j}$ are positive definite. The RM in the case of an even number of moments is a $2q \times 2q$ matrix polynomial,

$$
U_j(z) := \begin{pmatrix} U_{11,j}(z) & U_{12,j}(z) \\ U_{21,j}(z) & U_{22,j}(z) \end{pmatrix}, \; z \in \mathbb{C} \quad 0 \leq j \leq n
\tag{A.6}
$$

with

$$
\begin{aligned}
U_{11,j}(z) &:= I_q - (z-a)u_{2,j}^*(R_j(\bar{z}))^* H_{2,j}^{-1} R_j(a)v_j, \\
U_{12,j}(z) &:= u_{1,j}^*(R_j(\bar{z}))^* H_{1,j}^{-1} R_j(a)u_{1,j}, \\
U_{21,j}(z) &:= -(b-z)(z-a)v_j^*(R_j(\bar{z}))^* H_{2,j}^{-1} R_j(a)v_j
\end{aligned}
$$

and

$$
U_{22,j}(z) := I_q + (z-a)v_j^*(R_j(\bar{z}))^* H_{1,j}^{-1} R_j(a)u_{1,j}.
\tag{A.7}
$$

Further, for $1 \leq j \leq n$, we write

$$
Y_{1,j} := -a \begin{pmatrix} s_j \\ s_{j+1} \\ \vdots \\ s_{2j-1} \end{pmatrix} + \begin{pmatrix} s_{j+1} \\ s_{j+2} \\ \vdots \\ s_{2j} \end{pmatrix}, \quad Y_{2,j} := b \begin{pmatrix} s_j \\ s_{j+1} \\ \vdots \\ s_{2j-1} \end{pmatrix} - \begin{pmatrix} s_{j+1} \\ s_{j+2} \\ \vdots \\ s_{2j} \end{pmatrix}, \quad \text{(A.8)}
$$

$$
D_{1,j} := -as_{2j} + s_{2j+1}, \qquad D_{2,j} := bs_{2j} - s_{2j+1}, \qquad \text{(A.9)}
$$

$$
w_{1,j}(a) := -s_j - Y_{1,j}^* H_{1,j-1}^{-1} R_{j-1}(a) u_{1,j-1}, \qquad \text{(A.10)}
$$

$$
w_{2,j}(a) := -w_{1,j}(a), \qquad \text{(A.11)}
$$

$$
\widehat{v}_j(a) := a^j I_q - Y_{2,j}^* H_{2,j-1}^{-1} R_{j-1}(a) v_{j-1}, \qquad \text{(A.12)}
$$

and

$$
\widehat{H}_{k,j} := D_{k,j} - Y_{k,j}^* H_{k,j-1}^{-1} Y_{k,j}, \quad k \in \{1,2\},
$$

for $1 \leq j \leq n$.

Note that $\widehat{H}_{k,j} > 0$ since $H_{k,j} > 0$.

Definition A.1. Define

$$
b_0(z) := U_0(z), \qquad \text{(A.13)}
$$

and

$$
b_j(z) := \begin{pmatrix} I_q - (z-a)w_{2,j}^*(a)\widehat{H}_{2,j}^{-1}\widehat{v}_j(a) & w_{1,j}^*(a)\widehat{H}_{1,j}^{-1}w_{1,j}(a) \\ (z-a)(z-b)\widehat{v}_j^*(a)\widehat{H}_{2,j}^{-1}\widehat{v}_j(a) & I_q + (z-a)\widehat{v}_j^*(a)\widehat{H}_{1,j}^{-1}w_{1,j}(a) \end{pmatrix}
$$

$$
\text{(A.14)}
$$

for all $z \in \mathbb{C}$ and all $1 \leq j \leq n$.

The matrix function b_j is called the Blaschke–Potapov factor of the HMM problem with an even number of moments.

Remark A.2. In [12] the matrix function b_0 was erroneously defined through the formula (A.14).

Theorem A.3 (Corollary of Theorem 2, [12]). *Let the elementary factors of the Blaschke–Potapov product be given by (A.13) and (A.14). Then the resolvent matrix (A.6) of the HMM problem in the case of an even number of moments admits a Blaschke–Potapov multiplicative representation of the form*

$$
U_n(z) = b_0 \times \cdots \times b_n. \qquad \text{(A.15)}
$$

Acknowledgment

The author wishes to thank Prof. Bernd Kirstein of Leipzig University for helpful suggestions. Many thanks to Dr. Conrad Mädler of Leipzig University for valuable comments.

References

[1] A.E. Choque Rivero, Yu.M. Dyukarev, B. Fritzsche, and B. Kirstein, *A truncated matricial moment problem on a finite interval.* Interpolation, Schur Functions and Moment Problems. *Oper. Theory: Adv. Appl.* **165** (2006), 121–173.

[2] A.E. Choque Rivero, Yu.M. Dyukarev, B. Fritzsche, and B. Kirstein, *A Truncated Matricial Moment Problem on a Finite Interval. The Case of an Odd Number of Prescribed Moments.* System Theory, Schur Algorithm and Multidimensional Analysis. *Oper. Theory: Adv. Appl.* **176** (2007), 99–174.

[3] A.E. Choque Rivero, A.E. Merzon; *The completely indeterminate Caratheodory matrix problem in the* $\mathcal{R}_q[a,b]$ *class,* Analysis 28 (2008), 177–207.

[4] A.E. Choque Rivero, *The Caratheodory Matrix Problem in the class* $S[a,b]$, Russian Mathematics (Iz. VUZ), Vol. 50, No.11, (2006), 58–73.

[5] A.E. Choque Rivero, *The resolvent matrix for the Hausdorff matrix moment problem expressed in terms of orthogonal matrix polynomials,* Complex Analysis and Operator Theory, to be submitted.

[6] V.K. Dubovoj, B. Fritzsche, B. Kirstein, *Matricial Version of the Classical Schur Problem* Teubner-Texte zur Mathematik, Bb. 129, B.G. Teubner, Stuttgart – Leipzig, 1992.

[7] Yu.M. Dyukarev, *The multiplicative structure of resolvent matrices of interpolation problems in the Stieltjes class,* Vestnik Kharkov Univ. Ser. Mat. Prikl. Mat. i Mekh. (1999), no. 458, 143–153.

[8] Yu.M. Dyukarev, V.E. Katsnelson, *Multiplicative and additive classes of analytic matrix functions of Stieltjes type and associated interpolation problems* (in Russian), Teor. Funkcii, Funkcional. Anal. i Prilozen., Part I: **36** (1981), 13–27; Part III: **41** (1984), 64–70.

[9] A.V. Efimov, V.P. Potapov, *J-expansive matrix-valued functions and their role in the analytical theory of electrical circuits,* Russian Math. Surveys, **28** (1973), 69–140.

[10] I.V. Kovalishina, *Analytic theory of a class of interpolation problems* Izv. Akad. Nauk SSSR Ser. Mat., Volume 47, Issue 3, (1983), 455–497.

[11] M.G. Krein and A.A. Nudelman, *The Markov Moment Problem and Extremal Problems.* Translations of Mathematical Monographs **50**, AMS, Providence, RI 1977.

[12] I.Yu. Serikova, *The multiplicative structure of resolvent matrix of the moment problem on the kompact interval (case of even numbers of moments),* Vestnik Kharkov Univ. Ser. Mat. Prikl. Mat. i Mekh. (2007), no. 790, 132–139.

[13] Yu.M. Dyukarev, *Multiplicative and additive classes of analytic matrix functions of Stieltjes type and associated interpolation problems, Part II* (in Russian), Teor. Funkcii. Funkcional. Anal. i Prilozen. 38 (1982), 40–48.

Abdon Eddy Choque Rivero
Instituto de Física y Matemáticas
Universidad Michoacana de San Nicolás de Hidalgo
Ciudad Universitaria, Morelia, Mich.
C.P. 58048, México
e-mail: abdon@ifm.umich.mx

Operator Theory:
Advances and Applications, Vol. 226, 211–250
© 2012 Springer Basel

On a Special Parametrization of Matricial α-Stieltjes One-sided Non-negative Definite Sequences

Bernd Fritzsche, Bernd Kirstein and Conrad Mädler

Abstract. The characterization of the solvability of matrix versions of truncated Stieltjes-type moment problems led to the class of one-sided α-Stieltjes non-negative definite sequences of complex $q \times q$ matrices. The study of this class and some of its important subclasses is the central theme of this paper. We introduce an inner parametrization for sequences of complex matrices, which is particularly well-suited to the class of sequences under consideration. Furthermore, several interrelations between this parametrization and the canonical Hankel parametrization are indicated.

Mathematics Subject Classification (2010). Primary 44A60; Secondary 47A57.

Keywords. Truncated matricial Stieltjes moment problem, one-sided Stieltjes non-negative definite sequences, one-sided Stieltjes non-negative definite extendable sequences, one-sided α-Stieltjes parametrization, canonical Hankel parametrization.

1. Introduction

This paper is a direct continuation of work done in the paper [12], where two truncated matricial generalized Stieltjes power moment problems made up one of the main topics. In order to properly formulate these problems, we first review some notation. Let $\mathbb{C}$, $\mathbb{R}$, $\mathbb{N}_0$, and $\mathbb{N}$ be the set of all complex numbers, the set of all real numbers, the set of all non-negative integers, and the set of all positive integers, respectively. Further, for every choice of $\alpha \in \mathbb{R} \cup \{-\infty\}$ and $\beta \in \mathbb{R} \cup \{+\infty\}$, let $\mathbb{Z}_{\alpha,\beta}$ be the set of all integers k for which $\alpha \leq k \leq \beta$ holds. Throughout this paper, let $p \in \mathbb{N}$ and let $q \in \mathbb{N}$. If $\mathcal{X}$ is a non-empty set, then $\mathcal{X}^{p \times q}$ stands for the set of all $p \times q$ matrices each entry of which belongs to $\mathcal{X}$, and $\mathcal{X}^p$ is short for $\mathcal{X}^{p \times 1}$. If $(\mathcal{X}, \mathfrak{A})$ is a measurable space, then each countably additive mapping whose domain is $\mathfrak{A}$ and whose values belong to the set $\mathbb{C}_{\geq}^{q \times q}$ of all non-negative

Hermitian complex $q \times q$ matrices is called a *non-negative Hermitian $q \times q$ measure on* $(\mathcal{X}, \mathfrak{A})$.

Let $\mathfrak{B}_{\mathbb{R}}$ be the σ-algebra of all Borel subsets of $\mathbb{R}$. For all $\Omega \in \mathfrak{B}_{\mathbb{R}} \setminus \{\emptyset\}$, let $\mathfrak{B}_{\Omega}$ be the σ-algebra of all Borel subsets of Ω, let $\mathcal{M}_{\geq}^q(\Omega)$ be the set of all non-negative Hermitian $q \times q$ measures on $(\Omega, \mathfrak{B}_{\Omega})$ and, for all $\kappa \in \mathbb{N}_0 \cup \{+\infty\}$, let $\mathcal{M}_{\geq,\kappa}^q(\Omega)$ be the set of all $\sigma \in \mathcal{M}_{\geq}^q(\Omega)$ such that the integral

$$s_j^{(\sigma)} := \int_{\Omega} t^j \sigma(dt) \tag{1.1}$$

exists for all $j \in \mathbb{Z}_{0,\kappa}$.

We now come to the formulation of two matricial power moment problems, which lie in the background of our considerations:

M$[\Omega; (s_j)_{j=0}^{\kappa}, =]$: Let $\Omega \in \mathfrak{B}_{\mathbb{R}} \setminus \{\emptyset\}$, let $\kappa \in \mathbb{N}_0 \cup \{+\infty\}$, and let $(s_j)_{j=0}^{\kappa}$ be a sequence from $\mathbb{C}^{q \times q}$. Describe the set $\mathcal{M}_{\geq}^q[\Omega; (s_j)_{j=0}^{\kappa}, =]$ of all $\sigma \in \mathcal{M}_{\geq,\kappa}^q(\Omega)$ for which $s_j^{(\sigma)} = s_j$ is fulfilled for each $j \in \mathbb{Z}_{0,\kappa}$.

The second matricial moment problem under consideration is a truncated one:

M$[\Omega; (s_j)_{j=0}^{m}, \leq]$: Let $\Omega \in \mathfrak{B}_{\mathbb{R}} \setminus \{\emptyset\}$, let $m \in \mathbb{N}_0$, and let $(s_j)_{j=0}^{m}$ be a sequence from $\mathbb{C}^{q \times q}$. Describe the set $\mathcal{M}_{\geq}^q[\Omega; (s_j)_{j=0}^{m}, \leq]$ of all $\sigma \in \mathcal{M}_{\geq,m}^q(\Omega)$ for which $s_m - s_m^{(\sigma)}$ is non-negative Hermitian and, in the case $m > 0$, moreover $s_j^{(\sigma)} = s_j$ is fulfilled for each $j \in \mathbb{Z}_{0,m-1}$.

The considerations of this paper are mostly concentrated on the case that the set Ω is a one-sided closed infinite interval of the real axis. To get deeper insight into this situation, we start with some observations on the case $\Omega = \mathbb{R}$, which is connected with matricial versions of the classical Hamburger moment problem. In this case, the above formulated matricial moment problems have been intensively investigated since the 1980's (see, e.g., Bolotnikov [4], Chen/Hu [5,6], Dym [10], Kovalishina [20,21], and [13]).

To explain some criteria of solvability in the case $\Omega = \mathbb{R}$, we introduce certain sets of sequences from $\mathbb{C}^{q \times q}$, which are determined by the properties of particular Hankel matrices built of them. For each $n \in \mathbb{N}_0$, let $\mathcal{H}_{q,2n}^{\geq}$ (respectively, $\mathcal{H}_{q,2n}^{>}$) be the set of all sequences $(s_j)_{j=0}^{2n}$ from $\mathbb{C}^{q \times q}$ such that the block Hankel matrix

$$H_n := [s_{j+k}]_{j,k=0}^{n} \tag{1.2}$$

is non-negative Hermitian (resp. positive Hermitian). Furthermore, let $\mathcal{H}_{q,\infty}^{\geq}$ (resp. $\mathcal{H}_{q,\infty}^{>}$) be the set of all sequences $(s_j)_{j=0}^{\infty}$ from $\mathbb{C}^{q \times q}$ such that, for all $n \in \mathbb{N}_0$, the sequence $(s_j)_{j=0}^{2n}$ belongs to $\mathcal{H}_{q,2n}^{\geq}$ (resp. $\mathcal{H}_{q,2n}^{>}$). For each $n \in \mathbb{N}_0$, the elements of the set $\mathcal{H}_{q,2n}^{\geq}$ (resp. $\mathcal{H}_{q,2n}^{>}$) are called *Hankel non-negative definite* (resp. *Hankel positive definite*) sequences. For each $n \in \mathbb{N}_0$, let $\mathcal{H}_{q,2n}^{\geq,e}$ be the set of all sequences $(s_j)_{j=0}^{2n}$ from $\mathbb{C}^{q \times q}$ for which there are matrices $s_{2n+1} \in \mathbb{C}^{q \times q}$ and $s_{2n+2} \in \mathbb{C}^{q \times q}$ such that $(s_j)_{j=0}^{2(n+1)}$ belongs to $\mathcal{H}_{q,2(n+1)}^{\geq}$. Furthermore, for each $n \in \mathbb{N}_0$, we will use $\mathcal{H}_{q,2n+1}^{\geq,e}$ to denote the set of sequences $(s_j)_{j=0}^{2n+1}$ from $\mathbb{C}^{q \times q}$ for which there is some $s_{2n+2} \in \mathbb{C}^{q \times q}$ such that $(s_j)_{j=0}^{2(n+1)}$ belongs to $\mathcal{H}_{q,2(n+1)}^{\geq}$. For each $m \in \mathbb{N}_0$, the elements of the set $\mathcal{H}_{q,m}^{\geq,\mathrm{e}}$ are called *Hankel non-negative definite extendable* sequences. For technical reasons, we set $\mathcal{H}_{q,\infty}^{\geq,\mathrm{e}} := \mathcal{H}_{q,\infty}^{\geq}$. The solvability of the Hamburger moment problems under consideration can be characterized as follows.

Theorem 1.1 (see [13, Theorem 4.17], [15, Theorem 6.6]). *Let $\kappa \in \mathbb{N}_0 \cup \{+\infty\}$ and let $(s_j)_{j=0}^{\kappa}$ be a sequence from $\mathbb{C}^{q \times q}$. Then $\mathcal{M}_{\geq}^q[\mathbb{R}; (s_j)_{j=0}^{\kappa}, =] \neq \emptyset$ if and only if $(s_j)_{j=0}^{\kappa} \in \mathcal{H}_{q,\kappa}^{\geq,\mathrm{e}}$.*

Theorem 1.2. *Let $n \in \mathbb{N}_0$ and let $(s_j)_{j=0}^{2n}$ be a sequence from $\mathbb{C}^{q \times q}$. Then the set $\mathcal{M}_{\geq}^q[\mathbb{R}; (s_j)_{j=0}^{2n}, \leq]$ is non-empty if and only if $(s_j)_{j=0}^{2n} \in \mathcal{H}_{q,2n}^{\geq}$.*

For different proofs of Theorem 1.2, we refer to [5, Theorem 3.2], [13, Theorem 4.16] and [23, Satz 9.20].

Now let $\Omega \in \mathfrak{B}_{\mathbb{R}} \setminus \{\emptyset\}$, $\kappa \in \mathbb{N}_0 \cup \{+\infty\}$ and $\sigma \in \mathcal{M}_{\geq,\kappa}^q(\Omega)$. If the mapping $\mu \colon \mathfrak{B}_{\mathbb{R}} \to \mathbb{C}^{q \times q}$ is defined by $\mu(B) := \sigma(B \cap \Omega)$, then clearly $\mu \in \mathcal{M}_{\geq,\kappa}^q(\mathbb{R})$, and taking (1.1) into account, we see that, for each $k \in \mathbb{Z}_{0,\kappa}$, the equation

$$s_k^{(\mu)} = s_k^{(\sigma)}$$

holds. Thus, we see that if $\sigma \in \mathcal{M}_{\geq}^q[\Omega; (s_j)_{j=0}^{\kappa}, =]$ (resp. $\sigma \in \mathcal{M}_{\geq}^q[\Omega; (s_j)_{j=0}^{2n}, \leq]$), then $\mu \in \mathcal{M}_{\geq}^q[\mathbb{R}; (s_j)_{j=0}^{\kappa}, =]$ (resp. $\mu \in \mathcal{M}_{\geq}^q[\mathbb{R}; (s_j)_{j=0}^{2n}, \leq]$). Consequently, if the set $\mathcal{M}_{\geq}^q[\Omega; (s_j)_{j=0}^{\kappa}, =]$ (resp. $\mathcal{M}_{\geq}^q[\Omega; (s_j)_{j=0}^{2n}, \leq]$) is non-empty, then $(s_j)_{j=0}^{\kappa} \in \mathcal{H}_{q,\kappa}^{\geq,\mathrm{e}}$ (resp. $(s_j)_{j=0}^{2n} \in \mathcal{H}_{q,2n}^{\geq}$). The choice of $\Omega \in \mathfrak{B}_{\mathbb{R}} \setminus \{\emptyset\}$ in the problems $\mathsf{M}[\Omega; (s_j)_{j=0}^{\kappa}, =]$ and $\mathsf{M}[\Omega; (s_j)_{j=0}^{m}, \leq]$ leads us consequently to special subclasses of the classes $\mathcal{H}_{q,2n}^{\geq}$ and $\mathcal{H}_{q,\kappa}^{\geq,\mathrm{e}}$. In this paper, we continue the investigations in [12], where the case of an interval $[\alpha, +\infty)$ with arbitrarily given $\alpha \in \mathbb{R}$ was considered. For the convenience of the reader, we are going to recall the characterizations of solvability of our moment problems which were obtained in [12]. First we introduce corresponding subclasses of the classes $\mathcal{H}_{q,2n}^{\geq}$ and $\mathcal{H}_{q,\kappa}^{\geq,\mathrm{e}}$, which depend on a given real number α.

Definition 1.3.

(a) For each $\alpha \in \mathbb{R}$, let $\mathcal{K}_{q,0,\alpha}^{\geq} := \mathcal{H}_{q,0}^{\geq}$, and, for each $\alpha \in \mathbb{R}$ and each $n \in \mathbb{N}$, let

$$\mathcal{K}_{q,2n,\alpha}^{\geq} := \left\{ (s_j)_{j=0}^{2n} \in \mathcal{H}_{q,2n}^{\geq} \colon (-\alpha s_j + s_{j+1})_{j=0}^{2(n-1)} \in \mathcal{H}_{q,2(n-1)}^{\geq} \right\}$$

and, for each $\alpha \in \mathbb{R}$ and each $n \in \mathbb{N}_0$, furthermore, let $\mathcal{K}_{q,2n+1,\alpha}^{\geq}$ be the set of all sequences $(s_j)_{j=0}^{2n+1}$ from $\mathbb{C}^{q \times q}$ such that $\{(s_j)_{j=0}^{2n}, (-\alpha s_j + s_{j+1})_{j=0}^{2n}\} \subseteq \mathcal{H}_{q,2n}^{\geq}$. Furthermore, for each $\alpha \in \mathbb{R}$ and each $m \in \mathbb{N}_0$, let $\mathcal{K}_{q,m,\alpha}^{\geq,\mathrm{e}}$ be the set of all sequences $(s_j)_{j=0}^{m}$ from $\mathbb{C}^{q \times q}$ for which there is some $s_{m+1} \in \mathbb{C}^{q \times q}$ such that $(s_j)_{j=0}^{m+1}$ belongs to $\mathcal{K}_{q,m+1,\alpha}^{\geq}$. For each $\alpha \in \mathbb{R}$ and each $m \in \mathbb{N}_0$, we call a sequence $(s_j)_{j=0}^{m}$ α-*Stieltjes right-sided non-negative definite* (resp. α-*Stieltjes right-sided non-negative definite extendable*) if it belongs to $\mathcal{K}_{q,m,\alpha}^{\geq}$ (resp. to $\mathcal{K}_{q,m,\alpha}^{\geq,\mathrm{e}}$). In the case $\alpha = 0$, the sequence $(s_j)_{j=0}^{m}$ is also called *Stieltjes non-negative definite* (resp. *Stieltjes non-negative definite extendable*).

(b) For each $\alpha \in \mathbb{R}$, let $\mathcal{L}^{\geq}_{q,0,\alpha} := \mathcal{H}^{\geq}_{q,0}$, and, for each $\alpha \in \mathbb{R}$ and each $n \in \mathbb{N}$, let

$$\mathcal{L}^{\geq}_{q,2n,\alpha} := \left\{ (s_j)_{j=0}^{2n} \in \mathcal{H}^{\geq}_{q,2n} : (\alpha s_j - s_{j+1})_{j=0}^{2(n-1)} \in \mathcal{H}^{\geq}_{q,2(n-1)} \right\}$$

and, for each $\alpha \in \mathbb{R}$ and each $n \in \mathbb{N}_0$, furthermore, let $\mathcal{L}^{\geq}_{q,2n+1,\alpha}$ be the set of all sequences $(s_j)_{j=0}^{2n+1}$ from $\mathbb{C}^{q \times q}$ such that $\{(s_j)_{j=0}^{2n}, (\alpha s_j - s_{j+1})_{j=0}^{2n}\} \subseteq \mathcal{H}^{\geq}_{q,2n}$. Furthermore, for each $\alpha \in \mathbb{R}$ and each $m \in \mathbb{N}_0$, let $\mathcal{L}^{\geq,e}_{q,m,\alpha}$ be the set of all sequences $(s_j)_{j=0}^{m}$ from $\mathbb{C}^{q \times q}$ for which there is some $s_{m+1} \in \mathbb{C}^{q \times q}$ such that $(s_j)_{j=0}^{m+1}$ belongs to $\mathcal{L}^{\geq}_{q,m+1,\alpha}$. For each $\alpha \in \mathbb{R}$ and each $m \in \mathbb{N}_0$, we call a sequence $(s_j)_{j=0}^{m}$ α-Stieltjes left-sided non-negative definite (resp. α-Stieltjes left-sided non-negative definite extendable) if it belongs to $\mathcal{L}^{\geq}_{q,m,\alpha}$ (resp. to $\mathcal{L}^{\geq,e}_{q,m,\alpha}$).

The classes of sequences introduced in part (b) of Definition 1.3 do not appear in [12]. We will see soon that these classes are convenient tools to handle the corresponding moment problems for intervals of the type $(-\infty, \alpha]$, where $\alpha \in \mathbb{R}$.

Remark 1.4. Let $\alpha \in \mathbb{R}$, $n \in \mathbb{N}_0$, and $(s_j)_{j=0}^{n} \in \mathcal{K}^{\geq}_{q,n,\alpha}$ (resp. $\mathcal{K}^{\geq,e}_{q,n,\alpha}$). Then we easily see that $(s_j)_{j=0}^{m} \in \mathcal{K}^{\geq}_{q,m,\alpha}$ (resp. $\mathcal{K}^{\geq,e}_{q,m,\alpha}$) for each $m \in \mathbb{Z}_{0,n}$.

In view of Remark 1.4, for each $\alpha \in \mathbb{R}$, let $\mathcal{K}^{\geq}_{q,\infty,\alpha}$ be the set of all sequences $(s_j)_{j=0}^{\infty}$ from $\mathbb{C}^{q \times q}$ such that $(s_j)_{j=0}^{m} \in \mathcal{K}^{\geq}_{q,m,\alpha}$ for all $m \in \mathbb{N}_0$. Further, let $\mathcal{K}^{\geq,e}_{q,\infty,\alpha} := \mathcal{K}^{\geq}_{q,\infty,\alpha}$. The notation I_q stands for the identity matrix in $\mathbb{C}^{q \times q}$.

The following result describes the interrelations between the classes of sequences from $\mathbb{C}^{q \times q}$, which were introduced in parts (a) and (b) of Definition 1.3.

Lemma 1.5. *Let* $\alpha \in \mathbb{R}$, *let* $\kappa \in \mathbb{N}_0 \cup \{+\infty\}$, *and let* $(s_j)_{j=0}^{\kappa}$ *be a sequence from* $\mathbb{C}^{q \times q}$. *For* $j \in \mathbb{Z}_{0,\kappa}$, *let* $u_j := (-1)^j s_j$.

(a) $(s_j)_{j=0}^{\kappa} \in \mathcal{K}^{\geq}_{q,\kappa,\alpha}$ *if and only if* $(u_j)_{j=0}^{\kappa} \in \mathcal{L}^{\geq}_{q,\kappa,-\alpha}$.
(b) $(s_j)_{j=0}^{\kappa} \in \mathcal{K}^{\geq,e}_{q,\kappa,\alpha}$ *if and only if* $(u_j)_{j=0}^{\kappa} \in \mathcal{L}^{\geq,e}_{q,\kappa,-\alpha}$.

Proof. For all $n \in \mathbb{N}_0$, the matrix $V_n := \operatorname{diag}[(-1)^j I_q]_{j=0}^{n}$ is obviously unitary. For all $n \in \mathbb{N}_0$ with $2n \leq \kappa$, the equation $H_n^{\langle u \rangle} = V_n^* H_n V_n$ holds and, for all $n \in \mathbb{N}_0$ with $2n + 1 \leq \kappa$, we have $(-\alpha)H_n^{\langle u \rangle} - K_n^{\langle u \rangle} = V_n^*(-\alpha H_n + K_n)V_n$. $\qquad\square$

Lemma 1.5 shows that it is justified for us to concentrate our further considerations mainly on the right-sided case.

Using the just introduced sets of sequences from $\mathbb{C}^{q \times q}$, we recall the solvability criteria of the problems $\mathsf{M}[[\alpha, +\infty); (s_j)_{j=0}^{m}, =]$ and $\mathsf{M}[[\alpha, +\infty); (s_j)_{j=0}^{m}, \leq]$:

Theorem 1.6. *Let* $\alpha \in \mathbb{R}$, *let* $\kappa \in \mathbb{N}_0 \cup \{+\infty\}$, *and let* $(s_j)_{j=0}^{\kappa}$ *be a sequence from* $\mathbb{C}^{q \times q}$. *Then* $\mathcal{M}^q_{\geq}[[\alpha, +\infty); (s_j)_{j=0}^{\kappa}, =] \neq \emptyset$ *if and only if* $(s_j)_{j=0}^{\kappa} \in \mathcal{K}^{\geq,e}_{q,\kappa,\alpha}$.

In the case $\kappa \in \mathbb{N}_0$, a proof of Theorem 1.6 is given in [12, Theorem 1.3]. In the case $\kappa = +\infty$, the asserted equivalence can be proved using the equation $\mathcal{M}^q_{\geq}[[\alpha, +\infty); (s_j)_{j=0}^{\infty}, =] = \bigcap_{m=0}^{\infty} \mathcal{M}^q_{\geq}[[\alpha, +\infty); (s_j)_{j=0}^{m}, =]$ and the matricial version of the Helly-Prohorov theorem (see [14, Satz 9]). We omit the details of the proof, the essential idea of which originated in [1].

For Problem $M[[\alpha, +\infty); (s_j)_{j=0}^m, \leq]$, the solvability criterion is the following:

Theorem 1.7 ([12, Theorem 1.4]). *Let* $\alpha \in \mathbb{R}$, *let* $m \in \mathbb{N}_0$, *and let* $(s_j)_{j=0}^m$ *be a sequence from* $\mathbb{C}^{q \times q}$. *Then* $\mathcal{M}_{\geq}^q[[\alpha, +\infty); (s_j)_{j=0}^m, \leq] \neq \emptyset$ *if and only if* $(s_j)_{j=0}^m \in \mathcal{K}_{q,m,\alpha}^{\geq}$.

Now we turn our attention to the interval $(-\infty, \alpha]$, where $\alpha \in \mathbb{R}$ is arbitrarily given.

Theorem 1.8. *Let* $\alpha \in \mathbb{R}$, *let* $\kappa \in \mathbb{N}_0 \cup \{+\infty\}$, *and let* $(s_j)_{j=0}^{\kappa}$ *be a sequence from* $\mathbb{C}^{q \times q}$. *Then* $\mathcal{M}_{\geq}^q[(-\infty, \alpha]; (s_j)_{j=0}^{\kappa}, =] \neq \emptyset$ *if and only if* $(s_j)_{j=0}^{\kappa} \in \mathcal{L}_{q,\kappa,\alpha}^{\geq,e}$.

Proof. Combine Theorem 1.6, part (b) of Lemma 1.5 and Lemma B.3. □

Theorem 1.9. *Let* $\alpha \in \mathbb{R}$, *let* $n \in \mathbb{N}_0$, *and let* $(s_j)_{j=0}^{2n}$ *be a sequence from* $\mathbb{C}^{q \times q}$. *Then* $\mathcal{M}_{\geq}^q[(-\infty, \alpha]; (s_j)_{j=0}^{2n}, \leq] \neq \emptyset$ *if and only if* $(s_j)_{j=0}^{2n} \in \mathcal{L}_{q,2n,\alpha}^{\geq}$.

Proof. Combine Lemma B.3, part (a) of Lemma 1.5, and Theorem 1.7. □

A characterization of the case $\mathcal{M}_{\geq}^q[(-\infty, \alpha]; (s_j)_{j=0}^{2n+1}, \leq] \neq \emptyset$, where $\alpha \in \mathbb{R}$, $n \in \mathbb{N}_0$, and a sequence $(s_j)_{j=0}^{2n+1}$ from $\mathbb{C}^{q \times q}$ are given, is more difficult, so that we want to discuss a necessary and sufficient condition for that somewhere else.

In the scalar situation, the particular case of the interval $[0, +\infty)$ played an important role in the history of the power moment problem. Namely, the classical investigations of Stieltjes [25, 26] are considered now as a cornerstone in the early period of moment problems. For a modern summary of the state of affairs on the Stieltjes moment problem and its relations to operator theory, we refer to the paper by Simon [24].

In the matrix case, several authors treated the power moment problem for the interval $[0, +\infty)$. Parametrizations of the set $\mathcal{M}_{\geq}^q[[0, +\infty); (s_j)_{j=0}^m, \leq]$ are given, under a certain nondegeneracy condition, by Dyukarev [11]. In the general case, descriptions of $\mathcal{M}_{\geq}^q[[0, +\infty); (s_j)_{j=0}^m, \leq]$ are stated by Bolotnikov [2], [3, Theorem 1.5] and by Chen/Hu in [7, Theorem 2.4], whereas in the case $m \in \mathbb{N}_0$ descriptions of the set $\mathcal{M}_{\geq}^q[[0, +\infty); (s_j)_{j=0}^m, =]$ are given by Hu/Chen in [19, Theorem 4.1, Lemmas 2.3 and 2.4]. Moreover, for each $m \in \mathbb{N}_0$ and arbitrary real numbers α, the cases $\mathcal{M}_{\geq}^q[[\alpha, +\infty); (s_j)_{j=0}^m, =] \neq \emptyset$ and $\mathcal{M}_{\geq}^q[[\alpha, +\infty); (s_j)_{j=0}^m, \leq] \neq \emptyset$ are characterized in [12, Theorems 1.3 and 1.4].

The importance of Theorems 1.6 and 1.7 led us in [12] to a closer look at the properties of sequences from $\mathbb{C}^{q \times q}$, which are α-Stieltjes right-sided non-negative definite or α-Stieltjes right-sided non-negative definite extendable. Guided by our former investigations on Hankel non-negative definite sequences and Hankel non-negative definite extendable sequences, which were done in [13], we started in [12, Section 4] a thorough study of the structure of α-Stieltjes right-sided non-negative definite sequences and α-Stieltjes right-sided non-negative definite extendable sequences. The central theme of this paper is to continue these investigations in order to gain deeper insights into the structure of the elements of the set of Hankel non-negative definite sequences and distinguished subsets, of which we systematically used the canonical Hankel parametrization of sequences (see [13, 15, 18]). The first main goal of this paper is to find a convenient parametrization of sequences from $\mathbb{C}^{q \times q}$, which is particularly well adapted to the class of α-Stieltjes right-sided non-negative definite sequences and its important subclasses. In order to realize this aim, we introduce the right-sided α-Stieltjes parametrization of sequences from $\mathbb{C}^{p \times q}$ (see Section 4).

This paper is organized as follows.

In Section 2, we summarize some basic facts on α-Stieltjes right-sided non-negative definite sequences and α-Stieltjes right-sided non-negative definite extendable sequences. This material is mostly taken from [12].

In Section 3, we recall the canonical Hankel parametrization of sequences of matrices from $\mathbb{C}^{p \times q}$.

In Section 4, we introduce the right-sided α-Stieltjes parametrization of sequences from $\mathbb{C}^{p \times q}$. The main result of Section 4 is Theorem 4.12 which contains characterizations of the class of α-Stieltjes right-sided non-negative definite sequences and their interesting subclasses in terms of the right-sided α-Stieltjes parametrization. As an example the right-sided α-Stieltjes parametrization of the sequence of Catalan numbers is computed.

In Section 5, we continue our former investigations on completely degenerate Hankel non-negative definite sequences. For arbitrarily given $\alpha \in \mathbb{R}$, we study the subclass of this class of sequences which consists of α-Stieltjes right-sided non-negative definite sequences.

In Section 6, we present some interrelations between the right-sided α-Stieltjes parametrization and the canonical Hankel parametrization of α-Stieltjes right-sided non-negative definite sequences. The main results are Theorems 6.8 and 6.20.

In Section 7, we fix numbers $\alpha \in \mathbb{R}$ and $\beta \in [\alpha, +\infty)$. It is shown that a β-Stieltjes right-sided non-negative definite sequence is also α-Stieltjes right-sided non-negative definite (see Remark 7.1). Furthermore, some interrelations between the α- and right-sided β-Stieltjes parametrization of this sequence are indicated (see Propositions 7.2, 7.5, and 7.6).

Inspired by investigations of Chen/Hu [5], we constructed in [18] a Schur-type algorithm for sequences of matrices from $\mathbb{C}^{p \times q}$. This algorithm turned out to be intimately connected with the canonical Hankel parametrization. More precisely, it was shown in [18, Theorem 9.15] that the canonical Hankel parametrization of a Hankel non-negative definite extendable sequence can be generated by the Schur-type algorithm under consideration. In a forthcoming paper, we will develop an α-Stieltjes version of a Schur-type algorithm for sequences of matrices from $\mathbb{C}^{p \times q}$. An important feature of this algorithm will be the fact that the right-sided α-Stieltjes parametrization of an α-Stieltjes right-sided non-negative definite extendable sequence can be generated by it.

2. Some facts on the classes of α-Stieltjes one-sided non-negative definite sequences and some of its subclasses

The central theme of this section is to study the classes of sequences of complex matrices which were given in Definition 1.3. It should be mentioned that the classes introduced in part (a) of Definition 1.3 have been discussed already in [12]. Since Lemma 1.5 shows that the left-sided classes are connected with the corresponding right-sided classes by a simple reflection principle, we will mainly concentrate on the right case.

Definition 2.1. For each $\alpha \in \mathbb{C}$, each $\kappa \in \mathbb{N} \cup \{+\infty\}$, and each sequence $(s_j)_{j=0}^{\kappa}$ from $\mathbb{C}^{p \times q}$, let the sequences $(s_{\alpha \triangleright j})_{j=0}^{\kappa-1}$ and $(s_{\alpha \triangleleft j})_{j=0}^{\kappa-1}$ be given by

$$s_{\alpha \triangleright j} := -\alpha s_j + s_{j+1} \qquad \text{and} \qquad s_{\alpha \triangleleft j} := \alpha s_j - s_{j+1} \qquad (2.1)$$

for each $j \in \mathbb{Z}_{0,\kappa-1}$. Then $(s_{\alpha \triangleright j})_{j=0}^{\kappa-1}$ (resp. $(s_{\alpha \triangleleft j})_{j=0}^{\kappa-1}$) is called the *sequence generated from* $(s_j)_{j=0}^{\kappa}$ by *right-sided* (resp. *left-sided*) α-*shifting*.

Remark 2.2. Let $\alpha \in \mathbb{C}$, let $\kappa \in \mathbb{N} \cup \{+\infty\}$, and let $(s_j)_{j=0}^{\kappa}$ be a sequence from $\mathbb{C}^{p \times q}$. In view of Definition 2.1, then $s_{\alpha \triangleleft j} = -s_{\alpha \triangleright j}$ for all $j \in \mathbb{Z}_{0,\kappa-1}$.

In view of Lemma 1.5, we add the following:

Remark 2.3. Let $\alpha \in \mathbb{C}$, $\kappa \in \mathbb{N} \cup \{+\infty\}$ and let $(s_j)_{j=0}^{\kappa}$ be a sequence from $\mathbb{C}^{p \times q}$. Let $\beta := -\alpha$ and $(u_j)_{j=0}^{\kappa}$ be defined by $u_j := (-1)^j s_j$. It can be easily checked that $u_{\beta \triangleleft j} = (-1)^j s_{\alpha \triangleright j}$ for each $j \in \mathbb{Z}_{0,\kappa-1}$.

Remarks 2.2 and 2.3 allow us to concentrate mainly on the right-sided case.
Theorems 1.6, 1.7, and 1.8 show that both sequences $(s_j)_{j=0}^{\kappa}$ and $(s_{\alpha \triangleright j})_{j=0}^{\kappa-1}$ (resp. $(s_{\alpha \triangleleft j})_{j=0}^{\kappa-1}$) together contain essential information about the considered moment problems connected with the interval $[\alpha, +\infty)$ (resp. $(-\infty, \alpha]$).

It should be mentioned that the construction in Definition 2.1 was introduced for $\alpha \in \mathbb{C}$ and rectangular matrices, though the generic case in this paper will be connected with real numbers α and square matrices.

Remark 2.4. Let $\kappa \in \mathbb{N} \cup \{+\infty\}$ and $(s_j)_{j=0}^{\kappa}$ be a sequence from $\mathbb{C}^{p \times q}$. In view of (2.1), then $s_{0 \triangleright j} = s_{j+1}$ for all $j \in \mathbb{Z}_{0,\kappa-1}$.

Obviously, for each $\alpha \in \mathbb{R}$ and each $n \in \mathbb{N}$, we have

$$\mathcal{K}_{q,2n,\alpha}^{\geq} = \left\{ (s_j)_{j=0}^{2n} \in \mathcal{H}_{q,2n}^{\geq} : (s_{\alpha \triangleright j})_{j=0}^{2(n-1)} \in \mathcal{H}_{q,2(n-1)}^{\geq} \right\},$$

$$\mathcal{K}_{q,2n,\alpha}^{\geq,e} = \left\{ (s_j)_{j=0}^{2n} \in \mathcal{H}_{q,2n}^{\geq} : (s_{\alpha \triangleright j})_{j=0}^{2n-1} \in \mathcal{H}_{q,2n-1}^{\geq,e} \right\},$$

and, for each $\alpha \in \mathbb{R}$ and each $n \in \mathbb{N}_0$, we see that $\mathcal{K}_{q,2n+1,\alpha}^{\geq}$ is the set of all sequences $(s_j)_{j=0}^{2n+1}$ from $\mathbb{C}^{q \times q}$ such that $\{(s_j)_{j=0}^{2n}, (s_{\alpha \triangleright j})_{j=0}^{2n}\} \subseteq \mathcal{H}_{q,2n}^{\geq}$ and that

$$\mathcal{K}_{q,2n+1,\alpha}^{\geq,e} = \left\{ (s_j)_{j=0}^{2n+1} \in \mathcal{H}_{q,2n+1}^{\geq,e} : (s_{\alpha \triangleright j})_{j=0}^{2n} \in \mathcal{H}_{q,2n}^{\geq} \right\}.$$

It seems to be useful to state some simple, but useful interrelations between the sequences $(s_j)_{j=0}^{\kappa}$ and $(s_{\alpha \triangleright j})_{j=0}^{\kappa-1}$. Here and in the following, for each $A \in \mathbb{C}^{p \times q}$, let $\mathcal{N}(A)$ be the null space of A and let $\mathcal{R}(A)$ be the column space of A.

Remark 2.5. Let $\alpha \in \mathbb{C}$, $\kappa \in \mathbb{N} \cup \{+\infty\}$, $(s_j)_{j=0}^{\kappa}$ be a sequence from $\mathbb{C}^{p \times q}$ and $m \in \mathbb{Z}_{1,\kappa}$. Then it is readily checked that

$$s_m = \alpha^m s_0 + \sum_{j=0}^{m-1} \alpha^{m-1-j} s_{\alpha \triangleright j}.$$

In particular, the inclusions $\mathcal{N}(s_0) \cap [\bigcap_{j=0}^{m-1} \mathcal{N}(s_{\alpha \triangleright j})] \subseteq \mathcal{N}(s_m)$ and $\mathcal{R}(s_m) \subseteq \mathcal{R}(s_0) + \sum_{j=0}^{m-1} \mathcal{R}(s_{\alpha \triangleright j})$ hold true.

Now we give some more or less simple observations on the arithmetics of several sets introduced in Definition 1.3.

Remark 2.6. Let $\alpha \in \mathbb{R}$, $\kappa \in \mathbb{N}_0 \cup \{+\infty\}$ and $r \in \mathbb{N}$. For all $v \in \mathbb{Z}_{1,r}$, let $(s_j^{(v)})_{j=0}^{\kappa} \in \mathcal{K}_{q,\kappa,\alpha}^{\geq}$ (resp. $\mathcal{K}_{q,\kappa,\alpha}^{\geq,e}$). Then we easily see that $(\sum_{v=1}^{r} s_j^{(v)})_{j=0}^{\kappa} \in \mathcal{K}_{q,\kappa,\alpha}^{\geq}$ (resp. $\mathcal{K}_{q,\kappa,\alpha}^{\geq,e}$).

Remark 2.7. Let $\alpha \in \mathbb{R}$, $\kappa \in \mathbb{N}_0 \cup \{+\infty\}$, $(s_j)_{j=0}^{\kappa} \in \mathcal{K}_{q,\kappa,\alpha}^{\geq}$ (resp. $\mathcal{K}_{q,\kappa,\alpha}^{\geq,e}$), and $A \in \mathbb{C}^{q \times p}$. Then it is readily checked that $(A^* s_j A)_{j=0}^{\kappa} \in \mathcal{K}_{p,\kappa,\alpha}^{\geq}$ (resp. $\mathcal{K}_{p,\kappa,\alpha}^{\geq,e}$).

Remark 2.8. Let $\alpha \in \mathbb{R}$, $\kappa \in \mathbb{N}_0 \cup \{+\infty\}$ and $n \in \mathbb{N}$. For all $l \in \mathbb{Z}_{1,n}$, let $q_l \in \mathbb{N}$ and $(s_j^{(l)})_{j=0}^{\kappa}$ be a sequence from $\mathbb{C}^{q_l \times q_l}$. Then it is readily checked that $(s_j^{(l)})_{j=0}^{\kappa} \in \mathcal{K}_{q_l,\kappa,\alpha}^{\geq}$ (resp. $\mathcal{K}_{q_l,\kappa,\alpha}^{\geq,e}$) for all $l \in \mathbb{Z}_{1,n}$ if and only if $(\operatorname{diag}[s_j^{(l)}]_{l=1}^{n})_{j=0}^{\kappa} \in \mathcal{K}_{\sum_{l=1}^{n} q_l,\kappa,\alpha}^{\geq}$ (resp. $\mathcal{K}_{\sum_{l=1}^{n} q_l,\kappa,\alpha}^{\geq,e}$).

We will write $\mathbb{C}_{\mathrm{H}}^{q \times q}$ for the set of all Hermitian complex $q \times q$ matrices. We will use the Löwner semi-odering in $\mathbb{C}_{\mathrm{H}}^{q \times q}$, i.e., we write $A \geq B$ (resp. $A > B$) in order to indicate that A and B are Hermitian complex matrices such that $A - B$ is non-negative (resp. positive) Hermitian. For each $\beta \in \mathbb{R}$, let $\lfloor \beta \rfloor := \max\{k \in \mathbb{Z} : k \leq \beta\}$, i.e., $\lfloor \beta \rfloor$ is the integer part of β.

Lemma 2.9. *Let* $\alpha \in \mathbb{R}$, $\kappa \in \mathbb{N}_0 \cup \{+\infty\}$ *and* $(s_j)_{j=0}^{\kappa} \in \mathcal{K}_{q,\kappa,\alpha}^{\geq}$. *Then:*

(a) $s_j \in \mathbb{C}_{\mathrm{H}}^{q \times q}$ *for all* $j \in \mathbb{Z}_{0,\kappa}$ *and* $s_{\alpha \triangleright j} \in \mathbb{C}_{\mathrm{H}}^{q \times q}$ *for all* $j \in \mathbb{Z}_{0,\kappa-1}$.
(b) $s_{2k} \in \mathbb{C}_{\geq}^{q \times q}$ *for all* $k \in \mathbb{N}_0$ *with* $2k \leq \kappa$ *and* $s_{\alpha \triangleright 2k} \in \mathbb{C}_{\geq}^{q \times q}$ *for all* $k \in \mathbb{N}_0$ *with* $2k + 1 \leq \kappa$.
(c) $\mathcal{N}(s_{2k}) \subseteq \mathcal{N}(s_j)$ *and* $\mathcal{R}(s_j) \subseteq \mathcal{R}(s_{2k})$ *for each* $k \in \mathbb{Z}_{0,\frac{\kappa}{2}}$ *and each* $j \in \mathbb{Z}_{2k,2\lfloor \frac{\kappa}{2} \rfloor - 1}$.
(d) $\mathcal{N}(s_{\alpha \triangleright 2k}) \subseteq \mathcal{N}(s_{\alpha \triangleright j})$ *and* $\mathcal{R}(s_{\alpha \triangleright j}) \subseteq \mathcal{R}(s_{\alpha \triangleright 2k})$ *for each* $k \in \mathbb{Z}_{0,\frac{\kappa-1}{2}}$ *and each* $j \in \mathbb{Z}_{2k,2\lfloor \frac{\kappa-1}{2} \rfloor - 1}$.

Proof. Combine Definition 1.3 with [18, Lemma 3.2]. $\qquad\square$

To obtain similar statements analogous to parts (c) and (d) for the class $\mathcal{K}^{\geq,e}_{q,m,\alpha}$, we recall the following result:

Lemma 2.10 ([12, Lemma 4.7, Lemma 4.11]). *Let* $\alpha \in \mathbb{R}$ *and* $n \in \mathbb{N}_0$. *Then* $\mathcal{K}^{\geq,e}_{q,2n,\alpha} \subseteq \mathcal{H}^{\geq,e}_{q,2n}$. *Furthermore, if* $(s_j)^{2n+1}_{j=0} \in \mathcal{K}^{\geq,e}_{q,2n+1,\alpha}$, *then* $(s_{\alpha \triangleright j})^{2n}_{j=0} \in \mathcal{H}^{\geq,e}_{q,2n}$.

Lemma 2.11. *Let* $\alpha \in \mathbb{R}$, $m \in \mathbb{N}_0$ *and* $(s_j)^m_{j=0} \in \mathcal{K}^{\geq,e}_{q,m,\alpha}$. *Then:*

(a) $\mathcal{N}(s_{2k}) \subseteq \mathcal{N}(s_j)$ *and* $\mathcal{R}(s_j) \subseteq \mathcal{R}(s_{2k})$ *for each* $k \in \mathbb{Z}_{0,\frac{m}{2}}$ *and each* $j \in \mathbb{Z}_{2k,m}$.
(b) $\mathcal{N}(s_{\alpha \triangleright 2k}) \subseteq \mathcal{N}(s_{\alpha \triangleright j})$ *and* $\mathcal{R}(s_{\alpha \triangleright j}) \subseteq \mathcal{R}(s_{\alpha \triangleright 2k})$ *for each* $k \in \mathbb{Z}_{0,\frac{m-1}{2}}$ *and each* $j \in \mathbb{Z}_{2k,m-1}$.

Proof. Combine Definition 1.3, Lemma 2.10 and [18, Lemma 3.1]. $\square$

At the end of this section, we turn our attention to some subclass of $\mathcal{K}^{\geq}_{q,\kappa,\alpha}$, which is characterized by stronger positivity properties. For each $n \in \mathbb{N}_0$, let $\mathcal{H}^{>}_{q,2n}$ be the set of sequences from $\mathbb{C}^{q \times q}$ such that the block Hankel matrix H_n given by (1.2) is positive Hermitian. If $\alpha \in \mathbb{R}$, then we set $\mathcal{K}^{>}_{q,0,\alpha} := \mathcal{H}^{>}_{q,0}$. For each $\alpha \in \mathbb{R}$ and each $n \in \mathbb{N}$, let

$$\mathcal{K}^{>}_{q,2n,\alpha} := \left\{ (s_j)^{2n}_{j=0} \in \mathcal{H}^{>}_{q,2n} : (s_{\alpha \triangleright j})^{2n-2}_{j=0} \in \mathcal{H}^{>}_{q,2n-2} \right\}.$$

Furthermore, if $\alpha \in \mathbb{R}$ and if $n \in \mathbb{N}_0$, then let $\mathcal{K}^{>}_{q,2n+1,\alpha}$ be the set of all sequences $(s_j)^{2n+1}_{j=0}$ from $\mathbb{C}^{q \times q}$ such that $\{(s_j)^{2n}_{j=0}, (s_{\alpha \triangleright j})^{2n}_{j=0}\} \subseteq \mathcal{H}^{>}_{q,2n}$.

Remark 2.12. Let $q \in \mathbb{N}$, $\alpha \in \mathbb{R}$, $n \in \mathbb{N}_0$, and $(s_j)^n_{j=0} \in \mathcal{K}^{>}_{q,n,\alpha}$. For each $m \in \mathbb{Z}_{0,n}$, then, $(s_j)^m_{j=0} \in \mathcal{K}^{>}_{q,m,\alpha}$.

In view of Remark 2.12, it is natural to denote by $\mathcal{K}^{>}_{q,\infty,\alpha}$ the set of all sequences $(s_j)^\infty_{j=0}$ from $\mathbb{C}^{q \times q}$ such that $(s_j)^m_{j=0} \in \mathcal{K}^{>}_{q,m,\alpha}$ holds for all $m \in \mathbb{N}_0$. For each $\alpha \in \mathbb{R}$ and each $\kappa \in \mathbb{N}_0 \cup \{+\infty\}$, a sequence $(s_j)^\kappa_{j=0} \in \mathcal{K}^{>}_{q,\kappa,\alpha}$ is called α-*Stieltjes right-sided positive definite*. In the special case $\alpha = 0$, it is also called *Stieltjes positive definite*. We state some first obvious observations on α-Stieltjes right-sided positive definite sequences.

Remark 2.13. Let $\alpha \in \mathbb{R}$ and $\kappa \in \mathbb{N}_0 \cup \{+\infty\}$. Then $\mathcal{K}^{>}_{q,\kappa,\alpha} \subseteq \mathcal{K}^{\geq}_{q,\kappa,\alpha}$.

Remark 2.14. Let $\alpha \in \mathbb{R}$, $\kappa \in \mathbb{N}_0 \cup \{+\infty\}$ and $r \in \mathbb{N}$. For all $v \in \mathbb{Z}_{1,r}$, let $(s_j^{(v)})^\kappa_{j=0} \in \mathcal{K}^{\geq}_{q,\kappa,\alpha}$. If there is some $v_0 \in \mathbb{Z}_{1,r}$ such that $(s_j^{(v_0)})^\kappa_{j=0} \in \mathcal{K}^{>}_{q,\kappa,\alpha}$, then $(\sum^r_{v=1} s_j^{(v)})^\kappa_{j=0} \in \mathcal{K}^{>}_{q,\kappa,\alpha}$.

Remark 2.15. Let $\alpha \in \mathbb{R}$, $\kappa \in \mathbb{N}_0 \cup \{+\infty\}$, $(s_j)^\kappa_{j=0} \in \mathcal{K}^{>}_{q,\kappa,\alpha}$ and $A \in \mathbb{C}^{q \times q}$. If A is non-singular, then $(A^* s_j A)^\kappa_{j=0} \in \mathcal{K}^{>}_{q,\kappa,\alpha}$, and if A is singular, then $(A^* s_j A)^\kappa_{j=0} \in \mathcal{K}^{\geq}_{q,\kappa,\alpha} \setminus \mathcal{K}^{>}_{q,\kappa,\alpha}$.

Remark 2.16. Let $\alpha \in \mathbb{R}$, $\kappa \in \mathbb{N}_0 \cup \{+\infty\}$ and $n \in \mathbb{N}$. For all $l \in \mathbb{Z}_{1,n}$, let $q_l \in \mathbb{N}$ and $(s_j^{(l)})^\kappa_{j=0}$ be a sequence from $\mathbb{C}^{q_l \times q_l}$. Then $(s_j^{(l)})^\kappa_{j=0} \in \mathcal{K}^{>}_{q_l,\kappa,\alpha}$ for all $l \in \mathbb{Z}_{1,n}$ if and only if $(\text{diag}[s_j^{(l)}]^n_{l=1})^\kappa_{j=0} \in \mathcal{K}^{>}_{\sum^n_{l=1} q_l,\kappa,\alpha}$.

In the sequel, we will need the Moore-Penrose inverse $A^\dagger$ of a complex $p \times q$ matrix A (see Appendix A).

In order to formulate some essential observations on the above introduced sets, it is useful to introduce now some further constructions of matrices notation, which will play a central role in the following.

Let $\kappa \in \mathbb{N}_0 \cup \{+\infty\}$ and a sequence $(s_j)_{j=0}^{\kappa}$ from $\mathbb{C}^{p \times q}$ be given. Then, for every choice of $j, k \in \mathbb{N}_0$ with $j \leq k \leq \kappa$, let

$$y_{j,k}^{\langle s \rangle} := \begin{bmatrix} s_j \\ s_{j+1} \\ \vdots \\ s_k \end{bmatrix} \qquad \text{and} \qquad z_{j,k}^{\langle s \rangle} := [s_j, s_{j+1}, \ldots, s_k]. \qquad (2.2)$$

For each $n \in \mathbb{N}_0$ with $2n \leq \kappa$, let $H_n^{\langle s \rangle} := [s_{j+k}]_{j,k=0}^n$. If $n \in \mathbb{N}_0$ is such that $2n + 1 \leq \kappa$, then let

$$K_n^{\langle s \rangle} := [s_{j+k+1}]_{j,k=0}^n, \qquad (2.3)$$

and for each $n \in \mathbb{N}_0$ with $2n + 2 \leq \kappa$, let

$$G_n^{\langle s \rangle} := [s_{j+k+2}]_{j,k=0}^n. \qquad (2.4)$$

We use the notation $L_0^{\langle s \rangle} := s_0$, and, for each $n \in \mathbb{N}$ with $2n \leq \kappa$, let

$$L_n^{\langle s \rangle} := s_{2n} - z_{n,2n-1}^{\langle s \rangle}(H_{n-1}^{\langle s \rangle})^\dagger y_{n,2n-1}^{\langle s \rangle}. \qquad (2.5)$$

In situations in which it is clear which sequence $(s_j)_{j=0}^{\kappa}$ of complex matrices is meant, we will write $y_{j,k}$, $z_{j,k}$, H_n, K_n, G_n, and L_n instead of $y_{j,k}^{\langle s \rangle}$, $z_{j,k}^{\langle s \rangle}$, $H_n^{\langle s \rangle}$, $K_n^{\langle s \rangle}$, $G_n^{\langle s \rangle}$, and $L_n^{\langle s \rangle}$, respectively. Furthermore, if $\alpha \in \mathbb{C}$, $\kappa \in \mathbb{N} \cup \{+\infty\}$, and a sequence $(s_j)_{j=0}^{\kappa}$ from $\mathbb{C}^{p \times q}$ are given, then let $(t_j)_{j=0}^{\kappa-1}$ be defined by $t_j := s_{\alpha \triangleright j}$ and (2.1) for each $j \in \mathbb{Z}_{0,\kappa-1}$, and let

$$H_{\alpha \triangleright n} := H_n^{\langle t \rangle} \qquad \text{and} \qquad L_{\alpha \triangleright n} := L_n^{\langle t \rangle} \qquad (2.6)$$

for each $n \in \mathbb{N}_0$ with $2n+1 \leq \kappa$, let $K_{\alpha \triangleright n} := K_n^{\langle t \rangle}$ for each $n \in \mathbb{N}_0$ with $2n+2 \leq \kappa$, and, for every choice of integers j, k with $0 \leq j \leq k \leq \kappa - 1$, let

$$y_{\alpha \triangleright j,k} := y_{j,k}^{\langle t \rangle} \qquad \text{and} \qquad z_{\alpha \triangleright j,k} := z_{j,k}^{\langle t \rangle}. \qquad (2.7)$$

Now we recall important characterizations of the set $\mathcal{K}_{q,m,\alpha}^{\geq,\mathrm{e}}$.

Lemma 2.17 ([12, Lemma 4.15]). *Let* $\alpha \in \mathbb{R}$, $n \in \mathbb{N}_0$ *and* $(s_j)_{j=0}^{2n+1} \in \mathcal{K}_{q,2n+1,\alpha}^{\geq}$. *Then* $(s_j)_{j=0}^{2n+1} \in \mathcal{K}_{q,2n+1,\alpha}^{\geq,\mathrm{e}}$ *if and only if* $\mathcal{N}(L_n) \subseteq \mathcal{N}(L_{\alpha \triangleright n})$.

Lemma 2.18 ([12, Lemma 4.16]). *Let* $\alpha \in \mathbb{R}$, $n \in \mathbb{N}$ *and* $(s_j)_{j=0}^{2n} \in \mathcal{K}_{q,2n,\alpha}^{\geq}$. *Then* $(s_j)_{j=0}^{2n} \in \mathcal{K}_{q,2n,\alpha}^{\geq,\mathrm{e}}$ *if and only if* $\mathcal{N}(L_{\alpha \triangleright n-1}) \subseteq \mathcal{N}(L_n)$.

Lemmas 2.17 and 2.18 have the following important consequences:

Proposition 2.19. *Let* $\alpha \in \mathbb{R}$.

(a) *Let* $n \in \mathbb{N}_0$ *and* $(s_j)_{j=0}^{2n+1} \in \mathcal{K}_{q,2n+1,\alpha}^{\geq,e}$. *Then*

$$\mathcal{N}(L_0) \subseteq \mathcal{N}(L_{\alpha \triangleright 0}) \subseteq \mathcal{N}(L_1) \subseteq \cdots \subseteq \mathcal{N}(L_{\alpha \triangleright n-1}) \subseteq \mathcal{N}(L_n) \subseteq \mathcal{N}(L_{\alpha \triangleright n})$$

and

$$\mathcal{R}(L_0) \supseteq \mathcal{R}(L_{\alpha \triangleright 0}) \supseteq \mathcal{R}(L_1) \supseteq \cdots \supseteq \mathcal{R}(L_{\alpha \triangleright n-1}) \supseteq \mathcal{R}(L_n) \supseteq \mathcal{R}(L_{\alpha \triangleright n}).$$

(b) *Let* $n \in \mathbb{N}$ *and* $(s_j)_{j=0}^{2n} \in \mathcal{K}_{q,2n,\alpha}^{\geq,e}$. *Then*

$$\mathcal{N}(L_0) \subseteq \mathcal{N}(L_{\alpha \triangleright 0}) \subseteq \mathcal{N}(L_1) \subseteq \cdots \subseteq \mathcal{N}(L_{\alpha \triangleright n-1}) \subseteq \mathcal{N}(L_n)$$

and

$$\mathcal{R}(L_0) \supseteq \mathcal{R}(L_{\alpha \triangleright 0}) \supseteq \mathcal{R}(L_1) \supseteq \cdots \supseteq \mathcal{R}(L_{\alpha \triangleright n-1}) \supseteq \mathcal{R}(L_n).$$

Proof. Use Remark 1.4 and Lemmas 2.17 and 2.18. □

Now we verify that α-Stieltjes right-sided positive definite sequences are α-Stieltjes right-sided non-negative definite extendable.

Proposition 2.20. *Let* $\alpha \in \mathbb{R}$ *and* $\kappa \in \mathbb{N}_0 \cup \{+\infty\}$. *Then* $\mathcal{K}_{q,\kappa,\alpha}^{>} \subseteq \mathcal{K}_{q,\kappa,\alpha}^{\geq,e}$.

Proof. Because of the definition of the sets in question, the cases $\kappa = 0$ and $\kappa = +\infty$ are trivial. Suppose now that $\kappa = 2n+1$ with some $n \in \mathbb{N}_0$. We consider an arbitrary sequence $(s_j)_{j=0}^{2n+1} \in \mathcal{K}_{q,2n+1,\alpha}^{>}$. Then $(s_j)_{j=0}^{2n+1}$ belongs to $\mathcal{K}_{q,2n+1,\alpha}^{\geq}$ and $(s_j)_{j=0}^{2n}$ belongs to $\mathcal{H}_{q,2n}^{>}$. Thus, [13, Remark 2.1] shows in particular that $\operatorname{rank} L_n = q$, i.e., $\mathcal{N}(L_n) = \{0_{q \times 1}\}$. Lemma 2.17 yields then $(s_j)_{j=0}^{2n+1} \in \mathcal{K}_{q,2n+1,\alpha}^{\geq,e}$. It remains to prove the case that $\kappa = 2n$ with some $n \in \mathbb{N}$. Let $(s_j)_{j=0}^{2n} \in \mathcal{K}_{q,2n,\alpha}^{>}$. Then $(s_j)_{j=0}^{2n} \in \mathcal{K}_{q,2n,\alpha}^{\geq}$ and $(s_{\alpha \triangleright j})_{j=0}^{2n-2} \in \mathcal{H}_{q,2n-2}^{>}$. Consequently, [13, Remark 2.1] implies $\operatorname{rank} L_{\alpha \triangleright n-1} = q$, i.e., $\mathcal{N}(L_{\alpha \triangleright n-1}) = \{0_{q \times 1}\}$. Using Lemma 2.18, we get $(s_j)_{j=0}^{2n} \in \mathcal{K}_{q,2n,\alpha}^{\geq,e}$. The proof is complete. □

3. Canonical Hankel parametrization

With later applications to the matrix version of the Hamburger moment problem in mind, a particular inner parametrization, called canonical Hankel parametrization, for sequences of complex matrices was developed in [13, 15, 18]. We recall the definition of the canonical Hankel parametrization of a sequence from $\mathbb{C}^{p \times q}$. To explain this notion, we need some further matrices built from the given data. Let $\kappa \in \mathbb{N}_0 \cup \{+\infty\}$ and $(s_j)_{j=0}^{\kappa}$ be a sequence from $\mathbb{C}^{p \times q}$. For all $k \in \mathbb{N}$ with $2k - 1 \leq \kappa$, let

$$\begin{aligned}
\Theta_k^{\langle s \rangle} &:= z_{k,2k-1}^{\langle s \rangle} (H_{k-1}^{\langle s \rangle})^{\dagger} y_{k,2k-1}^{\langle s \rangle}, \\
\Sigma_k^{\langle s \rangle} &:= z_{k,2k-1}^{\langle s \rangle} (H_{k-1}^{\langle s \rangle})^{\dagger} K_{k-1}^{\langle s \rangle} (H_{k-1}^{\langle s \rangle})^{\dagger} y_{k,2k-1}^{\langle s \rangle}
\end{aligned} \tag{3.1}$$

and, for all $k \in \mathbb{N}$ with $2k \leq \kappa$, let

$$M_k^{\langle s \rangle} := z_{k,2k-1}^{\langle s \rangle}(H_{k-1}^{\langle s \rangle})^{\dagger}y_{k+1,2k}^{\langle s \rangle}, \qquad N_k^{\langle s \rangle} := z_{k+1,2k}^{\langle s \rangle}(H_{k-1}^{\langle s \rangle})^{\dagger}y_{k,2k-1}^{\langle s \rangle}, \qquad (3.2)$$

and let

$$\Lambda_k^{\langle s \rangle} := M_k^{\langle s \rangle} + N_k^{\langle s \rangle} - \Sigma_k^{\langle s \rangle}. \qquad (3.3)$$

Moreover, let $0_{p \times q}$ be the null matrix in $\mathbb{C}^{p \times q}$ and let

$$\Theta_0^{\langle s \rangle} := 0_{p \times q}, \quad \Sigma_0^{\langle s \rangle} := 0_{p \times q}, \quad M_0^{\langle s \rangle} := 0_{p \times q}, \quad N_0^{\langle s \rangle} := 0_{p \times q}, \quad \text{and} \quad \Lambda_0^{\langle s \rangle} := 0_{p \times q}. \qquad (3.4)$$

If a sequence $(s_j)_{j=0}^{\kappa}$ from $\mathbb{C}^{p \times q}$ is given, then we will write Θ_k, Σ_k, M_k, N_k and Λ_k for $\Theta_k^{\langle s \rangle}$, $\Sigma_k^{\langle s \rangle}$, $M_k^{\langle s \rangle}$, $N_k^{\langle s \rangle}$ and $\Lambda_k^{\langle s \rangle}$, respectively. Furthermore, if $\alpha \in \mathbb{C}$, $\kappa \in \mathbb{N} \cup \{+\infty\}$, and a sequence $(s_j)_{j=0}^{\kappa}$ from $\mathbb{C}^{p \times q}$ are given, then let

$$\Theta_{\alpha \triangleright k} := \Theta_k^{\langle t \rangle} \qquad \text{for all } k \in \mathbb{N}_0 \text{ with } 2k \leq \kappa \qquad (3.5)$$

and let

$$M_{\alpha \triangleright k} := M_k^{\langle t \rangle} \quad \text{and} \quad \Lambda_{\alpha \triangleright k} := \Lambda_k^{\langle t \rangle} \quad \text{for all } k \in \mathbb{N}_0 \text{ with } 2k \leq \kappa - 1, \qquad (3.6)$$

where $(t_j)_{j=0}^{\kappa-1}$ is defined by $t_j := s_{\alpha \triangleright j}$ and (2.1) for all $j \in \mathbb{Z}_{0,\kappa-1}$.

Remark 3.1. Let $\kappa \in \mathbb{N} \cup \{+\infty\}$ and let $(s_j)_{j=0}^{2\kappa}$ be a sequence from $\mathbb{C}^{p \times q}$. Then one can easily see that there are unique sequences $(C_k)_{k=1}^{\kappa}$ and $(D_k)_{k=0}^{\kappa}$ from $\mathbb{C}^{p \times q}$ such that $s_{2k} = \Theta_k + D_k$ for all $k \in \mathbb{Z}_{0,\kappa}$ and $s_{2k-1} = \Lambda_{k-1} + C_k$ for all $k \in \mathbb{Z}_{1,\kappa}$, where Θ_k and Λ_{k-1} are given by (3.1) and (3.3), respectively. In particular, we see that $C_k = s_{2k-1} - \Lambda_{k-1}$ for all $k \in \mathbb{Z}_{1,\kappa}$ and moreover $D_k = L_k$ for all $k \in \mathbb{Z}_{0,\kappa}$.

Remark 3.1 leads to the notion of canonical Hankel parametrization, which was introduced in [13, Definition 2.28] and [18, Definition 2.3].

Definition 3.2. Let $\kappa \in \mathbb{N} \cup \{+\infty\}$ and let $(s_j)_{j=0}^{2\kappa}$ be a sequence from $\mathbb{C}^{p \times q}$. For all $k \in \mathbb{Z}_{1,\kappa}$, let $C_k := s_{2k-1} - \Lambda_{k-1}$, where Λ_{k-1} was introduced in (3.4) and (3.3), and, for all $k \in \mathbb{Z}_{0,\kappa}$, let $D_k := L_k$, where L_k was introduced in (2.5). Then the pair $[(C_k)_{k=1}^{\kappa}, (D_k)_{k=0}^{\kappa}]$ is called the *canonical Hankel parametrization of* $(s_j)_{j=0}^{2\kappa}$.

Remark 3.3. Let $\kappa \in \mathbb{N} \cup \{+\infty\}$ and let $(C_k)_{k=1}^{\kappa}$ and $(D_k)_{k=0}^{\kappa}$ be sequences from $\mathbb{C}^{p \times q}$. Then it can be immediately checked by induction that there is a unique sequence $(s_j)_{j=0}^{2\kappa}$ from $\mathbb{C}^{p \times q}$ such that $[(C_k)_{k=1}^{\kappa}, (D_k)_{k=0}^{\kappa}]$ is the canonical Hankel parametrization of $(s_j)_{j=0}^{2\kappa}$, namely the sequence $(s_j)_{j=0}^{2\kappa}$ recursively given by $s_{2k} = \Theta_k + D_k$ for all $k \in \mathbb{Z}_{0,\kappa}$ and $s_{2k+1} = \Lambda_k + C_{k+1}$ for all $k \in \mathbb{Z}_{0,\kappa-1}$.

In view of Definition 3.2, [15, Proposition 2.10 (b)] admits the following reformulation:

Proposition 3.4. *Let $n \in \mathbb{N}$ and let $(s_j)_{j=0}^{2n}$ be a sequence from $\mathbb{C}^{q \times q}$ with canonical Hankel parametrization $[(C_k)_{k=1}^{n}, (D_k)_{k=0}^{n}]$. Then $(s_j)_{j=0}^{2n} \in \mathcal{H}_{q,2n}^{\geq}$ if and only if the following three conditions hold:*

(I) *$(C_k)_{k=1}^{n}$ is a sequence of Hermitian matrices.*

(II) *$(D_k)_{k=0}^{n}$ is a sequence of non-negative Hermitian matrices.*

(III) *All the inclusions*

$$\mathcal{N}(D_0) \subseteq \mathcal{N}(D_1) \subseteq \cdots \subseteq \mathcal{N}(D_{n-2}) \subseteq \mathcal{N}(D_{n-1})$$

$$\mathcal{N}(C_1) \quad \mathcal{N}(C_2) \qquad\qquad \mathcal{N}(C_{n-1}) \quad \mathcal{N}(C_n)$$

hold true.

Proposition 3.4 shows that the membership of a sequence to $\mathcal{H}^{\geq}_{q,2n}$ can be nicely expressed in terms of its canonical Hankel parametrization. Taking Remark 3.3 into account, we see from Proposition 3.4 that Hankel non-negative definite sequences with prescribed canonical Hankel parametrization can be constructed. In the next section, we will introduce a particular parametrization for sequences from $\mathbb{C}^{p\times q}$ which plays a similar role concerning the membership of a sequence to the set $\mathcal{K}^{\geq}_{q,\kappa,\alpha}$ where $\alpha \in \mathbb{R}$.

At the end of this section, we express the rank and the determinant of the matrix H_n built from a sequence $(s_j)_{j=0}^{2n} \in \mathcal{H}^{\geq}_{q,2n}$ in terms of its canonical Hankel parametrization.

Lemma 3.5. *Let $\kappa \in \mathbb{N} \cup \{+\infty\}$ and let $(s_j)_{j=0}^{2\kappa} \in \mathcal{H}^{\geq}_{q,2\kappa}$ with canonical Hankel parametrization $[(C_k)_{k=1}^{\kappa}, (D_k)_{k=0}^{\kappa}]$. Then $\operatorname{rank} H_n = \sum_{k=0}^{n} \operatorname{rank} D_k$ and $\det H_n = \prod_{k=0}^{n} \det D_k$ for all $n \in \mathbb{Z}_{0,\kappa}$.*

Proof. According to [15, Proposition 4.17], for all $n \in \mathbb{N}$, there exists a complex $(n+1)q \times (n+1)q$ matrix $\mathbb{F}_n$ with $\det \mathbb{F}_n = 1$ such that $\mathbb{F}_n^* H_n \mathbb{F}_n = \operatorname{diag}[L_k]_{k=0}^{n}$. In view of the definition of the set $\mathcal{H}^{\geq}_{q,2\kappa}$ and Definition 3.2, this implies the assertions. $\qquad\square$

4. One-sided α-Stieltjes parametrizations

Our current focus on matricial moment problems for intervals of the type $[\alpha, +\infty)$ or $(-\infty, \alpha]$, where α is an arbitrary real number, motivates us to look for corresponding one-sided α-analogues of the canonical Hankel parametrization of seqeunces from $\mathbb{C}^{p\times q}$. Our above considerations show that, for reasons of symmetry, we can mainly concentrate on the right-sided case. In this case, we use the matrices defined in (3.1), (2.5) and (2.6) to introduce the following notion which will turn out to be one of the central objects of this paper.

Remark 4.1. Let $\alpha \in \mathbb{C}$, let $\kappa \in \mathbb{N}_0 \cup \{+\infty\}$, and let $(s_j)_{j=0}^{\kappa}$ be a sequence from $\mathbb{C}^{p\times q}$. Then one can easily see that there is a unique sequence $(Q_j)_{j=0}^{\kappa}$ from $\mathbb{C}^{p\times q}$ such that $s_{2k} = \Theta_k + Q_{2k}$ for all $k \in \mathbb{N}_0$ with $2k \leq \kappa$ and $s_{2k+1} = \alpha s_{2k} + \Theta_{\alpha\triangleright k} + Q_{2k+1}$ for all $k \in \mathbb{N}_0$ with $2k+1 \leq \kappa$. In particular, we see that $Q_{2k} = L_k$ for all $k \in \mathbb{N}_0$ with $2k \leq \kappa$ and moreover $Q_{2k+1} = L_{\alpha\triangleright k}$ for all $k \in \mathbb{N}_0$ with $2k+1 \leq \kappa$.

Remark 4.1 leads us to the following notion.

Definition 4.2. Let $\alpha \in \mathbb{C}$, let $\kappa \in \mathbb{N}_0 \cup \{+\infty\}$, and let $(s_j)_{j=0}^{\kappa}$ be a sequence from $\mathbb{C}^{p\times q}$. Then the sequence $(Q_j)_{j=0}^{\kappa}$ given by $Q_{2k} := L_k$ for all $k \in \mathbb{N}_0$ with $2k \leq \kappa$

and by $Q_{2k+1} := L_{\alpha \rhd k}$ for all $k \in \mathbb{N}_0$ with $2k + 1 \leq \kappa$ is called the *right-sided* α-*Stieltjes parametrization of* $(s_j)_{j=0}^{\kappa}$. In the case $\alpha = 0$ the sequence $(Q_j)_{j=0}^{\kappa}$ is simply called the *right-sided Stieltjes parametrization of* $(s_j)_{j=0}^{\kappa}$.

Remark 4.3. Let $\alpha \in \mathbb{C}$, let $\kappa \in \mathbb{N}_0 \cup \{+\infty\}$, and let $(Q_j)_{j=0}^{\kappa}$ be a sequence from $\mathbb{C}^{p \times q}$. Then it can be immediately checked by induction that there is a unique sequence $(s_j)_{j=0}^{\kappa}$ from $\mathbb{C}^{p \times q}$ such that $(Q_j)_{j=0}^{\kappa}$ is the right-sided α-Stieltjes parametrization of $(s_j)_{j=0}^{\kappa}$, namely the sequence $(s_j)_{j=0}^{\kappa}$ recursively given by $s_{2k} = \Theta_k + Q_{2k}$ for all $k \in \mathbb{N}_0$ with $2k \leq \kappa$ and $s_{2k+1} = \alpha s_{2k} + \Theta_{\alpha \rhd k} + Q_{2k+1}$ for all $k \in \mathbb{N}_0$ with $2k + 1 \leq \kappa$.

Let us add the left-sided analogue of Definition 4.2.

Remark 4.4. Let $\alpha \in \mathbb{C}$, let $\kappa \in \mathbb{N}_0 \cup \{+\infty\}$, and let $(s_j)_{j=0}^{\kappa}$ be a sequence from $\mathbb{C}^{p \times q}$. Then there is a unique sequence $(\tilde{Q}_j)_{j=0}^{\kappa}$ from $\mathbb{C}^{p \times q}$ such that $s_{2k} = \Theta_k + \tilde{Q}_{2k}$ for all $k \in \mathbb{N}_0$ with $2k \leq \kappa$ and in the case $\kappa \geq 1$ moreover $s_{2k+1} = \alpha s_{2k} - \Theta_k^{\langle t \rangle} - \tilde{Q}_{2k+1}$ for all $k \in \mathbb{N}_0$ with $2k + 1 \leq \kappa$, where the sequence $(t_j)_{j=0}^{\kappa-1}$ is given by $t_j := s_{\alpha \lhd j}$ and (2.1) for all $j \in \mathbb{Z}_{0,\kappa-1}$. In particular, we see that $\tilde{Q}_{2k} = L_k$ for all $k \in \mathbb{N}_0$ with $2k \leq \kappa$ and $\tilde{Q}_{2k+1} = L_k^{\langle t \rangle}$ for all $k \in \mathbb{N}_0$ with $2k + 1 \leq \kappa$.

Definition 4.5. Let $\alpha \in \mathbb{C}$, $\kappa \in \mathbb{N}_0 \cup \{+\infty\}$ and let $(s_j)_{j=0}^{\kappa}$ be a sequence from $\mathbb{C}^{p \times q}$. Let $(t_j)_{j=0}^{\kappa-1}$ be defined by $t_j := s_{\alpha \lhd j}$ and (2.1) for all $j \in \mathbb{Z}_{0,\kappa-1}$. Using (2.5), let $\tilde{Q}_{2k} := L_k$ for all $k \in \mathbb{N}_0$ with $2k \leq \kappa$ and $\tilde{Q}_{2k+1} := L_k^{\langle t \rangle}$ for all $k \in \mathbb{N}_0$ with $2k + 1 \leq \kappa$. Then the sequence $(\tilde{Q}_j)_{j=0}^{\kappa}$ is called the *left-sided* α-*Stieltjes parametrization of* $(s_j)_{j=0}^{\kappa}$. In the particular case $\alpha = 0$, the sequence $(\tilde{Q}_j)_{j=0}^{\kappa}$ is simply called the *left-sided Stieltjes parametrization of* $(s_j)_{j=0}^{\kappa}$.

Remark 4.6. Let $\alpha \in \mathbb{C}$, let $\kappa \in \mathbb{N}_0 \cup \{+\infty\}$, and let $(\tilde{Q}_j)_{j=0}^{\kappa}$ be a sequence from $\mathbb{C}^{p \times q}$. Then it can be immediately checked by induction that there is a unique sequence $(s_j)_{j=0}^{\kappa}$ from $\mathbb{C}^{p \times q}$ such that $(\tilde{Q}_j)_{j=0}^{\kappa}$ is the left-sided α-Stieltjes parametrization of $(s_j)_{j=0}^{\kappa}$, namely the sequence $(s_j)_{j=0}^{\kappa}$ given by $s_{2k} = \Theta_k + \tilde{Q}_{2k}$ for all $k \in \mathbb{N}_0$ with $2k \leq \kappa$ and, in the case $\kappa \geq 1$, moreover by $s_{2k+1} = \alpha s_{2k} - \Theta_k^{\langle t \rangle} - \tilde{Q}_{2k+1}$ for all $k \in \mathbb{N}_0$ with $2k + 1 \leq \kappa$, where the sequence $(t_j)_{j=0}^{\kappa-1}$ is given by $t_j := s_{\alpha \lhd j}$ and (2.1) for all $j \in \mathbb{Z}_{0,\kappa-1}$.

Remark 4.7. Let $\alpha \in \mathbb{C}$, $\kappa \in \mathbb{N}_0 \cup \{+\infty\}$ and let $(s_j)_{j=0}^{\kappa}$ be a sequence from $\mathbb{C}^{p \times q}$. Denote by $(Q_j)_{j=0}^{\kappa}$ and $(\tilde{Q}_j)_{j=0}^{\kappa}$ the right and left-sided α-Stieltjes parametrization of $(s_j)_{j=0}^{\kappa}$, respectively. Then from the definition of the corresponding matrices it is clear that $\tilde{Q}_j = (-1)^j Q_j$ for each $j \in \mathbb{Z}_{0,\kappa}$.

In view of Remark 4.7, we can mainly concentrate on the investigation of the right-sided α-Stieltjes parametrization.

Remark 4.8. Let $\alpha \in \mathbb{C}$, let $\kappa \in \mathbb{N}_0 \cup \{+\infty\}$, and let $(s_j)_{j=0}^{\kappa}$ be a sequence from $\mathbb{C}^{p \times q}$. Further, let $(Q_j)_{j=0}^{\kappa}$ be the right-sided α-Stieltjes parametrization of $(s_j)_{j=0}^{\kappa}$.

Keeping in mind Definition 4.2 and the definition of the matrices L_k, $k \in \mathbb{Z}_{0,\frac{\kappa}{2}}$, and $L_{\alpha \rhd k}$, $k \in \mathbb{Z}_{0,\frac{\kappa-1}{2}}$, it is readily checked that, for each $m \in \mathbb{Z}_{0,\kappa}$, the sequence $(Q_j)_{j=0}^m$ is the right-sided α-Stieltjes parametrization of $(s_j)_{j=0}^m$.

We list now some properties of the right-sided α-Stieltjes parametrization, which can be easily proved using basic results on Moore-Penrose inverse of complex matrices, which are especially summarized in [17, Remark A.2, Lemma A.3]. We omit the details of the proofs.

Remark 4.9. Let $\alpha \in \mathbb{C}$, $\kappa \in \mathbb{N}_0 \cup \{+\infty\}$, $(s_j)_{j=0}^\kappa$ be a sequence from $\mathbb{C}^{p \times q}$ and $(Q_j)_{j=0}^\kappa$ be the right-sided α-Stieltjes parametrization of $(s_j)_{j=0}^\kappa$. Then:

(a) $(Q_j^{\mathrm{T}})_{j=0}^\kappa$ is the right-sided α-Stieltjes parametrization of $(s_j^{\mathrm{T}})_{j=0}^\kappa$.
(b) $(Q_j^*)_{j=0}^\kappa$ is the right-sided $\bar{\alpha}$-Stieltjes parametrization of $(s_j^*)_{j=0}^\kappa$.
(c) For each choice of $m, n \in \mathbb{N}$, $U \in \mathbb{C}^{m \times p}$ with $U^*U = I_p$ and $V \in \mathbb{C}^{q \times n}$ with $VV^* = I_q$, the sequence $(UQ_jV)_{j=0}^\kappa$ is the right-sided α-Stieltjes parametrization of $(Us_jV)_{j=0}^\kappa$.

Remark 4.10. Let $\alpha \in \mathbb{C}$, $\kappa \in \mathbb{N}_0 \cup \{+\infty\}$ and $n \in \mathbb{N}$. For all $m \in \mathbb{Z}_{1,n}$, let $p_m, q_m \in \mathbb{N}$, $(s_j^{(m)})_{j=0}^\kappa$ be a sequence from $\mathbb{C}^{p_m \times q_m}$, and $(Q_j^{(m)})_{j=0}^\kappa$ be the right-sided α-Stieltjes parametrization of $(s_j^{(m)})_{j=0}^\kappa$. Then it is readily checked that $(\mathrm{diag}[Q_j^{(m)}]_{m=1}^n)_{j=0}^\kappa$ is the right-sided α-Stieltjes parametrization of

$$(\mathrm{diag}[s_j^{(m)}]_{m=1}^n)_{j=0}^\kappa.$$

Lemma 4.11. *Let* $\alpha \in \mathbb{R}$, *let* $\kappa \in \mathbb{N}_0 \cup \{+\infty\}$, *and let* $(s_j)_{j=0}^\kappa \in \mathcal{K}_{q,\kappa,\alpha}^\geq$ *with right-sided* α-*Stieltjes parametrization* $(Q_j)_{j=0}^\kappa$. *Then:*

(a) $\mathrm{rank}\, H_n = \sum_{k=0}^n \mathrm{rank}\, Q_{2k}$ *and* $\det H_n = \prod_{k=0}^n \det Q_{2k}$ *for all* $n \in \mathbb{N}_0$ *with* $2n \leq \kappa$.
(b) *If* $\kappa \geq 1$, *then* $\mathrm{rank}\, H_{\alpha \rhd n} = \sum_{k=0}^n \mathrm{rank}\, Q_{2k+1}$ *and* $\det H_{\alpha \rhd n} = \prod_{k=0}^n \det Q_{2k+1}$ *for all* $n \in \mathbb{N}_0$ *with* $2n + 1 \leq \kappa$

Proof. Combine the Definition of $\mathcal{K}_{q,\kappa,\alpha}^\geq$ with Definition 4.2 and Lemma 3.5. $\square$

Guided by Proposition 3.4, now we are looking for characterizations of particular classes of sequences from $\mathbb{C}^{q \times q}$ in terms of their canonical right-sided α-Stieltjes parametrization.

Theorem 4.12. *Let* $\alpha \in \mathbb{R}$, $\kappa \in \mathbb{N}_0 \cup \{+\infty\}$, $(s_j)_{j=0}^\kappa$ *be a sequence from* $\mathbb{C}^{q \times q}$ *and* $(Q_j)_{j=0}^\kappa$ *be the right-sided* α-*Stieltjes parametrization of* $(s_j)_{j=0}^\kappa$. *Then:*

(a) $(s_j)_{j=0}^\kappa$ *is a sequence from* $\mathbb{C}_{\mathrm{H}}^{q \times q}$ *if and only if* $(Q_j)_{j=0}^\kappa$ *is a sequence from* $\mathbb{C}_{\mathrm{H}}^{q \times q}$.
(b) *The following statements are equivalent:*
   (i) $(s_j)_{j=0}^\kappa \in \mathcal{K}_{q,\kappa,\alpha}^\geq$.
   (ii) $Q_j \in \mathbb{C}_\geq^{q \times q}$ *for each* $j \in \mathbb{Z}_{0,\kappa}$ *and, in the case* $\kappa \geq 2$, *the inclusion* $\mathcal{N}(Q_j) \subseteq \mathcal{N}(Q_{j+1})$ *holds for each* $j \in \mathbb{Z}_{0,\kappa-2}$.

(iii) $Q_j \in \mathbb{C}_\geq^{q \times q}$ for each $j \in \mathbb{Z}_{0,\kappa}$ and, in the case $\kappa \geq 2$, the equations
$$Q_{j+1}Q_j^\dagger Q_j = Q_{j+1} \text{ holds for each } j \in \mathbb{Z}_{0,\kappa-2}.$$

(c) *The following statements are equivalent:*

(iv) $(s_j)_{j=0}^\kappa \in \mathcal{K}_{q,\kappa,\alpha}^{\geq,e}$.

(v) $Q_j \in \mathbb{C}_\geq^{q \times q}$ for each $j \in \mathbb{Z}_{0,\kappa}$ and, in the case $\kappa \geq 1$, furthermore $\mathcal{N}(Q_j) \subseteq \mathcal{N}(Q_{j+1})$ for all $j \in \mathbb{Z}_{0,\kappa-1}$.

(vi) $Q_j \in \mathbb{C}_\geq^{q \times q}$ for each $j \in \mathbb{Z}_{0,\kappa}$ and, in the case $\kappa \geq 1$, furthermore
$$Q_{j+1}Q_j^\dagger Q_j = Q_{j+1} \text{ for all } j \in \mathbb{Z}_{0,\kappa-1}.$$

(d) *The following statements are equivalent:*

(vii) $(s_j)_{j=0}^\kappa \in \mathcal{K}_{q,\kappa,\alpha}^{>}$.

(viii) $Q_j \in \mathbb{C}_>^{q \times q}$ for each $j \in \mathbb{Z}_{0,\kappa}$.

Proof. (a) This equivalence is readily checked by induction.

(b) "(i)$\Rightarrow$(ii)": Since (i) is supposed, $(Q_j)_{j=0}^\kappa$ is a sequence of Schur complements of non-negative Hermitian block Hankel matrices. Hence, $Q_j \in \mathbb{C}_\geq^{q \times q}$ for each $j \in \mathbb{Z}_{0,\kappa}$ (see also [12, Propositions 4.9 and 4.13]). Obviously, if $\kappa \geq 2$, then $(s_j)_{j=0}^{\kappa-1}$ belongs to $\mathcal{K}_{q,\kappa-1,\alpha}^{\geq,e}$, and, thus, Remark 4.8, Definition 4.2, and Proposition 2.19 yield $\mathcal{N}(Q_j) \subseteq \mathcal{N}(Q_{j+1})$ for each $j \in \mathbb{Z}_{0,\kappa-2}$.

"(ii)$\Rightarrow$(i)": Because of $Q_0 \in \mathbb{C}_\geq^{q \times q}$, we have $(s_j)_{j=0}^0 \in \mathcal{H}_{q,0}^\geq = \mathcal{K}_{q,0,\alpha}^{\geq,e}$. If $\kappa = 0$, thus (i) holds. If $\kappa \geq 1$, then $Q_0, Q_1 \in \mathbb{C}_\geq^{q \times q}$ imply immediately $(s_j)_{j=0}^1 \in \mathcal{K}_{q,1,\alpha}^\geq$. Now let $\kappa \geq 2$. We know that there is an $m \in \mathbb{Z}_{1,\kappa-1}$ such that $(s_j)_{j=0}^m \in \mathcal{K}_{q,m,\alpha}^\geq$. Because of $\mathcal{N}(Q_{m-1}) \subseteq \mathcal{N}(Q_m)$, Lemmas 2.17 and 2.18 yield $(s_j)_{j=0}^m \in \mathcal{K}_{q,m,\alpha}^{\geq,e}$. Thus, from $Q_m \in \mathbb{C}_\geq^{q \times q}$ and [12, Proposition 4.9 and Proposition 4.13] we see that $(s_j)_{j=0}^m$ belongs to $\mathcal{K}_{q,m,\alpha}^\geq$. Consequently, (i) is inductively proved.

"(ii)$\Leftrightarrow$(iii)": Use Lemma A.1.

(c) "(iv)$\Rightarrow$(v)": If $\kappa = +\infty$, then (v) follows from part (b). The case $\kappa = 0$ is trivial. Suppose that $\kappa \in \mathbb{N}$. According to (iv), there is an $s_{\kappa+1} \in \mathbb{C}^{q \times q}$ such that $(s_j)_{j=0}^{\kappa+1} \in \mathcal{K}_{q,\kappa+1,\alpha}^\geq$. Thus, (v) follows from part (b).

"(v)$\Rightarrow$(iv)": If $\kappa = +\infty$, then this implication is true because of (b). If $\kappa = 2n$ with some $n \in \mathbb{N}_0$, then, setting $s_{2n+1} := \alpha s_{2n} + z_{\alpha \triangleright n, 2n-1} H_{\alpha \triangleright n-1}^\dagger y_{\alpha \triangleright n, 2n-1}$, Definition 4.2 shows $Q_{2n+1} = 0_{q \times q}$, so that part (b) implies $(s_j)_{j=0}^{2n+1} \in \mathcal{K}_{q,2n+1,\alpha}^\geq$. If $\kappa = 2n-1$ with some $n \in \mathbb{N}$, then with $s_{2n} := z_{n,2n-1} H_{n-1}^\dagger y_{n,2n-1}$ we have $Q_{2n} = 0_{q \times q}$ and part (b) yields $(s_j)_{j=0}^{2n} \in \mathcal{K}_{q,2n,\alpha}^\geq$. In each case, (iv) holds true.

"(v)$\Leftrightarrow$(vi)": Use Lemma A.1.

(d) "(vii)$\Rightarrow$(viii)": Because of (vii), the sequence $(s_j)_{j=0}^\kappa$ belongs to $\mathcal{K}_{q,\kappa,\alpha}^\geq$. Thus, from (b) it follows that $Q_j \in \mathbb{C}_\geq^{q \times q}$ for each $j \in \mathbb{Z}_{0,\kappa}$. Since (vii) holds, we know that

$$\operatorname{rank} H_n = (n+1)q \qquad \text{for each } n \in \mathbb{N}_0 \text{ with } 2n \leq \kappa \qquad (4.1)$$

and that

$$\operatorname{rank} H_{\alpha \triangleright n} = (n+1)q \qquad \text{for each } n \in \mathbb{N}_0 \text{ with } 2n+1 \leq \kappa. \qquad (4.2)$$

Consequently, Lemma 4.11 yields (viii).

"(viii)⇒(vii)": Part (b) shows that (vii) implies $(s_j)_{j=0}^{\kappa} \in \mathcal{K}_{q,\kappa,\alpha}^{\geq}$. Because of Lemma 4.11, we get then (4.1) and (4.2). Thus, (vii) holds true. □

Taking Remark 4.3 into account, the application of Theorem 4.12 provides us the possibility to construct, for arbitrarily given $\alpha \in \mathbb{R}$, sequences from $\mathbb{C}^{q \times q}$ with prescribed right α-Stieltjes properties.

As a first application of Theorem 4.12 we present now a complete answer to the completion problem for α-Stieltjes right-sided positive definite sequences.

Proposition 4.13. *Let $\alpha \in \mathbb{R}$, let $n \in \mathbb{N}_0$, and let $(s_j)_{j=0}^{2n+1}$ be a sequence from $\mathbb{C}^{q \times q}$. Then $(s_j)_{j=0}^{2n+1} \in \mathcal{K}_{q,2n+1,\alpha}^{>}$ if and only if $(s_j)_{j=0}^{2n} \in \mathcal{K}_{q,2n,\alpha}^{>}$ and $s_{2n+1} > \alpha s_{2n} + \Theta_{\alpha \triangleright n}$, where $\Theta_{\alpha \triangleright n}$ is given by (3.5).*

Proof. Use part (d) of Theorem 4.12, Definition 4.2, (2.6), (2.5), and (2.1). □

Proposition 4.14. *Let $\alpha \in \mathbb{R}$, let $n \in \mathbb{N}$, and let $(s_j)_{j=0}^{2n}$ be a sequence from $\mathbb{C}^{q \times q}$. Then $(s_j)_{j=0}^{2n} \in \mathcal{K}_{q,2n,\alpha}^{>}$ if and only if $(s_j)_{j=0}^{2n-1} \in \mathcal{K}_{q,2n-1,\alpha}^{>}$ and $s_{2n} > \Theta_n$, where Θ_n is given by (3.1).*

Proof. Use part (d) of Theorem 4.12, Definition 4.2, and (2.5). □

It should be mentioned, that Propositions 4.13 and 4.14 allow us an alternate approach to Proposition 2.20. The combination of Propositions 4.13 and 4.14 even produces an improvement of Proposition 2.20. Namely, we see now that an α-Stieltjes right-sided positive definite sequence admits always an α-Stieltjes right-sided positive definite extension.

Example 4.15 (Catalan numbers). Let $s_j := \frac{1}{j+1}\binom{2j}{j}$ for all $j \in \mathbb{N}_0$ and let $(Q_j)_{j=0}^{\infty}$ be the right-sided 0-Stieltjes parametrization of $(s_j)_{j=0}^{\infty}$. Then $(s_j)_{j=0}^{\infty} \in \mathcal{K}_{1,\infty,0}^{>}$ and $Q_j = 1$ for all $j \in \mathbb{N}_0$. Indeed, $\det H_n = 1$ for all $n \in \mathbb{N}_0$ (see, e.g., [22]). Thus, from [15, Remark 3.1 (a)] we see that $D_k = 1$ for all $k \in \mathbb{N}_0$, where $[(C_k)_{k=1}^{\infty}, (D_k)_{k=0}^{\infty}]$ is the canonical Hankel parametrization of $(s_j)_{j=0}^{\infty}$. In view of Definitions 4.2 and 3.2 we get that $Q_{2k} = L_k = D_k = 1$ for all $k \in \mathbb{N}_0$. For all $j \in \mathbb{N}_0$, let $t_j := s_{j+1}$. Then $\det H_n^{\langle t \rangle} = 1$ for all $n \in \mathbb{N}_0$ (see, e.g., [22]). Thus, from [15, Remark 3.1 (a)] we see that $F_k = 1$ for all $k \in \mathbb{N}_0$, where $[(E_k)_{k=1}^{\infty}, (F_k)_{k=0}^{\infty}]$ is the canonical Hankel parametrization of $(t_j)_{j=0}^{\infty}$. In view of Definition 4.2, (2.6), Remark 2.4 and Definition 3.2, we get that $Q_{2k+1} = L_{0 \triangleright k} = L_k^{\langle t \rangle} = F_k = 1$ for all $k \in \mathbb{N}_0$. Finally, part (d) of Theorem 4.12 yields $(s_j)_{j=0}^{\infty} \in \mathcal{K}_{1,\infty,0}^{>}$.

5. Completely degenerate α-Stieltjes non-negative definite sequences

In this section, we continue our former investigations on completely degenerate Hankel non-negative definite sequences, which were done in [12, 13, 15, 18]. First we recall the corresponding notions. Based on the matrices defined via (2.5), we now introduce an important subclass of $\mathcal{H}^{\geq}_{q,2n}$. If $n \in \mathbb{N}_0$ and if $(s_j)_{j=0}^{2n} \in \mathcal{H}^{\geq}_{q,2n}$, then $(s_j)_{j=0}^{2n}$ is called *completely degenerate* if $L_n = 0$. For each $n \in \mathbb{N}_0$, the set $\mathcal{H}^{\geq,\mathrm{cd}}_{q,2n}$ of all completely degenerate sequences belonging to $\mathcal{H}^{\geq}_{q,2n}$ is a subset of $\mathcal{H}^{\geq,\mathrm{e}}_{q,2n}$ (see [13, Corollary 2.14]). If $m \in \mathbb{N}_0$ and $(s_j)_{j=0}^{2m} \in \mathcal{H}^{\geq,\mathrm{e}}_{q,2m}$ are given, then from [13, Proposition 2.13] one can easily see that $(s_j)_{j=0}^{2m}$ belongs to $\mathcal{H}^{\geq,\mathrm{cd}}_{q,2m}$ if and only if there is some $n \in \mathbb{Z}_{0,m}$ such that $L_n = 0_{q \times q}$. A Hankel non-negative definite sequence $(s_j)_{j=0}^{\infty}$ from $\mathbb{C}^{q \times q}$ is said to be *completely degenerate* if there is some non-negative integer n such that $(s_j)_{j=0}^{2n}$ is a completely degenerate Hankel non-negative definite sequence. By $\mathcal{H}^{\geq,\mathrm{cd}}_{q,\infty}$ we denote the set of all completely degenerate Hankel non-negative definite sequences $(s_j)_{j=0}^{\infty}$ from $\mathbb{C}^{q \times q}$. A Hankel non-negative definite sequence $(s_j)_{j=0}^{\infty}$ from $\mathbb{C}^{q \times q}$ is called *completely degenerate of order n* if the sequence $(s_j)_{j=0}^{2n}$ is completely degenerate. By $\mathcal{H}^{\geq,\mathrm{cd},n}_{q,\infty}$ we denote the set of all Hankel non-negative definite sequences $(s_j)_{j=0}^{\infty}$ from $\mathbb{C}^{q \times q}$ which are completely degenerate of order n. If $n \in \mathbb{N}_0$ and $(s_j)_{j=0}^{\infty} \in \mathcal{H}^{\geq,\mathrm{cd},n}_{q,\infty}$, then $(s_j)_{j=0}^{2m} \in \mathcal{H}^{\geq,\mathrm{cd}}_{q,2m}$ for each $m \in \mathbb{Z}_{n,\infty}$.

Given a fixed number $\alpha \in \mathbb{R}$, now we introduce important subclasses of the class of completely degenerate Hankel non-negative definite sequences. Let $n \in \mathbb{N}_0$. Then we define

$$\mathcal{K}^{\geq,\mathrm{cd}}_{q,2n,\alpha} := \mathcal{K}^{\geq}_{q,2n,\alpha} \cap \mathcal{H}^{\geq,\mathrm{cd}}_{q,2n}, \quad \mathcal{L}^{\geq,\mathrm{cd}}_{q,2n,\alpha} := \mathcal{L}^{\geq}_{q,2n,\alpha} \cap \mathcal{H}^{\geq,\mathrm{cd}}_{q,2n},$$

$$\mathcal{K}^{\geq,\mathrm{cd}}_{q,2n+1,\alpha} := \left\{ (s_j)_{j=0}^{2n+1} \in \mathcal{K}^{\geq}_{q,2n+1,\alpha} \colon (s_{\alpha \triangleright j})_{j=0}^{2n} \in \mathcal{H}^{\geq,\mathrm{cd}}_{q,2n} \right\},$$

and

$$\mathcal{L}^{\geq,\mathrm{cd}}_{q,2n+1,\alpha} := \left\{ (s_j)_{j=0}^{2n+1} \in \mathcal{L}^{\geq}_{q,2n+1,\alpha} \colon (s_{\alpha \triangleleft j})_{j=0}^{2n} \in \mathcal{H}^{\geq,\mathrm{cd}}_{q,2n} \right\}.$$

Furthermore, let $\mathcal{K}^{\geq,\mathrm{cd}}_{q,\infty,\alpha}$ be the set of all sequences $(s_j)_{j=0}^{\infty} \in \mathcal{K}^{\geq}_{q,\infty,\alpha}$ for which there exists some $m \in \mathbb{N}_0$ such that $(s_j)_{j=0}^{m} \in \mathcal{K}^{\geq,\mathrm{cd}}_{q,m,\alpha}$ and let $\mathcal{L}^{\geq,\mathrm{cd}}_{q,\infty,\alpha}$ be the set of all sequences $(s_j)_{j=0}^{\infty} \in \mathcal{L}^{\geq}_{q,\infty,\alpha}$ for which there exists some $m \in \mathbb{N}_0$ such that $(s_j)_{j=0}^{m} \in \mathcal{L}^{\geq,\mathrm{cd}}_{q,m,\alpha}$.

Example 5.1. Let $m \in \mathbb{N}$, let $(\xi_l)_{l=1}^{m}$ be a sequence from $\mathbb{R}$, and let $(A_l)_{l=1}^{m}$ be a sequence from $\mathbb{C}^{q \times q}_{\geq}$. For all $j \in \mathbb{N}_0$, let $s_j := \sum_{l=1}^{m} \xi_l^j A_l$. Let $\alpha, \beta \in \mathbb{R}$ with $\alpha \leq \min\{\xi_1, \xi_2, \ldots, \xi_m\}$ and $\max\{\xi_1, \xi_2, \ldots, \xi_m\} \leq \beta$. Then $(s_j)_{j=0}^{\infty} \in \mathcal{L}^{\geq,\mathrm{cd}}_{q,\infty,\beta} \cap \mathcal{K}^{\geq,\mathrm{cd}}_{q,\infty,\alpha}$. Indeed, [13, Lemma 2.40 (b)] yields $(s_j)_{j=0}^{\infty} \in \mathcal{H}^{\geq,\mathrm{cd},u}_{q,\infty}$ for some $u \in \mathbb{N}$. Hence, $(s_j)_{j=0}^{\infty} \in \mathcal{H}^{\geq}_{q,\infty}$ and $(s_j)_{j=0}^{2u} \in \mathcal{H}^{\geq,\mathrm{cd}}_{q,2u}$. For all $n \in \mathbb{N}_0$, we have, in view of (2.1), furthermore

$$
U_n^* \left(\mathrm{diag}[A_l]_{l=1}^m \right) U_n = \left[\sum_{l=1}^m (\xi_l^j \sqrt{\xi_l - \alpha I_q}) A_l (\xi_l^k \sqrt{\xi_l - \alpha I_q}) \right]_{j,k=0}^n
$$

$$
= \left[\sum_{l=1}^m (\xi_l - \alpha) \xi_l^{j+k} A_l \right]_{j,k=0}^n
$$

$$
= \left[-\alpha \sum_{l=1}^m \xi_l^{j+k} A_l + \sum_{l=1}^m \xi_l^{j+k+1} A_l \right]_{j,k=0}^n
$$

$$
= [-\alpha s_{j+k} + s_{j+k+1}]_{j,k=0}^n = [s_{\alpha \triangleright j+k}]_{j,k=0}^n,
$$

where $U_n := [\xi_l^k \sqrt{\xi_l - \alpha I_q}]_{\substack{l=1,\ldots,m \\ k=0,\ldots,n}}$, and

$$
V_n^* \left(\mathrm{diag}[A_l]_{l=1}^m \right) V_n = \left[\sum_{l=1}^m (\xi_l^j \sqrt{\beta - \xi_l I_q}) A_l (\xi_l^k \sqrt{\beta - \xi_l I_q}) \right]_{j,k=0}^n
$$

$$
= \left[\sum_{l=1}^m (\beta - \xi_l) \xi_l^{j+k} A_l \right]_{j,k=0}^n
$$

$$
= \left[\beta \sum_{l=1}^m \xi_l^{j+k} A_l - \sum_{l=1}^m \xi_l^{j+k+1} A_l \right]_{j,k=0}^n
$$

$$
= [\beta s_{j+k} - s_{j+k+1}]_{j,k=0}^n = [s_{\beta \triangleleft j+k}]_{j,k=0}^n,
$$

where $V_n := [\xi_l^k \sqrt{\beta - \xi_l I_q}]_{\substack{l=1,\ldots,m \\ k=0,\ldots,n}}$. Hence, since $(A_l)_{l=1}^m$ is a sequence of non-negative Hermitian matrices, we have $\{(s_{\alpha \triangleright j})_{j=0}^\infty, (s_{\beta \triangleleft j})_{j=0}^\infty\} \subseteq \mathcal{H}_{q,\infty}^\geq$, which, in view of $(s_j)_{j=0}^\infty \in \mathcal{H}_{q,\infty}^\geq$ and $(s_j)_{j=0}^{2u} \in \mathcal{H}_{q,2u}^{\geq,\mathrm{cd}}$, yields the assertion.

The membership of a sequence to $\mathcal{H}_{q,2n}^{\geq,\mathrm{cd}}$ can be characterized in the following way.

Remark 5.2. Let $n \in \mathbb{N}$ and $(s_j)_{j=0}^{2n} \in \mathcal{H}_{q,2n}^\geq$ with canonical Hankel parametrization $[(C_k)_{k=1}^n, (D_k)_{k=0}^n]$. Then $(s_j)_{j=0}^{2n} \in \mathcal{H}_{q,2n}^{\geq,\mathrm{cd}}$ if and only if $D_n = 0_{q \times q}$ (see Definition 3.2).

The right α-analogue of Remark 5.2 looks as follows.

Proposition 5.3. *Let* $\alpha \in \mathbb{R}$, $m \in \mathbb{N}_0$ *and* $(s_j)_{j=0}^m \in \mathcal{K}_{q,m,\alpha}^\geq$ *with right-sided* α-*Stieltjes parametrization* $(Q_j)_{j=0}^m$. *Then* $(s_j)_{j=0}^m \in \mathcal{K}_{q,m,\alpha}^{\geq,\mathrm{cd}}$ *if and only if* $Q_m = 0_{q \times q}$.

Proof. Taking the definition of the set $\mathcal{K}_{q,m,\alpha}^{\geq,\mathrm{cd}}$ into account, the assertion follows immediately by combining Definition 4.2 and Remark 5.2.   □

Corollary 5.4. *Let $\alpha \in \mathbb{R}$ and $(s_j)_{j=0}^{\infty} \in \mathcal{K}_{q,\infty,\alpha}^{\geq}$ with right-sided α-Stieltjes parametrization $(Q_j)_{j=0}^{\infty}$. Then $(s_j)_{j=0}^{\infty} \in \mathcal{K}_{q,\infty,\alpha}^{\geq,cd}$ if and only if there exists some $m \in \mathbb{N}_0$ with $Q_m = 0_{q \times q}$.*

Proof. Use Proposition 5.3. $\square$

Lemma 5.5. *Let $\alpha \in \mathbb{R}$, let $\kappa \in \mathbb{N}_0 \cup \{+\infty\}$, let $(s_j)_{j=0}^{\kappa} \in \mathcal{K}_{q,\kappa,\alpha}^{\geq,e}$, and let $m \in \mathbb{Z}_{0,\kappa}$ be such that $(s_j)_{j=0}^{m} \in \mathcal{K}_{q,m,\alpha}^{\geq,cd}$. Then $(s_j)_{j=0}^{n} \in \mathcal{K}_{q,n,\alpha}^{\geq,cd}$ for all $n \in \mathbb{Z}_{m,\kappa}$.*

Proof. Use Proposition 5.3 and part (c) of Theorem 4.12. $\square$

Lemma 5.6. *Let $\alpha \in \mathbb{R}$, $\kappa \in \mathbb{N}_0 \cup \{+\infty\}$ and $r \in \mathbb{N}$. For all $v \in \mathbb{Z}_{1,r}$, let $(s_j^{(v)})_{j=0}^{\kappa} \in \mathcal{K}_{q,\kappa,\alpha}^{\geq}$. Let $m \in \mathbb{Z}_{0,\kappa}$ and $(\sum_{v=1}^{r} s_j^{(v)})_{j=0}^{\kappa} \in \mathcal{K}_{q,\kappa,\alpha}^{\geq,cd}$. Then $(s_j^{(v)})_{j=0}^{\kappa} \in \mathcal{K}_{q,\kappa,\alpha}^{\geq,cd}$ for all $v \in \mathbb{Z}_{1,r}$.*

Proof. For each $j \in \mathbb{Z}_{0,\kappa}$, let $t_j := \sum_{v=1}^{r} s_j^{(v)}$. By assumption, then $(t_j)_{j=0}^{\kappa}$ belongs to $\mathcal{K}_{q,\kappa,\alpha}^{\geq,cd}$ and, in particular, to $\mathcal{K}_{q,\kappa,\alpha}^{\geq}$. Remark 1.4 shows that $(t_j)_{j=0}^{n} \in \mathcal{K}_{q,n,\alpha}^{\geq}$ for each $n \in \mathbb{Z}_{0,\kappa}$. Let $(R_j)_{j=0}^{\kappa}$ be the right-sided α-Stieltjes parametrization of $(t_j)_{j=0}^{\kappa}$. Furthermore, for each $v \in \mathbb{Z}_{1,r}$, let $(Q_j^{(v)})_{j=0}^{\kappa}$ be the right-sided α-Stieltjes parametrization of $(s_j^{(v)})_{j=0}^{\kappa}$.

(I) First we consider the case $\kappa \in \mathbb{N}_0$. Since $(s_j^{(v)})_{j=0}^{\kappa} \in \mathcal{K}_{q,\kappa,\alpha}^{\geq}$ is supposed for each $v \in \mathbb{Z}_{1,r}$, from part (b) of Theorem 4.12 we see that

$$Q_{\kappa}^{(v)} \in \mathbb{C}_{\geq}^{q \times q} \tag{5.1}$$

holds true for each $v \in \mathbb{Z}_{1,r}$. Because of the assumption $(t_j)_{j=0}^{\kappa} \in \mathcal{K}_{q,\kappa,\alpha}^{\geq,cd}$, we have $R_{\kappa} = 0_{q \times q}$. Using a result on Schur complements of sums of non-negative Hermitian block matrices (see [15, Lemma 4.21]), we conclude that

$$-\sum_{v=1}^{r} Q_{\kappa}^{(v)} = R_{\kappa} - \sum_{v=1}^{r} Q_{\kappa}^{(v)} \in \mathbb{C}_{\geq}^{q \times q}.$$

Consequently, because of (5.1), for each $v \in \mathbb{Z}_{1,r}$, we get $-Q_{\kappa}^{(v)} \in \mathbb{C}_{\geq}^{q \times q}$ and, using (5.1) again, then $Q_{\kappa}^{(v)} = 0_{q \times q}$. Thus, in the case $\kappa \in \mathbb{N}_0$, the proof is complete.

(II) Now let $\kappa = +\infty$. Then there is an $m \in \mathbb{N}_0$ such that $(t_j)_{j=0}^{m} \in \mathcal{K}_{q,m,\alpha}^{\geq,cd}$. Part (I) of this proof yields that $(s_j^{(v)})_{j=0}^{m} \in \mathcal{K}_{q,m,\alpha}^{\geq,cd}$ for each $v \in \mathbb{Z}_{1,r}$. This implies $(s_j^{(v)})_{j=0}^{\infty} \in \mathcal{K}_{q,\infty,\alpha}^{\geq,cd}$ for each $v \in \mathbb{Z}_{1,r}$. $\square$

Lemma 5.7. *Let $\alpha \in \mathbb{R}$, $\kappa \in \mathbb{N}_0 \cup \{+\infty\}$, $(s_j)_{j=0}^{\kappa} \in \mathcal{K}_{q,\kappa,\alpha}^{\geq,cd}$, $p \in \mathbb{N}$ and $A \in \mathbb{C}^{p \times q}$. If $\kappa \geq 1$, then suppose that*

$$\mathcal{N}(A) \subseteq \bigcap_{j=0}^{\kappa-1} \mathcal{N}(s_j). \tag{5.2}$$

Then $(As_j A^*)_{j=0}^{\kappa} \in \mathcal{K}_{p,\kappa,\alpha}^{\geq,cd}$.

Proof. Remark 2.7 shows that $(t_j)_{j=0}^{\kappa}$ defined by $t_j := As_jA^*$, $j \in \mathbb{Z}_{0,\kappa}$, belongs to $\mathcal{K}_{p,\kappa,\alpha}^{\geq}$. Let $(R_j)_{j=0}^{\kappa}$ be the right-sided α-Stieltjes parametrization of $(t_j)_{j=0}^{\kappa}$.

(I) The case $\kappa = 0$ is trivial.

(II) Now we consider the case $\kappa \in \mathbb{N}$. If $\kappa \geq 2$, then we see from (2.1) and (5.2) that

$$\bigcap_{j=0}^{\kappa-1} \mathcal{N}(s_j) \subseteq \bigcap_{j=0}^{\kappa-2} \mathcal{N}(s_{\alpha \triangleright j}). \tag{5.3}$$

Because of $(s_j)_{j=0}^{\kappa} \in \mathcal{K}_{q,\kappa,\alpha}^{\geq,\text{cd}}$, we have $Q_\kappa = 0_{q\times q}$. From a result on Schur complements of non-negative Hermitian block Hankel matrices (see [18, Proposition 3.11]), and the inclusions (5.2) and (5.3) we get that $R_\kappa = AQ_\kappa A^* = A \cdot 0_{q\times q} \cdot A^* = 0_{p\times p}$. Thus, $(t_j)_{j=0}^{\kappa}$ belongs to $\mathcal{K}_{p,\kappa,\alpha}^{\geq,\text{cd}}$.

(III) Now let $\kappa = +\infty$. Then there is an $m \in \mathbb{N}_0$ such that $(s_j)_{j=0}^{m} \in \mathcal{K}_{q,m,\alpha}^{\geq,\text{cd}}$. Parts (I) and (II) yield that $(t_j)_{j=0}^{m} \in \mathcal{K}_{p,m,\alpha}^{\geq,\text{cd}}$. Thus, $(t_j)_{j=0}^{\infty}$ belongs to $\mathcal{K}_{p,\infty,\alpha}^{\geq,\text{cd}}$. $\square$

Remark 5.8. Let $\alpha \in \mathbb{R}$, $\kappa \in \mathbb{N}_0 \cup \{+\infty\}$ and $n \in \mathbb{N}$. For all $l \in \mathbb{Z}_{1,n}$, let $q_l \in \mathbb{N}$ and $(s_j^{(l)})_{j=0}^{\kappa}$ be a sequence from $\mathbb{C}^{q_l \times q_l}$. Further, let $m \in \mathbb{Z}_{0,\kappa}$. In view of Remark 2.16 and elementary properties of the Moore-Penrose inverse of block diagonal matrices, we easily see then that $(s_j^{(l)})_{j=0}^{\kappa} \in \mathcal{K}_{q_l,\kappa,\alpha}^{\geq,\text{cd},m}$ for all $l \in \mathbb{Z}_{1,n}$ if and only if $(\text{diag}[s_j^{(l)}]_{l=1}^{n})_{j=0}^{\kappa} \in \mathcal{K}_{\sum_{l=1}^{n} q_l,\kappa,\alpha}^{\geq,\text{cd},m}$.

Proposition 5.9. *Let $\alpha \in \mathbb{R}$ and let $\kappa \in \mathbb{N}_0 \cup \{+\infty\}$. Then $\mathcal{K}_{q,\kappa,\alpha}^{\geq,\text{cd}} \subseteq \mathcal{K}_{q,\kappa,\alpha}^{\geq,\text{e}}$.*

Proof. Let $(s_j)_{j=0}^{\kappa} \in \mathcal{K}_{q,\kappa,\alpha}^{\geq,\text{cd}}$. Then $(s_j)_{j=0}^{\kappa}$ belongs in particular to $\mathcal{K}_{q,\kappa,\alpha}^{\geq}$. Since the cases $\kappa = 0$ and $\kappa = +\infty$ are trivial, it remains to consider the case $\kappa \in \mathbb{N}$. Then $Q_\kappa = 0_{q\times q}$, which implies $\mathcal{N}(Q_{\kappa-1}) \subseteq \mathbb{C}^q = \mathcal{N}(Q_\kappa)$. Thus Lemmas 2.17 and 2.18 yield $(s_j)_{j=0}^{\kappa} \in \mathcal{K}_{q,\kappa,\alpha}^{\geq,\text{e}}$. $\square$

Proposition 5.10. *Let $\alpha \in \mathbb{R}$, let $m \in \mathbb{N}_0$, let $(s_j)_{j=0}^{m}$ be a sequence from $\mathbb{C}^{q\times q}$, and let $(Q_j)_{j=0}^{m}$ be the right-sided α-Stieltjes parametrization of $(s_j)_{j=0}^{m}$. Then the following statements are equivalent:*

(i) $(s_j)_{j=0}^{m} \in \mathcal{K}_{q,m,\alpha}^{\geq,\text{cd}}$.
(ii) *The following three conditions are fulfilled:*
  (ii-1) $Q_m = 0_{q\times q}$.
  (ii-2) *If $m \geq 1$, then $Q_j \in \mathbb{C}_{\succeq}^{q\times q}$ for each $j \in \mathbb{Z}_{0,m-1}$.*
  (ii-3) *If $m \geq 2$, then $\mathcal{N}(Q_j) \subseteq \mathcal{N}(Q_{j+1})$ for each $j \in \mathbb{Z}_{0,m-2}$.*

Proof. "(i)$\Rightarrow$(ii)": Because of (i), we have $(s_j)_{j=0}^{m} \in \mathcal{K}_{q,m,\alpha}^{\geq}$ and (ii-1). Proposition 5.9 shows that $(s_j)_{j=0}^{m}$ belongs to $\mathcal{K}_{q,m,\alpha}^{\geq,\text{e}}$. Thus, part (c) of Theorem 4.12 yields (ii-2) and (ii-3).

"(ii)$\Rightarrow$(i)": In view of (ii), from part (c) of Theorem 4.12, we obtain $(s_j)_{j=0}^{m} \in \mathcal{K}_{q,m,\alpha}^{\geq,\text{e}}$ and, hence, $(s_j)_{j=0}^{m} \in \mathcal{K}_{q,m,\alpha}^{\geq}$. Thus, (ii-1) implies (i). $\square$

Proposition 5.11. *Let $\alpha \in \mathbb{R}$, let $(s_j)_{j=0}^\infty$ be a sequence from $\mathbb{C}^{q \times q}$, and let $(Q_j)_{j=0}^\infty$ be the right-sided α-Stieltjes parametrization of $(s_j)_{j=0}^\infty$. Then $(s_j)_{j=0}^\infty$ belongs to $\mathcal{K}_{q,\infty,\alpha}^{\geq,\mathrm{cd}}$ if and only if there is an $m \in \mathbb{N}_0$ such that the following statements hold true:*

(i) *$Q_k = 0_{q \times q}$ for each $k \in \mathbb{Z}_{m,\infty}$.*

(ii) *If $m \geq 1$, then $Q_l \in \mathbb{C}_\geq^{q \times q}$ for each $l \in \mathbb{Z}_{0,m-1}$.*

(iii) *If $m \geq 2$, then $\mathcal{N}(Q_j) \subseteq \mathcal{N}(Q_{j+1})$ for each $j \in \mathbb{Z}_{0,m-2}$.*

Proof. First suppose $(s_j)_{j=0}^\infty \in \mathcal{K}_{q,\infty,\alpha}^{\geq,\mathrm{cd}}$. Then $(s_j)_{j=0}^\infty \in \mathcal{K}_{q,\infty,\alpha}^\geq = \mathcal{K}_{q,\infty,\alpha}^{\geq,\mathrm{e}}$ and there is an $m \in \mathbb{N}_0$ such that $Q_m = 0_{q \times q}$. Thus, part (c) of Theorem 4.12 yields that $Q_j \in \mathbb{C}_\geq^{q \times q}$ and $\mathcal{N}(Q_j) \subseteq \mathcal{N}(Q_{j+1})$ hold true for each $j \in \mathbb{N}_0$. Since $\mathcal{N}(Q_m) = \mathcal{N}(0_{q \times q}) = \mathbb{C}^q$, the conditions (i), (ii), and (iii) are fulfilled.

Conversely, now suppose that (i), (ii), and (iii) are true. Then part (b) of Theorem 4.12 provides us $(s_j)_{j=0}^\infty \in \mathcal{K}_{q,\infty,\alpha}^\geq$ and, consequently, (i) implies $(s_j)_{j=0}^\infty \in \mathcal{K}_{q,\infty,\alpha}^{\geq,\mathrm{cd}}$. □

The structure of the completely degenerate Hankel non-negative definite sequences was completely described in [13, Proposition 4.9 (c)]. A closer look at this result yields the following observation on the structure of the elements of the set $\mathcal{K}_{q,\infty,\alpha}^{\geq,\mathrm{cd}}$.

Proposition 5.12. *Let $\alpha \in \mathbb{R}$, let $(s_j)_{j=0}^\infty \in \mathcal{K}_{q,\infty,\alpha}^{\geq,\mathrm{cd}}$, and let $n \in \mathbb{N}$ be such that $s_0 \neq 0_{q \times q}$ and $(s_j)_{j=0}^n \in \mathcal{K}_{q,n,\alpha}^{\geq,\mathrm{cd}}$. Then there exist a number $m \in \mathbb{Z}_{1,\lfloor \frac{n+1}{2} \rfloor q}$, a sequence $(\xi_l)_{l=1}^m$ of pairwise different numbers from $[\alpha, +\infty)$ and a sequence $(A_l)_{l=1}^m$ from $\mathbb{C}_\geq^{q \times q} \setminus \{0_{q \times q}\}$ such that for all $j \in \mathbb{N}_0$ the representation*

$$s_j = \sum_{l=1}^m \xi_l^j A_l$$

holds true.

Proof. We first consider the case $n = 2k$ with some $k \in \mathbb{N}$. We have then $(s_j)_{j=0}^\infty \in \mathcal{H}_{q,\infty}^{\geq,\mathrm{cd},k}$. Hence, from [13, Proposition 4.9 (a), (b)] we obtain the existence of a number $m \in \mathbb{Z}_{1,kq}$, a sequence $(\xi_l)_{l=1}^m$ of pairwise different numbers from $\mathbb{R}$ and a sequence $(A_l)_{l=1}^m$ from $\mathbb{C}_\geq^{q \times q}$ such that $\mathcal{M}_\geq^q[\mathbb{R}; (s_j)_{j=0}^\infty, =] = \{\sum_{l=1}^m \delta_{\xi_l} A_l\}$, where $\delta_{\xi_l} \colon \mathfrak{B}_\mathbb{R} \to [0, +\infty)$ is the Dirac measure with unit mass in ξ_l for all $l \in \mathbb{Z}_{1,m}$. In view of $s_0 \neq 0_{q \times q}$, we assume without loss of generality that $A_l \neq 0_{q \times q}$ for all $l \in \mathbb{Z}_{1,m}$. Because of $(s_j)_{j=0}^\infty \in \mathcal{K}_{q,\infty,\alpha}^{\geq,\mathrm{cd}}$, we have $(s_j)_{j=0}^\infty \in \mathcal{K}_{q,\infty,\alpha}^{\geq,\mathrm{e}}$, which in view of Theorem 1.6 implies $\mathcal{M}_\geq^q[[\alpha, +\infty); (s_j)_{j=0}^\infty, =] \neq \emptyset$. Let $\mu \in \mathcal{M}_\geq^q[[\alpha, +\infty); (s_j)_{j=0}^\infty, =]$ and let $\nu \colon \mathfrak{B}_\mathbb{R} \to \mathbb{C}^{q \times q}$ be defined by $\nu(B) := \mu(B \cap [\alpha, +\infty))$. According to [12, Remark 2.25 (b)], we have then $\nu \in \mathcal{M}_\geq^q[\mathbb{R}; (s_j)_{j=0}^\infty, =]$. Thus, $\nu = \sum_{l=1}^m \delta_{\xi_l} A_l$. For all $j \in \mathbb{N}_0$, we have then $s_j = s_j^{(\nu)} = \sum_{l=1}^m \xi_l^j A_l$ and for all $l \in \mathbb{Z}_{1,m}$ furthermore $0_{q \times q} \neq A_l = \nu(\{\xi_l\}) = \mu(\{\xi_l\} \cap [\alpha, +\infty))$, which implies $\xi_l \in [\alpha, +\infty)$.

Now we consider the case $n = 2k+1$ with some $k \in \mathbb{N}_0$. Because of $(s_j)_{j=0}^\infty \in \mathcal{K}_{q,\infty,\alpha}^{\geq,\mathrm{cd}}$ we have $(s_j)_{j=0}^\infty \in \mathcal{K}_{q,\infty,\alpha}^{\geq,\mathrm{e}}$. In view of $(s_j)_{j=0}^n \in \mathcal{K}_{q,n,\alpha}^{\geq,\mathrm{cd}}$, Lemma 5.5 yields then $(s_j)_{j=0}^{2(k+1)} \in \mathcal{K}_{q,2(k+1),\alpha}^{\geq,\mathrm{cd}}$. Hence, the assertion follows in this case from the already proved assertion for the case of an even number n. $\square$

6. Connection between α-Stieltjes and canonical Hankel parametrization

The central theme in this section is to investigate, in the case of given α-Stieltjes right-sided non-negative definite sequences, certain interrelations between the right-sided α-Stieltjes parametrization introduced in Definition 4.2 and the canonical Hankel parametrization introduced in Definition 3.2.

Remark 6.1. Let $\alpha \in \mathbb{C}$, $\kappa \in \mathbb{Z}_{2,+\infty} \cup \{+\infty\}$ and $(s_j)_{j=0}^\kappa$ be a sequence from $\mathbb{C}^{p \times q}$ with right-sided α-Stieltjes parametrization $(Q_j)_{j=0}^\kappa$. Then we easily see that:

(a) $Q_{2k} = D_k$ for all $k \in \mathbb{N}_0$ with $2k \leq \kappa$, where $[(C_k)_{k=1}^{\lfloor \frac{\kappa}{2} \rfloor}, (D_k)_{k=0}^{\lfloor \frac{\kappa}{2} \rfloor}]$ is the canonical Hankel parametrization of $(s_j)_{j=0}^{2\lfloor \frac{\kappa}{2} \rfloor}$.

(b) If $\kappa \geq 3$, then $Q_{2k+1} = F_k$ for every choice of $k \in \mathbb{N}_0$ with $2k+1 \leq \kappa$, where $[(E_k)_{k=1}^{\lfloor \frac{\kappa-1}{2} \rfloor}, (F_k)_{k=0}^{\lfloor \frac{\kappa-1}{2} \rfloor}]$ is the canonical Hankel parametrization of $(s_{\alpha \triangleright j})_{j=0}^{2\lfloor \frac{\kappa-1}{2} \rfloor}$.

Remark 6.2. Let $\kappa \in \mathbb{Z}_{2,+\infty} \cup \{+\infty\}$ and let $(s_j)_{j=0}^\kappa$ be a sequence from $\mathbb{C}^{p \times q}$. For each $k \in \mathbb{N}$ with $2k \leq \kappa$, then:

(a) Taking (2.1), (2.2), (2.3) and (2.4) into account the matrix H_k admits the block representations

$$H_k = \begin{bmatrix} H_{k-1} & y_{k,2k-1} \\ z_{k,2k-1} & s_{2k} \end{bmatrix}, \qquad H_k = \begin{bmatrix} y_{0,k-1} & K_{k-1} \\ s_k & z_{k+1,2k} \end{bmatrix}, \qquad (6.1)$$

and

$$H_k = \begin{bmatrix} z_{0,k-1} & s_k \\ K_{k-1} & y_{k+1,2k} \end{bmatrix}, \qquad H_k = \begin{bmatrix} s_0 & z_{1,k} \\ y_{1,k} & G_{k-1} \end{bmatrix}.$$

(b) Taking the block decompositions (6.1) into account, from [13, Remark 2.1] and [17, Lemma A.3] we see that $(s_j)_{j=0}^{2k}$ belongs to $\mathcal{H}_{q,2k}^\geq$ if and only if the four conditions $(s_j)_{j=0}^{2(k-1)} \in \mathcal{H}_{q,k-1}^\geq$, $H_{k-1} H_{k-1}^\dagger y_{k,2k-1} = y_{k,2k-1}$, $z_{k,2k-1} = y_{k,2k-1}^*$ and $L_k \in \mathbb{C}_\geq^{q \times q}$ hold true. Furthermore, if $(s_j)_{j=1}^{2k} \in \mathcal{H}_{q,2k}^\geq$, then $s_j^* = s_j$ for each $j \in \mathbb{Z}_{0,2k}$ and the equations $z_{k,2k-1} H_{k-1}^\dagger H_{k-1} = z_{k,2k-1}$ and $\operatorname{rank} H_k = \sum_{j=0}^k \operatorname{rank} L_j$ are true.

(c) From [13, Proposition 2.13] and [17, Lemma A.3] we know that $(s_j)_{j=0}^{2k} \in \mathcal{H}_{q,2k}^{\geq,\mathrm{e}}$ if and only if $(s_j)_{j=0}^{2k} \in \mathcal{H}_{q,2k}^\geq$ and $L_k L_{k-1}^\dagger L_{k-1} = L_k$ hold true. If $(s_j)_{j=0}^{2k} \in \mathcal{H}_{q,2k}^{\geq,\mathrm{e}}$, then $L_{k-1} L_{k-1}^\dagger L_k = L_k$.

Remark 6.3 ([13, Lemma 3.14]). Let $n \in \mathbb{N}$ and let $(s_j)_{j=0}^{2n} \in \mathcal{H}_{q,2n}^{\geq,\mathrm{e}}$. Then the matrices introduced in (3.2), (3.3), (3.4) and (2.5) satisfy

$$\Lambda_n = M_n + L_n L_{n-1}^{\dagger}(s_{2n-1} - M_{n-1}).$$

For fixed $\alpha \in \mathbb{R}$, we indicate now some interrelations between the matrices introduced in (2.5) and (3.6). To describe these connections, we will use the matrices given via (3.2), (3.4), and (3.6).

Proposition 6.4. *Let* $n \in \mathbb{N}$, *let* $\alpha \in \mathbb{R}$ *and let* $(s_j)_{j=0}^{2n} \in \mathcal{K}_{q,2n,\alpha}^{\geq}$. *Then*

$$L_n = (s_{\alpha \triangleright 2n-1} - M_{\alpha \triangleright n-1}) - L_{\alpha \triangleright n-1} L_{n-1}^{\dagger}(s_{2n-1} - M_{n-1}). \tag{6.2}$$

Proof. Since $(s_j)_{j=0}^{2n}$ belongs to $\mathcal{K}_{q,2n,\alpha}^{\geq}$, we see that

$$(s_j)_{j=0}^{2n} \in \mathcal{H}_{q,2n}^{\geq}, \tag{6.3}$$

$$\left\{ (s_j)_{j=0}^{2(n-1)}, (s_{\alpha \triangleright j})_{j=0}^{2(n-1)} \right\} \subseteq \mathcal{H}_{q,2(n-1)}^{\geq} \tag{6.4}$$

and

$$(s_j)_{j=0}^{2n-1} \in \mathcal{K}_{q,2n-1,\alpha}^{\geq,\mathrm{e}}. \tag{6.5}$$

Since, in view of (6.3), (1.2), and (2.2), part (b) of Remark 6.2 shows that $s_0 s_0^{\dagger} s_1 = s_1$ is valid. Thus, in the case $n = 1$, equation (6.2) obviously holds true.

Now we consider the case $n \geq 2$. We have then

$$s_{2n-1} - M_{n-1} = s_{2n-1} - z_{n-1,2n-3} H_{n-2}^{\dagger} y_{n,2n-2} = [-z_{n-1,2n-3} H_{n-2}^{\dagger}, I_q] y_{n,2n-1}$$

$$= [-z_{n-1,2n-3} H_{n-2}^{\dagger}, I_q] H_{n-1} H_{n-1}^{\dagger} y_{n,2n-1}$$

$$= [-z_{n-1,2n-3} H_{n-2}^{\dagger}, I_q] \begin{bmatrix} H_{n-2} & y_{n-1,2n-3} \\ z_{n-1,2n-3} & s_{2n-2} \end{bmatrix} H_{n-1}^{\dagger} y_{n,2n-1}$$

$$= [0_{q \times (n-1)q}, L_{n-1}] H_{n-1}^{\dagger} y_{n,2n-1}, \tag{6.6}$$

where the 1st equation is due to (3.2), the 2nd equation is due to (2.2), the 3rd equation is due to (6.3) and part (b) of Remark 6.2, the 4th equation is due to $n-1 \geq 1$ and part (a) of Remark 6.2 and the 5th equation is due to $n-1 \geq 1$, (6.4), part (b) of Remark 6.2 and (2.5). Moreover, we conclude

$$s_{\alpha \triangleright 2n-1} - M_{\alpha \triangleright n-1} = s_{\alpha \triangleright 2n-1} - z_{\alpha \triangleright n-1,2n-3} H_{\alpha \triangleright n-2}^{\dagger} y_{\alpha \triangleright n,2n-2}$$

$$= [-z_{\alpha \triangleright n-1,2n-3} H_{\alpha \triangleright n-2}^{\dagger}, I_q] y_{\alpha \triangleright n,2n-1}, \tag{6.7}$$

where the 1st equation is due to (3.2) and the 2nd equation is due to (2.2). Because of (6.5), Lemma 2.17 yields $\mathcal{N}(L_{n-1}) \subseteq \mathcal{N}(L_{\alpha \triangleright n-1})$, which, in view of part (b) of Lemma A.1, implies

$$L_{\alpha \triangleright n-1} L_{n-1}^{\dagger} L_{n-1} = L_{\alpha \triangleright n-1}. \tag{6.8}$$

In view of $n-1 \geq 1$, (6.4) and part (b) of Remark 6.2, we have

$$z_{\alpha \triangleright n-1,2n-3} H_{\alpha \triangleright n-2}^{\dagger} H_{\alpha \triangleright n-2} = z_{\alpha \triangleright n-1,2n-3}.$$

Taking additionally into account (2.5), we get

$$[-z_{\alpha\triangleright n-1,2n-3}H^{\dagger}_{\alpha\triangleright n-2}, I_q] \begin{bmatrix} H_{\alpha\triangleright n-2} & y_{\alpha\triangleright n-1,2n-3} \\ z_{\alpha\triangleright n-1,2n-3} & s_{\alpha\triangleright 2n-2} \end{bmatrix} = [0_{q\times(n-1)q}, L_{\alpha\triangleright n-1}].$$

(6.9)

We have then

$$L_{\alpha\triangleright n-1}L^{\dagger}_{n-1}(s_{2n-1} - M_{n-1}) = [0_{q\times(n-1)q}, L_{\alpha\triangleright n-1}L^{\dagger}_{n-1}L_{n-1}]H^{\dagger}_{n-1}y_{n,2n-1}$$

$$= [0_{q\times(n-1)q}, L_{\alpha\triangleright n-1}]H^{\dagger}_{n-1}y_{n,2n-1}$$

$$= [-z_{\alpha\triangleright n-1,2n-3}H^{\dagger}_{\alpha\triangleright n-2}, I_q] \begin{bmatrix} H_{\alpha\triangleright n-2} & y_{\alpha\triangleright n-1,2n-3} \\ z_{\alpha\triangleright n-1,2n-3} & s_{\alpha\triangleright 2n-2} \end{bmatrix} H^{\dagger}_{n-1}y_{n,2n-1}$$

$$= [-z_{\alpha\triangleright n-1,2n-3}H^{\dagger}_{\alpha\triangleright n-2}, I_q]H_{\alpha\triangleright n-1}H^{\dagger}_{n-1}y_{n,2n-1},$$

(6.10)

where the 1st equation is due to (6.6), the 2nd equation is due to (6.8), the 3rd equation is due to (6.9) and the 4th equation is due to (2.6), (2.7), $n-1 \geq 1$ and part (a) of Remark 6.2. Taking (6.7) and (6.10) into account, we see that

$$(s_{\alpha\triangleright 2n-1} - M_{\alpha\triangleright n-1}) - L_{\alpha\triangleright n-1}L^{\dagger}_{n-1}(s_{2n-1} - M_{n-1})$$

$$= [-z_{\alpha\triangleright n-1,2n-3}H^{\dagger}_{\alpha\triangleright n-2}, I_q](y_{\alpha\triangleright n,2n-1} - H_{\alpha\triangleright n-1}H^{\dagger}_{n-1}y_{n,2n-1}). \quad (6.11)$$

We also obtain from (2.6), (1.2), (2.1), (2.3), (2.7), (2.2) and part (a) of Remark 6.2 that

$$[H_{\alpha\triangleright n-1}, y_{\alpha\triangleright n,2n-1}] = [-\alpha H_{n-1} + K_{n-1}, -\alpha y_{n,2n-1} + y_{n+1,2n}]$$

$$= \left(-\alpha[I_{nq}, 0_{nq\times q}] \begin{bmatrix} H_{n-1} & y_{n,2n-1} \\ z_{n,2n-1} & s_{2n} \end{bmatrix} + [0_{nq\times q}, I_{nq}] \begin{bmatrix} z_{0,n-1} & s_n \\ K_{n-1} & y_{n+1,2n} \end{bmatrix}\right)$$

$$= (-\alpha[I_{nq}, 0_{nq\times q}]H_n + [0_{nq\times q}, I_{nq}]H_n) = (-\alpha[I_{nq}, 0_{nq\times q}] + [0_{nq\times q}, I_{nq}])H_n$$

and, due to part (a) of Remark 6.2, (6.3), part (b) of Remark 6.2, and (2.5), consequently,

$$y_{\alpha\triangleright n,2n-1} - H_{\alpha\triangleright n-1}H^{\dagger}_{n-1}y_{n,2n-1} = [H_{\alpha\triangleright n-1}, y_{\alpha\triangleright n,2n-1}] \begin{bmatrix} -H^{\dagger}_{n-1}y_{n,2n-1} \\ I_q \end{bmatrix}$$

$$= (-\alpha[I_{nq}, 0_{nq\times q}] + [0_{nq\times q}, I_{nq}]) \begin{bmatrix} H_{n-1} & y_{n,2n-1} \\ z_{n,2n-1} & s_{2n} \end{bmatrix} \begin{bmatrix} -H^{\dagger}_{n-1}y_{n,2n-1} \\ I_q \end{bmatrix}$$

(6.12)

$$= (-\alpha[I_{nq}, 0_{nq\times q}] + [0_{nq\times q}, I_{nq}]) \begin{bmatrix} 0_{nq\times q} \\ L_n \end{bmatrix} = \begin{bmatrix} 0_{(n-1)q\times q} \\ L_n \end{bmatrix}.$$

From (6.11) and (6.12) then (6.2) follows.   □

Proposition 6.5. Let $\alpha \in \mathbb{R}$, $n \in \mathbb{N}$ and $(s_j)^{2n+1}_{j=0} \in \mathcal{K}^{\geq}_{q,2n+1,\alpha}$. Then

$$L_{\alpha\triangleright n} = (s_{2n+1} - M_n) - L_n \left[\alpha I_q + L^{\dagger}_{\alpha\triangleright n-1}(s_{\alpha\triangleright 2n-1} - M_{\alpha\triangleright n-1})\right]. \quad (6.13)$$

Proof. Because $(s_j)_{j=0}^{2n+1}$ belongs to $\mathcal{K}_{q,2n+1,\alpha}^{\geq}$, we have $\{(s_j)_{j=0}^{2n}, (s_{\alpha \rhd j})_{j=0}^{2n}\} \subseteq$
$\mathcal{H}_{q,2n}^{\geq}$, $(s_{\alpha \rhd j})_{j=0}^{2(n-1)} \in \mathcal{H}_{q,2(n-1)}^{\geq}$, and $(s_j)_{j=0}^{2n} \in \mathcal{K}_{q,2n,\alpha}^{\geq,e}$. If $n \geq 2$, then, similar
to (6.6), from Remark 6.2 we get

$$s_{\alpha \rhd 2n-1} - M_{\alpha \rhd n-1} = [0_{q \times (n-1)q}, L_{\alpha \rhd n-1}]H_{\alpha \rhd n-1}^{\dagger}y_{\alpha \rhd n, 2n-1}. \tag{6.14}$$

Thus, if $n \geq 2$, then using (6.14), Proposition 2.19, and part (b) of Lemma A.1, we conclude

$$L_n L_{\alpha \rhd n-1}^{\dagger}(s_{\alpha \rhd 2n-1} - M_{\alpha \rhd n-1}) = [0_{q \times (n-1)q}, L_n]H_{\alpha \rhd n-1}^{\dagger}y_{\alpha \rhd n, 2n-1}$$

and, consequently,

$$(s_{2n+1} - M_n) - L_n\left[\alpha I_q + L_{\alpha \rhd n-1}^{\dagger}(s_{\alpha \rhd n-1} - M_{\alpha \rhd n-1})\right]$$
$$= (s_{2n+1} - M_n) - \alpha L_n - [0_{q \times (n-1)q}, L_n]H_{\alpha \rhd n-1}^{\dagger}y_{\alpha \rhd n, 2n-1}. \tag{6.15}$$

If $n = 1$, then it is similarly checked that

$$(s_{2n+1} - M_n) - L_n\left[\alpha I_q + L_{\alpha \rhd n-1}^{\dagger}(s_{\alpha \rhd n-1} - M_{\alpha \rhd n-1})\right]$$
$$= (s_{2n+1} - M_n) - \alpha L_n - L_n H_{\alpha \rhd n-1}^{\dagger}y_{\alpha \rhd n, 2n-1} \tag{6.16}$$

holds true. Hence, in the case $n = 1$ as well as in the case $n \geq 2$, from (6.15) and (6.16) it follows that

$$(s_{2n+1} - M_n) - L_n\left[\alpha I_q + L_{\alpha \rhd n-1}^{\dagger}(s_{\alpha \rhd n-1} - M_{\alpha \rhd n-1})\right] \tag{6.17}$$
$$= [0_{q \times nq}, L_n, s_{2n+1} - M_n]\left(\begin{bmatrix} -\alpha I_{(n+1)q} \\ 0_{q \times (n+1)q} \end{bmatrix} + \begin{bmatrix} 0_{q \times (n+1)q} \\ I_{(n+1)q} \end{bmatrix}\right)\begin{bmatrix} -H_{\alpha \rhd n-1}^{\dagger}y_{\alpha \rhd n, 2n-1} \\ I_q \end{bmatrix}.$$

Taking into account (2.5), (3.2), and part (b) of Remark 6.2, we have

$$[0_{q \times nq}, L_n, s_{2n+1} - M_n]$$
$$= [0_{q \times nq}, s_{2n} - z_{n,2n-1}H_{n-1}^{\dagger}y_{n,2n-1}, s_{2n+1} - z_{n,2n-1}H_{n-1}^{\dagger}y_{n+1,2n}] \tag{6.18}$$
$$= [-z_{n,2n-1}H_{n-1}^{\dagger}, I_q]\begin{bmatrix} H_{n-1} & y_{n,2n-1} & y_{n+1,2n} \\ z_{n,2n-1} & s_{2n} & s_{2n+1} \end{bmatrix}.$$

Thus, from (6.17), (6.18) and part (a) of Remark 6.2 we obtain

$$(s_{2n+1} - M_n) - L_n\left[\alpha I_q + L_{\alpha \rhd n-1}^{\dagger}(s_{\alpha \rhd n-1} - M_{\alpha \rhd n-1})\right]$$
$$= [-z_{n,2n-1}H_{n-1}^{\dagger}, I_q](-\alpha H_n + K_n)\begin{bmatrix} -H_{\alpha \rhd n-1}^{\dagger}y_{\alpha \rhd n, 2n-1} \\ I_q \end{bmatrix}. \tag{6.19}$$

Since $(s_{\alpha\triangleright j})_{j=0}^{2n}$ belongs to $\mathcal{H}_{q,2n}^{\geq}$, part (b) of Remark 6.2 yields

$$(-\alpha H_n + K_n)\begin{bmatrix} -H_{\alpha\triangleright n-1}^{\dagger} y_{\alpha\triangleright n,2n-1} \\ I_q \end{bmatrix}$$

$$= \begin{bmatrix} H_{\alpha\triangleright n-1} & y_{\alpha\triangleright n,2n-1} \\ z_{\alpha\triangleright n,2n-1} & s_{\alpha\triangleright 2n} \end{bmatrix} \begin{bmatrix} -H_{\alpha\triangleright n-1}^{\dagger} y_{\alpha\triangleright n,2n-1} \\ I_q \end{bmatrix} = \begin{bmatrix} 0_{nq\times q} \\ L_{\alpha\triangleright n} \end{bmatrix}. \quad (6.20)$$

Finally, (6.19) and (6.20) imply (6.13). □

Corollary 6.6. *Let $n \in \mathbb{N}$.*

(a) *If $(s_j)_{j=0}^{2n} \in \mathcal{K}_{q,2n,0}^{\geq}$, then*

$$L_n = (s_{2n} - M_{0\triangleright n-1}) - L_{0\triangleright n-1} L_{n-1}^{\dagger}(s_{2n-1} - M_{n-1}).$$

(b) *If $(s_j)_{j=0}^{2n+1} \in \mathcal{K}_{q,2n+1,0}^{\geq}$, then*

$$L_{0\triangleright n} = (s_{2n+1} - M_n) - L_n L_{0\triangleright n-1}^{\dagger}(s_{2n} - M_{0\triangleright n-1}).$$

Proof. (a) Use Proposition 6.4 and Remark 2.4.

(b) This follows from Proposition 6.5 and Remark 2.4. □

Lemma 6.7. *Let $\alpha \in \mathbb{R}$, $\kappa \in \mathbb{N} \cup \{+\infty\}$ and $(s_j)_{j=0}^{\kappa} \in \mathcal{K}_{q,\kappa,\alpha}^{\geq}$. Further, let $(Q_j)_{j=0}^{\kappa}$ be the right-sided α-Stieltjes parametrization of $(s_j)_{j=0}^{\kappa}$. Then*

$$s_{2k+1} - \Lambda_k = Q_{2k+1} + (\alpha I_q + Q_{2k} Q_{2k-1}^{\dagger}) Q_{2k}$$

for all $k \in \mathbb{N}_0$ with $2k+1 \leq \kappa$, where Λ_k is given via (3.3), where $Q_{-1} := 0_{q\times q}$.

Proof. Obviously, from Definition 4.2 we see that

$$s_1 - \Lambda_0 = s_1 = -\alpha s_0 + s_1 + \alpha s_0 = L_{\alpha\triangleright 0} + \alpha L_0 = Q_1 + \alpha Q_0.$$

Now let $\kappa \geq 3$ and $k \in \mathbb{N}$ be such that $2k+1 \leq \kappa$. Since $(s_j)_{j=0}^{\kappa}$ belongs to $\mathcal{K}_{q,\kappa,\alpha}^{\geq}$, we have then $(s_j)_{j=0}^{2k+1} \in \mathcal{K}_{q,2k+1,\alpha}^{\geq}$ and hence $(s_j)_{j=0}^{2k} \in \mathcal{H}_{q,2k}^{\geq,\mathrm{e}} \cap \mathcal{K}_{q,2k,\alpha}^{\geq,\mathrm{e}} \cap \mathcal{K}_{q,2k,\alpha}^{\geq}$. Consequently, we get

$$s_{2k+1} - \Lambda_k = (s_{2k+1} - M_k) - (\Lambda_k - M_k)$$

$$= (s_{2k+1} - M_k) - L_k L_{k-1}^{\dagger}(s_{2k-1} - M_{k-1})$$

$$= (s_{2k+1} - M_k) - L_k L_{\alpha\triangleright k-1}^{\dagger} L_{\alpha\triangleright k-1} L_{k-1}^{\dagger}(s_{2k-1} - M_{k-1})$$

$$= (s_{2k+1} - M_k) + L_k L_{\alpha\triangleright k-1}^{\dagger}[L_k - (s_{\alpha\triangleright 2k-1} - M_{\alpha\triangleright k-1})]$$

$$= (s_{2k+1} - M_k) - L_k L_{\alpha\triangleright k-1}^{\dagger}(s_{\alpha\triangleright 2k-1} - M_{\alpha\triangleright k-1}) + L_k L_{\alpha\triangleright k-1}^{\dagger} L_k$$

$$= L_{\alpha\triangleright k} + \alpha L_k + L_k L_{\alpha\triangleright k-1}^{\dagger} L_k = L_{\alpha\triangleright k} + (\alpha I_q + L_k L_{\alpha\triangleright k-1}^{\dagger}) L_k$$

$$= Q_{2k+1} + (\alpha I_q + Q_{2k} Q_{2k-1}^{\dagger}) Q_{2k},$$

where the 2nd equation is due to $(s_j)_{j=0}^{2k} \in \mathcal{H}_{q,2k}^{\geq,\mathrm{e}}$ and Remark 6.3, the 3rd equation is due to $(s_j)_{j=0}^{2k} \in \mathcal{K}_{q,2k,\alpha}^{\geq}$, part (b) of Proposition 2.19 and part (b) of Lemma A.1, the 4th equation is due to $(s_j)_{j=0}^{2k} \in \mathcal{K}_{q,2k,\alpha}^{\geq}$ and Proposition 6.4, the 6th equation is due to $(s_j)_{j=0}^{2k+1} \in \mathcal{K}_{q,2k+1,\alpha}^{\geq}$ and Proposition 6.5, and the 8th equation is due to Definition 4.2. The proof is complete. $\qquad\square$

Now we are able to state the interrelation between right α-Stieltjes parametrization and canonical Hankel parametrization for sequences of the class $\mathcal{K}_{q,\kappa,\alpha}^{\geq}$.

Theorem 6.8. *Let $\alpha \in \mathbb{R}$, $\kappa \in \mathbb{Z}_{2,+\infty} \cup \{+\infty\}$, and $(s_j)_{j=0}^{\kappa} \in \mathcal{K}_{q,\kappa,\alpha}^{\geq}$. Let $(Q_j)_{j=0}^{\kappa}$ be the right-sided α-Stieltjes parametrization of $(s_j)_{j=0}^{\kappa}$ and let $[(C_k)_{k=1}^{\lfloor \frac{\kappa}{2} \rfloor}, (D_k)_{k=0}^{\lfloor \frac{\kappa}{2} \rfloor}]$ be the canonical Hankel parametrization of $(s_j)_{j=0}^{2\lfloor \frac{\kappa}{2} \rfloor}$. Let $Q_{-1} := 0_{q\times q}$. Then*

$$C_k = Q_{2k-1} + (\alpha I_q + Q_{2k-2} Q_{2k-3}^{\dagger}) Q_{2k-2}$$

for all $k \in \mathbb{N}$ with $2k \leq \kappa$ and $D_k = Q_{2k}$ for all $k \in \mathbb{N}_0$ with $2k \leq \kappa$.

Proof. Use Definition 3.2, Lemma 6.7, and Remark 6.1. $\qquad\square$

If the canonical Hankel parametrization of $(s_j)_{j=0}^{2\kappa} \in \mathcal{K}_{q,2\kappa,\alpha}^{\geq}$ is given, then, we are going to show that the right-sided α-Stieltjes parametrization of $(s_j)_{j=0}^{2\kappa}$ can be computed recursively. To derive this we still need the following auxiliary result.

Lemma 6.9. *Let $\alpha \in \mathbb{R}$, $\kappa \in \mathbb{N} \cup \{+\infty\}$, and $(s_j)_{j=0}^{\kappa} \in \mathcal{K}_{q,\kappa,\alpha}^{\geq}$. Let $(Q_j)_{j=0}^{\kappa}$ be the right-sided α-Stieltjes parametrization of $(s_j)_{j=0}^{\kappa}$. Let $Q_{-1} := 0_{q\times q}$. For all $k \in \mathbb{N}_0$ with $2k+1 \leq \kappa$, then*

$$Q_{2k+1} = s_{2k+1} - \Lambda_k - (\alpha I_q + L_k Q_{2k-1}^{\dagger}) L_k. \qquad (6.21)$$

Proof. Use Lemma 6.7 and Definition 4.2. $\qquad\square$

Proposition 6.10. *Let $\alpha \in \mathbb{R}$, let $\kappa \in \mathbb{N} \cup \{+\infty\}$, and let $(s_j)_{j=0}^{2\kappa} \in \mathcal{K}_{q,2\kappa,\alpha}^{\geq}$. Let $[(C_k)_{k=1}^{\kappa}, (D_k)_{k=0}^{\kappa}]$ be the canonical Hankel parametrization of $(s_j)_{j=0}^{2\kappa}$ and let $Q_{-1} := 0_{q\times q}$. Then the right-sided α-Stieltjes parametrization $(Q_j)_{j=0}^{2\kappa}$ of $(s_j)_{j=0}^{2\kappa}$ is given by $Q_{2k} = D_k$ for all $k \in \mathbb{Z}_{0,\kappa}$ and by the recurrence formulas*

$$Q_{2k+1} = C_{k+1} - (\alpha I_q + D_k Q_{2k-1}^{\dagger}) D_k$$

for all $k \in \mathbb{Z}_{0,\kappa-1}$.

Proof. Use Remark 6.1, Lemma 6.9 and Definition 3.2. $\qquad\square$

Corollary 6.11. *Let $\alpha \in \mathbb{R}$, let $\kappa \in \mathbb{N} \cup \{+\infty\}$, and let $(s_j)_{j=0}^{2\kappa} \in \mathcal{K}_{q,2\kappa,\alpha}^{\geq}$ with canonical Hankel parametrization $[(C_k)_{k=1}^{\kappa}, (D_k)_{k=0}^{\kappa}]$ and right-sided α-Stieltjes parametrization $(Q_j)_{j=0}^{2\kappa}$. Then $(C_k)_{k=1}^{\kappa}$ and $(D_k)_{k=0}^{\kappa}$ are sequences of Hermitian matrices. Then:*

(a) $C_k - \alpha D_{k-1} \geq Q_{2k-1} \geq 0_{q\times q}$ *for all $k \in \mathbb{Z}_{1,\kappa}$.*

(b) $\mathcal{N}(C_k - \alpha D_{k-1}) \subseteq \mathcal{N}(Q_{2k-1})$ *and* $\mathcal{R}(Q_{2k-1}) \subseteq \mathcal{R}(C_k - \alpha D_{k-1})$ *for all $k \in \mathbb{Z}_{1,\kappa}$.*

(c) $0 \leq \det Q_{2k-1} \leq \det(C_k - \alpha D_{k-1})$ *and* $\operatorname{rank} Q_{2k-1} \leq \operatorname{rank}(C_k - \alpha D_{k-1})$ *for all $k \in \mathbb{Z}_{1,\kappa}$.*

(d) *For all* $n \in \mathbb{Z}_{0,\kappa-1}$ *the inequalities* $0 \leq \det H_{\alpha \rhd n} \leq \prod_{k=1}^{n+1} \det(C_k - \alpha D_{k-1})$ *and* $\operatorname{rank} H_{\alpha \rhd n} \leq \sum_{k=1}^{n+1} \operatorname{rank}(C_k - \alpha D_{k-1})$ *hold.*

Proof. From [13, Proposition 2.30] and [15, Proposition 2.15] we know that $(C_k)_{k=1}^{\kappa}$ and $(D_k)_{k=0}^{\kappa}$ are sequences of Hermitian matrices.

(a) According to part (b) of Theorem 4.12, we have $Q_j \in \mathbb{C}_{\geq}^{q \times q}$ for all $j \in \mathbb{Z}_{0,2\kappa}$, which, in view of Proposition 6.10, implies

$$C_k - \alpha D_{k-1} = Q_{2k-1} + D_{k-1} Q_{2k-3}^{\dagger} D_{k-1} = Q_{2k-1} + Q_{2k-2} Q_{2k-3}^{\dagger} Q_{2k-2}$$

$$= Q_{2k-1} + Q_{2k-2}^* Q_{2k-3}^{\dagger} Q_{2k-2} \geq Q_{2k-1} \geq 0_{q \times q}$$

for all $k \in \mathbb{Z}_{1,\kappa}$, where $Q_{-1} := 0_{q \times q}$.

(b), (c) Use (a).

(d) This follows from Lemma 4.11 and (c).   □

Corollary 6.12. *Let* $\alpha \in [0, +\infty)$, *let* $\kappa \in \mathbb{N} \cup \{+\infty\}$, *and let* $(s_j)_{j=0}^{2\kappa} \in \mathcal{K}_{q,2\kappa,\alpha}^{\geq}$ *with canonical Hankel parametrization* $[(C_k)_{k=1}^{\kappa}, (D_k)_{k=0}^{\kappa}]$ *and right-sided* α*-Stieltjes parametrization* $(Q_j)_{j=0}^{2\kappa}$. *Then:*

(a) $C_k \geq Q_{2k-1} \geq 0_{q \times q}$ *for all* $k \in \mathbb{Z}_{1,\kappa}$.
(b) $\mathcal{N}(C_k) \subseteq \mathcal{N}(Q_{2k-1})$ *and* $\mathcal{R}(Q_{2k-1}) \subseteq \mathcal{R}(C_k)$ *for all* $k \in \mathbb{Z}_{1,\kappa}$.
(c) $0 \leq \det Q_{2k-1} \leq \det C_k$ *and* $\operatorname{rank} Q_{2k-1} \leq \operatorname{rank} C_k$ *for all* $k \in \mathbb{Z}_{1,\kappa}$.
(d) $0 \leq \det H_{\alpha \rhd n} \leq \prod_{k=1}^{n+1} \det C_k$ *and* $\operatorname{rank} H_{\alpha \rhd n} \leq \sum_{k=1}^{n+1} \operatorname{rank} C_k$ *for all* $n \in \mathbb{Z}_{0,\kappa-1}$.

Proof. (a) According to part (b) of Theorem 4.12, we have $Q_j \in \mathbb{C}_{\geq}^{q \times q}$ for all $j \in \mathbb{Z}_{0,2\kappa}$, which, in view of Proposition 6.10 and $\alpha \in [0, +\infty)$, implies

$$C_k = Q_{2k-1} + (\alpha I_q + D_{k-1} Q_{2k-3}^{\dagger}) D_{k-1}$$

$$= Q_{2k-1} + \alpha Q_{2k-2} + Q_{2k-2}^* Q_{2k-3}^{\dagger} Q_{2k-2} \geq Q_{2k-1} \geq 0_{q \times q}$$

for all $k \in \mathbb{Z}_{1,\kappa}$, where $Q_{-1} := 0_{q \times q}$.

(b), (c) Use (a).

(d) This follows from Lemma 4.11 and (c).   □

The next result contains interesting monotonicity results for the canonical Hankel parameters of a sequence belonging to $\mathcal{K}_{q,2\kappa,\alpha}^{\geq}$.

Corollary 6.13. *Let* $\alpha \in \mathbb{R}$, *let* $\kappa \in \mathbb{N} \cup \{+\infty\}$, *and let* $(s_j)_{j=0}^{2\kappa} \in \mathcal{K}_{q,2\kappa,\alpha}^{\geq}$ *with canonical Hankel parametrization* $[(C_k)_{k=1}^{\kappa}, (D_k)_{k=0}^{\kappa}]$. *Then:*

(a) $C_k \geq \alpha D_{k-1}$ *for all* $k \in \mathbb{Z}_{1,\kappa}$.
(b) *If* $\kappa \geq 2$, *then* $\mathcal{N}(C_k - \alpha D_{k-1}) \subseteq \mathcal{N}(D_k)$ *and* $\mathcal{R}(D_k) \subseteq \mathcal{R}(C_k - \alpha D_{k-1})$ *for all* $k \in \mathbb{Z}_{1,\kappa-1}$.

Proof. (a) Use part (a) of Corollary 6.11.

(b) According to part (b) of Corollary 6.11, we have $\mathcal{N}(C_k - \alpha D_{k-1}) \subseteq \mathcal{N}(Q_{2k-1})$ and $\mathcal{R}(Q_{2k-1}) \subseteq \mathcal{R}(C_k - \alpha D_{k-1})$ for all $k \in \mathbb{Z}_{1,\kappa}$. Part (b) of Theorem 4.12 yields $Q_j \in \mathbb{C}_{\geq}^{q \times q}$ for all $j \in \mathbb{Z}_{0,2\kappa}$ and $\mathcal{N}(Q_j) \subseteq \mathcal{N}(Q_{j+1})$ for all

$j \in \mathbb{Z}_{0,2\kappa-2}$. From Proposition 6.10 we obtain $Q_{2k} = D_k$ for all $k \in \mathbb{Z}_{0,\kappa}$. Thus, we get

$$\mathcal{N}(C_k - \alpha D_{k-1}) \subseteq \mathcal{N}(Q_{2k-1}) \subseteq \mathcal{N}(Q_{2k}) = \mathcal{N}(D_k)$$

and hence $\mathcal{R}(D_k) \subseteq \mathcal{R}(C_k - \alpha D_{k-1})$ for all $k \in \mathbb{Z}_{1,\kappa-1}$. $\square$

The following result should be compared with Proposition 3.4.

Corollary 6.14. *Let* $\alpha \in [0, +\infty)$, *let* $\kappa \in \mathbb{N} \cup \{+\infty\}$, *and let* $(s_j)_{j=0}^{2\kappa} \in \mathcal{K}_{q,2\kappa,\alpha}^{\geq}$ *with canonical Hankel parametrization* $[(C_k)_{k=1}^{\kappa}, (D_k)_{k=0}^{\kappa}]$. *Then* $(C_k)_{k=1}^{\kappa}$ *and* $(D_k)_{k=0}^{\kappa}$ *are sequences of non-negative Hermitian matrices. Further,* $\mathcal{N}(D_{k-1}) \subseteq \mathcal{N}(C_k)$ *for all* $k \in \mathbb{Z}_{1,\kappa}$ *and, in the case* $\kappa \geq 2$, *moreover* $\mathcal{N}(C_k) \subseteq \mathcal{N}(D_k)$ *for all* $k \in \mathbb{Z}_{1,\kappa-1}$.

Proof. Because of [15, Propositions 2.10 and 2.15], we know that $D_k \in \mathbb{C}_{\geq}^{q \times q}$ for all $k \in \mathbb{Z}_{0,\kappa}$ and $\mathcal{N}(D_{k-1}) \subseteq \mathcal{N}(C_k)$ for all $k \in \mathbb{Z}_{1,\kappa}$. Since $\alpha \in [0, +\infty)$ is assumed, part (a) of Corollary 6.13 yields then $C_k \in \mathbb{C}_{\geq}^{q \times q}$ for all $k \in \mathbb{Z}_{1,\kappa}$ and, furthermore, $\mathcal{N}(C_k) \subseteq \mathcal{N}(\alpha D_{k-1})$, which implies $\mathcal{N}(C_k) \subseteq \mathcal{N}(C_k - \alpha D_{k-1})$ for all $k \in \mathbb{Z}_{1,\kappa}$. Thus, in view of part (b) of Corollary 6.13, the proof is complete. $\square$

The following example shows that the converse statement to the first part of Corollary 6.14 is false. More precisely, we will construct a sequence $(s_j)_{j=0}^4 \in \mathcal{H}_{1,4}^{\geq} \setminus \mathcal{K}_{1,4,0}^{\geq}$ the canonical Hankel parametrization of which consists of positive real numbers.

Example 6.15. Let $s_0 := 2$, let $s_1 := 2$, let $s_2 := 6$, let $s_3 := 12$, and let $s_4 := 29$. Let $(Q_j)_{j=0}^4$ be the right-sided 0-Stieltjes parametrization of $(s_j)_{j=0}^4$ and let $[(C_k)_{k=1}^2, (D_k)_{k=0}^2]$ be the canonical Hankel parametrization of $(s_j)_{j=0}^4$. Then $D_0 = C_1 = 2$, $D_1 = 4$, and $C_2 = D_2 = 2$, which in view of Proposition 3.4 implies $(s_j)_{j=0}^4 \in \mathcal{H}_{1,4}^{\geq}$. Furthermore, $Q_0 = Q_1 = 2$, $Q_2 = 4$, $Q_3 = -6$, and $Q_4 = 2$, which in view of part (b) of Theorem 4.12 implies $(s_j)_{j=0}^4 \notin \mathcal{K}_{1,4,0}^{\geq}$ although $C_1 = C_2 = 2 > 0$.

Example 6.16. Let $\alpha \in (-\infty, 0)$ and let $\kappa \in \mathbb{Z}_{2,+\infty} \cup \{+\infty\}$. Let $Q_0 := I_3$, let $Q_1 := -\alpha \begin{bmatrix} 2 & 0 & 0 \\ 0 & 1 & 0 \\ 0 & 0 & 0 \end{bmatrix}$, let $Q_2 := \begin{bmatrix} 0 & 0 & 0 \\ 0 & 1 & 0 \\ 0 & 0 & 0 \end{bmatrix}$, and let $Q_j := 0_{3 \times 3}$ for all $j \in \mathbb{Z}_{3,2\kappa}$. Let $(s_j)_{j=0}^{2\kappa}$ be the sequence from $\mathbb{C}^{3 \times 3}$ with right-sided α-Stieltjes parametrization $(Q_j)_{j=0}^{2\kappa}$ and let $[(C_k)_{k=1}^{\kappa}, (D_k)_{k=0}^{\kappa}]$ be the canonical Hankel parametrization of $(s_j)_{j=0}^{2\kappa}$. In view of $-\alpha \in (0, +\infty)$ and part (b) of Theorem 4.12, then $(s_j)_{j=0}^{2\kappa} \in \mathcal{K}_{3,2\kappa,\alpha}^{\geq}$. According to Theorem 6.8, furthermore, $C_1 = Q_1 + \alpha Q_0 = -\alpha \begin{bmatrix} 1 & 0 & 0 \\ 0 & 0 & 0 \\ 0 & 0 & -1 \end{bmatrix}$ and $D_1 = Q_2 = \begin{bmatrix} 0 & 0 & 0 \\ 0 & 1 & 0 \\ 0 & 0 & 0 \end{bmatrix}$, which implies $\{-C_1, C_1\} \cap \mathbb{C}_{\geq}^{3 \times 3} = \emptyset$, $\mathcal{N}(C_1) \setminus \mathcal{N}(D_1) \neq \emptyset$ and $\mathcal{N}(D_1) \setminus \mathcal{N}(C_1) \neq \emptyset$.

Corollary 6.17. *Let* $\alpha \in (0, +\infty)$, *let* $\kappa \in \mathbb{N} \cup \{+\infty\}$, *and let* $(s_j)_{j=0}^{2\kappa} \in \mathcal{K}_{q,2\kappa,\alpha}^{\geq}$ *with canonical Hankel parametrization* $[(C_k)_{k=1}^{\kappa}, (D_k)_{k=0}^{\kappa}]$. *Then* $(C_k)_{k=1}^{\kappa}$ *and* $(D_k)_{k=0}^{\kappa}$ *are sequences of non-negative Hermitian matrices. Further,* $\mathcal{N}(D_{k-1}) = \mathcal{N}(C_k)$ *for all* $k \in \mathbb{Z}_{1,\kappa}$ *and, in the case* $\kappa \geq 2$, *moreover* $\mathcal{N}(D_{k-1}) \subseteq \mathcal{N}(D_k)$ *for all* $k \in \mathbb{Z}_{1,\kappa-1}$.

Proof. Because of Corollary 6.14, we know that $D_k \in \mathbb{C}_\geq^{q \times q}$ for all $k \in \mathbb{Z}_{0,\kappa}$. Since $\alpha \in (0, +\infty)$ is assumed, part (a) of Corollary 6.13 yields then $\mathcal{N}(C_k) \subseteq \mathcal{N}(\alpha D_{k-1}) = \mathcal{N}(D_{k-1})$ for all $k \in \mathbb{Z}_{1,\kappa}$. Thus, in view of Corollary 6.14, the proof is complete. $\qquad\square$

Example 6.18. Let $\alpha \in (-\infty, 0]$ and let $\kappa \in \mathbb{N} \cup \{+\infty\}$. Let $Q_0 \in \mathbb{C}_\geq^{q \times q} \setminus \{0_{q \times q}\}$, let $Q_1 := -\alpha Q_0$, and let $Q_j := 0_{q \times q}$ for all $j \in \mathbb{Z}_{2,2\kappa}$. Let $(s_j)_{j=0}^{2\kappa}$ be the unique sequence from $\mathbb{C}^{q \times q}$ with right-sided α-Stieltjes parametrization $(Q_j)_{j=0}^{2\kappa}$ (see Remark 4.3) and let $[(C_k)_{k=1}^\kappa, (D_k)_{k=0}^\kappa]$ be the canonical Hankel parametrization of $(s_j)_{j=0}^{2\kappa}$. In view of $-\alpha \in [0, +\infty)$ and part (b) of Theorem 4.12, then $(s_j)_{j=0}^{2\kappa} \in \mathcal{K}_{q,2\kappa,\alpha}^\geq$. According to Theorem 6.8, furthermore, $D_0 = Q_0 \neq 0_{q \times q}$ and $C_1 = Q_1 + \alpha Q_0 = 0_{q \times q}$, which in particular implies $\mathcal{N}(D_0) \neq \mathcal{N}(C_1)$.

Now we derive a further interrelation between right-sided α-Stieltjes parametrization and canonical Hankel parametrization. For this reason, we still need the following auxiliary result.

Lemma 6.19. *Let* $\alpha \in \mathbb{R}$, $\kappa \in \mathbb{Z}_{2,+\infty} \cup \{+\infty\}$ *and* $(s_j)_{j=0}^\kappa \in \mathcal{K}_{q,\kappa,\alpha}^\geq$ *with right-sided α-Stieltjes parametrization* $(Q_j)_{j=0}^\kappa$. *For all* $k \in \mathbb{N}_0$ *with* $2k + 1 \leq \kappa - 1$,

$$s_{\alpha \triangleright 2k+1} - \Lambda_{\alpha \triangleright k} = Q_{2k+2} + (\alpha I_q + Q_{2k+1} Q_{2k}^\dagger) Q_{2k+1}.$$

Proof. We have

$$Q_2 + (\alpha I_q + Q_1 Q_0^\dagger) Q_1 = L_1 + (\alpha I_q + L_{\alpha \triangleright 0} L_0^\dagger) L_{\alpha \triangleright 0}$$

$$= (s_2 - z_{1,1} H_0^\dagger y_{1,1}) + (\alpha I_q + s_{\alpha \triangleright 0} s_0^\dagger) s_{\alpha \triangleright 0}$$

$$= (s_2 - s_1 s_0^\dagger s_1) + \left[\alpha I_q + (-\alpha s_0 + s_1) s_0^\dagger \right] (-\alpha s_0 + s_1)$$

$$= s_2 - s_1 s_0^\dagger s_1 - \alpha^2 s_0 + \alpha s_1 + \alpha^2 s_0 s_0^\dagger s_0 - \alpha s_0 s_0^\dagger s_1 - \alpha s_1 s_0^\dagger s_0 + s_1 s_0^\dagger s_1$$

$$= s_2 - s_1 s_0^\dagger s_1 - \alpha^2 s_0 + \alpha s_1 + \alpha^2 s_0 - \alpha s_1 - \alpha s_1 + s_1 s_0^\dagger s_1 = -\alpha s_1 + s_2 = s_{\alpha \triangleright 1}$$

$$= s_{\alpha \triangleright 1} - \Lambda_{\alpha \triangleright 0},$$

where the 1st equation is due to Definition 4.2, the 2nd equation is due to (2.5), the 3rd equation is due to (2.2), (1.2) and (2.1), the 6th equation is due to $(s_j)_{j=0}^\kappa \in \mathcal{K}_{q,\kappa,\alpha}^\geq$, $\kappa \geq 2$, part (c) of Lemma 2.9 and parts (c) and (b) of Lemma A.1, the 7th equation is due to (2.1), and the 8th equation is due to (3.4). Now suppose $\kappa \geq 4$ and let $k \in \mathbb{N}$ with $2k+2 \leq \kappa$. Since $(s_j)_{j=0}^\kappa$ belongs to $\mathcal{K}_{q,\kappa,\alpha}^\geq$, we have $(s_j)_{j=0}^{2k+2} \in \mathcal{K}_{q,2k+2,\alpha}^\geq$ and hence $(s_j)_{j=0}^{2k+1} \in \mathcal{K}_{q,2k+1,\alpha}^{\geq,e}$, which, because of Lemma 2.10, implies $(s_{\alpha \triangleright j})_{j=0}^{2k} \in \mathcal{H}_{q,2k}^{\geq,e}$. From Remark 6.3 we obtain then

$$\Lambda_{\alpha \triangleright k} = M_{\alpha \triangleright k} + L_{\alpha \triangleright k} L_{\alpha \triangleright k-1}^\dagger (s_{\alpha \triangleright 2k-1} - M_{\alpha \triangleright k-1}). \tag{6.22}$$

Because of $(s_j)_{j=0}^{2k+2} \in \mathcal{K}_{q,2k+2,\alpha}^\geq$ we have $(s_j)_{j=0}^{2k+1} \in \mathcal{K}_{q,2k+1,\alpha}^\geq \cap \mathcal{K}_{q,2k+1,\alpha}^{\geq,e}$. From Lemma 2.17 we obtain that $\mathcal{N}(L_k) \subseteq \mathcal{N}(L_{\alpha \triangleright k})$, which, in view of part (b) of Lemma A.1, implies

$$L_{\alpha \triangleright k} L_k^\dagger L_k = L_{\alpha \triangleright k}. \tag{6.23}$$

Because of $(s_j)_{j=0}^{2k+2} \in \mathcal{K}_{q,2k+2,\alpha}^{\geq}$, we have $(s_j)_{j=0}^{2k+1} \in \mathcal{K}_{q,2k+1,\alpha}^{\geq}$, which, because of Proposition 6.5, implies

$$L_{\alpha \triangleright k} = (s_{2k+1} - M_k) - L_k \left[\alpha I_q + L_{\alpha \triangleright k-1}^\dagger (s_{\alpha \triangleright 2k-1} - M_{\alpha \triangleright k-1}) \right]. \tag{6.24}$$

Since $(s_j)_{j=0}^{2k+2}$ belongs to $\mathcal{K}_{q,2k+2,\alpha}^{\geq}$, from Proposition 6.4 we have

$$L_{k+1} = (s_{\alpha \triangleright 2k+1} - M_{\alpha \triangleright k}) - L_{\alpha \triangleright k} L_k^\dagger (s_{2k+1} - M_k). \tag{6.25}$$

Finally, we obtain

$$\begin{aligned}
s_{\alpha \triangleright 2k+1} - \Lambda_{\alpha \triangleright k} &= (s_{\alpha \triangleright 2k+1} - M_{\alpha \triangleright k}) - (\Lambda_{\alpha \triangleright k} - M_{\alpha \triangleright k}) \\
&= (s_{\alpha \triangleright 2k+1} - M_{\alpha \triangleright k}) - L_{\alpha \triangleright k} L_{\alpha \triangleright k-1}^\dagger (s_{\alpha \triangleright 2k-1} - M_{\alpha \triangleright k-1}) \\
&= (s_{\alpha \triangleright 2k+1} - M_{\alpha \triangleright k}) - L_{\alpha \triangleright k} L_k^\dagger L_k L_{\alpha \triangleright k-1}^\dagger (s_{\alpha \triangleright 2k-1} - M_{\alpha \triangleright k-1}) \\
&= (s_{\alpha \triangleright 2k+1} - M_{\alpha \triangleright k}) + L_{\alpha \triangleright k} L_k^\dagger [L_{\alpha \triangleright k} - (s_{2k+1} - M_k) + \alpha L_k] \\
&= (s_{\alpha \triangleright 2k+1} - M_{\alpha \triangleright k}) - L_{\alpha \triangleright k} L_k^\dagger (s_{2k+1} - M_k) + \alpha L_{\alpha \triangleright k} L_k^\dagger L_k + L_{\alpha \triangleright k} L_k^\dagger L_{\alpha \triangleright k} \\
&= L_{k+1} + \alpha L_{\alpha \triangleright k} + L_{\alpha \triangleright k} L_k^\dagger L_{\alpha \triangleright k} = L_{k+1} + (\alpha I_q + L_{\alpha \triangleright k} L_k^\dagger) L_{\alpha \triangleright k} \\
&= Q_{2k+2} + (\alpha I_q + Q_{2k+1} Q_{2k}^\dagger) Q_{2k+1},
\end{aligned}$$

where the 2nd equation is due to (6.22), the 3rd equation is due to (6.23), the 4th equation is due to (6.24), the 6th equation is due to (6.25) and (6.23), and the 8th equation is due to Definition 4.2. $\qquad\square$

Theorem 6.20. Let $\alpha \in \mathbb{R}$, $\kappa \in \mathbb{Z}_{3,+\infty} \cup \{+\infty\}$, and $(s_j)_{j=0}^{\kappa} \in \mathcal{K}_{q,\kappa,\alpha}^{\geq}$. Let $(Q_j)_{j=0}^{\kappa}$ be the right-sided α-Stieltjes parametrization of $(s_j)_{j=0}^{\kappa}$ and let $[(E_k)_{k=1}^{\lfloor \frac{\kappa-1}{2} \rfloor}, (F_k)_{k=0}^{\lfloor \frac{\kappa-1}{2} \rfloor}]$ be the canonical Hankel parametrization of $(s_{\alpha \triangleright j})_{j=0}^{2\lfloor \frac{\kappa-1}{2} \rfloor}$. Then

$$E_k = Q_{2k} + (\alpha I_q + Q_{2k-1} Q_{2k-2}^\dagger) Q_{2k-1}$$

for all $k \in \mathbb{N}$ with $2k+1 \leq \kappa$ and $F_k = Q_{2k+1}$ for all $k \in \mathbb{N}_0$ with $2k+1 \leq \kappa$.

Proof. Use Definition 3.2, Lemma 6.19, and Definition 4.2. $\qquad\square$

Let s_0 and the canonical Hankel parametrization of $(s_{\alpha \triangleright j})_{j=0}^{2\kappa}$ be given, then we will show that the right-sided α-Stieltjes parametrization of $(s_j)_{j=0}^{2\kappa+1} \in \mathcal{K}_{q,2\kappa+1,\alpha}^{\geq}$ can be computed recursively. To realize this aim, we need the following auxiliary result.

Lemma 6.21. Let $\alpha \in \mathbb{R}$, $\kappa \in \mathbb{N}_0 \cup \{+\infty\}$, and $(s_j)_{j=0}^{\kappa} \in \mathcal{K}_{q,\kappa,\alpha}^{\geq}$. Let $(Q_j)_{j=0}^{\kappa}$ be the right-sided α-Stieltjes parametrization of $(s_j)_{j=0}^{\kappa}$. Then

$$Q_{2k} = \begin{cases} s_0 & \text{if } k = 0 \\ s_{\alpha \triangleright 2k-1} - \Lambda_{\alpha \triangleright k-1} - (\alpha I_q + L_{\alpha \triangleright k-1} Q_{2k-2}^\dagger) L_{\alpha \triangleright k-1} & \text{if } k \geq 1 \end{cases}$$

for all $k \in \mathbb{N}_0$ with $2k \leq \kappa$.

Proof. Use Definition 4.2 and Lemma 6.19. $\qquad\square$

Proposition 6.22. *Let* $\alpha \in \mathbb{R}$, *let* $\kappa \in \mathbb{N} \cup \{+\infty\}$, *let* $(s_j)_{j=0}^{2\kappa+1} \in \mathcal{K}_{q,2\kappa+1,\alpha}^{\geq}$, *and let* $[(E_k)_{k=1}^{\kappa}, (F_k)_{k=0}^{\kappa}]$ *be the canonical Hankel parametrization of* $(s_{\alpha\triangleright j})_{j=0}^{2\kappa}$. *Then the right-sided* α-*Stieltjes parametrization* $(Q_j)_{j=0}^{2\kappa+1}$ *of* $(s_j)_{j=0}^{2\kappa+1}$ *is given by the recurrence formulas*

$$Q_{2k} = \begin{cases} s_0 & \text{if } k = 0 \\ E_k - (\alpha I_q + F_{k-1}Q_{2k-2}^{\dagger})F_{k-1} & \text{if } k \geq 1 \end{cases}$$

for all $k \in \mathbb{Z}_{0,\kappa}$ *and by* $Q_{2k+1} = F_k$ *for all* $k \in \mathbb{Z}_{0,\kappa}$.

Proof. Use Lemma 6.21, Definition 3.2 and Remark 6.1. □

Corollary 6.23. *Let* $\alpha \in \mathbb{R}$, *let* $\kappa \in \mathbb{N} \cup \{+\infty\}$, *and let* $(s_j)_{j=0}^{2\kappa+1} \in \mathcal{K}_{q,2\kappa+1,\alpha}^{\geq}$ *with right-sided* α-*Stieltjes parametrization* $(Q_j)_{j=0}^{2\kappa+1}$. *Let* $[(E_k)_{k=1}^{\kappa}, (F_k)_{k=0}^{\kappa}]$ *be the canonical Hankel parametrization of* $(s_{\alpha\triangleright j})_{j=0}^{2\kappa}$. *Then:*

(a) $E_k - \alpha F_{k-1} \geq Q_{2k} \geq 0_{q\times q}$ *for all* $k \in \mathbb{Z}_{1,\kappa}$.
(b) $\mathcal{N}(E_k - \alpha F_{k-1}) \subseteq \mathcal{N}(Q_{2k})$ *and* $\mathcal{R}(Q_{2k}) \subseteq \mathcal{R}(E_k - \alpha F_{k-1})$ *for all* $k \in \mathbb{Z}_{1,\kappa}$.
(c) $0 \leq \det Q_{2k} \leq \det(E_k - \alpha F_{k-1})$ *and* $\operatorname{rank} Q_{2k} \leq \operatorname{rank}(E_k - \alpha F_{k-1})$ *for all* $k \in \mathbb{Z}_{1,\kappa}$.
(d) $0 \leq \det H_n \leq (\det s_0) \prod_{k=1}^{n} \det(E_k - \alpha F_{k-1})$ *and* $\operatorname{rank} H_n \leq \operatorname{rank} s_0 + \sum_{k=1}^{n} \operatorname{rank}(E_k - \alpha F_{k-1})$ *for all* $n \in \mathbb{Z}_{1,\kappa}$.

Proof. From [13, Proposition 2.30] and [15, Proposition 2.15] we know that $(E_k)_{k=1}^{\kappa}$ and $(F_k)_{k=0}^{\kappa}$ are sequences of Hermitian matrices.

(a) According to part (b) of Theorem 4.12, we have $Q_j \in \mathbb{C}_{\geq}^{q\times q}$ for all $j \in \mathbb{Z}_{0,2\kappa+1}$, which, in view of Proposition 6.22, for all $k \in \mathbb{Z}_{1,\kappa}$, implies

$$E_k - \alpha F_{k-1} = Q_{2k} + F_{k-1}Q_{2k-2}^{\dagger}F_{k-1}$$

$$= Q_{2k} + Q_{2k-1}Q_{2k-2}^{\dagger}Q_{2k-1}$$

$$= Q_{2k} + Q_{2k-1}^{*}Q_{2k-2}^{\dagger}Q_{2k-1} \geq Q_{2k} \geq 0_{q\times q}.$$

(b), (c) Use (a).
(d) Use Lemma 4.11, Definition 4.2, and (c). □

Corollary 6.24. *Let* $\alpha \in [0, +\infty)$, *let* $\kappa \in \mathbb{N} \cup \{+\infty\}$, *and let* $(s_j)_{j=0}^{2\kappa+1} \in \mathcal{K}_{q,2\kappa+1,\alpha}^{\geq}$ *with right-sided* α-*Stieltjes parametrization* $(Q_j)_{j=0}^{2\kappa+1}$. *Further, let* $[(E_k)_{k=1}^{\kappa}, (F_k)_{k=0}^{\kappa}]$ *be the canonical Hankel parametrization of* $(s_{\alpha\triangleright j})_{j=0}^{2\kappa}$. *Then:*

(a) $E_k \geq Q_{2k} \geq 0_{q\times q}$ *for all* $k \in \mathbb{Z}_{1,\kappa}$.
(b) $\mathcal{N}(E_k) \subseteq \mathcal{N}(Q_{2k})$ *and* $\mathcal{R}(Q_{2k}) \subseteq \mathcal{R}(E_k)$ *for all* $k \in \mathbb{Z}_{1,\kappa}$.
(c) $0 \leq \det Q_{2k} \leq \det E_k$ *and* $\operatorname{rank} Q_{2k} \leq \operatorname{rank} E_k$ *for all* $k \in \mathbb{Z}_{1,\kappa}$.
(d) $0 \leq \det H_n \leq (\det s_0) \prod_{k=1}^{n} \det E_k$ *and* $\operatorname{rank} H_n \leq \operatorname{rank} s_0 + \sum_{k=1}^{n} \operatorname{rank} E_k$ *for all* $n \in \mathbb{Z}_{1,\kappa}$.

Proof. (a) According to part (b) of Theorem 4.12, we have $Q_j \in \mathbb{C}_\succeq^{q \times q}$ for all $j \in \mathbb{Z}_{0,2\kappa+1}$, which, in view of Proposition 6.22 and $\alpha \in [0, +\infty)$, implies

$$E_k = Q_{2k} + (\alpha I_q + F_{k-1}Q_{2k-2}^\dagger)F_{k-1}$$
$$= Q_{2k} + \alpha Q_{2k-1} + Q_{2k-1}^* Q_{2k-2}^\dagger Q_{2k-1} \geq Q_{2k} \geq 0_{q \times q}$$

for all $k \in \mathbb{Z}_{1,\kappa}$.

 (b), (c) Use (a).

 (d) Use Lemma 4.11, Definition 4.2, and (c).   □

7. Connection between right α- and right β-Stieltjes parametrizations

In this section, we consider a fixed $\alpha \in \mathbb{R}$, a fixed $\beta \in [\alpha, +\infty)$ and a sequence $(s_j)_{j=0}^\kappa \in \mathcal{K}_{q,\kappa,\beta}^\geq$. First we will verify that $(s_j)_{j=0}^\kappa$ belongs to $\mathcal{K}_{q,\kappa,\alpha}^\geq$, too. Afterwards, we will study interrelations between the right-sided α-Stieltjes parametrization and the right-sided β-Stieltjes parametrization of $(s_j)_{j=0}^\kappa$.

Remark 7.1. Let $\alpha \in \mathbb{R}$, $\beta \in [\alpha, +\infty)$, and $\kappa \in \mathbb{N}_0 \cup \{+\infty\}$. Then by inspection of the corresponding block Hankel matrices it is readily checked that $\mathcal{K}_{q,\kappa,\beta}^\geq \subseteq \mathcal{K}_{q,\kappa,\alpha}^\geq$.

If the given real numbers α and β fulfill $\alpha \leq \beta$ and $(s_j)_{j=0}^\kappa \in \mathcal{K}_{q,\kappa,\beta}^\geq$, then the right-sided β-Stieltjes parametrization of $(s_j)_{j=0}^\kappa$ and the right-sided α-Stieltjes parametrization of $(s_j)_{j=0}^\kappa$ are connected recursively.

Proposition 7.2. *Let* $\alpha \in \mathbb{R}$, $\beta \in [\alpha, +\infty)$, $\kappa \in \mathbb{N}_0 \cup \{+\infty\}$ *and* $(s_j)_{j=0}^\kappa \in \mathcal{K}_{q,\kappa,\beta}^\geq$ *with right-sided α-Stieltjes parametrization* $(Q_j)_{j=0}^\kappa$ *and right-sided β-Stieltjes parametrization* $(R_j)_{j=0}^\kappa$. *Then* $Q_{2k} = R_{2k}$ *for all* $k \in \mathbb{N}_0$ *with* $2k \leq \kappa$ *and*

$$Q_{2k+1} \geq (\beta - \alpha)R_{2k} + R_{2k+1}$$

for all $k \in \mathbb{N}_0$ *with* $2k + 1 \leq \kappa$.

Proof. From Definition 4.2 we get $Q_{2k} = L_k = R_{2k}$ for all $k \in \mathbb{N}_0$ with $2k \leq \kappa$. Now suppose $\kappa \geq 1$ and let $k \in \mathbb{N}_0$ with $2k + 1 \leq \kappa$. For all $j \in \mathbb{Z}_{0,2k}$, let $t_j := (\beta - \alpha)s_j$. According to (2.1), we get for all $j \in \mathbb{Z}_{0,2k}$ that

$$s_{\alpha \triangleright j} = -\alpha s_j + s_{j+1} = (\beta - \alpha)s_j + (-\beta s_j + s_{j+1}) = t_j + s_{\beta \triangleright j},$$

which in view of (1.2) implies

$$H_{\alpha \triangleright k} = H_k^{\langle t \rangle} + H_{\beta \triangleright k}. \tag{7.1}$$

Because of $(s_j)_{j=0}^\kappa \in \mathcal{K}_{q,\kappa,\beta}^\geq$ and $2k + 1 \leq \kappa$, we have $(s_j)_{j=0}^{2k+1} \in \mathcal{K}_{q,2k+1,\beta}^\geq$ and hence $\{(s_j)_{j=0}^{2k}, (s_{\beta \triangleright j})_{j=0}^{2k}\} \subseteq \mathcal{H}_{q,2k}^\geq$, i.e., $H_k \in \mathbb{C}_\succeq^{(k+1)q \times (k+1)q}$ and

$$H_{\beta \triangleright k} \in \mathbb{C}_\succeq^{(k+1)q \times (k+1)q}. \tag{7.2}$$

According to (1.2), we get $H_k^{\langle t \rangle} = (\beta - \alpha)H_k$, which, in view of $\beta - \alpha \in [0, +\infty)$ and $H_k \in \mathbb{C}_{\geq}^{(k+1)q \times (k+1)q}$, implies

$$H_k^{\langle t \rangle} \in \mathbb{C}_{\geq}^{(k+1)q \times (k+1)q}. \tag{7.3}$$

In view of (7.3), (7.2), and (7.1), the application of [15, Lemma 4.21 (b)] yields $L_{\alpha \triangleright k} \geq L_k^{\langle t \rangle} + L_{\beta \triangleright k}$. Moreover, from (7.3) we know $L_k^{\langle t \rangle} \in \mathbb{C}_{\geq}^{q \times q}$, and from (2.5) we get $L_k^{\langle t \rangle} = (\beta - \alpha)L_k$. Finally, we obtain $L_{\alpha \triangleright k} \geq L_k^{\langle t \rangle} + L_{\beta \triangleright k} = (\beta - \alpha)L_k + L_{\beta \triangleright k} \geq L_{\beta \triangleright k}$, which in view of Definition 4.2 implies the asserted inequalities. $\qquad \square$

Corollary 7.3. *Let $\alpha \in \mathbb{R}$, $\beta \in (\alpha, +\infty)$, $\kappa \in \mathbb{N} \cup \{+\infty\}$ and $(s_j)_{j=0}^{\kappa} \in \mathcal{K}_{q,\kappa,\beta}^{\geq}$ with right-sided α-Stieltjes parametrization $(Q_j)_{j=0}^{\kappa}$ and right-sided β-Stieltjes parametrization $(R_j)_{j=0}^{\kappa}$. Then:*

(a) $Q_{2k+1} \geq (\beta - \alpha)R_{2k} \geq 0_{q \times q}$ *and* $Q_{2k+1} \geq R_{2k+1} \geq 0_{q \times q}$ *for all $k \in \mathbb{N}_0$ with $2k + 1 \leq \kappa$.*

(b) $\mathcal{N}(Q_{2k+1}) \subseteq \mathcal{N}(R_{2k}) \cap \mathcal{N}(R_{2k+1})$ *and* $\mathcal{R}(R_{2k1}) \cup \mathcal{R}(R_{2k+1}) \subseteq \mathcal{R}(Q_{2k+1})$ *for all $k \in \mathbb{N}_0$ with $2k + 1 \leq \kappa$.*

(c) $0 \leq (\beta - \alpha)^q \det R_{2k} \leq \det Q_{2k+1}$, $0 \leq \det R_{2k+1} \leq \det Q_{2k+1}$, $\operatorname{rank} R_{2k} \leq \operatorname{rank} Q_{2k+1}$, *and* $\operatorname{rank} R_{2k+1} \leq \operatorname{rank} Q_{2k+1}$ *for all $k \in \mathbb{N}_0$ with $2k + 1 \leq \kappa$.*

(d) $H_{\alpha \triangleright n} \geq (\beta - \alpha)H_n \geq 0_{q \times q}$ *and* $H_{\alpha \triangleright n} \geq H_{\beta \triangleright n} \geq 0_{q \times q}$ *for all $n \in \mathbb{N}_0$ with $2n + 1 \leq \kappa$.*

Proof. (a) According to part (b) of Theorem 4.12, we have $R_j \in \mathbb{C}_{\geq}^{q \times q}$ for all $j \in \mathbb{Z}_{0,\kappa}$, which, in view of Proposition 7.2 and $\beta - \alpha \in (0, +\infty)$, implies (a).

Parts (b) and (c) follow from (a). Part (d) is obvious. $\qquad \square$

Corollary 7.4. *Let $\alpha \in \mathbb{R}$, let $\kappa \in \mathbb{N} \cup \{+\infty\}$, let $(s_j)_{j=0}^{\kappa} \in \mathcal{K}_{q,\kappa,\alpha}^{\geq}$ with right-sided α-Stieltjes parametrization $(Q_j)_{j=0}^{\kappa}$, and let $\epsilon \in (0, +\infty)$ be such that $(s_j)_{j=0}^{\kappa} \in \mathcal{K}_{q,\kappa,\alpha+\epsilon}^{\geq}$. Then:*

(a) $Q_{2k+1} \geq \epsilon Q_{2k} \geq 0_{q \times q}$ *for all $k \in \mathbb{N}_0$ with $2k + 1 \leq \kappa$.*

(b) $\mathcal{N}(Q_{2k}) = \mathcal{N}(Q_{2k+1})$ *and* $\mathcal{R}(Q_{2k+1}) = \mathcal{R}(Q_{2k})$ *for all $k \in \mathbb{N}_0$ with $2k + 1 \leq \kappa - 1$.*

(c) $0 \leq \det Q_{2k} \leq \epsilon^{-q} \det Q_{2k+1}$ *for all $k \in \mathbb{N}_0$ with $2k + 1 \leq \kappa$.*

(d) $\operatorname{rank} Q_{2k} = \operatorname{rank} Q_{2k+1}$ *for all $k \in \mathbb{N}_0$ with $2k + 1 \leq \kappa - 1$.*

(e) $0 \leq \det H_n \leq \epsilon^{\frac{-(n+1)(n+2)q}{2}} \det H_{\alpha \triangleright n}$ *for all $n \in \mathbb{N}_0$ with $2n + 1 \leq \kappa$.*

(f) $\operatorname{rank} H_n = \operatorname{rank} H_{\alpha \triangleright n}$ *for all $n \in \mathbb{N}_0$ with $2n + 1 \leq \kappa - 1$.*

Proof. (a) Let $\beta := \alpha + \epsilon$. From Proposition 7.2 (with the notation used there) we get that $Q_{2k} = R_{2k}$ for all $k \in \mathbb{N}_0$ with $2k \leq \kappa$, which, in view of part (a) of Corollary 7.3, implies (a).

(b) Since $\epsilon \in (0, +\infty)$ is assumed, (a) yields $\mathcal{N}(Q_{2k+1}) \subseteq \mathcal{N}(\epsilon Q_{2k}) = \mathcal{N}(Q_{2k})$ for all $k \in \mathbb{N}_0$ with $2k + 1 \leq \kappa$, which, in view of part (b) of Theorem 4.12, implies (b).

(c) Use (a).

(d) This follows from (b).

(e) Use Lemma 4.11 and (c).

(f) This is a consequence of Lemma 4.11 and (d).   □

Proposition 7.5. *Let $\alpha \in \mathbb{R}$, $\beta \in [\alpha, +\infty)$, $\kappa \in \mathbb{N}_0 \cup \{+\infty\}$ and $(s_j)_{j=0}^{\kappa} \in \mathcal{K}_{q,\kappa,\beta}^{\geq}$. Then the right-sided α-Stieltjes parametrization $(Q_j)_{j=0}^{\kappa}$ of $(s_j)_{j=0}^{\kappa}$ is given in terms of the right-sided β-Stieltjes parametrization $(R_j)_{j=0}^{\kappa}$ of $(s_j)_{j=0}^{\kappa}$ by $Q_{2k} = R_{2k}$ for all $k \in \mathbb{N}_0$ with $2k \leq \kappa$ and, in the case $\kappa \geq 1$, by the recurrence formulas*

$$Q_{2k+1} = R_{2k+1} + \left[(\beta - \alpha)I_q + R_{2k}(R_{2k-1}^\dagger - Q_{2k-1}^\dagger) \right] R_{2k} \qquad (7.4)$$

for all $k \in \mathbb{N}_0$ with $2k + 1 \leq \kappa$, where $Q_{-1} := 0_{q \times q}$ and $R_{-1} := 0_{q \times q}$.

Proof. From Definition 4.2 we get $Q_{2k} = L_k = R_{2k}$ for all $k \in \mathbb{N}_0$ with $2k \leq \kappa$. Now suppose $\kappa \geq 1$ and let $k \in \mathbb{N}_0$ be such that $2k + 1 \leq \kappa$. From Remark 7.1 we get $(s_j)_{j=0}^{\kappa} \in \mathcal{K}_{q,\kappa,\alpha}^{\geq}$. Because of Lemma 6.9, we have then (6.21). According to $(s_j)_{j=0}^{\kappa} \in \mathcal{K}_{q,\kappa,\beta}^{\geq}$ and Lemma 6.7, we get $s_1 - \Lambda_0 = R_1 + \beta R_0$ and, if $k \geq 1$, furthermore, $s_{2k+1} - \Lambda_k = R_{2k+1} + (\beta I_q + R_{2k}R_{2k-1}^\dagger)R_{2k}$. Definition 4.2 yields $R_{2k} = L_k$. Thus, (7.4) is obtained by substituting the last equations into (6.21). □

Proposition 7.6. *Let $\alpha \in \mathbb{R}$, $\beta \in [\alpha, +\infty)$, $\kappa \in \mathbb{N}_0 \cup \{+\infty\}$ and $(s_j)_{j=0}^{\kappa} \in \mathcal{K}_{q,\kappa,\beta}^{\geq}$. Then the right-sided β-Stieltjes parametrization $(R_j)_{j=0}^{\kappa}$ of $(s_j)_{j=0}^{\kappa}$ is given in terms of the right-sided α-Stieltjes parametrization $(Q_j)_{j=0}^{\kappa}$ of $(s_j)_{j=0}^{\kappa}$ by $R_{2k} = Q_{2k}$ for all $k \in \mathbb{N}_0$ with $2k \leq \kappa$ and in the case $\kappa \geq 1$ by the recurrence formulas*

$$R_{2k+1} = Q_{2k+1} - \left[(\beta - \alpha)I_q + Q_{2k}(R_{2k-1}^\dagger - Q_{2k-1}^\dagger) \right] Q_{2k}$$

for all $k \in \mathbb{N}_0$ with $2k + 1 \leq \kappa$, where $Q_{-1} := 0_{q \times q}$ and $R_{-1} := 0_{q \times q}$.

Proof. Use Proposition 7.5.   □

Appendix A. Moore-Penrose inverse

If $A \in \mathbb{C}^{p \times q}$, then the Moore-Penrose inverse $A^\dagger$ of A is the unique complex $q \times p$ matrix G which fulfills the four equations $AGA = A$, $GAG = G$, $(AG)^* = AG$, and $(GA)^* = GA$.

Lemma A.1. *Let $A \in \mathbb{C}^{p \times q}$ and $r, s \in \mathbb{N}$. Then:*

(a) $(A^\dagger)^\dagger = A$, $(A^\dagger)^* = (A^*)^\dagger$, $\mathcal{R}(A^\dagger) = \mathcal{R}(A^*)$ and $\mathcal{N}(A^\dagger) = \mathcal{N}(A^*)$.

(b) *Let $B \in \mathbb{C}^{r \times q}$. Then $\mathcal{N}(A) \subseteq \mathcal{N}(B)$ if and only if $BA^\dagger A = B$.*

(c) *Let $C \in \mathbb{C}^{p \times s}$. Then $\mathcal{R}(C) \subseteq \mathcal{R}(A)$ if and only if $AA^\dagger C = C$.*

(d) *If $U \in \mathbb{C}^{r \times p}$ with $U^*U = I_p$ and if $V \in \mathbb{C}^{q \times s}$ with $VV^* = I_q$, then $(UAV)^\dagger = V^* A^\dagger U^*$.*

Lemma A.1 is well known and the corresponding proof is straightforward.

Appendix B. Some considerations on non-negative Hermitian measures

Let $\mathfrak{B}_{\mathbb{C}}$ be the set of all Borel subsets of $\mathbb{C}$. If $(\Omega, \mathfrak{A})$ is a measurable space, then let $\mathcal{M}_{\geq}^q(\Omega, \mathfrak{A})$ be the set of all non-negative Hermitian $q \times q$ measures on $(\Omega, \mathfrak{A})$.

Proposition B.1. *Let* $(\Omega, \mathfrak{A})$ *and* $(\tilde{\Omega}, \tilde{\mathfrak{A}})$ *be measurable spaces and* $\mu \in \mathcal{M}_{\geq}^q(\Omega, \mathfrak{A})$. *Furthermore, let* $T \colon \Omega \to \tilde{\Omega}$ *be an* $\mathfrak{A}$-$\tilde{\mathfrak{A}}$-*measurable mapping. Then* $T(\mu) \colon \tilde{\mathfrak{A}} \to \mathbb{C}^{q \times q}$ *defined by* $[T(\mu)](\tilde{A}) := \mu(T^{-1}(\tilde{A}))$ *is a non-negative Hermitian measure which belongs to* $\mathcal{M}_{\geq}^q(\tilde{\Omega}, \tilde{\mathfrak{A}})$. *Furthermore, if* $\tilde{f} \colon \tilde{\Omega} \to \mathbb{C}$ *is an* $\tilde{\mathfrak{A}}$-$\mathfrak{B}_{\mathbb{C}}$-*measurable mapping, then*

$$\tilde{f} \in \mathcal{L}^1\left(\tilde{\Omega}, \tilde{\mathfrak{A}}, T(\mu); \mathbb{C}\right) \tag{B.1}$$

if and only if

$$\tilde{f} \circ T \in \mathcal{L}^1(\Omega, \mathfrak{A}, \mu; \mathbb{C}). \tag{B.2}$$

If $\tilde{f}$ *belongs to* $\mathcal{L}^1(\tilde{\Omega}, \tilde{\mathfrak{A}}, T(\mu); \mathbb{C})$, *then*

$$\int_{\tilde{A}} \tilde{f} \, \mathrm{d}\left[T(\mu)\right] = \int_{T^{-1}(\tilde{A})} (\tilde{f} \circ T) \mathrm{d}\mu \qquad \textit{for all } \tilde{A} \in \tilde{\mathfrak{A}}. \tag{B.3}$$

Proof. Obviously, $\tilde{\mu} := T(\mu)$ belongs to $\mathcal{M}_{\geq}^q(\tilde{\Omega}, \tilde{\mathfrak{A}})$. Let $\tilde{f} \colon \tilde{\Omega} \to \mathbb{C}$ be an $\tilde{\mathfrak{A}}$-$\mathfrak{B}_{\mathbb{C}}$-measurable mapping. According to [16, Lemma B.1], condition (B.1) holds true if and only if the following statement is fulfilled:

(I) For each $v \in \mathbb{C}^q$, the function $\tilde{f}$ belongs to $\mathcal{L}^1(\tilde{\Omega}, \tilde{\mathfrak{A}}, v^*[T(\mu)]v; \mathbb{C})$.

Since

$$T(v^*\mu v) = v^* \left[T(\mu)\right] v \tag{B.4}$$

holds true for each $v \in \mathbb{C}^q$, from [8, Proposition 2.6.5] we see that (I) is equivalent to:

(II) For each $v \in \mathbb{C}^q$, the function $\tilde{f} \circ T$ belongs to $\mathcal{L}^1(\Omega, \mathfrak{A}, v^*\mu v; \mathbb{C})$.

Using [16, Lemma B.1], we get that (II) is fulfilled if and only if (B.2) is true.

For each $j \in \mathbb{Z}_{1,q}$, let $e_j := [\delta_{j1}, \delta_{j2}, \ldots, \delta_{jq}]$, where $\delta_{jk} := 1$ for $j = k$ and $\delta_{jk} := 0$ for $j \neq k$. Furthermore, for each $j \in \mathbb{Z}_{1,q}$ and each $k \in \mathbb{Z}_{1,q}$, let $v_{jk}^{(0)} := \frac{1}{2}(e_j + e_k)$, $v_{jk}^{(1)} := \frac{1}{2}(e_j - \mathrm{i}e_k)$, $v_{jk}^{(2)} := \frac{1}{2}(e_j - e_k)$, and $v_{jk}^{(3)} := \frac{1}{2}(e_j + \mathrm{i}e_k)$. We consider an arbitrary $\tilde{f} \in \mathcal{L}^1(\tilde{\Omega}, \tilde{\mathfrak{A}}, T(\mu); \mathbb{C})$ and an arbitrary $\tilde{A} \in \tilde{\mathfrak{A}}$. For every choice of j and k in $\mathbb{Z}_{1,q}$, from [9, Remark 1.1.1], [16, Lemma B.3], (B.4), and [8, Proposition 2.6.5] we then get

$$
\begin{aligned}
e_j^* \left(\int_{T^{-1}(\tilde{A})} \tilde{f}\mathrm{d}\,[T(\mu)] \right) e_k &= \sum_{l=0}^{3} \mathrm{i}^l (v_{jk}^{(l)})^* \left(\int_{T^{-1}(\tilde{A})} \tilde{f}\mathrm{d}\,[T(\mu)] \right) v_{jk}^{(l)} \\
&= \sum_{l=0}^{3} \mathrm{i}^l \int_{T^{-1}(\tilde{A})} \tilde{f}\mathrm{d}\left((v_{jk}^{(l)})^* [T(\mu)]\, v_{jk}^{(l)} \right) \\
&= \sum_{l=0}^{3} \mathrm{i}^l \int_{T^{-1}(\tilde{A})} \tilde{f}\left[T\left((v_{jk}^{(l)})^* \mu v_{jk}^{(l)} \right) \right] \\
&= \sum_{l=0}^{3} \mathrm{i}^l \int_{\tilde{A}} (\tilde{f}\circ T)\mathrm{d}\left[(v_{jk}^{(l)})^* \mu v_{jk}^{(l)} \right] \\
&= \sum_{l=0}^{3} \mathrm{i}^l (v_{jk}^{(l)})^* \left[\int_{\tilde{A}} (\tilde{f}\circ T)\mathrm{d}\mu \right] v_{jk}^{(l)} \\
&= e_j^* \left[\int_{\tilde{A}} (\tilde{f}\circ T)\mathrm{d}\mu \right] e_k.
\end{aligned}
$$

Consequently, (B.3) is also proved. □

For all subsets Ω of $\mathbb{C}$, let $\check{\Omega} := \{-\omega \colon \omega \in \Omega\}$. Furthermore for all $\Omega \in \mathfrak{B}_{\mathbb{R}} \setminus \{\emptyset\}$ and all $\mu \in \mathcal{M}_{\geq}^{q}(\Omega)$, let $\check{\mu}$ be the image measure of μ under the mapping $r \colon \Omega \to \check{\Omega}$ defined by $r(t) := -t$.

Lemma B.2. *Let Ω be a non-empty subset of $\mathbb{R}$, let $\mu \in \mathcal{M}_{\geq}^{q}(\Omega)$, and let $\kappa \in \mathbb{N}_0 \cup \{+\infty\}$. Then $\mu \in \mathcal{M}_{\geq,\kappa}^{q}(\Omega)$ if and only if $\check{\mu} \in \mathcal{M}_{\geq,\kappa}^{q}(\check{\Omega})$ and, in this case, $s_j^{(\check{\mu})} = (-1)^j s_j^{(\mu)}$ for all $j \in \mathbb{Z}_{0,\kappa}$.*

Proof. Use Proposition B.1. □

Lemma B.3. *Let $\alpha \in \mathbb{R}$, let $\kappa \in \mathbb{N}_0 \cup \{+\infty\}$, let $(t_j)_{j=0}^{\kappa}$ be a sequence from $\mathbb{C}^{q \times q}$, and let $\tau \in \mathcal{M}_{\geq}^{q}[(-\infty, \alpha]; (t_j)_{j=0}^{\kappa}, =]$. Then $\check{\tau} \in \mathcal{M}_{\geq}^{q}[[-\alpha, +\infty); ((-1)^j t_j)_{j=0}^{\kappa}, =]$.*

Proof. The proof consists of an application of Lemma B.2. □

References

[1] N.I. Akhiezer, *Klassicheskaya problema momentov i nekotorye voprosy analiza, svyazannye s neyu*, Gosudarstv. Izdat. Fiz.-Mat. Lit., Moscow, 1961 (Russian); English transl., *The classical moment problem and some related questions in analysis*, Translated by N. Kemmer, Hafner Publishing Co., New York, 1965. MR0154069 (27 #4028)

[2] V.A. Bolotnikov, *Descriptions of solutions of a degenerate moment problem on the axis and the halfaxis*, Teor. Funktsiĭ Funktsional. Anal. i Prilozhen. **50** (1988), 25–31, i (Russian); English transl., J. Soviet Math. **49** (1990), no. 6, 1253–1258. MR975671 (90g:30037)

[3] ———, *Degenerate Stieltjes moment problem and associated J-inner polynomials*, Z. Anal. Anwendungen **14** (1995), no. 3, 441–468. MR1362524 (96k:47020)

[4] V.A. Bolotnikov, *On degenerate Hamburger moment problem and extensions of non-negative Hankel block matrices*, Integral Equations Operator Theory **25** (1996), no. 3, 253–276, DOI 10.1007/BF01262294. MR1395706 (97k:44011)

[5] G.-N. Chen and Y.-J. Hu, *The truncated Hamburger matrix moment problems in the nondegenerate and degenerate cases, and matrix continued fractions*, Linear Algebra Appl. **277** (1998), no. 1-3, 199–236, DOI 10.1016/S0024-3795(97)10076-3. MR1624548 (99j:44015)

[6] ———, *The Nevanlinna-Pick interpolation problems and power moment problems for matrix-valued functions. III. The infinitely many data case*, Linear Algebra Appl. **306** (2000), no. 1-3, 59–86, DOI 10.1016/S0024-3795(99)00241-4. MR1740433 (2001g:47029)

[7] ———, *A unified treatment for the matrix Stieltjes moment problem in both nondegenerate and degenerate cases*, J. Math. Anal. Appl. **254** (2001), no. 1, 23–34, DOI 10.1006/jmaa.2000.7195. MR1807884 (2001k:44008)

[8] D.L. Cohn, *Measure theory*, Birkhäuser Boston, Mass., 1980. MR578344 (81k:28001)

[9] V.K. Dubovoj, B. Fritzsche, and B. Kirstein, *Matricial version of the classical Schur problem*, Teubner-Texte zur Mathematik [Teubner Texts in Mathematics], vol. 129, B.G. Teubner Verlagsgesellschaft mbH, Stuttgart, 1992. With German, French and Russian summaries. MR1152328 (93e:47021)

[10] H. Dym, *On Hermitian block Hankel matrices, matrix polynomials, the Hamburger moment problem, interpolation and maximum entropy*, Integral Equations Operator Theory **12** (1989), no. 6, 757–812, DOI 10.1007/BF01196878. MR1018213 (91c:30065)

[11] Yu.M. Dyukarev, *Indeterminacy criteria for the Stieltjes matrix moment problem*, Mat. Zametki **75** (2004), no. 1, 71–88, DOI 10.1023/B:MATN.0000015022.02925.bd (Russian, with Russian summary); English transl., Math. Notes **75** (2004), no. 1-2, 66–82. MR2053150 (2005c:47016)

[12] Yu.M. Dyukarev, B. Fritzsche, B. Kirstein, and C. Mädler, *On truncated matricial Stieltjes type moment problems*, Complex Anal. Oper. Theory **4** (2010), no. 4, 905–951, DOI 10.1007/s11785-009-0002-8. MR2735313 (2011i:44009)

[13] Yu.M. Dyukarev, B. Fritzsche, B. Kirstein, C. Mädler, and H.C. Thiele, *On distinguished solutions of truncated matricial Hamburger moment problems*, Complex Anal. Oper. Theory **3** (2009), no. 4, 759–834, DOI 10.1007/s11785-008-0061-2. MR2570113

[14] B. Fritzsche and B. Kirstein, *Schwache Konvergenz nichtnegativ hermitescher Borelmaße*, Wiss. Z. Karl-Marx-Univ. Leipzig Math.-Natur. Reihe **37** (1988), no. 4, 375–398 (German, with English summary). MR975253 (90a:28015)

[15] B. Fritzsche, B. Kirstein, and C. Mädler, *On Hankel nonnegative definite sequences, the canonical Hankel parametrization, and orthogonal matrix polynomials*, Complex Anal. Oper. Theory **5** (2011), no. 2, 447–511, DOI 10.1007/s11785-010-0054-9. MR2805417

[16] ———, *On Matrix-Valued Herglotz-Nevanlinna Functions with an Emphasis on Particular Subclasses*, to appear in Math. Nachr.

[17] B. Fritzsche, B. Kirstein, C. Mädler, and T. Schwarz, *On the Concept of Invertibility for Sequences of Complex p × q-Matrices and its Application to Holomorphic p × q-Matrix-Valued Functions*. See this volume.

[18] ———, *On a Schur-Type Algorithm for Sequences of Complex p × q-Matrices and its Interrelations with the Canonical Hankel Parametrization*. See this volume.

[19] Y.-J. Hu and G.-N. Chen, *A unified treatment for the matrix Stieltjes moment problem*, Linear Algebra Appl. **380** (2004), 227–239, DOI 10.1016/j.laa.2003.10.012. MR2038751 (2004j:47029)

[20] I.V. Kovalishina, *J-expansive matrix-valued functions, and the classical problem of moments*, Akad. Nauk Armjan. SSR Dokl. **60** (1975), no. 1, 3–10 (Russian, with Armenian summary). MR0396962 (53 #822)

[21] ———, *Analytic theory of a class of interpolation problems*, Izv. Akad. Nauk SSSR Ser. Mat. **47** (1983), no. 3, 455–497 (Russian). MR703593 (84i:30043)

[22] OEIS Foundation Inc., *The On-Line Encyclopedia of Integer Sequences*, 2012. http://oeis.org/A000108.

[23] A. Rösch, *Über das Hamburger Momentenproblem im matriziellen Fall*, Diplomarbeit, Universität Leipzig, Leipzig, Nov. 2003.

[24] B. Simon, *The classical moment problem as a self-adjoint finite difference operator*, Adv. Math. **137** (1998), no. 1, 82–203, DOI 10.1006/aima.1998.1728. MR1627806 (2001e:47020)

[25] T.-J. Stieltjes, *Recherches sur les fractions continues*, Ann. Fac. Sci. Toulouse Sci. Math. Sci. Phys. **8** (1894), no. 4, J1–J122 (French). MR1508159

[26] ———, *Recherches sur les fractions continues [Suite et fin]*, Ann. Fac. Sci. Toulouse Sci. Math. Sci. Phys. **9** (1895), no. 1, A5–A47 (French). MR1508160

Bernd Fritzsche, Bernd Kirstein and Conrad Mädler
Mathematisches Institut
Universität Leipzig
Augustusplatz 10/11
D-04109 Leipzig, Germany
e-mail: fritzsche@math.uni-leipzig.de
       kirstein@math.uni-leipzig.de
       maedler@math.uni-leipzig.de

Operator Theory:
Advances and Applications, Vol. 226, 251–300
© 2012 Springer Basel

On Maximal Weight Solutions of a Moment Problem for Rational Matrix-valued Functions

Bernd Fritzsche, Bernd Kirstein and Andreas Lasarow

Abstract. We study some special solutions of a finite moment problem for rational matrix functions. Roughly speaking, we discuss a family of molecular non-negative Hermitian matrix-valued Borel measures on the unit circle with a special structure. As one of the main results, we will see that each member of this family offers an extremal property in the solution set of the moment problem concerning the weight assigned to some point of the open unit disk. Our approach uses the theory of orthogonal rational matrix functions.

Mathematics Subject Classification (2010). 30E05, 42C05, 44A60, 47A56.

Keywords. Non-negative Hermitian matrix measures, matrix moment problem, matricial Carathéodory functions, orthogonal rational matrix functions.

0. Introduction

The study of a moment problem for rational matrix functions, called Problem (R), underlies the present paper. This problem can be regarded as a generalization of the truncated trigonometric matrix moment problem (see Section 2 for the exact formulation). In a certain way, we continue in this paper the investigations of [23] and [24], where we began to explore some extremal questions within the solution set of Problem (R) in the nondegenerate case. For an introduction to Problem (R) and related topics we refer the reader to [17]–[19].

The considerations on Problem (R) are motivated by an extension of the theory of orthogonal rational functions on the unit circle drawn up by Bultheel, González-Vera, Hendriksen, and Njåstad in [8] (see also [5]–[7]) for the matrix case. Thus, the investigations here are connected with those in [20], [21], and [34] as well.

As explained in [5], para-orthogonal rational (complex-valued) functions can be used to obtain quadrature formulas on the unit circle. This can be done quite similar to the classical case of polynomials pointed out by Jones, Njåstad, and

The third author's research for this paper was supported by the German Research Foundation (Deutsche Forschungsgemeinschaft) on badge LA 1386/3–1.

Thron in [30]. In a sense, the treatments in [25] can be regarded as a first step toward an implementation of this method to the case of rational matrix-valued functions and the present work continues to build on this approach.

One of the central aims in [23] was to extend the construction of a particular solution of Problem (R) stated in [17, Theorem 31] to a whole family of solutions. In doing so, we have focussed on the nondegenerate case. Roughly speaking, this means that the given moment matrix $\mathbf{G}$ in Problem (R) has to satisfy an additional condition of regularity. In the present paper we will mostly assume that condition as well. The set of particular solutions studied here features then formal structural similarities to the family introduced in [23].

The family of solutions in [23] is parametrized by points w of the open unit disk of the complex plane which are not poles of the underlying rational matrix functions. In [23] and [24] we have shown that each member $F_{n,w}^{(\alpha)}$ of this family is extremal in several directions with respect to that point w which plays the role of the parameter. Moreover, in [24, Section 6] we have verified that the associated Riesz–Herglotz transform of the non-negative Hermitian matrix-valued Borel measure $F_{n,w}^{(\alpha)}$ is a rational matrix function which can be expressed in terms of orthogonal rational matrix functions on the unit circle. The paper on hand ties directly in with this insight in a certain way. In fact, for the nondegenerate case, the particular solutions analyzed below form a family of solutions of Problem (R), where the elements of the associated family of Riesz–Herglotz transforms admit similar representations. In contrast to [24, Section 6], instead of strictly contractive matrices, now some special unitary matrices appear in these descriptions as parameters. In particular, we will determine which extremal properties these special solutions of Problem (R) comprise.

An application of the theory of orthogonal rational matrix functions with respect to a non-negative Hermitian matrix Borel measure on the unit circle will be the basic strategy. The considerations below can be regarded as a generalization of the investigations in [22, Section 9], where the Szegő theory of orthogonal matrix polynomials is used to explore some extremal solutions of the matricial Carathéodory problem. Effectively, we will see that larger parts of the results presented there can be extended to the studied moment problem for rational matrix functions here. Beyond that, inspired by some thoughts in [12] and [35], we get somewhat more insight on the extremality as in [22, Section 9].

Concerning the truncated trigonometric moment problem in the scalar case, a result closely related to the extremal feature studied here was obtained by Geronimus (see [27, Theorem 20.1]). From today's point of view, these treatments of Geronimus are connected with those of para-orthogonal polynomials on the unit circle (see, e.g., [9], [28], [30], [40], and [41]). For a comprehensive exposition of the discussion of extremal questions (similar to [27, Theorem 20.1]) associated with several scalar power moment problems we refer to Kreĭn [32, Section 2 in Chapter I] as well as to Kreĭn and Nudelman [33, Section 3 in Chapter III].

With a view to former considerations of extremal questions in the matrix case like those in the present paper we turn to Arov [2], where underlying sets given by some specific linear fractional transformations were analyzed in this regard (see also Sakhnovich [39], where another method is applied).

In comparison with the classical investigations, the matrix case is somewhat more complicated. One of the main difficulties is as follows. The extremal problems in question offer in the scalar case solutions which are uniquely determined (see, e.g., [22, Proposition 9.17]). This fact can be deduced, for instance, from elementary results on para-orthogonal polynomials on the unit circle. In the matrix case, it is not that simple to see whether this uniqueness is available or not. In fact, this matter is not discussed in [2] (or in [39] and [12]) and in [22] only some special situations are mentioned, where uniqueness can be met. However, in [35] it is shown that the uniqueness relating to this kind of extremal questions within the solution set of the truncated trigonometric matrix moment problem for the nondegenerate case applies. In keeping with that, in the present paper we will verify a similar condition for uniqueness with respect to Problem (R).

The paper is organized as follows. For the reader's convenience we recall in Section 1 some notation which we have already used in previous papers on rational matrix functions and which we will use in the following as well. We will also review a characterization of the nondegenerate case from our former work (see Theorem 1.1) which is fundamental for that which follows.

In Section 2 we turn to canonical solutions of Problem (R) which were introduced in [25]. We first give in Section 2 the exact formulation of Problem (R) and then we recall the notion of canonical solutions of Problem (R). Here, we will concentrate on the nondegenerate case. In this case (cf. Theorem 2.1), the canonical solutions of Problem (R) can be parametrized by the set of unitary matrices (of appropriate size). The solutions $F_{n,u}^{(\alpha)}$ of Problem (R) which are of particular interest in the present paper form a subclass of this parametrized family, where points u of the unit circle come into consideration as parameters.

The special solutions $F_{n,u}^{(\alpha)}$ concerning points u of the unit circle will be introduced in Section 3. By definition we will see that the family of these matrix measures have formal structural similarities to the family of extremal solutions of Problem (R) discussed in [24] and complete this family to a certain extent. We will present some basic facts on $F_{n,u}^{(\alpha)}$ which are closely related to some on the extremal solutions in [23] and [24]. However, there are also some essential contrasts (see, e.g., Theorem 3.9 and Proposition 3.13).

In Section 4, we single out some features of $F_{n,u}^{(\alpha)}$ given by the exceptional position in the whole family of canonical solutions of Problem (R). The main result here (see Theorem 4.1) reveals conditions which can be used to characterize $F_{n,u}^{(\alpha)}$. We will also point out (cf. Proposition 4.15) that the solution $F_{n,u}^{(\alpha)}$ can be represented as the limit with respect to weak convergence of non-negative Hermitian matrix measures and the family of extremal solutions studied in [23] and [24].

Finally, for the nondegenerate case, we explain in Sections 5 and 6 that canonical solutions of Problem (R) comprise an extremal property within the solution set concerning the values of singletons. In this context, we will verify that the measure $F_{n,u}^{(\alpha)}$ has an exceptional position with respect to the point u which plays the role of the parameter. More precisely, we will see that for each point u of the unit circle the measure $F_{n,u}^{(\alpha)}$ maximizes the value of the matrix $F(\{u\})$ with respect to the Löwner semiordering of Hermitian matrices when F varies over the solution set of Problem (R) for the nondegenerate case. In the process, we will also realize that $F_{n,u}^{(\alpha)}$ is the unique solution of this extremal question (see Theorem 5.5).

Since an application of results on orthogonal rational matrix functions is the basic strategy, in keeping with [20] and [21] (see also [24]), we mostly focus on the situation that the poles of the underlying rational matrix functions must be in a sense in good position with respect to the unit circle. In particular, the measure $F_{n,u}^{(\alpha)}$ is exclusively well defined under this assumption. However, in Section 6 we will also discuss the somewhat more general case, where it is only assumed that the poles are not located on the unit circle. In fact, we will see that the maximal weight task is essentially the same (see, e.g., Theorem 6.1 and Proposition 6.3). We will also derive some conclusions from the above-mentioned extremal property of canonical solutions of Problem (R). For instance, we will obtain a result which shows some kind of universality concerning the pole location of the underlying rational matrix functions (see Proposition 6.6).

1. Preliminaries

Let $\mathbb{N}_0$ and $\mathbb{N}$ be the set of all non-negative integers and the set of all positive integers, respectively. For each $k \in \mathbb{N}_0$ and each $\tau \in \mathbb{N}_0$ or $\tau = +\infty$, let $\mathbb{N}_{k,\tau}$ be the set of all integers n for which $k \leq n \leq \tau$ holds. Furthermore, let

$$\mathbb{D} := \{w \in \mathbb{C} : |w| < 1\} \quad \text{and} \quad \mathbb{T} := \{z \in \mathbb{C} : |z| = 1\}$$

be the unit disk and the unit circle of the complex plane $\mathbb{C}$. The extended complex plane $\mathbb{C} \cup \{\infty\}$ will be designated by $\mathbb{C}_0$.

Throughout this paper, let p and q be positive integers. If $\mathfrak{X}$ is a nonempty set, then $\mathfrak{X}^{p \times q}$ stands for the set of all $p \times q$ matrices each entry of which belongs to $\mathfrak{X}$. If $\mathbf{A} \in \mathbb{C}^{p \times q}$, then the null space (resp., the range) of a $\mathbf{A}$ will be designated by $\mathcal{N}(\mathbf{A})$ (resp., $\mathcal{R}(\mathbf{A})$), the notation $\mathbf{A}^*$ means the adjoint matrix of $\mathbf{A}$, and $\mathbf{A}^+$ stands for the Moore–Penrose inverse of $\mathbf{A}$. For the null matrix which belongs to $\mathbb{C}^{p \times q}$ we will write $0_{p \times q}$. The identity matrix that belongs to $\mathbb{C}^{q \times q}$ will be denoted by $\mathbf{I}_q$. If $\mathbf{A} \in \mathbb{C}^{q \times q}$, then $\det \mathbf{A}$ is the determinant of $\mathbf{A}$ and $\operatorname{Re} \mathbf{A}$ (resp., $\operatorname{Im} \mathbf{A}$) stands for the real (resp., imaginary) part of $\mathbf{A}$, i.e., $\operatorname{Re} \mathbf{A} := \frac{1}{2}(\mathbf{A} + \mathbf{A}^*)$ (resp., $\operatorname{Im} \mathbf{A} := \frac{1}{2i}(\mathbf{A} - \mathbf{A}^*)$). We will write $\mathbf{A} \geq \mathbf{B}$ (resp., $\mathbf{A} > \mathbf{B}$) when $\mathbf{A}$ and $\mathbf{B}$ are Hermitian matrices (square and of the same size) such that $\mathbf{A} - \mathbf{B}$ is a non-negative (resp., positive) Hermitian matrix. Recall that a complex $p \times q$ matrix $\mathbf{A}$ is contractive (resp., strictly contractive) in the case of $\mathbf{I}_q \geq \mathbf{A}^* \mathbf{A}$ (resp., $\mathbf{I}_q > \mathbf{A}^* \mathbf{A}$). If $\mathbf{A}$ is a non-negative Hermitian matrix, then $\sqrt{\mathbf{A}}$ stands for the (unique) non-negative Hermitian matrix $\mathbf{B}$ given by $\mathbf{B}^2 = \mathbf{A}$.

Let $\tau \in \mathbb{N}$ or $\tau = +\infty$. Suppose that $(\alpha_j)_{j=1}^{\tau}$ is a sequence of numbers belonging to $\mathbb{C} \setminus \mathbb{T}$ and $n \in \mathbb{N}_{0,\tau}$. For $n = 0$, let $\pi_{\alpha,0}$ be the constant function on $\mathbb{C}_0$ with value 1 and $\mathcal{R}_{\alpha,0}$ be the set of all constant complex-valued functions defined on $\mathbb{C}_0$. Let $\mathbb{P}_{\alpha,0} := \emptyset$ and $\mathbb{Z}_{\alpha,0} := \emptyset$. If $n \in \mathbb{N}$, then let $\pi_{\alpha,n} : \mathbb{C} \to \mathbb{C}$ be defined by

$$\pi_{\alpha,n}(u) := \prod_{j=1}^{n} (1 - \overline{\alpha_j} u)$$

and let $\mathcal{R}_{\alpha,n}$ denote the set of all rational functions f which admit a representation

$$f = \frac{p_n}{\pi_{\alpha,n}}$$

with some polynomial $p_n : \mathbb{C} \to \mathbb{C}$ of degree not greater than n. Furthermore (using the convention $\frac{1}{0} := \infty$), let

$$\mathbb{P}_{\alpha,n} := \bigcup_{j=1}^{n} \left\{ \frac{1}{\overline{\alpha_j}} \right\} \qquad \text{and} \qquad \mathbb{Z}_{\alpha,n} := \bigcup_{j=1}^{n} \{\alpha_j\} .$$

Let $F \in \mathcal{M}_{\geq}^{q}(\mathbb{T}, \mathcal{B}_{\mathbb{T}})$, where $\mathcal{M}_{\geq}^{q}(\mathbb{T}, \mathcal{B}_{\mathbb{T}})$ stands for the set of all non-negative Hermitian $q \times q$ measures defined on the σ-algebra $\mathcal{B}_{\mathbb{T}}$ of all Borel subsets of $\mathbb{T}$. As already in [17], the right (resp., left) $\mathbb{C}^{q \times q}$-module $\mathcal{R}_{\alpha,n}^{q \times q}$ will be equipped by

$$\left(X, Y \right)_{F,r} := \int_{\mathbb{T}} (X(z))^* F(\mathrm{d}z) \, Y(z)$$

$$\left(\text{resp.,} \quad \left(X, Y \right)_{F,l} := \int_{\mathbb{T}} X(z) \, F(\mathrm{d}z) \, (Y(z))^* \right)$$

for all $X, Y \in \mathcal{R}_{\alpha,n}^{q \times q}$ with a matrix-valued inner product. (For details on the integration theory with respect to non-negative Hermitian $q \times q$ measures, we refer to Kats [31] and Rosenberg [36]–[38].) Moreover, if $(X_k)_{k=0}^{n}$ is a sequence of matrix-valued functions which belong to the right (resp., left) $\mathbb{C}^{q \times q}$-module $\mathcal{R}_{\alpha,n}^{q \times q}$, then we associate the non-negative Hermitian matrix

$$\mathbf{G}_{X,n}^{(F)} := \left(\int_{\mathbb{T}} (X_j(z))^* F(\mathrm{d}z) \, X_k(z) \right)_{j,k=0}^{n}$$

$$\left(\text{resp.,} \quad \mathbf{H}_{X,n}^{(F)} := \left(\int_{\mathbb{T}} X_j(z) \, F(\mathrm{d}z) \, (X_k(z))^* \right)_{j,k=0}^{n} \right).$$

Particular attention will be payed to the situation that some nondegeneracy condition holds. Recall that a matrix measure $F \in \mathcal{M}_{\geq}^{q}(\mathbb{T}, \mathcal{B}_{\mathbb{T}})$ is called *nondegenerate of order n* if the block Toeplitz matrix

$$\mathbf{T}_n^{(F)} := \left(\mathbf{c}_{j-k}^{(F)} \right)_{j,k=0}^{n}$$

is non-singular, where

$$\mathbf{c}_{\ell}^{(F)} := \int_{\mathbb{T}} z^{-\ell} F(\mathrm{d}z)$$

for some integer ℓ. We will write $\mathcal{M}_{\geq}^{q,n}(\mathbb{T}, \mathcal{B}_{\mathbb{T}})$ for the set of all $F \in \mathcal{M}_{\geq}^{q}(\mathbb{T}, \mathcal{B}_{\mathbb{T}})$ which are nondegenerate of order n.

The condition that $F \in \mathcal{M}_{\geq}^{q}(\mathbb{T}, \mathcal{B}_{\mathbb{T}})$ belongs to $\mathcal{M}_{\geq}^{q,n}(\mathbb{T}, \mathcal{B}_{\mathbb{T}})$ can be also expressed in terms of $\mathbf{G}_{X,n}^{(\alpha,F)}$ (resp., $\mathbf{H}_{X,n}^{(\alpha,F)}$) for some $X_0, X_1, \ldots, X_n \in \mathcal{R}_{\alpha,n}^{q \times q}$. In this regard, we recall now a result which is taken from [18] (there, Theorem 5.6).

Theorem 1.1. *Let $F \in \mathcal{M}_{\geq}^{q}(\mathbb{T}, \mathfrak{B}_{\mathbb{T}})$ and let $n \in \mathbb{N}_0$. Then the following statements are equivalent:*

(i) *$F \in \mathcal{M}_{\geq}^{q,n}(\mathbb{T}, \mathfrak{B}_{\mathbb{T}})$.*

(ii) *There exist a sequence $(\alpha_j)_{j=1}^{\infty}$ with $\alpha_j \in \mathbb{C} \setminus \mathbb{T}$, $j \in \mathbb{N}$, and a basis $(X_k)_{k=0}^{n}$ of the right $\mathbb{C}^{q \times q}$-module $\mathcal{R}_{\alpha,n}^{q \times q}$ such that the matrix $\mathbf{G}_{X,n}^{(\alpha,F)}$ is non-singular.*

(iii) *There exist a sequence $(\alpha_j)_{j=1}^{\infty}$ with $\alpha_j \in \mathbb{C} \setminus \mathbb{T}$, $j \in \mathbb{N}$, and a basis $(Y_k)_{k=0}^{n}$ of the left $\mathbb{C}^{q \times q}$-module $\mathcal{R}_{\alpha,n}^{q \times q}$ such that the matrix $\mathbf{H}_{Y,n}^{(\alpha,F)}$ is non-singular.*

(iv) *For each sequence $(\alpha_j)_{j=1}^{\infty}$ with $\alpha_j \in \mathbb{C} \setminus \mathbb{T}$, $j \in \mathbb{N}$, every basis $(X_k)_{k=0}^{n}$ of the right $\mathbb{C}^{q \times q}$-module $\mathcal{R}_{\alpha,n}^{q \times q}$, and every basis $(Y_k)_{k=0}^{n}$ of the left $\mathbb{C}^{q \times q}$-module $\mathcal{R}_{\alpha,n}^{q \times q}$, the complex matrices $\mathbf{G}_{X,n}^{(\alpha,F)}$ and $\mathbf{H}_{Y,n}^{(\alpha,F)}$ is positive Hermitian.*

Let $F \in \mathcal{M}_{\geq}^{q,n}(\mathbb{T}, \mathfrak{B}_{\mathbb{T}})$. In view of Theorem 1.1 (cf. [18, Theorem 5.8]) and [17, Theorem 10] one can see that by $(\mathcal{R}_{\alpha,n}^{q \times q}, (\cdot, \cdot)_{F,r})$ a right (resp., by $(\mathcal{R}_{\alpha,n}^{q \times q}, (\cdot, \cdot)_{F,l})$ a left) $\mathbb{C}^{q \times q}$-Hilbert module with reproducing kernel $K_{n;r}^{(\alpha,F)}$ (resp., $K_{n;l}^{(\alpha,F)}$) is given. In doing so, along the lines of the classical theory of reproducing kernels which goes back to the landmark paper [1] by Aronszajn, the machinery happens here in the context of matrix functions (cf., e.g., [3], [4], [14], and [29]). The reproducing kernels with respect to rational matrix functions under consideration were intensively studied in [17], [19], and [20] (see also [23]). The relevant kernel is a mapping from $(\mathbb{C}_0 \setminus \mathbb{P}_{\alpha,n}) \times (\mathbb{C}_0 \setminus \mathbb{P}_{\alpha,n})$ into $\mathbb{C}^{q \times q}$. For each $w \in \mathbb{C}_0 \setminus \mathbb{P}_{\alpha,n}$, let the matrix function $A_{n,w}^{(\alpha,F)} : \mathbb{C}_0 \setminus \mathbb{P}_{\alpha,n} \to \mathbb{C}^{q \times q}$ (resp., $C_{n,w}^{(\alpha,F)} : \mathbb{C}_0 \setminus \mathbb{P}_{\alpha,n} \to \mathbb{C}^{q \times q}$) be defined by

$$A_{n,w}^{(\alpha,F)}(v) := K_{n;r}^{(\alpha,F)}(v, w) \quad \left(\text{resp.,} \quad C_{n,w}^{(\alpha,F)}(v) := K_{n;l}^{(\alpha,F)}(w, v) \right). \tag{1.1}$$

Here, $K_{n;r}^{(\alpha,F)}$ (resp., $K_{n;l}^{(\alpha,F)}$) is the reproducing kernel relating to $(\mathcal{R}_{\alpha,n}^{q \times q}, (\cdot, \cdot)_{F,r})$ (resp., $(\mathcal{R}_{\alpha,n}^{q \times q}, (\cdot, \cdot)_{F,l})$) means that $A_{n,w}^{(\alpha,F)} \in \mathcal{R}_{\alpha,n}^{q \times q}$ (resp., $C_{n,w}^{(\alpha,F)} \in \mathcal{R}_{\alpha,n}^{q \times q}$) and that

$$\left(A_{n,w}^{(\alpha,F)}, X \right)_{F,r} = X(w) \quad \left(\text{resp.,} \quad \left(X, C_{n,w}^{(\alpha,F)} \right)_{F,l} = X(w) \right), \quad X \in \mathcal{R}_{\alpha,n}^{q \times q},$$

for each $w \in \mathbb{C}_0 \setminus \mathbb{P}_{\alpha,n}$. Note that (cf. [17, Remark 12]), if $X_0, X_1, \ldots, X_n$ is a basis of the right $\mathbb{C}^{q \times q}$-module $\mathcal{R}_{\alpha,n}^{q \times q}$ (resp., $Y_0, Y_1, \ldots, Y_n$ is a basis of the left $\mathbb{C}^{q \times q}$-module $\mathcal{R}_{\alpha,n}^{q \times q}$), then this kernel can be represented via

$$K_{n;r}^{(\alpha,F)}(v, w) = \Xi_n(v) \left(\mathbf{G}_{X,n}^{(F)} \right)^{-1} \left(\Xi_n(w) \right)^{*}$$

$$\left(\text{resp.,} \quad K_{n;l}^{(\alpha,F)}(w, v) = \left(\Upsilon_n(w) \right)^{*} \left(\mathbf{H}_{Y,n}^{(F)} \right)^{-1} \Upsilon_n(v) \right)$$

for all $v, w \in \mathbb{C}_0 \setminus \mathbb{P}_{\alpha,n}$, where

$$\Xi_n := \left(X_0, X_1, \ldots, X_n \right) \quad \left(\text{resp.,} \quad \Upsilon_n := \begin{pmatrix} Y_0 \\ Y_1 \\ \vdots \\ Y_n \end{pmatrix} \right).$$

In keeping with the studies in [24], we mostly focus the considerations below on the situation that the elements of the underlying sequence $(\alpha_j)_{j=1}^n$ must be in a sense in good position with respect to $\mathbb{T}$. In doing so, the notation $\mathcal{T}_1$ stands for the set of all sequences $(\alpha_j)_{j=1}^\infty$ of complex numbers which satisfy $\overline{\alpha_j}\alpha_k \neq 1$ for all $j, k \in \mathbb{N}$. For example, if $(\alpha_j)_{j=1}^\infty$ is a sequence of numbers belonging to $\mathbb{D}$, then $(\alpha_j)_{j=1}^\infty \in \mathcal{T}_1$. Moreover, if $(\alpha_j)_{j=1}^\infty \in \mathcal{T}_1$, then obviously $\alpha_j \notin \mathbb{T}$ for all $j \in \mathbb{N}$. Effectively, we choose $(\alpha_j)_{j=1}^\infty \in \mathcal{T}_1$ in the following to angle for applying some results of [20] and [21] on orthogonal rational matrix functions on $\mathbb{T}$.

Let $(\alpha_j)_{j=1}^\infty \in \mathcal{T}_1$. Furthermore, for each $j \in \mathbb{N}$, let

$$\eta_j := \begin{cases} -1 & \text{if } \alpha_j = 0 \\ \dfrac{\alpha_j}{|\alpha_j|} & \text{if } \alpha_j \neq 0 \end{cases}$$

and let the function $b_{\alpha_j} : \mathbb{C}_0 \setminus \{\frac{1}{\alpha_j}\} \to \mathbb{C}$ be given by

$$b_{\alpha_j}(u) := \begin{cases} \eta_j \dfrac{\alpha_j - u}{1 - \overline{\alpha_j}u} & \text{if } u \in \mathbb{C} \setminus \{\frac{1}{\alpha_j}\} \\ \dfrac{1}{|\alpha_j|} & \text{if } u = \infty. \end{cases}$$

Observe that, if $B_{\alpha,0}^{(q)}$ stands for the constant function on $\mathbb{C}_0$ with value $\mathbf{I}_q$ and if

$$B_{\alpha,k}^{(q)} := \left(\prod_{j=1}^k b_{\alpha_j}\right)\mathbf{I}_q, \quad k \in \mathbb{N}_{1,n},$$

then the system $B_{\alpha,0}^{(q)}, B_{\alpha,1}^{(q)}, \ldots, B_{\alpha,n}^{(q)}$ forms a basis of the right (resp., left) $\mathbb{C}^{q \times q}$-module $\mathcal{R}_{\alpha,n}^{q \times q}$ (see, e.g., [18, Section 2]). Thus, if $X \in \mathcal{R}_{\alpha,n}^{q \times q}$, then there are uniquely determined matrices $\mathbf{A}_0, \mathbf{A}_1, \ldots, \mathbf{A}_n$ belonging to $\mathbb{C}^{q \times q}$ such that

$$X = \sum_{j=0}^n \mathbf{A}_j B_{\alpha,j}^{(q)}.$$

The *reciprocal rational (matrix-valued) function* $X^{[\alpha,n]}$ of X with respect to $(\alpha_j)_{j=1}^\infty$ and n is given by

$$X^{[\alpha,n]} := \sum_{j=0}^n \mathbf{A}_{n-j}^* B_{\beta,j}^{(q)},$$

where $(\beta_j)_{j=1}^\infty$ is defined by $\beta_j := \alpha_{n+1-j}$ for each $j \in \mathbb{N}_{1,n}$ and $\beta_j := \alpha_j$ otherwise (cf. [20, Section 2]). This transform of a function X belonging to $\mathcal{R}_{\alpha,n}^{q \times q}$ into another $X^{[\alpha,n]}$ belonging to $\mathcal{R}_{\alpha,n}^{q \times q}$ is an essential tool in the theory of orthogonal rational matrix functions on $\mathbb{T}$ (see, e.g., [20] and [21]). With respect to the preceding case of orthogonal matrix polynomials on $\mathbb{T}$, we refer to [10] (see also [11, Section 3.6]).

2. Some basics on canonical solutions of Problem (R)

As a continuation of the studies in [17] (see also [19], [23], [24], and [25]), we consider the following rational matrix moment problem, called Problem (R).

Problem (R): *Let* $n \in \mathbb{N}$ *and* $\alpha_1, \alpha_2, \ldots, \alpha_n \in \mathbb{C} \setminus \mathbb{T}$. *Let* $\mathbf{G} \in \mathbb{C}^{(n+1)q \times (n+1)q}$ *and suppose that* $X_0, X_1, \ldots, X_n$ *is a basis of the right* $\mathbb{C}^{q \times q}$-*module* $\mathcal{R}_{\alpha,n}^{q \times q}$. *Describe the set* $\mathcal{M}[(\alpha_j)_{j=1}^n, \mathbf{G}; (X_k)_{k=0}^n]$ *of all measures* $F \in \mathcal{M}_{\geq}^q(\mathbb{T}, \mathfrak{B}_{\mathbb{T}})$ *such that* $\mathbf{G}_{X,n}^{(F)} = \mathbf{G}$.

The case $n = 0$ which includes only a condition on the weight $F(\mathbb{T})$ of some $F \in \mathcal{M}_{\geq}^q(\mathbb{T}, \mathfrak{B}_{\mathbb{T}})$ does not enter into Problem (R). However, the considerations in previous papers and below are intrinsically practicable for that case as well.

If $\alpha_j = 0$ for each $j \in \mathbb{N}_{1,n}$, then $\mathcal{R}_{\alpha,n}^{q \times q}$ is the set of all complex $q \times q$ matrix polynomials of degree not greater than n. Thus (cf. [17, Section 2]), Problem (R) leads to the truncated trigonometric matrix moment problem choosing X_k as the complex $q \times q$ matrix polynomial $E_{k,q}$ given, for each $k \in \mathbb{N}_{0,n}$, by

$$E_{k,q}(u) := u^k \mathbf{I}_q, \quad u \in \mathbb{C}.$$

To single out this case, if $F \in \mathcal{M}_{\geq}^q(\mathbb{T}, \mathfrak{B}_{\mathbb{T}})$, then we will use the notation $\mathcal{M}[\mathbf{T}_n^{(F)}]$ instead of $\mathcal{M}[(\alpha_j)_{j=1}^n, \mathbf{T}_n^{(F)}; (E_{k,q})_{k=0}^n]$.

Unless otherwise indicated, let $n \in \mathbb{N}$ and let $\alpha_1, \alpha_2, \ldots, \alpha_n \in \mathbb{C} \setminus \mathbb{T}$ in the following. Also, in view of Problem (R), let $\mathbf{G} \in \mathbb{C}^{(n+1)q \times (n+1)q}$ and suppose that $X_0, X_1, \ldots, X_n$ is a basis of the right $\mathbb{C}^{q \times q}$-module $\mathcal{R}_{\alpha,n}^{q \times q}$. We call a measure $F \in \mathcal{M}[(\alpha_j)_{j=1}^n, \mathbf{G}; (X_k)_{k=0}^n]$ a *canonical* solution (cf. [15, Section 5] and [25]) when

$$\operatorname{rank} \mathbf{T}_{n+1}^{(F)} = \operatorname{rank} \mathbf{G}. \tag{2.1}$$

(Note that the size of the matrix $\mathbf{T}_{n+1}^{(F)}$ in (2.1) is $(n+2)q \times (n+2)q$, whereas the size of $\mathbf{G}$ is $(n+1)q \times (n+1)q$.) In the case that $\mathcal{M}[(\alpha_j)_{j=1}^n, \mathbf{G}; (X_k)_{k=0}^n]$ is nonempty, such special solution always exists (cf. [25, Theorem 2.6]). Moreover (cf. [25, Theorem 2.10]), a solution F is a canonical solution in $\mathcal{M}[(\alpha_j)_{j=1}^n, \mathbf{G}; (X_k)_{k=0}^n]$ if and only if there is a finite subset Δ of $\mathbb{T}$ such that $F(\mathbb{T} \setminus \Delta) = 0_{q \times q}$ and

$$\sum_{z \in \Delta} \operatorname{rank} F(\{z\}) = \operatorname{rank} \mathbf{G}. \tag{2.2}$$

We now turn our consideration to the nondegenerate case, i.e., we presume that $\mathcal{M}[(\alpha_j)_{j=1}^n, \mathbf{G}; (X_k)_{k=0}^n]$ is nonempty, where $\mathbf{G}$ is some non-singular matrix. In this case, the canonical solutions of Problem (R) form a family which can be parametrized by the set of unitary $q \times q$ matrices. In doing so, the investigations in the subsequent section are based on the concrete formulas in [25, Theorem 3.6]. These are expressed in terms of orthogonal rational matrix functions on $\mathbb{T}$. For a better grasp, we recall in the following the relevant objects and [25, Theorem 3.6].

Let $(\alpha_j)_{j=1}^{\infty} \in \mathcal{T}_1$ and $F \in \mathcal{M}_{\geq}^q(\mathbb{T}, \mathfrak{B}_{\mathbb{T}})$. Furthermore, let $\tau \in \mathbb{N}_0$ or $\tau = +\infty$. A sequence $(Y_k)_{k=0}^{\tau}$ of functions with $Y_k \in \mathcal{R}_{\alpha,k}^{q \times q}$ for each $k \in \mathbb{N}_{0,\tau}$ is called a *left* (resp., *right*) *orthonormal system corresponding to* $(\alpha_j)_{j=1}^{\infty}$ *and* F *in the case of*

$$\left(Y_m, Y_s\right)_{F,l} = \delta_{m,s} \mathbf{I}_q \quad \left(\text{resp.,} \ \left(Y_m, Y_s\right)_{F,r} = \delta_{m,s} \mathbf{I}_q\right), \quad m, s \in \mathbb{N}_{0,\tau},$$

where $\delta_{m,s} := 1$ if $m = s$ and $\delta_{m,s} := 0$ otherwise (cf. [20, Definition 3.3]). If $(L_k)_{k=0}^{\tau}$ is a left orthonormal system and if $(R_k)_{k=0}^{\tau}$ is a right orthonormal system, respectively, corresponding to $(\alpha_j)_{j=1}^{\infty}$ and F, then we call $[(L_k)_{k=0}^{\tau}, (R_k)_{k=0}^{\tau}]$ a *pair of orthonormal systems corresponding to* $(\alpha_j)_{j=1}^{\infty}$ *and* F.

In what follows, let $\mathbf{L}_0$ and $\mathbf{R}_0$ be non-singular complex $q \times q$ matrices fulfilling

$$\mathbf{L}_0^* \mathbf{L}_0 = \mathbf{R}_0 \mathbf{R}_0^* \tag{2.3}$$

and (with $\tau \geq 1$) let $(\mathbf{U}_j)_{j=1}^{\tau}$ be a sequence of complex $2q \times 2q$ matrices such that

$$\mathbf{U}_j^* \mathbf{j}_{qq} \mathbf{U}_j = \begin{cases} \mathbf{j}_{qq} & \text{if } (1 - |\alpha_{j-1}|)(1 - |\alpha_j|) > 0 \\ -\mathbf{j}_{qq} & \text{if } (1 - |\alpha_{j-1}|)(1 - |\alpha_j|) < 0 \end{cases} \tag{2.4}$$

for each $j \in \mathbb{N}_{1,\tau}$, where $\mathbf{j}_{qq}$ is the $2q \times 2q$ signature matrix given by

$$\mathbf{j}_{qq} := \begin{pmatrix} \mathbf{I}_q & 0_{q \times q} \\ 0_{q \times q} & -\mathbf{I}_q \end{pmatrix}$$

and where we use for technical reasons the setting $\alpha_0 := 0$. Besides, we set

$$\rho_j := \begin{cases} \sqrt{\dfrac{1 - |\alpha_j|^2}{1 - |\alpha_{j-1}|^2}} & \text{if } (1 - |\alpha_{j-1}|)(1 - |\alpha_j|) > 0 \\[4mm] -\sqrt{\dfrac{|\alpha_j|^2 - 1}{1 - |\alpha_{j-1}|^2}} & \text{if } (1 - |\alpha_{j-1}|)(1 - |\alpha_j|) < 0 \end{cases}$$

for each $j \in \mathbb{N}_{1,\tau}$. As in [21, Section 3], we define sequences of rational matrix functions $(L_k)_{k=0}^{\tau}$ and $(R_k)_{k=0}^{\tau}$ by the initial conditions

$$L_0(u) = \mathbf{L}_0 \quad \text{and} \quad R_0(u) = \mathbf{R}_0 \tag{2.5}$$

for each $u \in \mathbb{C}$ and recursively by

$$\begin{pmatrix} L_j(u) \\ R_j^{[\alpha,j]}(u) \end{pmatrix} = \rho_j \frac{1 - \overline{\alpha_{j-1}} u}{1 - \alpha_j u} \mathbf{U}_j \begin{pmatrix} b_{\alpha_{j-1}}(u)\mathbf{I}_q & 0_{q \times q} \\ 0_{q \times q} & \mathbf{I}_q \end{pmatrix} \begin{pmatrix} L_{j-1}(u) \\ R_{j-1}^{[\alpha,j-1]}(u) \end{pmatrix}$$

for each $j \in \mathbb{N}_{1,\tau}$ and each $u \in \mathbb{C} \setminus \mathbb{P}_{\alpha,j}$. The pair $[(L_k)_{k=0}^{\tau}, (R_k)_{k=0}^{\tau}]$ of rational matrix functions is called the *pair which is left-generated by* $[(\alpha_j)_{j=1}^{\tau}; (\mathbf{U}_j)_{j=1}^{\tau}; \mathbf{L}_0, \mathbf{R}_0]$. Based on this concept and the bijective correspondence between such kind of pairs and pairs of orthonormal systems of rational matrix functions stated in [21] we will use the notation dual pair of orthonormal systems as explained below.

Let $F \in \mathcal{M}_{\geq}^{q,\tau}(\mathbb{T}, \mathfrak{B}_{\mathbb{T}})$ and let $[(L_k)_{k=0}^{\tau}, (R_k)_{k=0}^{\tau}]$ be a pair of orthonormal systems corresponding to $(\alpha_j)_{j=1}^{\tau}$ and F. Obviously (cf. [20, Remark 5.3]), there are non-singular complex $q \times q$ matrices $\mathbf{L}_0$ and $\mathbf{R}_0$ satisfying (2.3) and (2.5). Recalling that (2.3) implies the identity

$$(\mathbf{L}_0^{-*})^* \mathbf{L}_0^{-*} = \mathbf{R}_0^{-*}(\mathbf{R}_0^{-*})^*, \tag{2.6}$$

the pair $[(L_k^{\#})_{k=0}^0, (R_k^{\#})_{k=0}^0]$ which is given, for each $u \in \mathbb{C}$, by the formulas

$$L_0^{\#}(u) = \mathbf{L}_0^{-*} \quad \text{and} \quad R_0^{\#}(u) = \mathbf{R}_0^{-*}$$

is called the *dual pair of orthonormal systems corresponding to* $[(L_k)_{k=0}^0, (R_k)_{k=0}^0]$.
Now, let $\tau \geq 1$ and (by virtue of [21, Remark 3.5, Definition 3.6, Proposition 3.14, and Theorem 4.12]) let $(\mathbf{U}_j)_{j=1}^{\tau}$ be the unique sequence of complex $2q \times 2q$ matrices fulfilling (2.4) for each $j \in \mathbb{N}_{1,\tau}$ such that $[(L_k)_{k=0}^{\tau}, (R_k)_{k=0}^{\tau}]$ is the pair which is left-generated by $[(\alpha_j)_{j=1}^{\tau}; (\mathbf{U}_j)_{j=1}^{\tau}; \mathbf{L}_0, \mathbf{R}_0]$ with some non-singular complex $q \times q$ matrices $\mathbf{L}_0$ and $\mathbf{R}_0$ satisfying (2.3) and (2.5). Taking (2.6) into account and that, by setting $\mathbf{V}_j := \mathbf{j}_{qq} \mathbf{U}_j \mathbf{j}_{qq}$ for each $j \in \mathbb{N}_{1,\tau}$, (2.4) yields

$$\mathbf{V}_j^* \mathbf{j}_{qq} \mathbf{V}_j = \begin{cases} \mathbf{j}_{qq} & \text{if } (1 - |\alpha_{j-1}|)(1 - |\alpha_j|) > 0 \\ -\mathbf{j}_{qq} & \text{if } (1 - |\alpha_{j-1}|)(1 - |\alpha_j|) < 0, \end{cases}$$

the pair $[(L_k^{\#})_{k=0}^{\tau}, (R_k^{\#})_{k=0}^{\tau}]$ which is left-generated by $[(\alpha_j)_{j=1}^{\tau}; (\mathbf{V}_j)_{j=1}^{\tau}; \mathbf{L}_0^{-*}, \mathbf{R}_0^{-*}]$ is called the *dual pair of orthonormal systems corresponding to* $[(L_k)_{k=0}^{\tau}, (R_k)_{k=0}^{\tau}]$.

Recalling [24, Remark 4.1], a pair of orthonormal systems $[(L_k)_{k=0}^n, (R_k)_{k=0}^n]$ corresponding to $(\alpha_j)_{j=1}^{\infty}$ and some $F \in \mathcal{M}[(\alpha_j)_{j=1}^n, \mathbf{G}; (X_k)_{k=0}^n]$ is called a *pair of orthonormal systems corresponding to* $\mathcal{M}[(\alpha_j)_{j=1}^n, \mathbf{G}; (X_k)_{k=0}^n]$. We also speak of the *dual pair of orthonormal systems corresponding to that pair* $[(L_k)_{k=0}^n, (R_k)_{k=0}^n]$.

Suppose that $[(L_k)_{k=0}^n, (R_k)_{k=0}^n]$ is a pair of orthonormal systems corresponding to $\mathcal{M}[(\alpha_j)_{j=1}^n, \mathbf{G}; (X_k)_{k=0}^n]$ and let $[(L_k^{\#})_{k=0}^n, (R_k^{\#})_{k=0}^n]$ be the dual pair of orthonormal systems corresponding to $[(L_k)_{k=0}^n, (R_k)_{k=0}^n]$. The matrix functions

$$P_{n;\mathbf{U}}^{(\alpha)} := R_n^{[\alpha,n]} + b_{\alpha_n} \mathbf{U} L_n \quad \text{and} \quad P_{n;\mathbf{U}}^{(\alpha,\#)} := (R_n^{\#})^{[\alpha,n]} - b_{\alpha_n} \mathbf{U} L_n^{\#}$$

$$\left(\text{resp.,} \quad Q_{n;\mathbf{U}}^{(\alpha)} := L_n^{[\alpha,n]} + b_{\alpha_n} R_n \mathbf{U} \quad \text{and} \quad Q_{n;\mathbf{U}}^{(\alpha,\#)} := (L_n^{\#})^{[\alpha,n]} - b_{\alpha_n} R_n^{\#} \mathbf{U} \right) \tag{2.7}$$

with some unitary $q \times q$ matrix $\mathbf{U}$ will be of particular interest in the following. Because of (2.7) and the unitarity of the matrix $\mathbf{U}$ it follows that the rational matrix function $\Psi_{n;\mathbf{U}}^{(\alpha)} := (P_{n;\mathbf{U}}^{(\alpha)})^{-1} P_{n;\mathbf{U}}^{(\alpha,\#)}$ admits also the representation

$$\Psi_{n;\mathbf{U}}^{(\alpha)} = \left(\frac{1}{b_{\alpha_n}} \mathbf{U}^* R_n^{[\alpha,n]} + L_n \right)^{-1} \left(\frac{1}{b_{\alpha_n}} \mathbf{U}^* (R_n^{\#})^{[\alpha,n]} - L_n^{\#} \right).$$

Thus, [24, Lemma 6.7] implies that $\Psi_{n;\mathbf{U}}^{(\alpha)}$ is given by $\Psi_{n;\mathbf{U}}^{(\alpha)} = Q_{n;\mathbf{U}}^{(\alpha,\#)} (Q_{n;\mathbf{U}}^{(\alpha)})^{-1}$ and

$$\Psi_{n;\mathbf{U}}^{(\alpha)} = \left(\frac{1}{b_{\alpha_n}} (L_n^{\#})^{[\alpha,n]} \mathbf{U}^* - R_n^{\#} \right) \left(\frac{1}{b_{\alpha_n}} L_n^{[\alpha,n]} \mathbf{U}^* + R_n \right)^{-1}$$

as well, where the complex $q \times q$ matrices $P_{n;\mathbf{U}}^{(\alpha)}(v)$, $\frac{1}{b_{\alpha_n}(v)} \mathbf{U}^* R_n^{[\alpha,n]}(v) + L_n(v)$, $Q_{n;\mathbf{U}}^{(\alpha)}(v)$, and $\frac{1}{b_{\alpha_n}(v)} L_n^{[\alpha,n]}(v) \mathbf{U}^* + R_n(v)$ are non-singular for each $v \in \mathbb{D} \setminus \mathbb{P}_{\alpha,n}$.

A function $\Omega : \mathbb{D} \to \mathbb{C}^{q \times q}$ which is holomorphic in $\mathbb{D}$ and for which $\operatorname{Re} \Omega(w)$ is non-negative Hermitian for all $w \in \mathbb{D}$ is a $q \times q$ *Carathéodory function* (in $\mathbb{D}$). In particular, if $F \in \mathcal{M}_{\geq}^q(\mathbb{T}, \mathfrak{B}_{\mathbb{T}})$, then $\Omega : \mathbb{D} \to \mathbb{C}^{q \times q}$ defined by

$$\Omega(w) := \int_{\mathbb{T}} \frac{z + w}{z - w} F(dz)$$

is a $q \times q$ Carathéodory function (see, e.g., [11, Theorem 2.2.2]). We call this Ω the *Riesz–Herglotz transform* of (the non-negative Hermitian $q \times q$ Borel measure) F.

With a view to the rational matrix function $\Psi^{(\alpha)}_{n;\mathbf{U}}$ for some unitary $q \times q$ matrix $\mathbf{U}$, the following characterization of canonical solutions of Problem (R) for the nondegenerate case is proven in [25, Theorem 3.6].

Theorem 2.1. *Let $(\alpha_j)^{\infty}_{j=1} \in \mathcal{T}_1$ and let $n \in \mathbb{N}$. Let $X_0, X_1, \ldots, X_n$ be a basis of the right $\mathbb{C}^{q \times q}$-module $\mathcal{R}^{q \times q}_{\alpha,n}$ and suppose that $\mathbf{G}$ is a non-singular matrix such that $\mathcal{M}[(\alpha_j)^{n}_{j=1}, \mathbf{G}; (X_k)^{n}_{k=0}] \neq \emptyset$ holds. Furthermore, let $F \in \mathcal{M}^{q}_{\geq}(\mathbb{T}, \mathfrak{B}_{\mathbb{T}})$ and let Ω be the Riesz–Herglotz transform of F. Then the following statements are equivalent:*

(i) *F is a canonical solution in $\mathcal{M}[(\alpha_j)^{n}_{j=1}, \mathbf{G}; (X_k)^{n}_{k=0}]$.*

(ii) *There is a unitary $q \times q$ matrix $\mathbf{U}$ such that $\Omega(v) = \Psi^{(\alpha)}_{n;\mathbf{U}}(v)$ for $v \in \mathbb{D} \setminus \mathbb{P}_{\alpha,n}$.*

Moreover, if (i) *holds, then the unitary $q \times q$ matrix $\mathbf{U}$ in* (ii) *is uniquely determined.*

In view of Theorem 2.1 and [11, Theorem 2.2.2], if $\mathbf{U}$ is a unitary $q \times q$ matrix, then we will use the notation $F^{(\alpha)}_{n;\mathbf{U}}$ and $\Omega^{(\alpha)}_{n;\mathbf{U}}$. Here, $F^{(\alpha)}_{n;\mathbf{U}}$ stands for the uniquely determined measure belonging to $\mathcal{M}^{q}_{\geq}(\mathbb{T}, \mathfrak{B}_{\mathbb{T}})$ such that its Riesz–Herglotz transform $\Omega^{(\alpha)}_{n;\mathbf{U}}$ satisfies $\Omega^{(\alpha)}_{n;\mathbf{U}}(v) = \Psi^{(\alpha)}_{n;\mathbf{U}}(v)$ for each $v \in \mathbb{D} \setminus \mathbb{P}_{\alpha,n}$.

The representation of $\Omega^{(\alpha)}_{n;\mathbf{U}}$ which appears in Theorem 2.1 depends on the concrete choice of the pair of orthonormal systems $[(L_k)^{n}_{k=0}, (R_k)^{n}_{k=0}]$ corresponding to the solution set $\mathcal{M}[(\alpha_j)^{n}_{j=1}, \mathbf{G}; (X_k)^{n}_{k=0}]$. However, by [24, Lemma 6.6] one can see that this is not so essential. If we choose another pair of orthonormal systems $[(\tilde{L}_k)^{n}_{k=0}, (\tilde{R}_k)^{n}_{k=0}]$ corresponding to $\mathcal{M}[(\alpha_j)^{n}_{j=1}, \mathbf{G}; (X_k)^{n}_{k=0}]$ (with associated dual pair $[(\tilde{L}^{\#}_k)^{n}_{k=0}, (\tilde{R}^{\#}_k)^{n}_{k=0}]$), then the only possible difference is that another unitary $q \times q$ matrix $\tilde{\mathbf{U}}$ occurs in that representation of $\Omega^{(\alpha)}_{n;\mathbf{U}}$.

At the end of the present section is a brief statement concerning the elementary case $n = 0$, based on a constant function X_0 defined on $\mathbb{C}_0$ with a non-singular $q \times q$ matrix $\mathbf{X}_0$ as value and a positive Hermitian $q \times q$ matrix $\mathbf{G}$. Then (cf. [23, Remark 2.2]) there is a measure $F \in \mathcal{M}^{q}_{\geq}(\mathbb{T}, \mathfrak{B}_{\mathbb{T}})$ such that the equality

$$\int_{\mathbb{T}} (X_0(z))^* F(\mathrm{d}z) X_0(z) = \mathbf{G} \tag{2.8}$$

holds. In analogy to Problem (R) and (2.1) with some $n \in \mathbb{N}$, we call a measure $F \in \mathcal{M}^{q}_{\geq}(\mathbb{T}, \mathfrak{B}_{\mathbb{T}})$ fulfilling (2.8) a *canonical* solution of that problem when

$$\operatorname{rank} \mathbf{T}^{(F)}_1 = \operatorname{rank} \mathbf{G}. \tag{2.9}$$

(If $\mathbf{X}_0 = \mathbf{I}_q$, then we will also speak of a canonical solution in $\mathcal{M}[\mathbf{G}]$.)

Suppose that $F \in \mathcal{M}^{q}_{\geq}(\mathbb{T}, \mathfrak{B}_{\mathbb{T}})$. Similar to Theorem 2.1 (cf. [22, Remark 6.8 and Example 9.11]), one can find that F is a canonical solution with respect to

(2.8) and (2.9) if and only if the Riesz–Herglotz transform Ω of F admits

$$\Omega(v) = \sqrt{\mathbf{X}_0 \mathbf{G}^{-1} \mathbf{X}_0^*}^{-1} \mathbf{V} \begin{pmatrix} \frac{z_1+v}{z_1-v} & 0 & \cdots & 0 \\ 0 & \frac{z_2+v}{z_2-v} & \ddots & \vdots \\ \vdots & \ddots & \ddots & 0 \\ 0 & \cdots & 0 & \frac{z_q+v}{z_q-v} \end{pmatrix} \mathbf{V}^* \sqrt{\mathbf{X}_0 \mathbf{G}^{-1} \mathbf{X}_0^*}^{-1}$$

for each $v \in \mathbb{D}$ with some unitary $q \times q$ matrix $\mathbf{V}$ and (not necessarily pairwise different) points $z_1, z_2, \ldots, z_q$ belonging to $\mathbb{T}$.

Having fixed a $z \in \mathbb{T}$, the notation ε_z stands for the Dirac measure defined on the σ-algebra $\mathfrak{B}_\mathbb{T}$ with unit mass located at z. Furthermore, $\mathfrak{o}$ denotes the zero measure in $\mathcal{M}^1_{\geq}(\mathbb{T}, \mathfrak{B}_\mathbb{T})$. Thus, it follows that F is a canonical solution with respect to (2.8) and (2.9) if and only if F admits the representation

$$F = \sqrt{\mathbf{X}_0 \mathbf{G}^{-1} \mathbf{X}_0^*}^{-1} \mathbf{V} \begin{pmatrix} \varepsilon_{z_1} & \mathfrak{o} & \cdots & \mathfrak{o} \\ \mathfrak{o} & \varepsilon_{z_2} & \ddots & \vdots \\ \vdots & \ddots & \ddots & \mathfrak{o} \\ \mathfrak{o} & \cdots & \mathfrak{o} & \varepsilon_{z_q} \end{pmatrix} \mathbf{V}^* \sqrt{\mathbf{X}_0 \mathbf{G}^{-1} \mathbf{X}_0^*}^{-1} \tag{2.10}$$

with a unitary $q \times q$ matrix $\mathbf{V}$ and $z_1, z_2, \ldots, z_q \in \mathbb{T}$. So, one can see that for a canonical solution in this context each case of r mass points with $r \in \mathbb{N}_{1,q}$ is possible.

3. A special family of solutions of Problem (R)

In a certain way, the studies in this paper can be regarded as a continuation of those in [23] and [24] on a class of extremal solutions of Problem (R). In particular, for the nondegenerate case, the set of special solutions which plays a key role in this paper features formal structural similarities to the family introduced in [23].

Unless otherwise indicated, let $n \in \mathbb{N}$ and let $\alpha_1, \alpha_2, \ldots, \alpha_n \in \mathbb{C} \setminus \mathbb{T}$ in the following. Furthermore, in view of Problem (R), let $X_0, X_1, \ldots, X_n$ be a basis of the right $\mathbb{C}^{q \times q}$-module $\mathcal{R}_{\alpha,n}^{q \times q}$ and suppose that $\mathbf{G}$ is a non-singular complex $(n+1)q \times (n+1)q$ matrix such that $\mathcal{M}[(\alpha_j)_{j=1}^n, \mathbf{G}; (X_k)_{k=0}^n]$ is nonempty. By [23, Theorem 3.4] we know that, if $w \in \mathbb{D} \setminus \mathbb{P}_{\alpha,n}$, then $F_{n,w}^{(\alpha)} : \mathfrak{B}_\mathbb{T} \to \mathbb{C}^{q \times q}$ defined by

$$F_{n,w}^{(\alpha)}(B) := \frac{1}{2\pi} \int_B \frac{1-|w|^2}{|z-w|^2} \left(A_{n,w}^{(\alpha)}(z)\right)^{-*} A_{n,w}^{(\alpha)}(w) \left(A_{n,w}^{(\alpha)}(z)\right)^{-1} \underline{\lambda}(dz)$$

belongs to $\mathcal{M}[(\alpha_j)_{j=1}^n, \mathbf{G}; (X_k)_{k=0}^n]$, where $\underline{\lambda}$ stands for the linear Lebesgue measure defined on $\mathfrak{B}_\mathbb{T}$ and where

$$A_{n,w}^{(\alpha)} := \left(X_0, X_1, \ldots, X_n\right) \mathbf{G}^{-1} \left(X_0(w), X_1(w), \ldots, X_n(w)\right)^*. \tag{3.1}$$

We also use the setting given by (3.1) in the general case that $w \in \mathbb{C}_0 \setminus \mathbb{P}_{\alpha,n}$. Note that, due to (1.1) and (3.1), if $F \in \mathcal{M}[(\alpha_j)_{j=1}^n, \mathbf{G}; (X_k)_{k=0}^n]$, then it follows that

$$A_{n,w}^{(\alpha,F)} = A_{n,w}^{(\alpha)}, \quad w \in \mathbb{C}_0 \setminus \mathbb{P}_{\alpha,n}. \tag{3.2}$$

In particular (cf. [17, Remark 14]), if $\mathcal{M}[(\alpha_j)_{j=1}^n, \mathbf{G}; (X_k)_{k=0}^n] \neq \emptyset$, then

$$A_{n,w}^{(\alpha)}(w) > 0_{q \times q}, \quad w \in \mathbb{C}_0 \setminus \mathbb{P}_{\alpha,n}. \tag{3.3}$$

In this paper, the value $A_{n,u}^{(\alpha)}(u)$ with $u \in \mathbb{T}$ will be a matter of particular interest.

By using the above settings for $n = 0$ with $w \in \mathbb{D}$ as well, $A_{0,w}^{(\alpha)}$ is the constant function with value $\mathbf{X}_0 \mathbf{G}^{-1} \mathbf{X}_0^*$ and the matrix measure $F_{0,w}^{(\alpha)}$ is given by

$$F_{0,w}^{(\alpha)}(B) = \frac{1}{2\pi} \int_B \frac{1 - |w|^2}{|z - w|^2} \mathbf{X}_0^{-*} \mathbf{G} \mathbf{X}_0^{-1} \lambda(\mathrm{d}z), \quad B \in \mathfrak{B}_{\mathbb{T}},$$

whereby (2.8) holds by choosing F as $F_{0,w}^{(\alpha)}$ (cf. [23, Remark 2.2 and Remark 3.5]).

Below, let $(\alpha_j)_{j=1}^\infty \in \mathcal{T}_1$ and $n \in \mathbb{N}$. As in Section 2, let $[(L_k)_{k=0}^n, (R_k)_{k=0}^n]$ be a pair of orthonormal systems corresponding to $\mathcal{M}[(\alpha_j)_{j=1}^n, \mathbf{G}; (X_k)_{k=0}^n]$ and let $[(L_k^\#)_{k=0}^n, (R_k^\#)_{k=0}^n]$ be the dual pair of orthonormal systems corresponding to $[(L_k)_{k=0}^n, (R_k)_{k=0}^n]$. Following [24, Lemma 3.11], special attention will be come up to the rational matrix function Θ_n which is defined by

$$\Theta_n := \begin{cases} b_{\alpha_n} (L_n^{[\alpha,n]})^{-1} R_n & \text{if } \alpha_n \in \mathbb{D} \\ \dfrac{1}{b_{\alpha_n}} R_n^{-1} L_n^{[\alpha,n]} & \text{if } \alpha_n \in \mathbb{C} \setminus \mathbb{D}. \end{cases} \tag{3.4}$$

Suppose that $w \in \mathbb{D} \setminus \mathbb{P}_{\alpha,n}$. Because of [24, Lemma 5.7 and Theorem 5.8] one can see that the Riesz–Herglotz transform $\Omega_{n,w}^{(\alpha)}$ of the matrix measure $F_{n,w}^{(\alpha)}$ admits, for each $v \in \mathbb{D} \setminus \mathbb{P}_{\alpha,n}$, the representations

$$\Omega_{n,w}^{(\alpha)}(v) = \left((L_n^\#)^{[\alpha,n]}(v) - b_{\alpha_n}(v) R_n^\#(v) \mathbf{W} \right) \left(L_n^{[\alpha,n]}(v) + b_{\alpha_n}(v) R_n(v) \mathbf{W} \right)^{-1},$$

$$\Omega_{n,w}^{(\alpha)}(v) = \left(R_n^{[\alpha,n]}(v) + b_{\alpha_n}(v) \mathbf{W} L_n(v) \right)^{-1} \left((R_n^\#)^{[\alpha,n]}(v) - b_{\alpha_n}(v) \mathbf{W} L_n^\#(v) \right)$$

in the case of $\alpha_n \in \mathbb{D}$ and otherwise

$$\Omega_{n,w}^{(\alpha)}(v) = \left(\frac{1}{b_{\alpha_n}(v)} (L_n^\#)^{[\alpha,n]}(v) \mathbf{W} - R_n^\#(v) \right) \left(\frac{1}{b_{\alpha_n}(v)} L_n^{[\alpha,n]}(v) \mathbf{W} + R_n(v) \right)^{-1},$$

$$\Omega_{n,w}^{(\alpha)}(v) = \left(\frac{1}{b_{\alpha_n}(v)} \mathbf{W} R_n^{[\alpha,n]}(v) + L_n(v) \right)^{-1} \left(\frac{1}{b_{\alpha_n}(v)} \mathbf{W} (R_n^\#)^{[\alpha,n]}(v) - L_n^\#(v) \right)$$

with $\mathbf{W} := -\left(\Theta_n(w) \right)^*$, wherein the involved inverses exist. Furthermore, for some $u \in \mathbb{T}$, from [24, Lemma 3.11] we know that the matrices $L_n(u)$, $R_n(u)$, $L_n^{[\alpha,n]}(u)$, and $R_n^{[\alpha,n]}(u)$ are non-singular and that $\Theta_n(u)$ is a unitary $q \times q$ matrix, where

$$\Theta_n = \begin{cases} b_{\alpha_n} L_n (R_n^{[\alpha,n]})^{-1} & \text{if } \alpha_n \in \mathbb{D} \\ \dfrac{1}{b_{\alpha_n}} R_n^{[\alpha,n]} L_n^{-1} & \text{if } \alpha_n \in \mathbb{C} \setminus \mathbb{D}. \end{cases}$$

This fact along with a comparison of the formulas of $\Omega_{n,w}^{(\alpha)}$ for some $w \in \mathbb{D} \setminus \mathbb{P}_{\alpha,n}$ mentioned above and the representations of the Riesz–Herglotz transform of a canonical solution of Problem (R) which are the result of Theorem 2.1 forms the starting point for the further considerations in the paper on hand.

In the following, the notation $F_{n,u}^{(\alpha)}$ stands for the uniquely determined measure belonging to $\mathcal{M}_{\geq}^q(\mathbb{T}, \mathfrak{B}_{\mathbb{T}})$ such that its Riesz–Herglotz transform $\Omega_{n,u}^{(\alpha)}$ satisfies $\Omega_{n,u}^{(\alpha)}(v) = \Psi_{n;\mathbf{U}}^{(\alpha)}(v)$ for each $v \in \mathbb{D} \setminus \mathbb{P}_{\alpha,n}$, where

$$\Psi_{n;\mathbf{U}}^{(\alpha)} := (P_{n;\mathbf{U}}^{(\alpha)})^{-1} P_{n;\mathbf{U}}^{(\alpha,\#)}, \quad \mathbf{U} := -(\Theta_n(u))^*, \tag{3.5}$$

is the rational matrix function defined based on (2.7) and some $u \in \mathbb{T}$. In particular, by Theorem 2.1 and [24, Lemma 3.11] we get the following.

Remark 3.1. If $u \in \mathbb{T}$, then $F_{n,u}^{(\alpha)}$ is a canonical solution in $\mathcal{M}[(\alpha_j)_{j=1}^n, \mathbf{G}; (X_k)_{k=0}^n]$ (for the nondegenerate case with $(\alpha_j)_{j=1}^\infty \in \mathcal{T}_1$).

Remark 3.2. Let $w \in \mathbb{D} \setminus \mathbb{P}_{\alpha,n}$ and let $u \in \mathbb{T}$. Recalling [20, Remark 2.6], in view of (3.2) and [17, Theorem 25] one can realize that $(A_{n,w}^{(\alpha)})^{[\alpha,n]}(u) \neq 0_{q \times q}$. This implies $\Theta_n(w) \neq \Theta_n(u)$ because of (3.4) and [24, Lemma 3.11]. Thus, by using [24, Theorem 5.8] it follows that $F_{n,w}^{(\alpha)} \neq F_{n,u}^{(\alpha)}$ (see also [24, Proposition 3.12]).

The family $(F_{n,u}^{(\alpha)})_{u \in \mathbb{T}}$ of canonical solutions will play a key role in what follows. Because of (3.5) and (3.4) there are structural similarities between the solutions $F_{n,u}^{(\alpha)}$ and $F_{n,w}^{(\alpha)}$ with $u \in \mathbb{T}$ and $w \in \mathbb{D} \setminus \mathbb{P}_{\alpha,n}$. In this regard, the following results are quite similar to some in [24] concerning $F_{n,w}^{(\alpha)}$ with $w \in \mathbb{D} \setminus \mathbb{P}_{\alpha,n}$.

Remark 3.3. Let $u_1, u_2 \in \mathbb{T}$. Recalling Theorem 2.1 and [24, Lemma 3.11], because of (3.5) and (3.4) one can see that $F_{n,u_1}^{(\alpha)} = F_{n,u_2}^{(\alpha)}$ is satisfied if and only if the equality $\Theta_n(u_1) = \Theta_n(u_2)$ holds (cf. [24, Proposition 3.12]).

From now on, for some $w \in \mathbb{C}_0 \setminus \mathbb{P}_{\alpha,n}$, starting from the given data in Problem (R) we use, besides the notation $A_{n,w}^{(\alpha)}$ given by (3.1), also

$$C_{n,w}^{(\alpha)} := \begin{pmatrix} X_0^{[\alpha,n]}(w) \\ X_1^{[\alpha,n]}(w) \\ \vdots \\ X_n^{[\alpha,n]}(w) \end{pmatrix}^* \mathbf{G}^{-1} \begin{pmatrix} X_0^{[\alpha,n]} \\ X_1^{[\alpha,n]} \\ \vdots \\ X_n^{[\alpha,n]} \end{pmatrix}. \tag{3.6}$$

Comparing now (3.6) with (1.1), by using [24, Remark 2.1] one can realize that, if $F \in \mathcal{M}[(\alpha_j)_{j=1}^n, \mathbf{G}; (X_k)_{k=0}^n]$, then

$$C_{n,w}^{(\alpha,F)} = C_{n,w}^{(\alpha)}, \quad w \in \mathbb{C}_0 \setminus \mathbb{P}_{\alpha,n}. \tag{3.7}$$

In particular (cf. (3.3) and [19, Proposition 11]), we have

$$C_{n,w}^{(\alpha)}(w) > 0_{q \times q}, \quad w \in \mathbb{C}_0 \setminus \mathbb{P}_{\alpha,n}. \tag{3.8}$$

Note also that (3.2) and (3.7) imply (cf. [19, Lemma 5]) the relation

$$C_{n,w}^{(\alpha)}(v) = (B_{\alpha,n}^{(q)}(w))^* A_{n,\frac{1}{v}}^{(\alpha)}(\tfrac{1}{w}) B_{\alpha,n}^{(q)}(v), \quad v, w \in \mathbb{C}_0 \setminus (\mathbb{P}_{\alpha,n} \cup \mathbb{Z}_{\alpha,n}). \tag{3.9}$$

Remark 3.4. Let $u_1, u_2 \in \mathbb{T}$. Recalling (3.2) and (3.7), from Remark 3.3 and [24, Lemma 3.11] it follows that $F_{n,u_1}^{(\alpha)} = F_{n,u_2}^{(\alpha)}$ holds (cf. [23, Proposition 6.4]) if and only if $u_1 = u_2$ or $\big(A_{n,u_1}^{(\alpha)}\big)^{[\alpha,n]}(u_2) = 0_{q \times q}$ (resp., $\big(C_{n,u_1}^{(\alpha)}\big)^{[\alpha,n]}(u_2) = 0_{q \times q}$). Hence, in view of $u_1, u_2 \in \mathbb{T}$ and [20, Remark 2.6], we get that $F_{n,u_1}^{(\alpha)} = F_{n,u_2}^{(\alpha)}$ is satisfied if and only if $u_1 = u_2$ or $A_{n,u_1}^{(\alpha)}(u_2) = 0_{q \times q}$ (resp., $C_{n,u_1}^{(\alpha)}(u_2) = 0_{q \times q}$).

Remark 3.5. By Remark 3.4 and the fundamental theorem of algebra we get that, for every sequence $(u_k)_{k=0}^{n+1}$ of pairwise different points belonging to $\mathbb{T}$, the sequence $(F_{n,u_k}^{(\alpha)})_{k=0}^{n+1}$ contains at least two different measures (cf. [23, Corollary 6.5]).

In a certain way, having fixed some $u \in \mathbb{T}$, the values of the $q \times q$ Carathéodory function $\Omega_{n,u}^{(\alpha)}$ are unique within the possible values of any Riesz–Herglotz transform associated with a solution of Problem (R). In fact, we get the following characterization (which is somewhat stronger than the corresponding result in [24, Corollary 5.9] with respect to $F_{n,w}^{(\alpha)}$ for some $w \in \mathbb{D} \setminus \mathbb{P}_{\alpha,n}$). In doing so, for some holomorphic function $\Omega : \mathbb{D} \to \mathbb{C}^{q \times q}$, the function $\widehat{\Omega} : \mathbb{C} \setminus \mathbb{T} \to \mathbb{C}^{q \times q}$ is defined by

$$\widehat{\Omega}(v) := \begin{cases} \Omega(v) & \text{if } v \in \mathbb{D} \\ -\big(\Omega(\tfrac{1}{\overline{v}})\big)^* & \text{if } v \in \mathbb{C} \setminus (\mathbb{D} \cup \mathbb{T}) \end{cases}$$

and $\widehat{\Omega}^{(t)}(v)$ means the value of the tth derivative of $\widehat{\Omega}$ at $v \in \mathbb{C} \setminus \mathbb{T}$ with $t \in \mathbb{N}_0$.

Proposition 3.6. *Let $u \in \mathbb{T}$ and suppose that Ω is the Riesz–Herglotz transform of some $F \in \mathcal{M}[(\alpha_j)_{j=1}^n, \mathbf{G}; (X_k)_{k=0}^n]$. Furthermore, let m be the number of pairwise different points amongst $(\alpha_j)_{j=0}^n$ with $\alpha_0 := 0$ and let $\gamma_1, \gamma_2, \ldots, \gamma_m$ denote these points, where l_k stands for the number of occurrences of γ_k in $(\alpha_j)_{j=0}^n$ for each $k \in \mathbb{N}_{1,m}$. Then the following statements are equivalent:*

(i) $F = F_{n,u}^{(\alpha)}$.

(ii) *There exists some $v \in \mathbb{C} \setminus (\mathbb{T} \cup \mathbb{P}_{\alpha,n} \cup \mathbb{Z}_{\alpha,n})$ such that $\widehat{\Omega}(v) = \widehat{\Omega_{n,u}^{(\alpha)}}(v)$.*

(iii) *There exists some $k \in \mathbb{N}_{1,m}$ such that $\widehat{\Omega}^{(l_k)}(\gamma_k) = \big(\widehat{\Omega_{n,u}^{(\alpha)}}\big)^{(l_k)}(\gamma_k)$.*

(iv) *For each $k \in \mathbb{N}_{1,m}$, the equality $\widehat{\Omega}^{(l_k)}(\gamma_k) = \big(\widehat{\Omega_{n,u}^{(\alpha)}}\big)^{(l_k)}(\gamma_k)$ holds.*

Proof. Taking Remark 3.1 into account, the assertion is an immediate consequence of [25, Proposition 2.8]. $\square$

Remark 3.7. Let $u \in \mathbb{T}$. Taking Theorem 2.1 and [24, Lemma 3.11] into account, because of (3.5) and (3.4) we obtain, for each $v \in \mathbb{D} \setminus \mathbb{P}_{\alpha,n}$, that

$$\Omega_{n,u}^{(\alpha)}(v) = \Big(\big(R_n^{[\alpha,n]}(u)\big)^* R_n^{[\alpha,n]}(v) - \overline{b_{\alpha_n}(u)} b_{\alpha_n}(v)\big(L_n(u)\big)^* L_n(v)\Big)^{-1}$$
$$\cdot \Big(\big(R_n^{[\alpha,n]}(u)\big)^* (R_n^\#)^{[\alpha,n]}(v) + \overline{b_{\alpha_n}(u)} b_{\alpha_n}(v)\big(L_n(u)\big)^* L_n^\#(v)\Big),$$

$$\Omega_{n,u}^{(\alpha)}(v) = \Big((L_n^\#)^{[\alpha,n]}(v)\big(L_n^{[\alpha,n]}(u)\big)^* + b_{\alpha_n}(v)\overline{b_{\alpha_n}(u)} R_n^\#(v)\big(R_n(u)\big)^*\Big)$$
$$\cdot \Big(L_n^{[\alpha,n]}(v)\big(L_n^{[\alpha,n]}(u)\big)^* - b_{\alpha_n}(v)\overline{b_{\alpha_n}(u)} R_n(v)\big(R_n(u)\big)^*\Big)^{-1},$$

wherein the involved inverse matrices exist. So, applying the Christoffel–Darboux formulas for orthogonal rational matrix functions (use [20, Corollary 5.5] and [34, Proposition 3.1]) yields, for each $v \in \mathbb{D} \setminus \mathbb{P}_{\alpha,n}$, the identities

$$\Omega_{n,u}^{(\alpha)}(v) = \left(\sum_{k=0}^{n} (L_k(u))^* L_k(v) \right)^{-1} \left(\frac{2}{1 - \overline{u}v} I_q - \sum_{k=0}^{n} (L_k(u))^* L_k^{\#}(v) \right),$$

$$\Omega_{n,u}^{(\alpha)}(v) = \left(\frac{2}{1 - v\overline{u}} I_q - \sum_{k=0}^{n} R_k^{\#}(v) (R_k(u))^* \right) \left(\sum_{k=0}^{n} R_k(v) (R_k(u))^* \right)^{-1}.$$

In contrast to a (general) canonical solution of Problem (R) in the nondegenerate case, by [24, Lemma 6.6] one can see that the representations due to Theorem 2.1 and Remark 3.7 of the Riesz–Herglotz transform $\Omega_{n,u}^{(\alpha)}$ of a measure $F_{n,u}^{(\alpha)}$ with some $u \in \mathbb{T}$ do not depend on the concrete choice of the pair of orthonormal systems $[(L_k)_{k=0}^{n}, (R_k)_{k=0}^{n}]$ corresponding to $\mathcal{M}[(\alpha_j)_{j=1}^{n}, \mathbf{G}; (X_k)_{k=0}^{n}]$.

We are going now to give some information on the structure of $F_{n,u}^{(\alpha)}$ with some $u \in \mathbb{T}$. But first, we present an auxiliary result (in a more general context) on the nondegeneracy of molecular non-negative Hermitian $q \times q$ matrix measures on $\mathfrak{B}_{\mathbb{T}}$. In doing so, we will recall that, having fixed a $z \in \mathbb{T}$, we use ε_z to denote the Dirac measure defined on the σ-algebra $\mathfrak{B}_{\mathbb{T}}$ with unit mass located at z.

Lemma 3.8. *Let* $m, r \in \mathbb{N}$*. Suppose that* $u_0, u_1, \ldots, u_r \in \mathbb{T}$ *are pairwise different and let* $(\mathbf{A}_s)_{s=0}^{r}$ *be a sequence of non-negative Hermitian* $q \times q$ *matrices, where* $\mathbf{A}_0$ *is non-singular. Furthermore, let* F *and* $\widetilde{F}$ *be the matrix measures fulfilling*

$$F = \sum_{s=0}^{r} \varepsilon_{u_s} \mathbf{A}_s \qquad and \qquad \widetilde{F} = \sum_{s=1}^{r} \varepsilon_{u_s} \mathbf{A}_s.$$

Then the following statements are equivalent:

(i) $\mathbf{G}_{X,m}^{(F)}$ *(resp.,* $\mathbf{H}_{X,m}^{(F)}$*) is non-singular for some* $(\alpha_j)_{j=1}^{\infty}$ *with* $\alpha_j \in \mathbb{C} \setminus \mathbb{T}$ *and basis* $X_0, X_1, \ldots, X_m$ *of the right (resp., left)* $\mathbb{C}^{q \times q}$*-module* $\mathcal{R}_{\alpha,m}^{q \times q}$*.*

(ii) $\mathbf{G}_{Y,m-1}^{(\widetilde{F})}$ *(resp.,* $\mathbf{H}_{Y,m-1}^{(\widetilde{F})}$*) is non-singular for some* $(\alpha_j)_{j=1}^{\infty}$ *with* $\alpha_j \in \mathbb{C} \setminus \mathbb{T}$ *and basis* $Y_0, Y_1, \ldots, Y_{m-1}$ *of the right (resp., left)* $\mathbb{C}^{q \times q}$*-module* $\mathcal{R}_{\alpha,m-1}^{q \times q}$*.*

Proof. Because of Theorem 1.1 one can see that (i) is equivalent to

$$\det \mathbf{T}_m^{(F)} \neq 0, \tag{3.10}$$

whereas (ii) is satisfied if and only if

$$\det \mathbf{T}_{m-1}^{(\widetilde{F})} \neq 0 \tag{3.11}$$

holds. Therefore, if we have shown that (3.10) is equivalent to (3.11), then the assertion follows. However, that (3.10) implies (3.11) is already proven with [35, Lemma 3.1]. It remains to be shown that (3.11) leads to (3.10) as well. Thus, we suppose now that (3.11) holds. Furthermore, we assume that

$$\det \mathbf{T}_m^{(F)} = 0$$

is fulfilled. Hence (see, e.g., [18, Theorem 5.8]), we find a complex $q \times q$ matrix polynomial P of degree not greater than m such that $P(v_0)$ is not equal to the zero matrix $0_{q \times q}$ for some $v_0 \in \mathbb{C}$, but

$$\int_{\mathbb{T}} \left(P(z)\right)^* F(\mathrm{d}z)\, P(z) = 0_{q \times q}.$$

By the choice of the matrix measure F, the outcome of this is

$$\sum_{s=0}^{r} \left(P(u_s)\right)^* \mathbf{A}_s P(u_s) = 0_{q \times q}.$$

Since the matrices $\mathbf{A}_0, \mathbf{A}_1, \dots, \mathbf{A}_r$ are non-negative Hermitian, one can conclude

$$\left(P(u_s)\right)^* \mathbf{A}_s P(u_s) = 0_{q \times q}$$

for each $s \in \mathbb{N}_{0,r}$. Consequently, taking into account that the matrix $\mathbf{A}_0$ is non-singular, it follows that $P(u_0) = 0_{q \times q}$. This yields that there is a complex $q \times q$ matrix polynomial Q of degree not greater than $m-1$ such that $Q(v_0) \neq 0_{q \times q}$ and

$$P(v) = (v - u_0)Q(v), \quad v \in \mathbb{C}.$$

So, recalling that the points $u_0, u_1, \dots, u_r \in \mathbb{T}$ are pairwise different, we obtain

$$\int_{\mathbb{T}} \left(Q(z)\right)^* \widetilde{F}(\mathrm{d}z)\, Q(z) = \sum_{s=1}^{r} \left(Q(u_s)\right)^* \mathbf{A}_s Q(u_s)$$

$$= \sum_{s=1}^{r} \frac{1}{|u_s - u_0|^2} \left(P(u_s)\right)^* \mathbf{A}_s P(u_s) = 0_{q \times q}.$$

Admittedly (see again [18, Theorem 5.8]), this is contrary to (3.11). Hence, (3.11) implicates also (3.10). $\qquad\square$

Recall that, for some $u \in \mathbb{T}$, from (3.3) we know that $A_{n,u}^{(\alpha)}(u)$ is a positive Hermitian $q \times q$ matrix, where the equality

$$A_{n,u}^{(\alpha)}(u) = C_{n,u}^{(\alpha)}(u) \tag{3.12}$$

(which follows from (3.1) and (3.6) along with $u \in \mathbb{T}$; see also (3.9)) is satisfied.

Theorem 3.9. *Let $(\alpha_j)_{j=1}^{\infty} \in \mathcal{T}_1$ and let $n \in \mathbb{N}$. Let $X_0, X_1, \dots, X_n$ be a basis of the right $\mathbb{C}^{q \times q}$-module $\mathcal{R}_{\alpha,n}^{q \times q}$ and suppose that $\mathbf{G}$ is a non-singular matrix such that $\mathcal{M}[(\alpha_j)_{j=1}^{n}, \mathbf{G}; (X_k)_{k=0}^{n}] \neq \emptyset$. Let $u \in \mathbb{T}$ and $\mathbf{U} := -\left(\Theta_n(u)\right)^*$, where $\Theta_n(u)$ is defined by (3.4). Furthermore, let $P_{n;\mathbf{U}}^{(\alpha)}$ and $Q_{n;\mathbf{U}}^{(\alpha)}$ be the functions given by (2.7). Then:*

(a) *There exist some $r \in \mathbb{N}_{n,nq}$, pairwise different points $z_1, z_2, \dots, z_r \in \mathbb{T} \setminus \{u\}$, and a sequence $(\mathbf{A}_s)_{s=1}^{r}$ of non-negative Hermitian $q \times q$ matrices each of which is not equal to the zero matrix such that*

$$F_{n,u}^{(\alpha)} = \varepsilon_u \left(A_{n,u}^{(\alpha)}(u)\right)^{-1} + \sum_{s=1}^{r} \varepsilon_{z_s} \mathbf{A}_s \quad \text{and} \quad \sum_{s=1}^{r} \mathrm{rank}\, \mathbf{A}_s = nq. \tag{3.13}$$

In particular, $F_{n,u}^{(\alpha)}(\{u\}) = \left(A_{n,u}^{(\alpha)}(u)\right)^{-1}$ and $F_{n,u}^{(\alpha)}(\{z_s\}) = \mathbf{A}_s$ for all $s \in \mathbb{N}_{1,r}$.

(b) *If $z \in \mathbb{T}$, then the relations*

$$\mathcal{N}(P_{n;\mathrm{U}}^{(\alpha)}(z)) = \mathcal{N}((Q_{n;\mathrm{U}}^{(\alpha)}(z))^*) = \mathcal{R}(F_{n,u}^{(\alpha)}(\{z\})) \qquad (3.14)$$

and

$$F_{n,u}^{(\alpha)}(\{z\})A_{n,z}^{(\alpha)}(z)\mathbf{x} = \mathbf{x}, \qquad \mathbf{x} \in \mathcal{R}(F_{n,u}^{(\alpha)}(\{z\})) \qquad (3.15)$$

hold, where $A_{n,z}^{(\alpha)}(z) = C_{n,z}^{(\alpha)}(z)$ and where $A_{n,z}^{(\alpha)}$ and $C_{n,z}^{(\alpha)}$ are the rational matrix functions given by (3.1) and (3.6) with $w = z$.

(c) *For $z \in \mathbb{C}_0 \setminus \mathbb{P}_{\alpha,n}$, $\det P_{n;\mathrm{U}}^{(\alpha)}(z) = 0$ holds if and only if $z \in \{u, z_1, z_2, \ldots, z_r\}$.*

(d) *Let $F_u := F_{n,u}^{(\alpha)} - \varepsilon_u \big(A_{n,u}^{(\alpha)}(u)\big)^{-1}$. Then F_u belongs to $\mathcal{M}_{\geq}^{q,n-1}(\mathbb{T}, \mathfrak{B}_{\mathbb{T}})$ and is a canonical solution in the set $\mathcal{M}[\mathbf{T}_{n-1}^{(F_u)}]$, i.e.,*

$$\operatorname{rank} \mathbf{T}_n^{(F_u)} = \operatorname{rank} \mathbf{T}_{n-1}^{(F_u)} = nq.$$

Proof. Taking Theorem 2.1 and Remark 3.1 into account, [25, parts (b) of Theorem 4.3] immediately imply the assertion of (b). Because of [25, part (a) of Theorem 4.3] a similar argumentation shows that there are some $\ell \in \mathbb{N}_{n+1,(n+1)q}$, pairwise different points $z_1, z_2, \ldots, z_\ell \in \mathbb{T}$, and a sequence $(\mathbf{A}_s)_{s=1}^{\ell}$ of non-negative Hermitian $q \times q$ matrices each of which is not equal to the zero matrix such that

$$F_{n,u}^{(\alpha)} = \sum_{s=1}^{\ell} \varepsilon_{z_s} \mathbf{A}_s \qquad \text{and} \qquad \sum_{s=1}^{\ell} \operatorname{rank} \mathbf{A}_s = (n+1)q. \qquad (3.16)$$

Hence, the Riesz–Herglotz transform $\Omega_{n,u}^{(\alpha)}$ of $F_{n,u}^{(\alpha)}$ admits the representation

$$\Omega_{n,u}^{(\alpha)}(v) = \sum_{s=1}^{\ell} \frac{z_s + v}{z_s - v} \mathbf{A}_s, \qquad v \in \mathbb{D}.$$

This gives rise to

$$\lim_{t \to 1-0} \frac{1-t}{2} \Omega_{n,u}^{(\alpha)}(tz) = F_{n,u}^{(\alpha)}(\{z\})$$

for each $z \in \mathbb{T}$ (see also [13, Lemma 8.1]). Consequently, recalling (3.2) and (3.3), by using Remark 3.7 and [20, Lemma 5.1] we get

$$F_{n,u}^{(\alpha)}(\{u\}) = \lim_{t \to 1-0} \frac{1-t}{2} \left(\frac{2}{1-tu\bar{u}}\mathbf{I}_q - \sum_{k=0}^{n} R_k^\#(tu)\big(R_k(u)\big)^* \right) \left(\sum_{k=0}^{n} R_k(tu)\big(R_k(u)\big)^* \right)^{-1}$$

$$= \left(\sum_{k=0}^{n} R_k(u)\big(R_k(u)\big)^* \right)^{-1} = \big(A_{n,u}^{(\alpha)}(u)\big)^{-1}.$$

Apparently (note (3.14) and the fundamental theorem of algebra), (3.16) leads to (3.13) with some integer $r \in \mathbb{N}_{n,nq}$, pairwise different points $z_1, z_2, \ldots, z_r$ belonging to $\mathbb{T}$, and a sequence $(\mathbf{A}_s)_{s=1}^{r}$ of non-negative Hermitian $q \times q$ matrices each of which is not equal to $0_{q \times q}$. Thus, we have verified (a). Based on that and Remark 3.1, the assertion of (c) is a consequence of [25, parts (c) of Theorem 4.3] and Theorem 2.1. Finally, it remains to be shown (d). In view of (3.13) we get

$$F_u = \sum_{s=1}^{r} \varepsilon_{z_s} \mathbf{A}_s.$$

Therefore, F_u belongs obviously to $\mathcal{M}^q_\geq(\mathbb{T}, \mathfrak{B}_\mathbb{T})$, where (3.13) in combination with [18, Remark 3.9 and Theorem 6.6] as well as [11, Lemmas 1.1.7 and 1.1.9] yields

$$\operatorname{rank} \mathbf{T}^{(F_u)}_{n-1} \leq \operatorname{rank} \mathbf{T}^{(F_u)}_n \leq \sum_{s=1}^r \operatorname{rank} \mathbf{A}_s = nq.$$

Hence, recalling Theorem 1.1 and (2.1), part (d) follows from (a) and Lemma 3.8 along with $\det \mathbf{G} \neq 0$ and $F^{(\alpha)}_{n,u} \in \mathcal{M}[(\alpha_j)^n_{j=1}, \mathbf{G}; (X_k)^n_{k=0}]$ (see Remark 3.1). □

Corollary 3.10. *Let $(\alpha_j)^\infty_{j=1} \in \mathcal{T}_1$ and let $n \in \mathbb{N}$. Let $X_0, X_1, \ldots, X_n$ be a basis of the right $\mathbb{C}^{q\times q}$-module $\mathcal{R}^{q\times q}_{\alpha,n}$ and suppose that $\mathbf{G}$ is a non-singular matrix such that the set $\mathcal{M}[(\alpha_j)^n_{j=1}, \mathbf{G}; (X_k)^n_{k=0}]$ is nonempty. Furthermore, let F be a solution in $\mathcal{M}[(\alpha_j)^n_{j=1}, \mathbf{G}; (X_k)^n_{k=0}]$ which admits the representation*

$$F = \sum_{s=1}^r \varepsilon_{u_s} \mathbf{A}_s \tag{3.17}$$

for some $r \in \mathbb{N}$, points $u_1, u_2, \ldots, u_r \in \mathbb{T}$, and a sequence $(\mathbf{A}_s)^r_{s=1}$ of non-negative Hermitian $q \times q$ matrices. Then $r \geq n+1$, where the points $u_1, u_2, \ldots, u_r$ can be chosen as pairwise different and the sequence $(\mathbf{A}_s)^r_{s=1}$ such that each element is not equal to the zero matrix. Moreover (in that context), if $r \geq nq + 2$, then the matrix measure F does not coincide with $F^{(\alpha)}_{n,u}$ for certain $u \in \mathbb{T}$.

Proof. Since $F \in \mathcal{M}[(\alpha_j)^n_{j=1}, \mathbf{G}; (X_k)^n_{k=0}]$ such that (3.17) holds and since the matrix $\mathbf{G}$ is non-singular, Theorem 1.1 and [18, Theorem 6.11] imply that $r \geq n+1$, where the points $u_1, u_2, \ldots, u_r$ can be chosen as pairwise different and the sequence $(\mathbf{A}_s)^r_{s=1}$ such that $\mathbf{A}_s \neq 0_{q\times q}$ for each $s \in \mathbb{N}_{1,r}$. Based on such representation, by using Theorem 3.9 one can see that, if $r \geq nq+2$ and if $u \in \mathbb{T}$, then $F \neq F^{(\alpha)}_{n,u}$. □

The matrix functions L_n, R_n, $L^\#_n$, and $R^\#_n$ (which are the basis for (3.4) and (3.5)) can be constructed from the given data in different ways. In view of the recurrence relations for orthogonal rational matrix functions one needs to determine the corresponding matrices, which realize those. Following this way, one can apply the formulas in [21]. Moreover, because of [24, Remark 6.4] one can use Szegő parameters to obtain representations of $\Omega^{(\alpha)}_{n;u}$. The functions $L^\#_n$ and $R^\#_n$ can be also extracted from L_n and R_n using the integral formulas in [34, Section 5].

In the following, we present a further alternative to get descriptions of the $q \times q$ Carathéodory function $\Omega^{(\alpha)}_{n,u}$ with $u \in \mathbb{T}$. In other words, we will reformulate the above representations in terms of reproducing kernels based on the procedure already used in [24, Section 6] (see also [25, Remark 3.9 and Theorem 3.10]).

If $F \in \mathcal{M}^q_\geq(\mathbb{T}, \mathfrak{B}_\mathbb{T})$ is such that the weight $F(\mathbb{T})$ is a non-singular matrix and if Ω stands for the Riesz–Herglotz transform of F, then $\Omega(w)$ is a non-singular matrix for each $w \in \mathbb{D}$ and Ω^{-1} is a $q \times q$ Carathéodory function, where $(\Omega(0))^{-1}$ is a positive Hermitian $q \times q$ matrix (see, e.g., [11, Proposition 3.6.8]). Taking this and the matricial version of the Riesz–Herglotz theorem (see [11, Theorem 2.2.2]) into account, we call the unique measure $F^{\#} \in \mathcal{M}^q_{\geq}(\mathbb{T}, \mathfrak{B}_{\mathbb{T}})$ fulfilling

$$(\Omega(w))^{-1} = \int_{\mathbb{T}} \frac{z+w}{z-w} F^{\#}(\mathrm{d}z), \quad w \in \mathbb{D},$$

the *reciprocal measure corresponding to* F (cf. [11, Definition 3.6.10] and [26]).
Based on [24, Remarks 6.1 and 6.2], similar to (3.1) and (3.6), we set

$$A_{n,w}^{(\alpha,\#)} := \left(X_0, X_1, \dots, X_n \right) (\mathbf{G}^{\#})^{-1} \left(X_0(w), X_1(w), \dots, X_n(w) \right)^*$$

and

$$C_{n,w}^{(\alpha,\#)} := \begin{pmatrix} X_0^{[\alpha,n]}(w) \\ X_1^{[\alpha,n]}(w) \\ \vdots \\ X_n^{[\alpha,n]}(w) \end{pmatrix}^* (\mathbf{G}^{\#})^{-1} \begin{pmatrix} X_0^{[\alpha,n]} \\ X_1^{[\alpha,n]} \\ \vdots \\ X_n^{[\alpha,n]} \end{pmatrix},$$

where $\mathbf{G}^{\#}$ stands for the $(n+1)q \times (n+1)q$ matrix $\mathbf{G}_{X,n}^{(F^{\#})}$ which is (uniquely) given by the reciprocal measure $F^{\#}$ corresponding to some $F \in \mathcal{M}[(\alpha_j)_{j=1}^n, \mathbf{G}; (X_k)_{k=0}^n]$.
Let $\mathbf{V}$ be a unitary $q \times q$ matrix. In view of [25, Theorem 3.10] the function

$$\Phi_{n;\mathbf{V}}^{(\alpha)} := \left(A_{n,\alpha_n}^{(\alpha,\#)} \, \mathbf{\Omega}_n^{-*} \sqrt{A_{n,\alpha_n}^{(\alpha)}(\alpha_n)}^{-1} - b_{\alpha_n}(C_{n,\alpha_n}^{(\alpha,\#)})^{[\alpha,n]} \mathbf{\Omega}_n^{-1} \sqrt{C_{n,\alpha_n}^{(\alpha)}(\alpha_n)}^{-1} \mathbf{V} \right)$$
$$\cdot \left(A_{n,\alpha_n}^{(\alpha)} \sqrt{A_{n,\alpha_n}^{(\alpha)}(\alpha_n)}^{-1} + b_{\alpha_n}(C_{n,\alpha_n}^{(\alpha)})^{[\alpha,n]} \sqrt{C_{n,\alpha_n}^{(\alpha)}(\alpha_n)}^{-1} \mathbf{V} \right)^{-1},$$

where $\mathbf{\Omega}_n = (L_n^{\#})^{[\alpha,n]}(\alpha_n)(L_n^{[\alpha,n]}(\alpha_n))^{-1}$, is of particular interest. (We marginally comment that [24, Lemma 6.4] points out some other possibilities to calculate the non-singular $q \times q$ matrix $\mathbf{\Omega}_n$.) Thus, recalling the represenations of $\Psi_{n;\mathbf{U}}^{(\alpha)}$ stated in Section 2, by using [24, Lemma 6.4] (note (3.3) and (3.8)) one can realize that

$$\Phi_{n;\mathbf{V}}^{(\alpha)} = \left(\sqrt{C_{n,\alpha_n}^{(\alpha)}(\alpha_n)}^{-1} C_{n,\alpha_n}^{(\alpha)} + b_{\alpha_n} \mathbf{V} \sqrt{A_{n,\alpha_n}^{(\alpha)}(\alpha_n)}^{-1} (A_{n,\alpha_n}^{(\alpha)})^{[\alpha,n]} \right)^{-1}$$
$$\cdot \left(\sqrt{C_{n,\alpha_n}^{(\alpha)}(\alpha_n)}^{-1} \mathbf{\Omega}_n^{-*} C_{n,\alpha_n}^{(\alpha,\#)} - b_{\alpha_n} \mathbf{V} \sqrt{A_{n,\alpha_n}^{(\alpha)}(\alpha_n)}^{-1} \mathbf{\Omega}_n^{-1}(A_{n,\alpha_n}^{(\alpha,\#)})^{[\alpha,n]} \right),$$

$$\Phi_{n;\mathbf{V}}^{(\alpha)} = \left(\frac{1}{b_{\alpha_n}} A_{n,\alpha_n}^{(\alpha,\#)} \mathbf{\Omega}_n^{-*} \sqrt{A_{n,\alpha_n}^{(\alpha)}(\alpha_n)}^{-1} \mathbf{V}^* - (C_{n,\alpha_n}^{(\alpha,\#)})^{[\alpha,n]} \mathbf{\Omega}_n^{-1} \sqrt{C_{n,\alpha_n}^{(\alpha)}(\alpha_n)}^{-1} \right)$$
$$\cdot \left(\frac{1}{b_{\alpha_n}} A_{n,\alpha_n}^{(\alpha)} \sqrt{A_{n,\alpha_n}^{(\alpha)}(\alpha_n)}^{-1} \mathbf{V}^* + (C_{n,\alpha_n}^{(\alpha)})^{[\alpha,n]} \sqrt{C_{n,\alpha_n}^{(\alpha)}(\alpha_n)}^{-1} \right)^{-1},$$

$$\Phi_{n;\mathbf{V}}^{(\alpha)} = \left(\frac{1}{b_{\alpha_n}} \mathbf{V}^* \sqrt{C_{n,\alpha_n}^{(\alpha)}(\alpha_n)}^{-1} C_{n,\alpha_n}^{(\alpha)} + \sqrt{A_{n,\alpha_n}^{(\alpha)}(\alpha_n)}^{-1} (A_{n,\alpha_n}^{(\alpha)})^{[\alpha,n]} \right)^{-1}$$
$$\cdot \left(\frac{1}{b_{\alpha_n}} \mathbf{V}^* \sqrt{C_{n,\alpha_n}^{(\alpha)}(\alpha_n)}^{-1} \mathbf{\Omega}_n^{-*} C_{n,\alpha_n}^{(\alpha,\#)} - \sqrt{A_{n,\alpha_n}^{(\alpha)}(\alpha_n)}^{-1} \mathbf{\Omega}_n^{-1}(A_{n,\alpha_n}^{(\alpha,\#)})^{[\alpha,n]} \right).$$

Remark 3.11. If $u \in \mathbb{T}$ and if (note (3.3) and (3.8))

$$\tilde{\Theta}_n(u) := b_{\alpha_n}(u) \sqrt{A_{n,\alpha_n}^{(\alpha)}(\alpha_n)} (A_{n,\alpha_n}^{(\alpha)}(u))^{-1} (C_{n,\alpha_n}^{(\alpha)})^{[\alpha,n]}(u) \sqrt{C_{n,\alpha_n}^{(\alpha)}(\alpha_n)}^{-1},$$

then by [24, Lemma 6.4] one can find that $\tilde{\Theta}_n(u)$ is a unitary $q \times q$ matrix, where

$$\tilde{\Theta}_n(u) = b_{\alpha_n}(u)\sqrt{A_{n,\alpha_n}^{(\alpha)}(\alpha_n)}^{-1}(A_{n,\alpha_n}^{(\alpha)})^{[\alpha,n]}(u)\left(C_{n,\alpha_n}^{(\alpha)}(u)\right)^{-1}\sqrt{C_{n,\alpha_n}^{(\alpha)}(\alpha_n)}.$$

Moreover, [24, Lemma 6.4] along with (3.5) and (3.4) implies that $F_{n,u}^{(\alpha)}$ is the uni-quely determined measure belonging to $\mathcal{M}_{\geq}^q(\mathbb{T}, \mathfrak{B}_{\mathbb{T}})$ such that its Riesz–Herglotz transform $\Omega_{n,u}^{(\alpha)}$ satisfies $\Omega_{n,u}^{(\alpha)}(v) = \Phi_{n;\mathbf{V}}^{(\alpha)}(v)$ for each $v \in \mathbb{D} \setminus \mathbb{P}_{\alpha,n}$, where $\Phi_{n;\mathbf{V}}^{(\alpha)}$ is given as above with $\mathbf{V} := -\left(\tilde{\Theta}_n(u)\right)^*$ (cf. [24, part (c) of Theorem 6.5]).

Due to [25, Remark 2.5] one can see that, roughly speaking, the concept of reciprocal measures harmonizes with the notion of a canonical solution (see also [26]). As the next proposition emphasized (cf. [24, Proposition 6.3]), having fixed some $u \in \mathbb{T}$, the situation concerning the solution $F_{n,u}^{(\alpha)}$ is not that simple. However, we get at least the following statement (which is similar to Theorem 3.9).

Remark 3.12. Let $(\alpha_j)_{j=1}^{\infty} \in \mathcal{T}_1$ and let $n \in \mathbb{N}$. Let $X_0, X_1, \ldots, X_n$ be a basis of the right $\mathbb{C}^{q \times q}$-module $\mathcal{R}_{\alpha,n}^{q \times q}$ and suppose that $\mathbf{G}$ is a non-singular matrix such that $\mathcal{M}[(\alpha_j)_{j=1}^n, \mathbf{G}; (X_k)_{k=0}^n] \neq \emptyset$. Let $u \in \mathbb{T}$ and $\mathbf{U} := -\left(\Theta_n(u)\right)^*$, where the value $\Theta_n(u)$ is defined by (3.4). Furthermore, let the rational matrix functions $P_{n;\mathbf{U}}^{(\alpha,\#)}$ and $Q_{n;\mathbf{U}}^{(\alpha,\#)}$ be given by (2.7). Denote by $F_{n,u}^{(\alpha,\#)}$ the reciprocal measure correspon-ding to $F_{n,u}^{(\alpha)}$. Then, recalling Theorem 2.1 and Remark 3.1, as a consequence of [25, Corollary 4.4] one can conclude the following:

(a) There exist some $\ell \in \mathbb{N}_{n+1,(n+1)q}$, pairwise different points $u_1, u_2, \ldots, u_\ell \in \mathbb{T}$, and a sequence $(\mathbf{A}_s^{\#})_{s=1}^{\ell}$ of non-negative Hermitian $q \times q$ matrices each of which is not equal to the zero matrix such that

$$F_{n,u}^{(\alpha,\#)} = \sum_{s=1}^{\ell} \varepsilon_{u_s} \mathbf{A}_s^{\#} \quad \text{and} \quad \sum_{s=1}^{\ell} \text{rank}\, \mathbf{A}_s^{\#} = (n+1)q.$$

In particular, $F_{n,u}^{(\alpha,\#)}(\{u_s\}) = \mathbf{A}_s^{\#}$ for each $s \in \mathbb{N}_{1,\ell}$.

(b) If $z \in \mathbb{T}$, then the relations

$$\mathcal{N}\left(P_{n;\mathbf{U}}^{(\alpha,\#)}(z)\right) = \mathcal{N}\left((Q_{n;\mathbf{U}}^{(\alpha,\#)}(z))^*\right) = \mathcal{R}\left(F_{n,u}^{(\alpha,\#)}(\{z\})\right)$$

and

$$F_{n,u}^{(\alpha,\#)}(\{z\})A_{n,z}^{(\alpha,\#)}(z)\mathbf{x} = \mathbf{x}, \quad \mathbf{x} \in \mathcal{R}\left(F_{n,u}^{(\alpha,\#)}(\{z\})\right),$$

are satisfied, where $A_{n,z}^{(\alpha,\#)}(z) = C_{n,z}^{(\alpha,\#)}(z)$ holds.

(c) For $z \in \mathbb{C}_0 \setminus \mathbb{P}_{\alpha,n}$, $\det P_{n;\mathbf{U}}^{(\alpha,\#)}(z) = 0$ holds if and only if $z \in \{u_1, u_2, \ldots, u_\ell\}$.

A comparison of Theorem 3.9 with Remark 3.12 shows that the reciprocal measure $F_{n,u}^{(\alpha,\#)}$ corresponding to $F_{n,u}^{(\alpha)}$ with some $u \in \mathbb{T}$ has a similar structure as the underlying measure $F_{n,u}^{(\alpha)}$, but not at all points. The following result contains some thought in this regard (see also (3.19) and Example 4.13). Here again, $\mathbf{G}^{\#}$ stands for the complex $(n+1)q \times (n+1)q$ matrix $\mathbf{G}_{X,n}^{(F^{\#})}$ which is (uniquely) given by the reciprocal measure $F^{\#}$ corresponding to some $F \in \mathcal{M}[(\alpha_j)_{j=1}^n, \mathbf{G}; (X_k)_{k=0}^n]$.

Proposition 3.13. *Let $(\alpha_j)_{j=1}^{\infty} \in \mathcal{T}_1$ and let $n \in \mathbb{N}$. Let $X_0, X_1, \ldots, X_n$ be a basis of the right $\mathbb{C}^{q \times q}$-module $\mathcal{R}_{\alpha,n}^{q \times q}$ and suppose that $\mathbf{G}$ is a non-singular matrix such that $\mathcal{M}[(\alpha_j)_{j=1}^n, \mathbf{G}; (X_k)_{k=0}^n] \neq \emptyset$. Furthermore, let $u \in \mathbb{T}$ and let $F_{n,u}^{(\alpha,\#)}$ be the reciprocal measure corresponding to the canonical solution $F_{n,u}^{(\alpha)}$. Then, for some $z \in \mathbb{T}$, the following statements are equivalent:*

(i) *$F_{n,u}^{(\alpha,\#)}$ coincides with the (special) canonical solution $(F^{\#})_{n,z}^{(\alpha)}$ which is given relating to the solution set $\mathcal{M}[(\alpha_j)_{j=1}^n, \mathbf{G}^{\#}; (X_k)_{k=0}^n]$ and the point z.*

(ii) $b_{\alpha_n}(u)(L_n^{\#})^{[\alpha,n]}(z)L_n(u) + b_{\alpha_n}(z)R_n^{\#}(z)R_n^{[\alpha,n]}(u) = 0_{q \times q}$.

(iii) $b_{\alpha_n}(z)L_n^{[\alpha,n]}(u)L_n^{\#}(z) + b_{\alpha_n}(u)R_n(u)(R_n^{\#})^{[\alpha,n]}(z) = 0_{q \times q}$.

(iv) $(1 - \overline{u}z) \sum_{k=0}^{n} \left(L_k(u)\right)^* L_k^{\#}(z) = 2\mathbf{I}_q$.

(v) $(1 - z\overline{u}) \sum_{k=0}^{n} R_k^{\#}(z)\left(R_k(u)\right)^* = 2\mathbf{I}_q$.

In particular, the measure $F_{n,u}^{(\alpha,\#)}$ does not coincide with the measure $(F^{\#})_{n,u}^{(\alpha)}$.

Proof. Let $z \in \mathbb{T}$. The reciprocal measure $F_{n,u}^{(\alpha,\#)}$ is well defined because of [24, Remark 6.2] and the fact that Remark 3.1 implies $F_{n,u}^{(\alpha)} \in \mathcal{M}[(\alpha_j)_{j=1}^n, \mathbf{G}; (X_k)_{k=0}^n]$. Moreover, in consequence of [24, Remarks 6.1 and 6.2] and the choice of $\mathbf{G}^{\#}$, the matrix measure $(F^{\#})_{n,z}^{(\alpha)}$ is also well defined. Let Ω_u and $\tilde{\Omega}_z$ be the Riesz–Herglotz transform of $F_{n,u}^{(\alpha,\#)}$ and $(F^{\#})_{n,z}^{(\alpha)}$, respectively. Furthermore, let $\Theta_n(u)$ be defined by (3.4). Thus, in view of (3.5) and (2.7), the representation

$$\Omega_u(v) = \left((R_n^{\#})^{[\alpha,n]}(v) + b_{\alpha_n}(v)\left(\Theta_n(u)\right)^* L_n^{\#}(v)\right)^{-1}\left(R_n^{[\alpha,n]}(v) - b_{\alpha_n}(v)\left(\Theta_n(u)\right)^* L_n(v)\right)$$

holds for each $v \in \mathbb{D} \setminus \mathbb{P}_{\alpha,n}$. Similarly, since [25, Remark 3.4] (see also [34, Theorem 4.6]) tells us that $[(L_k^{\#})_{k=0}^n, (R_k^{\#})_{k=0}^n]$ is a pair of orthonormal systems corresponding to $\mathcal{M}[(\alpha_j)_{j=1}^n, \mathbf{G}^{\#}; (X_k)_{k=0}^n]$, where $[(L_k)_{k=0}^n, (R_k)_{k=0}^n]$ is obviously the dual pair of orthonormal systems corresponding to $[(L_k^{\#})_{k=0}^n, (R_k^{\#})_{k=0}^n]$, we get

$$\tilde{\Omega}_z(v) = \left((R_n^{\#})^{[\alpha,n]}(v) - b_{\alpha_n}(v)\left(\Theta_n^{\#}(z)\right)^* L_n^{\#}(v)\right)^{-1}\left(R_n^{[\alpha,n]}(v) + b_{\alpha_n}(v)\left(\Theta_n^{\#}(z)\right)^* L_n(v)\right)$$

for each $v \in \mathbb{D} \setminus \mathbb{P}_{\alpha,n}$, where $\Theta_n^{\#}(z) := b_{\alpha_n}(z)\left((L_n^{\#})^{[\alpha,n]}(z)\right)^{-1}R_n^{\#}(z)$. Consequently, taking the matricial version of the Riesz–Herglotz theorem (see, e.g., [11, Theorem 2.2.2]) and (3.5) into account, by Theorem 2.1 and [24, Lemma 3.11] one can see that (i) is equivalent to (ii) (resp., to (iii)). Furthermore (note $u, z \in \mathbb{T}$ and the definition of b_{α_n}), an application of the Christoffel–Darboux formulas for orthogonal rational matrix functions stated in [34, Proposition 3.1] (cf. Remark 3.7) along with [20, Remark 2.5] yields that (ii) holds if and only if (iv) (resp., (v)) is satisfied. Thus, (i)–(v) are equivalent. Based on this fact (i.e., that the statements (i) and (v) are equivalent) and $u \in \mathbb{T}$ one can realize that $F_{n,u}^{(\alpha,\#)} \neq (F^{\#})_{n,u}^{(\alpha)}$. $\square$

Let $u \in \mathbb{T}$ and let us consider the settings (3.5) and (3.4) in the elementary case $n = 0$ (see also [22, Remark 6.8]). Applying a similar reasoning as in Remark 3.7 one can find that the Riesz–Herglotz transform $\Omega_{0,u}^{(\alpha)}$ of $F_{0,u}^{(\alpha)}$ admits

$$\Omega_{0,u}^{(\alpha)}(v) = \frac{u+v}{u-v} \mathbf{X}_0^{-*} \mathbf{G} \mathbf{X}_0^{-1}, \quad v \in \mathbb{D}.$$

From here one can read in a straightforward manner that appropriate statements are satisfied for that case as presented in this section for $n \in \mathbb{N}$. In particular (cf. (2.10)), in view of the matricial version of the Riesz–Herglotz theorem we see that

$$F_{0,u}^{(\alpha)} = \varepsilon_u \mathbf{X}_0^{-*} \mathbf{G} \mathbf{X}_0^{-1}. \tag{3.18}$$

Moreover, by using the notation of Proposition 3.13 for $n = 0$ as well it follows that

$$F_{0,u}^{(\alpha,\#)} = (F^\#)_{0,-u}^{(\alpha)}. \tag{3.19}$$

However, this relationship is in some contrast to the case $n \in \mathbb{N}$ (see Example 4.13).

4. Characterizations of $F_{n,u}^{(\alpha)}$ in the set of canonical solutions

The considerations in this section are attached to those in the previous one. In doing so, we will still use the notation fixed there. In particular, in view of (3.5) and (3.4), we will analyze the rational matrix functions given by (2.7) based on a point $u \in \mathbb{T}$ and a pair $[(L_k)_{k=0}^n, (R_k)_{k=0}^n]$ of orthonormal systems corresponding to $\mathcal{M}[(\alpha_j)_{j=1}^n, \mathbf{G}; (X_k)_{k=0}^n]$ with dual pair $[(L_k^\#)_{k=0}^n, (R_k^\#)_{k=0}^n]$ of orthonormal systems corresponding to $[(L_k)_{k=0}^n, (R_k)_{k=0}^n]$. Furthermore, the values $A_{n,u}^{(\alpha)}(u)$ and $C_{n,u}^{(\alpha)}(u)$ for $u \in \mathbb{T}$ will play a crucial part as well, where $A_{n,u}^{(\alpha)}$ and $C_{n,u}^{(\alpha)}$ are the rational matrix functions defined by (3.1) and (3.6) with $w = u$, respectively.

Let $u \in \mathbb{T}$. By the choice of the matrix measure $F_{n,u}^{(\alpha)}$ it is clear that this is a canonical solution of Problem (R) for the nondegenerate case (cf. Remark 3.1). Roughly speaking, recalling Theorems 2.1 and 3.9 (see also [25, Theorem 4.3]), we discuss below special features of the matrix measure $F_{n,u}^{(\alpha)}$ concerning its exceptional position in the whole family of canonical solutions of Problem (R). We begin directly with the main result of this section.

Theorem 4.1. *Let* $(\alpha_j)_{j=1}^\infty \in \mathcal{T}_1$ *and let* $n \in \mathbb{N}$. *Let* $X_0, X_1, \ldots, X_n$ *be a basis of the right* $\mathbb{C}^{q \times q}$-*module* $\mathcal{R}_{\alpha,n}^{q \times q}$ *and suppose that* $\mathbf{G}$ *is a non-singular matrix such that* $\mathcal{M}[(\alpha_j)_{j=1}^n, \mathbf{G}; (X_k)_{k=0}^n] \neq \emptyset$. *Let* $u \in \mathbb{T}$ *and let* $\mathbf{U}$ *be a unitary* $q \times q$ *matrix. Let* $P_{n;\mathbf{U}}^{(\alpha)}$ *and* $P_{n;\mathbf{U}}^{(\alpha,\#)}$ *(resp.,* $Q_{n;\mathbf{U}}^{(\alpha)}$ *and* $Q_{n;\mathbf{U}}^{(\alpha,\#)}$*) as well as* Θ_n *be the matrix functions given by (2.7) and (3.4). Furthermore, let* $\Omega_{n,u}^{(\alpha)}$ *and* $\Omega_{n;\mathbf{U}}^{(\alpha)}$ *be the Riesz–Herglotz transform of* $F_{n,u}^{(\alpha)}$ *and* $F_{n;\mathbf{U}}^{(\alpha)}$, *respectively. Let*

$$\widetilde{F}_u := F_{n;\mathbf{U}}^{(\alpha)} - \varepsilon_u F_{n;\mathbf{U}}^{(\alpha)}(\{u\}).$$

Then the following statements are equivalent:

(i) $F_{n;\mathbf{U}}^{(\alpha)} = F_{n,u}^{(\alpha)}$.

(ii) $F_{n;\mathbf{U}}^{(\alpha)}(\{u\}) = \left(A_{n,u}^{(\alpha)}(u)\right)^{-1}$ *(resp.,* $F_{n;\mathbf{U}}^{(\alpha)}(\{u\}) = \left(C_{n,u}^{(\alpha)}(u)\right)^{-1}$*).*

(iii) $F_{n;\mathbf{U}}^{(\alpha)}(\{u\})$ *is non-singular.*

(iv) $P_{n;\mathbf{U}}^{(\alpha)}(u) = 0_{q\times q}$ *(resp.,* $Q_{n;\mathbf{U}}^{(\alpha)}(u) = 0_{q\times q}$*).*

(v) $\mathbf{U} = -\big(\Theta_n(u)\big)^{*}$.

(vi) *There is some* $v \in \mathbb{C} \setminus (\mathbb{T} \cup \mathbb{P}_{\alpha,n} \cup \mathbb{Z}_{\alpha,n})$ *such that* $\widehat{\Omega}_{n;\mathbf{U}}^{(\alpha)}(v) = \widehat{\Omega}_{n,u}^{(\alpha)}(v)$.

(vii) *There is some* $k \in \mathbb{N}_{1,m}$ *such that* $\big(\widehat{\Omega}_{n;\mathbf{U}}^{(\alpha)}\big)^{(l_k)}(\gamma_k) = \big(\widehat{\Omega}_{n,u}^{(\alpha)}\big)^{(l_k)}(\gamma_k)$.

(viii) *For each* $k \in \mathbb{N}_{1,m}$, *the equality* $\big(\widehat{\Omega}_{n;\mathbf{U}}^{(\alpha)}\big)^{(l_k)}(\gamma_k) = \big(\widehat{\Omega}_{n,u}^{(\alpha)}\big)^{(l_k)}(\gamma_k)$ *holds.*

(ix) $\widetilde{F}_u$ *is a canonical solution in the set* $\mathcal{M}[\mathbf{T}_{n-1}^{(\widetilde{F}_u)}]$.

(x) $\operatorname{rank} \mathbf{T}_n^{(\widetilde{F}_u)} = nq$.

(xi) *The set* $\mathcal{M}[\mathbf{T}_n^{(\widetilde{F}_u)}]$ *is a singleton.*

In particular, if (i) *holds, then* $\det P_{n;\mathbf{U}}^{(\alpha,\#)}(u) \neq 0$ *(resp.,* $\det Q_{n;\mathbf{U}}^{(\alpha,\#)}(u) \neq 0$*) and* $F_{n;\mathbf{U}}^{(\alpha,\#)}(\{u\}) = 0_{q\times q}$, *where* $F_{n;\mathbf{U}}^{(\alpha,\#)}$ *is the reciprocal measure corresponding to* $F_{n;\mathbf{U}}^{(\alpha)}$.

Proof. In view of the definition of $F_{n;\mathbf{U}}^{(\alpha)}$ and Theorem 2.1 one can conclude the equivalence of (i), (vi), (vii), and (viii) directly from Proposition 3.6. Taking [24, Lemma 3.11] and (3.5) into account, the equivalence of (i) and (v) is a consequence of Theorem 2.1 and the special choice of $F_{n,u}^{(\alpha)}$. Recalling again [24, Lemma 3.11], because of (2.7) and (3.4) one can find that (v) holds if and only if (iv) is satisfied (cf. [22, part (b) of Proposition 9.8]). Since from [25, Theorem 4.3] we know that

$$\mathcal{N}\big(P_{n;\mathbf{U}}^{(\alpha)}(z)\big)\widetilde{F}_u = \mathcal{N}\big((Q_{n;\mathbf{U}}^{(\alpha)}(z))^{*}\big) = \mathcal{R}\big(F_{n;\mathbf{U}}^{(\alpha)}(\{z\})\big)$$

holds for each $z \in \mathbb{T}$, one can see that (iii) is equivalent to (iv). Furthermore, from (3.3) we know that $A_{n,u}^{(\alpha)}(u)$ is a positive Hermitian $q \times q$ matrix, where (3.12) holds (see also (3.9)). In particular, one can realize that (ii) leads to (iii). applying part (a) of Theorem 3.9 and (3.12), we get that (i) implies (ii). Keeping this in mind, from part (d) of Theorem 3.9 one can derive that (i) yields also (ix). Taking Theorem 2.1 into account and the fact that the matrix $\mathbf{G}$ is non-singular, by using Theorem 1.1 we have on the one hand

$$\operatorname{rank} \mathbf{T}_n^{(F_{n;\mathbf{U}}^{(\alpha)})} = \operatorname{rank} \mathbf{G} = (n+1)q$$

and on the other hand (note [18, Example 3.11]) the choice of $\widetilde{F}_u$ gives rise to

$$\operatorname{rank} \mathbf{T}_n^{(F_{n;\mathbf{U}}^{(\alpha)})} \leq \operatorname{rank} \mathbf{T}_n^{(\widetilde{F}_u)} + \operatorname{rank} \mathbf{T}_n^{(\varepsilon_u F_{n;\mathbf{U}}^{(\alpha)}(\{u\}))} = \operatorname{rank} \mathbf{T}_n^{(\widetilde{F}_u)} + \operatorname{rank} F_{n;\mathbf{U}}^{(\alpha)}(\{u\}).$$

Consequently, we obtain

$$(n+1)q \leq \operatorname{rank} \mathbf{T}_n^{(\widetilde{F}_u)} + \operatorname{rank} F_{n;\mathbf{U}}^{(\alpha)}(\{u\}). \tag{4.1}$$

In particular, because of (4.1) and $\operatorname{rank} F_{n;\mathbf{U}}^{(\alpha)}(\{u\}) \leq q$ it follows that

$$nq \leq \operatorname{rank} \mathbf{T}_n^{(\widetilde{F}_u)}. \tag{4.2}$$

We suppose now that (ix) is fulfilled. Thus, since the size of the matrix $\mathbf{T}_{n-1}^{(\tilde{F}_u)}$ is $nq \times nq$ and since (4.2) in combination with (ix) leads to

$$nq \leq \operatorname{rank} \mathbf{T}_n^{(\tilde{F}_u)} = \operatorname{rank} \mathbf{T}_{n-1}^{(\tilde{F}_u)},$$

one can see that (x) holds. Finally, we come from (x). By (4.1) along with (x) we get

$$(n+1)q \leq \operatorname{rank} \mathbf{T}_n^{(\tilde{F}_u)} + \operatorname{rank} F_{n;\mathrm{U}}^{(\alpha)}(\{u\}) = nq + \operatorname{rank} F_{n;\mathrm{U}}^{(\alpha)}(\{u\}).$$

Hence, the matrix $F_{n;\mathrm{U}}^{(\alpha)}(\{u\})$ is non-singular, i.e., (iii) holds. Finally, recalling a fundamental result on non-negative definite sequences of matrices (see, e.g., [11, Theorem 3.4.2]), the equivalence of (ix) and (xi) is an immediate consequence of [25, Proposition 1.12]. So, we have shown that the statements (i)–(xi) are equivalent. Based on that (more precisely, on the fact that (i) and (iii) are equivalent), the remaining part of the assertion can be deduced from [25, Corollary 4.5]. □

Corollary 4.2. *Let* $(\alpha_j)_{j=1}^\infty \in \mathcal{T}_1$ *and let* $n \in \mathbb{N}$*. Let* $X_0, X_1, \ldots, X_n$ *be a basis of the right* $\mathbb{C}^{q \times q}$*-module* $\mathcal{R}_{\alpha,n}^{q \times q}$ *and suppose that* $\mathbf{G}$ *is a non-singular matrix such that* $\mathcal{M}[(\alpha_j)_{j=1}^n, \mathbf{G}; (X_k)_{k=0}^n] \neq \emptyset$*. If* $u \in \mathbb{T}$ *and if* $F \in \mathcal{M}[(\alpha_j)_{j=1}^n, \mathbf{G}; (X_k)_{k=0}^n]$*, then the following statements are equivalent:*

(i) $F = F_{n,u}^{(\alpha)}$.
(ii) *F is a canonical solution in* $\mathcal{M}[(\alpha_j)_{j=1}^n, \mathbf{G}; (X_k)_{k=0}^n]$ *and* $\det F(\{u\}) \neq 0$.
(iii) *There is a finite subset* Δ *of* $\mathbb{T}$ *such that* $F(\mathbb{T} \backslash \Delta) = 0_{q \times q}$*,* $\det F(\{u\}) \neq 0$*, and*

$$\sum_{z \in \Delta \backslash \{u\}} \operatorname{rank} F(\{z\}) = nq. \tag{4.3}$$

(iv) *There is a finite subset* Δ *of* $\mathbb{T}$ *such that* $F(\mathbb{T} \backslash \Delta) = 0_{q \times q}$ *and* (4.3) *hold.*

Proof. Use Theorem 4.1 in combination with Theorem 2.1 and (2.2) (see also Remark 3.1 and [25, Remark 3.1]). □

Corollary 4.3. *Let* $(\alpha_j)_{j=1}^\infty \in \mathcal{T}_1$ *and let* $n \in \mathbb{N}$*. Let* $X_0, X_1, \ldots, X_n$ *be a basis of the right* $\mathbb{C}^{q \times q}$*-module* $\mathcal{R}_{\alpha,n}^{q \times q}$ *and suppose that* $\mathbf{G}$ *is a non-singular matrix such that* $\mathcal{M}[(\alpha_j)_{j=1}^n, \mathbf{G}; (X_k)_{k=0}^n] \neq \emptyset$*. Let* $u \in \mathbb{T}$*. If* $F \in \mathcal{M}[(\alpha_j)_{j=1}^n, \mathbf{G}; (X_k)_{k=0}^n]$ *and if* $F^{(\alpha,n)} : \mathcal{B}_{\mathbb{T}} \to \mathbb{C}^{q \times q}$ *is the matrix measure defined by*

$$F^{(\alpha,n)}(B) := \int_B \left(\frac{1}{\pi_{\alpha,n}(z)} \mathbf{I}_q \right)^* F(dz) \left(\frac{1}{\pi_{\alpha,n}(z)} \mathbf{I}_q \right), \tag{4.4}$$

then $F = F_{n,u}^{(\alpha)}$ *holds if and only if* $F^{(\alpha,n)}$ *coincides with the (special) canonical solution* $F_{n,u}$ *which is given relating to the set* $\mathcal{M}[\mathbf{T}_n^{(F^{(\alpha,n)})}]$ *and the point* u*.*

Proof. In view of Theorem 1.1 and [18, Remark 5.10], if $F^{(\alpha,n)} : \mathcal{B}_{\mathbb{T}} \to \mathbb{C}^{q \times q}$ is defined by (4.4) based on a matrix measure $F \in \mathcal{M}[(\alpha_j)_{j=1}^n, \mathbf{G}; (X_k)_{k=0}^n]$, then $F^{(\alpha,n)}$ belongs to $\mathcal{M}_{\geq}^{q,n}(\mathbb{T}, \mathcal{B}_{\mathbb{T}})$ since $\mathbf{G}$ is a non-singular matrix. Furthermore, because of [25, Remark 2.4] we already know that F is a canonical solution in $\mathcal{M}[(\alpha_j)_{j=1}^n, \mathbf{G}; (X_k)_{k=0}^n]$ if and only if $F^{(\alpha,n)}$ is a canonical solution with respect to the solution set $\mathcal{M}[\mathbf{T}_n^{(F^{(\alpha,n)})}]$. Thus, the assertion follows immediately from Theorem 4.1 along with Theorem 2.1 and the fact that the value $F(\{u\})$ is obviously non-singular if and only if $F^{(\alpha,n)}(\{u\})$ is non-singular. $\square$

Corollary 4.4. *Let $F \in \mathcal{M}_{\geq}^{q}(\mathbb{T}, \mathfrak{B}_{\mathbb{T}})$. Furthermore, let $n \in \mathbb{N}$ and let $u \in \mathbb{T}$. Then the following statements are equivalent:*

  (i) *F coincides with the canonical solution $F_{n,u}$ given relating to $\mathcal{M}[\mathbf{T}_n^{(F)}]$ and u.*
  (ii) *There exist a sequence $(\alpha_j)_{j=1}^{\infty} \in \mathcal{T}_1$ and a basis $X_0, X_1, \ldots, X_n$ of the right $\mathbb{C}^{q \times q}$-module $\mathcal{R}_{\alpha,n}^{q \times q}$ such that F coincides with the canonical solution $F_{n,u}^{(\alpha)}$ which is given relating to $\mathcal{M}[(\alpha_j)_{j=1}^{n}, \mathbf{G}_{X,n}^{(F)}; (X_k)_{k=0}^{n}]$ and the point u.*
  (iii) *F coincides with the canonical solution $F_{n,u}^{(\alpha)}$ in $\mathcal{M}[(\alpha_j)_{j=1}^{n}, \mathbf{G}_{X,n}^{(F)}; (X_k)_{k=0}^{n}]$ for every $(\alpha_j)_{j=1}^{\infty} \in \mathcal{T}_1$ and basis $X_0, X_1, \ldots, X_n$ of the right $\mathbb{C}^{q \times q}$-module $\mathcal{R}_{\alpha,n}^{q \times q}$.*

Proof. Applying a similar argumentation as in the proof of Corollary 4.3 (note here Theorem 1.1 and Theorem 2.1), the assertion follows from Theorem 4.1 along with a general result on the canonical solution stated in [25, Remark 2.3]. $\square$

Corollary 4.5. *Let $(\alpha_j)_{j=1}^{\infty} \in \mathcal{T}_1$ and let $n \in \mathbb{N}$. Let $X_0, X_1, \ldots, X_n$ be a basis of the right $\mathbb{C}^{q \times q}$-module $\mathcal{R}_{\alpha,n}^{q \times q}$ and suppose that $\mathbf{G}$ is a non-singular matrix such that $\mathcal{M}[(\alpha_j)_{j=1}^{n}, \mathbf{G}; (X_k)_{k=0}^{n}] \neq \emptyset$. Let F be a canonical solution in $\mathcal{M}[(\alpha_j)_{j=1}^{n}, \mathbf{G}; (X_k)_{k=0}^{n}]$ such that the value $F(\{u_s\})$ is a non-singular $q \times q$ matrix for each $s \in \mathbb{N}_{1,\ell}$ with some integer $\ell \in \mathbb{N}$ and pairwise different points $u_1, u_2, \ldots, u_\ell \in \mathbb{T}$.*

  (a) *Then F coincides with $F_{n,u_s}^{(\alpha)}$ and $F^{\#}(\{u_s\}) = 0_{q \times q}$ for all $s \in \mathbb{N}_{1,\ell}$, where $F^{\#}$ stands for the reciprocal measure corresponding to F. In particular, $\ell \leq n+1$.*
  (b) *The matrix measure F admits the representation*

$$F = \sum_{s=1}^{n+1} \varepsilon_{u_s} \left(A_{n,u_s}^{(\alpha)}(u_s) \right)^{-1} \qquad (4.5)$$

if $\ell = n + 1$ and otherwise there are some $r \in \mathbb{N}_{n+1-\ell, (n+1-\ell)q}$, pairwise different points $z_1, z_2, \ldots, z_r \in \mathbb{T}$, and a sequence $(\mathbf{A}_s)_{s=1}^{r}$ of non-negative Hermitian $q \times q$ matrices each of which is not equal to $0_{q \times q}$ such that

$$F = \sum_{s=1}^{\ell} \varepsilon_{u_s} \left(A_{n,u_s}^{(\alpha)}(u_s) \right)^{-1} + \sum_{s=1}^{r} \varepsilon_{z_s} \mathbf{A}_s \qquad and \qquad \sum_{s=1}^{r} \operatorname{rank} \mathbf{A}_s = (n+1-\ell)q.$$

  (c) *Suppose that $\ell \neq n + 1$ and let*

$$\widetilde{F} := F - \sum_{s=1}^{\ell} \varepsilon_{u_s} \left(A_{n,u_s}^{(\alpha)}(u_s) \right)^{-1}. \qquad (4.6)$$

Then $\widetilde{F} \in \mathcal{M}_{\geq}^{q,n-\ell}(\mathbb{T}, \mathfrak{B}_{\mathbb{T}})$ and $\widetilde{F}$ is a canonical solution in $\mathcal{M}[\mathbf{T}_{n-\ell}^{(\widetilde{F})}]$, i.e.,

$$\operatorname{rank} \mathbf{T}_{n+1-\ell}^{(\widetilde{F})} = \operatorname{rank} \mathbf{T}_{n-\ell}^{(\widetilde{F})} = (n+1-\ell)q.$$

In particular, for some $u \in \mathbb{T} \setminus \{u_1, u_2, \ldots, u_\ell\}$, the equality $F = F_{n,u}^{(\alpha)}$ holds if and only if $\widetilde{F}$ coincides with the (special) canonical solution $\widetilde{F}_{n-\ell,u}$ which is given relating to the set $\mathcal{M}[\mathbf{T}_{n-\ell}^{(\widetilde{F})}]$ and the point u.

Proof. Taking Theorems 1.1 and 2.1 into account, this is a simple consequence of Theorem 4.1 (note also Theorem 3.9). $\qquad\square$

Corollary 4.6. *Let* $(\alpha_j)_{j=1}^{\infty} \in \mathcal{T}_1$ *and let* $n \in \mathbb{N}$. *Let* $X_0, X_1, \ldots, X_n$ *be a basis of the right* $\mathbb{C}^{q \times q}$-*module* $\mathcal{R}_{\alpha,n}^{q \times q}$ *and suppose that* $\mathbf{G}$ *is a non-singular matrix such that* $\mathcal{M}[(\alpha_j)_{j=1}^n, \mathbf{G}; (X_k)_{k=0}^n] \neq \emptyset$. *Let* F *be a canonical solution in* $\mathcal{M}[(\alpha_j)_{j=1}^n, \mathbf{G}; (X_k)_{k=0}^n]$ *and let* $F^{\#}$ *be the reciprocal measure corresponding to* F. *Suppose that the value* $F^{\#}(\{z_s\})$ *is a non-singular* $q \times q$ *matrix for each* $s \in \mathbb{N}_{1,\ell}$ *with some integer* $\ell \in \mathbb{N}$ *and pairwise different points* $z_1, z_2, \ldots, z_\ell \in \mathbb{T}$.

(a) *Then* $F^{\#}$ *coincides with* $(F^{\#})_{n,z_s}^{(\alpha)}$ *and* $F(\{z_s\}) = 0_{q \times q}$ *for all* $s \in \mathbb{N}_{1,\ell}$, *where* $(F^{\#})_{n,z_s}^{(\alpha)}$ *stands for the (special) canonical solution which is given relating to* $\mathcal{M}[(\alpha_j)_{j=1}^n, \mathbf{G}^{\#}; (X_k)_{k=0}^n]$ *and* z_s. *In particular,* $\ell \leq n+1$.

(b) *The matrix measure* $F^{\#}$ *admits the representation*

$$F^{\#} = \sum_{s=1}^{n+1} \varepsilon_{z_s} \left(A_{n,z_s}^{(\alpha,\#)}(z_s) \right)^{-1} \tag{4.7}$$

if $\ell = n+1$ *and otherwise there are some* $r \in \mathbb{N}_{n+1-\ell,(n+1-\ell)q}$, *pairwise different points* $u_1, u_2, \ldots, u_r \in \mathbb{T}$, *and a sequence* $(\mathbf{A}_s^{\#})_{s=1}^r$ *of non-negative Hermitian* $q \times q$ *matrices each of which is not equal to* $0_{q \times q}$ *such that*

$$F^{\#} = \sum_{s=1}^{\ell} \varepsilon_{z_s} \left(A_{n,z_s}^{(\alpha,\#)}(z_s) \right)^{-1} + \sum_{s=1}^{r} \varepsilon_{u_s} \mathbf{A}_s^{\#} \quad and \quad \sum_{s=1}^{r} \operatorname{rank} \mathbf{A}_s^{\#} = (n+1-\ell)q.$$

(c) *Suppose that* $\ell \neq n+1$ *and let* $\widetilde{F^{\#}} := F^{\#} - \sum_{s=1}^{\ell} \varepsilon_{z_s} \left(A_{n,z_s}^{(\alpha,\#)}(z_s) \right)^{-1}$. *Then* $\widetilde{F^{\#}}$ *belongs to* $\mathcal{M}_{\geq}^{q,n-\ell}(\mathbb{T}, \mathfrak{B}_{\mathbb{T}})$ *and is a canonical solution in the set* $\mathcal{M}[\mathbf{T}_{n-\ell}^{(\widetilde{F^{\#}})}]$, *i.e.,*

$$\operatorname{rank} \mathbf{T}_{n+1-\ell}^{(\widetilde{F^{\#}})} = \operatorname{rank} \mathbf{T}_{n-\ell}^{(\widetilde{F^{\#}})} = (n+1-\ell)q.$$

In particular, for some $z \in \mathbb{T} \backslash \{z_1, z_2, \ldots, z_\ell\}$, *the equality* $F^{\#} = (F^{\#})_{n,z}^{(\alpha)}$ *holds if and only if the measure* $\widetilde{F^{\#}}$ *coincides with the (special) canonical solution* $(\widetilde{F^{\#}})_{n-\ell,u}$ *given relating to the set* $\mathcal{M}[\mathbf{T}_{n-\ell}^{(\widetilde{F^{\#}})}]$ *and the point* z.

Proof. Recalling that the matrix $\mathbf{G}$ is non-singular, by [24, Remarks 6.1 and 6.2] and the choice of $\mathbf{G}^{\#}$ we see that the matrix $\mathbf{G}^{\#}$ is non-singular as well. Furthermore, in view of [25, Remark 2.5] and the fact that the matrix measure F is a canonical solution in $\mathcal{M}[(\alpha_j)_{j=1}^n, \mathbf{G}; (X_k)_{k=0}^n]$ we get that $F^{\#}$ is a canonical solution in $\mathcal{M}[(\alpha_j)_{j=1}^n, \mathbf{G}^{\#}; (X_k)_{k=0}^n]$. Hence, the assertion follows from Corollary 4.5. $\qquad\square$

In view of (4.5) we still comment as follows (see also [25, Remark 3.2]).

Proposition 4.7. *Let* $(\alpha_j)_{j=1}^{\infty} \in \mathcal{T}_1$ *and let* $n \in \mathbb{N}$. *Let* $X_0, X_1, \ldots, X_n$ *be a basis of the right* $\mathbb{C}^{q \times q}$-*module* $\mathcal{R}_{\alpha,n}^{q \times q}$ *and suppose that* $\mathbf{G}$ *is a non-singular matrix such that the set* $\mathcal{M}[(\alpha_j)_{j=1}^n, \mathbf{G}; (X_k)_{k=0}^n]$ *is nonempty. Furthermore, let* F *be a solution in* $\mathcal{M}[(\alpha_j)_{j=1}^n, \mathbf{G}; (X_k)_{k=0}^n]$ *which admits (3.17) for some integer* $r \in \mathbb{N}_{1,n+1}$, points $u_1, u_2, \ldots, u_r \in \mathbb{T}$, and a sequence $(\mathbf{A}_s)_{s=1}^r$ of non-negative Hermitian $q \times q$ matrices. Then $r = n + 1$, the points $u_1, u_2, \ldots, u_r$ are pairwise different, and

$$\mathbf{A}_s = \left(A_{n,u_s}^{(\alpha)}(u_s) \right)^{-1}, \quad s \in \mathbb{N}_{1,r}.$$

In particular, F admits (4.5) and $F = F_{n,u_s}^{(\alpha)}$ as well as $F^{\#}(\{u_s\}) = 0_{q \times q}$ for each $s \in \mathbb{N}_{1,n+1}$, where $F^{\#}$ stands for the reciprocal measure corresponding to F.

Proof. Since F belongs to $\mathcal{M}[(\alpha_j)_{j=1}^n, \mathbf{G}; (X_k)_{k=0}^n]$, where (3.17) is satisfied with some $r \in \mathbb{N}_{1,n+1}$, and since the matrix $\mathbf{G}$ is non-singular, an application of Theorem 1.1 and [18, Theorem 6.11] implies that $r = n + 1$, that $u_1, u_2, \ldots, u_r$ are pairwise different points, and that $\mathbf{A}_s$ is a non-singular matrix for each $s \in \mathbb{N}_{1,r}$ (cf. Corollary 3.10). In particular, from (3.17) we see that there is a finite subset Δ of $\mathbb{T}$ such that $F(\mathbb{T} \setminus \Delta) = 0_{q \times q}$ and

$$\sum_{z \in \Delta} \operatorname{rank} F(\{z\}) = (n+1)q = \operatorname{rank} \mathbf{G}$$

hold. Therefore, in view of (2.2) one can conclude that the matrix measure F is a canonical solution in $\mathcal{M}[(\alpha_j)_{j=1}^n, \mathbf{G}; (X_k)_{k=0}^n]$. The remaining part of the assertion is then an easy consequence of Corollary 4.5 and the fact that the argumentation above implies that $F(\{u_s\})$ is non-singular for each $s \in \mathbb{N}_{1,n+1}$, where the points $u_1, u_2, \ldots, u_{n+1} \in \mathbb{T}$ are pairwise different. $\square$

Corollary 4.8. *Let* $(\alpha_j)_{j=1}^\infty \in \mathcal{T}_1$ *and let* $n \in \mathbb{N}$. *Let* $X_0, X_1, \ldots, X_n$ *be a basis of the right* $\mathbb{C}^{q \times q}$*-module* $\mathcal{R}_{\alpha,n}^{q \times q}$ *and suppose that* $\mathbf{G}$ *is a non-singular matrix such that the set* $\mathcal{M}[(\alpha_j)_{j=1}^n, \mathbf{G}; (X_k)_{k=0}^n]$ *is nonempty. Furthermore, let the reciprocal measure* $F^{\#}$ *corresponding to some* $F \in \mathcal{M}[(\alpha_j)_{j=1}^n, \mathbf{G}; (X_k)_{k=0}^n]$ *admit the representation*

$$F^{\#} = \sum_{s=1}^r \varepsilon_{z_s} \mathbf{A}_s^{\#} \tag{4.8}$$

for some $r \in \mathbb{N}_{1,n+1}$, *points* $z_1, z_2, \ldots, z_r \in \mathbb{T}$, *and a sequence* $(\mathbf{A}_s^{\#})_{s=1}^r$ *of non-negative Hermitian* $q \times q$ *matrices. Then* $r = n + 1$, *the points* $z_1, z_2, \ldots, z_r$ *are pairwise different, and* $\mathbf{A}_s = \left(A_{n,z_s}^{(\alpha,\#)}(z_s) \right)^{-1}$ *for all* $s \in \mathbb{N}_{1,r}$. *In particular,* $F^{\#}$ *admits (4.7) and* $F^{\#} = (F^{\#})_{n,z_s}$ *as well as* $F(\{z_s\}) = 0_{q \times q}$ *for all* $s \in \mathbb{N}_{1,n+1}$, *where* $(F^{\#})_{n,z_s}$ *stands for the (special) canonical solution which is given relating to the solution set* $\mathcal{M}[(\alpha_j)_{j=1}^n, \mathbf{G}^{\#}; (X_k)_{k=0}^n]$ *and the point* z_s.

Proof. Using a similar argumentation as in the proof of Corollary 4.6, the assertion can be seen from Proposition 4.7. $\square$

Corollary 4.9. *Let* $(\alpha_j)_{j=1}^\infty \in \mathcal{T}_1$ *and let* $n \in \mathbb{N}$. *Suppose that* $X_0, X_1, \ldots, X_n$ *is a basis of the right* $\mathbb{C}^{q \times q}$*-module* $\mathcal{R}_{\alpha,n}^{q \times q}$. *Furthermore, let* $u_1, u_2, \ldots, u_{n+1} \in \mathbb{T}$ *be pairwise different, let* $(\mathbf{A}_s)_{s=1}^{n+1}$ *be a sequence of positive Hermitian* $q \times q$ *matrices, and let* F *be the measure fulfilling (3.17) with* $r = n + 1$. *Then* $F \in \mathcal{M}_{\geq}^{q,n}(\mathbb{T}, \mathfrak{B}_{\mathbb{T}})$ *and* $\mathbf{G}_{X,n}^{(F)}$ *is a non-singular matrix. Moreover, for each* $s \in \mathbb{N}_{1,n+1}$, *the equality*

$$A_{n,u_s}^{(\alpha,F)}(u_s) = \mathbf{A}_s^{-1}$$

holds, the matrix measure F coincides with the (special) canonical solution $F_{n,u_s}^{(\alpha)}$ relating to the solution set $\mathcal{M}[(\alpha_j)_{j=1}^n, \mathbf{G}_{X,n}^{(F)}; (X_k)_{k=0}^n]$ and the point u_s as well as $F^{\#}(\{u_s\}) = 0_{q\times q}$, where $F^{\#}$ stands for the reciprocal measure corresponding to F.

Proof. Because of the choice of F and [18, part (c) of Theorem 6.11] one can see that the measure F belongs to $\mathcal{M}_{\geq}^{q,n}(\mathbb{T}, \mathfrak{B}_{\mathbb{T}})$. Therefore, Theorem 1.1 implies that $\mathbf{G}_{X,n}^{(F)}$ is a non-singular matrix. Taking this and $F \in \mathcal{M}[(\alpha_j)_{j=1}^n, \mathbf{G}_{X,n}^{(F)}; (X_k)_{k=0}^n]$ into account, for each $s \in \mathbb{N}_{1,n+1}$, from Proposition 4.7 along with (3.2) it follows that $A_{n,u_s}^{(\alpha,F)}(u_s) = \mathbf{A}_s^{-1}$ and $F = F_{n,u_s}^{(\alpha)}$ as well as $F^{\#}(\{u_s\}) = 0_{q\times q}$.   □

In view of the special case already discussed in [25, Corollaries 4.6 and 4.9], we can here conclude the following (note Theorem 2.1).

Corollary 4.10. *Let $(\alpha_j)_{j=1}^\infty \in \mathcal{T}_1$ and let $n \in \mathbb{N}$. Let $X_0, X_1, \ldots, X_n$ be a basis of the right $\mathbb{C}^{q\times q}$-module $\mathcal{R}_{\alpha,n}^{q\times q}$ and suppose that $\mathbf{G}$ is a non-singular matrix such that $\mathcal{M}[(\alpha_j)_{j=1}^n, \mathbf{G}; (X_k)_{k=0}^n] \neq \emptyset$. Furthermore, let F be a canonical solution in $\mathcal{M}[(\alpha_j)_{j=1}^n, \mathbf{G}; (X_k)_{k=0}^n]$ and let*

$$\tilde{F}_u := F - \varepsilon_u F(\{u\})$$

for some $u \in \mathbb{T}$. Then the following statements are equivalent:

(i) *For each $z \in \mathbb{T}$, the value $F(\{z\})$ is either a non-singular matrix or $0_{q\times q}$.*

(ii) *F admits (3.17) for some $r \in \mathbb{N}$, points $u_1, u_2, \ldots, u_r \in \mathbb{T}$, and a sequence $(\mathbf{A}_s)_{s=1}^r$ of positive Hermitian matrices.*

(iii) *F admits (4.5) with some pairwise different points $u_1, u_2, \ldots, u_{n+1} \in \mathbb{T}$.*

(iv) *F admits (3.17) for some $r \in \mathbb{N}_{1,n+1}$, points $u_1, u_2, \ldots, u_r \in \mathbb{T}$, and a sequence $(\mathbf{A}_s)_{s=1}^r$ of non-negative Hermitian $q \times q$ matrices.*

(v) *There is a solution $\tilde{F} \in \mathcal{M}[(\alpha_j)_{j=1}^n, \mathbf{G}; (X_k)_{k=0}^n]$ fulfilling the representation*

$$\tilde{F} = \sum_{s=1}^r \varepsilon_{u_s} \mathbf{A}_s$$

for some $r \in \mathbb{N}_{1,n+1}$, points $u_1, u_2, \ldots, u_r \in \mathbb{T}$, and a sequence $(\mathbf{A}_s)_{s=1}^r$ of non-negative Hermitian $q \times q$ matrices, where $\tilde{F}(\{u_1\}) = F(\{u_1\})$.

(vi) *For each $u \in \mathbb{T}$ with $F(\{u\}) \neq 0_{q\times q}$, the measure $\tilde{F}_u$ is a canonical solution in the set $\mathcal{M}[\mathbf{T}_{n-1}^{(\tilde{F}_u)}]$.*

(vii) *For each $u \in \mathbb{T}$ with $F(\{u\}) \neq 0_{q\times q}$, the identity $\operatorname{rank}\mathbf{T}_n^{(\tilde{F}_u)} = nq$ holds.*

(viii) *For each $u \in \mathbb{T}$ with $F(\{u\}) \neq 0_{q\times q}$, the set $\mathcal{M}[\mathbf{T}_n^{(\tilde{F}_u)}]$ is a singleton.*

Moreover, if (i) is satisfied and if $u \in \mathbb{T}$ is chosen such that $F(\{u\}) \neq 0_{q\times q}$ holds, then $F(\{u\}) = (A_{n,u}^{(\alpha)}(u))^{-1}$ and $F = F_{n,u}^{(\alpha)}$.

Proof. Taking Theorem 2.1 into account and the fact that F is a canonical solution in $\mathcal{M}[(\alpha_j)_{j=1}^n, \mathbf{G}; (X_k)_{k=0}^n]$, the equivalence of (i), (vi), (vii), and (viii) follows immediately from the equivalence of (iii), (ix), (x), and (xi) in Theorem 4.1 (note Corollary 4.4). Recalling that F is a canonical solution in $\mathcal{M}[(\alpha_j)_{j=1}^n, \mathbf{G}; (X_k)_{k=0}^n]$ and (2.2), one can see that the matrix measure F admits (3.17) for some $r \in \mathbb{N}$, pairwise different points $u_1, u_2, \ldots, u_r \in \mathbb{T}$, and a sequence $(\mathbf{A}_s)_{s=1}^r$ of non-negative Hermitian $q \times q$ matrices. Thus, the equivalence of (i) and (ii) is clear as well. Based on this reasoning and Corollary 4.5 one can also realize that (i) yields (iii). Furthermore, (iii) implies obviously (ii) (resp., (iv)). By choosing $\tilde{F} := F$, from (iv) it follows immediately (v). To complete the proof of the equivalence of (i)–(viii), it is enough to show that (v) leads to (iii). Certainly, this we get from Proposition 4.7 in combination with Theorems 4.1 and 2.1. Finally, the rest of the assertion we can conclude from Proposition 4.7. $\qquad\square$

The line of argument of Corollary 4.10 includes that, if (v) holds under the assumptions there, then the measure $\tilde{F}$ occurring in (v) coincides with F. In doing so, we will point out that the condition $\tilde{F}(\{u_1\}) = F(\{u_1\})$ in (v) of Corollary 4.10 can be also replaced by some constraint concerning the values of the associated Riesz–Herglotz transforms (similar to (vi) and (vii) in Theorem 4.1).

The next result is attached to the equivalence of (i) and (ix) in Theorem 4.1. Roughly speaking, we will add now some mass point to a solution of Problem (R) instead of removing one. In particular, we will see that every canonical solution of Problem (R) in the nondegenerate case can be extended to a special which is of the form introduced in Section 3 (see also Corollary 6.11 below).

Proposition 4.11. *Let* $(\alpha_j)_{j=1}^\infty \in \mathcal{T}_1$ *and let* $n \in \mathbb{N}$. *Let* $X_0, X_1, \ldots, X_n$ *be a basis of the right* $\mathbb{C}^{q \times q}$*-module* $\mathcal{R}_{\alpha,n}^{q \times q}$ *and suppose that* $\mathbf{G}$ *is a non-singular matrix such that* $\mathcal{M}[(\alpha_j)_{j=1}^n, \mathbf{G}; (X_k)_{k=0}^n] \neq \emptyset$. *Let* $F \in \mathcal{M}[(\alpha_j)_{j=1}^n, \mathbf{G}; (X_k)_{k=0}^n]$ *which admits* (3.17) *for some* $r \in \mathbb{N}$, *points* $u_1, u_2, \ldots, u_r \in \mathbb{T}$, *and a sequence* $(\mathbf{A}_s)_{s=1}^r$ *of nonnegative Hermitian* $q \times q$ *matrices. Furthermore, let*

$$F_{u,\mathbf{A}} := F + \varepsilon_u \mathbf{A},$$

where $u \in \mathbb{T} \setminus \{u_1, u_2, \ldots, u_r\}$ *and where* $\mathbf{A} > 0_{q \times q}$. *Suppose that* $Y_0, Y_1, \ldots, Y_{n+1}$ *is a basis of the right* $\mathbb{C}^{q \times q}$*-module* $\mathcal{R}_{\alpha,n+1}^{q \times q}$. *Then* $F_{u,\mathbf{A}}$ *belongs to* $\mathcal{M}_{\geq}^{q,n+1}(\mathbb{T}, \mathfrak{B}_\mathbb{T})$ *and the matrix* $\mathbf{G}_{Y,n+1}^{(F_{u,\mathbf{A}})}$ *is non-singular. Moreover,* $F_{u,\mathbf{A}}$ *coincides with the measure* $F_{n+1,u}^{(\alpha)}$ *which is defined relating to the solution set* $\mathcal{M}[(\alpha_j)_{j=1}^{n+1}, \mathbf{G}_{Y,n+1}^{(F_{u,\mathbf{A}})}; (Y_k)_{k=0}^{n+1}]$ *and the point* u *if and only if* F *is a canonical solution in* $\mathcal{M}[(\alpha_j)_{j=1}^n, \mathbf{G}; (X_k)_{k=0}^n]$. *In particular, if* $F_{u,\mathbf{A}} = F_{n+1,u}^{(\alpha)}$, *then*

$$A_{n+1,u}^{(\alpha, F_{u,\mathbf{A}})}(u) = \mathbf{A}^{-1}.$$

Proof. The taken requirements imply along with Lemma 3.8 that $F_{u,\mathbf{A}}$ belongs to $\mathcal{M}_{\geq}^q(\mathbb{T}, \mathfrak{B}_\mathbb{T})$ and that the matrix $\mathbf{G}_{Y,n+1}^{(F_{u,\mathbf{A}})}$ is non-singular. Thus, by Theorem 1.1 we get $F_{u,\mathbf{A}} \in \mathcal{M}_{\geq}^{q,n+1}(\mathbb{T}, \mathfrak{B}_\mathbb{T})$. Recalling (2.2), the remaining part of the assertion follows then from Theorem 4.1 in combination with Corollary 4.4. $\qquad\square$

In view of part (d) of Theorem 3.9 we know that, roughly speaking, for the (special) canonical solution $F_{n,u}^{(\alpha)}$ of Problem (R) in the nondegenerate case (with some $u \in \mathbb{T}$) the order n of the nondegeneracy will be reduced to $n-1$ by substracting the measure $\varepsilon_u F_{n,u}^{(\alpha)}(\{u\})$ from $F_{n,u}^{(\alpha)}$. In contrast to (x) in Theorem 4.1, this property cannot be used to characterize $F_{n,u}^{(\alpha)}$ in the whole set of canonical solutions of Problem (R). This will be emphasized by the following example.

Example 4.12. Let $\alpha_1 \in \mathbb{C} \setminus \mathbb{T}$ and suppose that X_0, X_1 is a basis of the right $\mathbb{C}^{q \times q}$-module $\mathcal{R}_{\alpha,1}^{3 \times 3}$. Let $\widetilde{F} := \varepsilon_1 \mathbf{A}_1 + \varepsilon_i \mathbf{A}_2$, where

$$\mathbf{A}_1 := \begin{pmatrix} 1 & 0 & 0 \\ 0 & 1 & 0 \\ 0 & 0 & 0 \end{pmatrix} \quad \text{and} \quad \mathbf{A}_2 := \begin{pmatrix} 0 & 0 & 0 \\ 0 & 1 & 0 \\ 0 & 0 & 1 \end{pmatrix}.$$

Furthermore, let $F := \varepsilon_1 \mathbf{A}_1 + \varepsilon_i \mathbf{A}_2 + \varepsilon_{-i} \mathbf{A}_3$, where

$$\mathbf{A}_3 := \begin{pmatrix} 1 & 0 & 0 \\ 0 & 0 & 0 \\ 0 & 0 & 1 \end{pmatrix}.$$

Then $F \in \mathcal{M}_{\geq}^{3,1}(\mathbb{T}, \mathfrak{B}_{\mathbb{T}})$ and $\widetilde{F} \in \mathcal{M}_{\geq}^{3,0}(\mathbb{T}, \mathfrak{B}_{\mathbb{T}})$ hold, where $\widetilde{F} = F - \varepsilon_{-i} \mathbf{A}_3$ and $\det \mathbf{G}_{X,1}^{(F)} \neq 0$. Moreover, F is a canonical solution in $\mathcal{M}[(\alpha_j)_{j=1}^1, \mathbf{G}_{X,1}^{(F)}; (X_k)_{k=0}^1]$, but there does not exist a $u \in \mathbb{T}$ such that F coincides with the measure $F_{1,u}^{(\alpha)}$ which is defined relating to the set $\mathcal{M}[(\alpha_j)_{j=1}^1, \mathbf{G}_{X,1}^{(F)}; (X_k)_{k=0}^1]$ and the point u.

Proof. Clearly, the matrices $\mathbf{A}_1$, $\mathbf{A}_2$, and $\mathbf{A}_3$ are non-negative Hermitian and each is singular, where $\operatorname{rank} \mathbf{A}_j = 2$ for $j \in \{1, 2, 3\}$. Thus, the choice of F and $\widetilde{F}$ implies $F, \widetilde{F} \in \mathcal{M}_{\geq}^3(\mathbb{T}, \mathfrak{B}_{\mathbb{T}})$, where $\widetilde{F} = F - \varepsilon_{-i} \mathbf{A}_3$. Moreover, we have

$$\mathbf{c}_0^{(F)} = \begin{pmatrix} 2 & 0 & 0 \\ 0 & 2 & 0 \\ 0 & 0 & 2 \end{pmatrix} > 0_{3 \times 3} \quad \text{and} \quad \mathbf{c}_1^{(F)} = \begin{pmatrix} 1+i & 0 & 0 \\ 0 & 1-i & 0 \\ 0 & 0 & 0 \end{pmatrix}$$

which implies that the block Toeplitz matrix $\mathbf{T}_1^{(F)}$ is positive Hermitian (for this use, e.g., [11, Lemma 1.1.9]). In other words, F belongs to $\mathcal{M}_{\geq}^{3,1}(\mathbb{T}, \mathfrak{B}_{\mathbb{T}})$. Therefore, from Theorem 1.1 we obtain that $\det \mathbf{G}_{X,1}^{(F)} \neq 0$. Consequently, in view of (2.2) we get that F is a canonical solution in $\mathcal{M}[(\alpha_j)_{j=1}^1, \mathbf{G}_{X,1}^{(F)}; (X_k)_{k=0}^1]$, whereas Corollary 4.2 shows that there is no $u \in \mathbb{T}$ such that F coincides with the special matrix measure $F_{1,u}^{(\alpha)}$. Furthermore, we get

$$\mathbf{c}_0^{(\widetilde{F})} = \begin{pmatrix} 1 & 0 & 0 \\ 0 & 2 & 0 \\ 0 & 0 & 1 \end{pmatrix} > 0_{3 \times 3}.$$

This yields that $\widetilde{F}$ belongs to $\mathcal{M}_{\geq}^{3,0}(\mathbb{T}, \mathfrak{B}_{\mathbb{T}})$.   □

Note that Example 4.12 along with Remark 3.1 shows that the set of canonical solutions of Problem (R) in this context with respect to $\mathcal{M}[(\alpha_j)_{j=1}^1, \mathbf{G}_{X,1}^{(F)}; (X_k)_{k=0}^1]$ (and the nondegenerate case) is more comprehensive as $\{F_{1,u}^{(\alpha)} : u \in \mathbb{T}\}$.

With a view to Proposition 3.13 and (3.19) we give now an example which clarifies that, if $u \in \mathbb{T}$, then there does not exist some $z \in \mathbb{T}$ in general such that the reciprocal measure $F_{n,u}^{(\alpha,\#)}$ corresponding to the canonical solution $F_{n,u}^{(\alpha)}$

coincides with the measure $(F^{\#})_{n,z}^{(\alpha)}$ which is given relating to the solution set $\mathcal{M}[(\alpha_j)_{j=1}^n, \mathbf{G}^{\#}; (X_k)_{k=0}^n]$ and the point z (not even under the strong specification that $F := F_{n,u}^{(\alpha)}$ admits (4.5) with pairwise different points $u_1, u_2, \ldots, u_{n+1} \in \mathbb{T}$).

Example 4.13. Let $\alpha_1 \in \mathbb{C} \setminus \mathbb{T}$ and let X_0, X_1 be a basis of the right $\mathbb{C}^{2 \times 2}$-module $\mathcal{R}_{\alpha,1}^{2 \times 2}$. Furthermore, let $F := \varepsilon_1 \mathbf{A}_1 + \varepsilon_{-1} \mathbf{A}_2$, where

$$\mathbf{A}_1 := \begin{pmatrix} 1 & 0 \\ 0 & 1 \end{pmatrix} \quad \text{and} \quad \mathbf{A}_2 := \begin{pmatrix} 1 & 1 \\ 1 & 2 \end{pmatrix}.$$

(a) The measure F belongs to $\mathcal{M}_{\geq}^{2,1}(\mathbb{T}, \mathfrak{B}_{\mathbb{T}})$ and $\mathbf{G}_{X,1}^{(F)}$, $\mathbf{A}_1$, and $\mathbf{A}_2$ are non-singular matrices, where $A_{1,1}^{(\alpha,F)}(1) = \mathbf{A}_1$ as well as $A_{1,-1}^{(\alpha,F)}(-1) = \mathbf{A}_2^{-1}$ and

$$\mathbf{A}_2^{-1} = \begin{pmatrix} 2 & -1 \\ -1 & 1 \end{pmatrix}.$$

Moreover, F coincides with the measure $F_{1,1}^{(\alpha)}$ (resp., $F_{1,-1}^{(\alpha)}$) which is defined relating to the set $\mathcal{M}[(\alpha_j)_{j=1}^1, \mathbf{G}_{X,1}^{(F)}; (X_k)_{k=0}^1]$ and the point 1 (resp., -1).

(b) The reciprocal measure $F^{\#}$ corresponding to F is given by

$$F^{\#} = \varepsilon_{z_1} \mathbf{B}_1 + \varepsilon_{\overline{z_1}} \mathbf{B}_1 + \varepsilon_{-z_1} \mathbf{B}_2 + \varepsilon_{-\overline{z_1}} \mathbf{B}_2,$$

where $z_1 := \frac{1}{\sqrt{5}}(1 + 2i)$ and where

$$\mathbf{B}_1 := \frac{1}{20} \begin{pmatrix} 3 - \sqrt{5} & \sqrt{5} - 1 \\ \sqrt{5} - 1 & 2 \end{pmatrix} \quad \text{and} \quad \mathbf{B}_2 := \frac{1}{20} \begin{pmatrix} 3 + \sqrt{5} & -\sqrt{5} - 1 \\ -\sqrt{5} - 1 & 2 \end{pmatrix}.$$

In particular, the matrix $F^{\#}(\{u\})$ is singular for each $u \in \mathbb{T}$ and there is no $z \in \mathbb{T}$ such that $F^{\#}$ coincides with the measure $(F^{\#})_{n,z}^{(\alpha)}$ which is given relating to the set $\mathcal{M}[(\alpha_j)_{j=1}^1, \mathbf{G}_{X,1}^{(F^{\#})}; (X_k)_{k=0}^1]$ and the point z.

Proof. Since 1 and -1 are pairwise different point belonging to $\mathbb{T}$ and since $\mathbf{A}_1$ and $\mathbf{A}_2$ are positive Hermitian 2×2 matrices, where $\mathbf{A}_1^{-1} = \mathbf{A}_1$ and where

$$\mathbf{A}_2^{-1} = \begin{pmatrix} 2 & -1 \\ -1 & 1 \end{pmatrix},$$

in view of Corollary 4.9 we get the assertion of (a). Moreover, the choice of F implies that the Riesz–Herglotz transform Ω of F satisfies the representation

$$\Omega(v) = \frac{1}{(1-v)(1+v)} \begin{pmatrix} (1+v)^2 + (1-v)^2 & (1-v)^2 \\ (1-v)^2 & (1+v)^2 + 2(1-v)^2 \end{pmatrix}$$

for each $v \in \mathbb{D}$. Thus, for each $v \in \mathbb{D}$, one can conclude that $\det \Omega(v) \neq 0$ and that

$$(\Omega(v))^{-1} = \frac{(1-v)(1+v)}{5v^4 + 6v^2 + 5} \begin{pmatrix} (1+v)^2 + 2(1-v)^2 & -(1-v)^2 \\ -(1-v)^2 & (1+v)^2 + (1-v)^2 \end{pmatrix}.$$

Consequently, by a straightforward calculation one can see that the reciprocal measure $F^{\#}$ corresponding to F is given by

$$F^{\#} = \varepsilon_{z_1} \mathbf{B}_1 + \varepsilon_{\overline{z_1}} \mathbf{B}_1 + \varepsilon_{-z_1} \mathbf{B}_2 + \varepsilon_{-\overline{z_1}} \mathbf{B}_2.$$

Since $z_1, \overline{z_1}, -z_1$, and $-\overline{z_1}$ are pairwise different and since $\mathbf{B}_1$ and $\mathbf{B}_2$ are singular matrices, Corollary 4.2 (or Corollary 3.10) yields that $F^{\#}(\{u\})$ is singular for each $u \in \mathbb{T}$ and that there is no $z \in \mathbb{T}$ such that $F^{\#}$ coincides with the measure $(F^{\#})_{n,z}^{(\alpha)}$ which is given relating to $\mathcal{M}[(\alpha_j)_{j=1}^1, \mathbf{G}_{X,1}^{(F^{\#})}; (X_k)_{k=0}^1]$ and z. $\square$

In the scalar case $q = 1$, the state of affairs is more straightforward. In particular, the set of canonical solutions of that rational (complex-valued) moment problem for the nondegenerate case is already given via $\{F_{n,u}^{(\alpha)} : u \in \mathbb{T}\}$. Furthermore, the structure discussed in Proposition 4.7 and Corollary 4.10 is significant for a canonical solution in this context. This will be emphasized by the following:

Remark 4.14. Let $(\alpha_j)_{j=1}^{\infty} \in \mathcal{T}_1$ and let $n \in \mathbb{N}$. Suppose that $X_0, X_1, \ldots, X_n$ is a basis of the linear space $\mathcal{R}_{\alpha,n}$ and let $\mathbf{G}$ be a non-singular matrix such that $\mathcal{M}[(\alpha_j)_{j=1}^n, \mathbf{G}; (X_k)_{k=0}^n] \neq \emptyset$. Let $F \in \mathcal{M}[(\alpha_j)_{j=1}^n, \mathbf{G}; (X_k)_{k=0}^n]$ and let $F^{\#}$ be the reciprocal measure corresponding to F. By Theorem 4.1 and Proposition 4.7 along with Corollaries 4.2 and 4.6 (where $q = 1$; see also [25, Remark 2.15]) one can conclude that the following statements are equivalent:

(i) F is a canonical solution in $\mathcal{M}[(\alpha_j)_{j=1}^n, \mathbf{G}; (X_k)_{k=0}^n]$.
(ii) $F = F_{n,u}^{(\alpha)}$ for some $u \in \mathbb{T}$.
(iii) F admits (4.5) with some pairwise different points $u_1, u_2, \ldots, u_{n+1} \in \mathbb{T}$.
(iv) F admits (3.17) for some $r \in \mathbb{N}_{1,n+1}$, points $u_1, u_2, \ldots, u_r \in \mathbb{T}$, and a sequence $(\mathbf{A}_s)_{s=1}^r$ of non-negative real numbers.
(v) $F^{\#}$ is a canonical solution in $\mathcal{M}[(\alpha_j)_{j=1}^n, \mathbf{G}^{\#}; (X_k)_{k=0}^n]$.
(vi) $F^{\#} = (F^{\#})_{n,z}^{(\alpha)}$ for some $z \in \mathbb{T}$.
(vii) $F^{\#}$ admits (4.7) with some pairwise different points $z_1, z_2, \ldots, z_{n+1} \in \mathbb{T}$.
(viii) $F^{\#}$ admits (4.8) for some $r \in \mathbb{N}_{1,n+1}$, points $z_1, z_2, \ldots, z_r \in \mathbb{T}$, and a sequence $(\mathbf{A}_s^{\#})_{s=1}^r$ of non-negative real numbers.

This reasoning yields also that, if (i) is satisfied and if $u \in \mathbb{T}$ is chosen such that $F(\{u\}) \neq 0$ holds, then $F(\{u\}) = (A_{n,u}^{(\alpha)}(u))^{-1}$ and $F = F_{n,u}^{(\alpha)}$, where $F^{\#}(\{u\}) = 0$.

We will now slightly modify the statement of [23, Proposition 6.6]. The result shows then that an element $F_{n,u}^{(\alpha)}$ with some $u \in \mathbb{T}$ can appear as the weak limit of some sequence of solutions of Problem (R) discussed in [23] and [24].

Recall that, if $(F_j)_{j=1}^{\infty}$ is a sequence with $F_j \in \mathcal{M}_{\geq}^q(\mathbb{T}, \mathfrak{B}_{\mathbb{T}})$ for $j \in \mathbb{N}$ and if $F \in \mathcal{M}_{\geq}^q(\mathbb{T}, \mathfrak{B}_{\mathbb{T}})$, then we say that $(F_j)_{j=1}^{\infty}$ *converges weakly* to F when

$$\lim_{j \to +\infty} \int_{\mathbb{T}} g(z)\, F_j(\mathrm{d}z) = \int_{\mathbb{T}} g(z)\, F(\mathrm{d}z)$$

holds for each real-valued, continuous, and bounded function g on $\mathbb{T}$ (see also [16]).

Proposition 4.15. *Let $(\alpha_j)_{j=1}^{\infty} \in \mathcal{T}_1$ and let $n \in \mathbb{N}$. Let $X_0, X_1, \ldots, X_n$ be a basis of the right $\mathbb{C}^{q \times q}$-module $\mathcal{R}_{\alpha,n}^{q \times q}$ and suppose that $\mathbf{G}$ is a non-singular matrix such that $\mathcal{M}[(\alpha_j)_{j=1}^n, \mathbf{G}; (X_k)_{k=0}^n] \neq \emptyset$. Furthermore, let $(w_j)_{j=1}^{\infty}$ be a sequence of points belonging to $(\mathbb{D} \setminus \mathbb{P}_{\alpha,n}) \cup \mathbb{T}$ which converges to some point $w_0 \in (\mathbb{D} \setminus \mathbb{P}_{\alpha,n}) \cup \mathbb{T}$. Then the sequence $(F_{n,w_j}^{(\alpha)})_{j=1}^{\infty}$ converges weakly to the measure $F_{n,w_0}^{(\alpha)}$.*

Proof. Our strategy is based on an application of [22, Proposition 10.7]. In doing so, for some $F \in \mathcal{M}[(\alpha_j)_{j=1}^n, \mathbf{G}; (X_k)_{k=0}^n]$, let $F^{(\alpha,n)} : \mathfrak{B}_{\mathbb{T}} \to \mathbb{C}^{q \times q}$ be the measure given via (4.4). In view of the concrete form of the measures under consideration, it is obvious that the sequence $(F_{n,w_j}^{(\alpha)})_{j=1}^\infty$ converges weakly to $F_{n,w_0}^{(\alpha)}$ if and only if the sequence $((F_{n,w_j}^{(\alpha)})^{(\alpha,n)})_{j=1}^\infty$ converges weakly to $(F_{n,w_0}^{(\alpha)})^{(\alpha,n)}$. Recalling the matricial version of the Riesz–Herglotz theorem (see, e.g., [11, Theorem 2.2.2] and note also [22, Lemma 3.3 and Remark 3.9]), a combination of [22, Proposition 10.7] with Corollary 4.3 and [23, Lemma 4.3] shows that the sequence $((F_{n,w_j}^{(\alpha)})^{(\alpha,n)})_{j=1}^\infty$ converges weakly to the measure $(F_{n,w_0}^{(\alpha)})^{(\alpha,n)}$. Thus, the proof is complete. □

Again, we make a brief statement concerning the elementary case $n = 0$ which is related to looking for measures $F \in \mathcal{M}_{\geq}^q(\mathbb{T}, \mathfrak{B}_{\mathbb{T}})$ fulfilling (2.8), based on a constant function X_0 defined on $\mathbb{C}_0$ with a non-singular $q \times q$ matrix $\mathbf{X}_0$ as value and a positive Hermitian $q \times q$ matrix $\mathbf{G}$. In view of (3.18) with (2.10) one can immediately find that appropriate statements are satisfied in that context as presented in this section concerning Problem (R). In particular, one can realize that the set of canonical solutions with respect to (2.8) and (2.9) is more comprehensive as $\{F_{0,u}^{(\alpha)} : u \in \mathbb{T}\}$ in the case of $q \geq 2$ (cf. [22, Example 9.11]).

5. An extremal property of $F_{n,u}^{(\alpha)}$ in the solution set of Problem (R)

The considerations in the present section are attached to (3.15). In fact, we will see that this relation results in an extremal property for canonical solutions of Problem (R) within the solution set in the nondegenerate case. Concerning this matter, based on Theorem 4.1 we will verify that the (special) canonical solution $F_{n,u}^{(\alpha)}$ of Problem (R) has an extra behavior with respect to the point $u \in \mathbb{T}$.

We first give some auxiliary results in view of (3.15).

Lemma 5.1. *Let* $\mathbf{A} \geq 0_{q \times q}$ *and let* $\mathbf{B} > 0_{q \times q}$. *Then:*

(a) *The following statements are equivalent:*
   (i) $\mathbf{B}^{-1} \geq \mathbf{A}$.
   (ii) $\begin{pmatrix} \mathbf{B}^{-1} & \mathbf{A} \\ \mathbf{A} & \mathbf{A} \end{pmatrix} \geq 0_{2q \times 2q}$.
   (iii) $\mathbf{A} \geq \mathbf{ABA}$.
   (iv) $\mathbf{A}^+ \geq \mathbf{A}^+ \mathbf{ABA}^+ \mathbf{A}$.
   (v) *The identity* $\mathbf{x}^* \mathbf{A}^+ \mathbf{x} \geq \mathbf{x}^* \mathbf{Bx}$ *holds for each* $\mathbf{x} \in \mathcal{R}(\mathbf{A})$.

(b) *The following statements are equivalent:*
   (vi) *The identity* $\mathbf{x}^* \mathbf{B}^{-1} \mathbf{x} \geq \mathbf{x}^* \mathbf{Ax}$ *holds for each* $\mathbf{x} \in \mathcal{R}(\mathbf{A})$.
   (vii) $\mathbf{A}^+ \mathbf{B}^{-1} \mathbf{A}^+ \geq \mathbf{A}^+$.
   (viii) $\mathbf{A}^+ \mathbf{AB}^{-1} \mathbf{A}^+ \mathbf{A} \geq \mathbf{A}$.
   (ix) $\begin{pmatrix} \mathbf{A}^+ \mathbf{AB}^{-1} \mathbf{A}^+ \mathbf{A} & \mathbf{A} \\ \mathbf{A} & \mathbf{A} \end{pmatrix} \geq 0_{2q \times 2q}$.

(c) *If* (i) *is satisfied, then it follows that* (vi) *holds.*

Proof. In view of the definition of the Moore–Penrose inverse $\mathbf{A}^+$ of $\mathbf{A}$ we have

$$\mathbf{A} = \mathbf{A}\mathbf{A}^+\mathbf{A} \qquad \text{and} \qquad \mathbf{A}^+ = \mathbf{A}^+\mathbf{A}\mathbf{A}^+.$$

Thus, taking some well-known results on the block decomposition of non-negative Hermitian matrices into account and the fact that the complex $q \times q$ matrix $\mathbf{A}$ is non-negative Hermitian (see, e.g., [11, Lemmas 1.1.6 and 1.1.9]), the assertions of (a) and (b) follow by a straightforward calculation. Furthermore, since (i) is satisfied if and only if the inequality

$$\mathbf{y}^*\mathbf{B}^{-1}\mathbf{y} \geq \mathbf{y}^*\mathbf{A}\mathbf{y}$$

holds for each $\mathbf{y} \in \mathbb{C}^{q\times1}$, obviously, (i) leads to (vi).   □

With a view to part (c) of Lemma 5.1 we give now a simple example (with $q = 2$) which shows that, conversely, (vi) does not generally imply (i).

Example 5.2. Let

$$\mathbf{A} := \begin{pmatrix} 1 & 0 \\ 0 & 0 \end{pmatrix} \qquad \text{and} \qquad \mathbf{B} := \frac{2}{3}\begin{pmatrix} 2 & -1 \\ -1 & 2 \end{pmatrix}.$$

Then $\mathbf{A} \geq 0_{2\times2}$ and $\mathbf{B} > 0_{2\times2}$ hold, where

$$\mathcal{R}(\mathbf{A}) = \left\{ \begin{pmatrix} x \\ 0 \end{pmatrix} : x \in \mathbb{C} \right\} \qquad \text{and} \qquad \mathbf{B}^{-1} = \frac{1}{2}\begin{pmatrix} 2 & 1 \\ 1 & 2 \end{pmatrix}. \qquad (5.1)$$

In particular, the identity $\mathbf{x}^*\mathbf{B}^{-1}\mathbf{x} = \mathbf{x}^*\mathbf{A}\mathbf{x}$ is satisfied for each $\mathbf{x} \in \mathcal{R}(\mathbf{A})$, but $\mathbf{B}^{-1} - \mathbf{A}$ is not a non-negative Hermitian 2×2 matrix.

Proof. It is easy to see that $\mathbf{A}$ is a non-negative Hermitian 2×2 matrix and that $\mathbf{B}$ is a positive Hermitian 2×2 matrix, where the equations in (5.1) are fulfilled. Therefore, we have on the one hand

$$\begin{pmatrix} x \\ 0 \end{pmatrix}^* \mathbf{B}^{-1} \begin{pmatrix} x \\ 0 \end{pmatrix} = |x|^2 = \begin{pmatrix} x \\ 0 \end{pmatrix}^* \mathbf{A} \begin{pmatrix} x \\ 0 \end{pmatrix}, \qquad x \in \mathbb{C},$$

which shows that $\mathbf{x}^*\mathbf{B}^{-1}\mathbf{x} = \mathbf{x}^*\mathbf{A}\mathbf{x}$ for each $\mathbf{x} \in \mathcal{R}(\mathbf{A})$. On the other hand, we get

$$\mathbf{B}^{-1} - \mathbf{A} = \frac{1}{2}\begin{pmatrix} 0 & 1 \\ 1 & 2 \end{pmatrix}$$

which reveals that $\mathbf{B}^{-1} - \mathbf{A}$ is not a non-negative Hermitian 2×2 matrix.   □

Lemma 5.3. *Let* $\mathbf{A},\mathbf{B} \in \mathbb{C}^{q\times q}$ *be Hermitian. The following statements are equivalent:*

(i) $\mathbf{A}\mathbf{B}\mathbf{x} = \mathbf{x}$ *for each* $\mathbf{x} \in \mathcal{R}(\mathbf{A})$.
(ii) $\mathbf{y}^*\mathbf{B}\mathbf{A} = \mathbf{y}^*$ *for each* $\mathbf{y} \in \mathcal{R}(\mathbf{A})$.
(iii) $\mathbf{y}^*\mathbf{A}\mathbf{B}\mathbf{x} = \mathbf{y}^*\mathbf{x}$ *for all* $\mathbf{x},\mathbf{y} \in \mathcal{R}(\mathbf{A})$.
(iv) $\mathbf{y}^*\mathbf{B}\mathbf{A}\mathbf{x} = \mathbf{y}^*\mathbf{x}$ *for all* $\mathbf{x},\mathbf{y} \in \mathcal{R}(\mathbf{A})$.
(v) $\mathbf{y}^*\mathbf{B}\mathbf{x} = \mathbf{y}^*\mathbf{A}^+\mathbf{x}$ *for all* $\mathbf{x},\mathbf{y} \in \mathcal{R}(\mathbf{A})$.
(vi) $\mathbf{A}\mathbf{B}\mathbf{A} = \mathbf{A}$.
(vii) $\mathbf{A}^+\mathbf{B}\mathbf{A} = \mathbf{A}^+$.
(viii) $\mathbf{A}\mathbf{B}\mathbf{A}^+ = \mathbf{A}^+$.

In particular, if $\mathbf{A}^+\mathbf{x} = \mathbf{B}\mathbf{x}$ *is fulfilled for each* $\mathbf{x} \in \mathcal{R}(\mathbf{A})$, *then* (i) *holds.*

Proof. Obviously, (i) holds if and only if (ii) is satisfied. Furthermore, (i) implies immediately (iii) as well as (vi) leads to (i). We are going now to the essential part of the proof. Similar to the argumentation of Lemma 5.1, this will be based on some fundamental properties of the Moore–Penrose inverse of a (Hermitian) matrix (see, e.g., [11, pages 13–17] for details). One can see that (iii) is equivalent to

$$\mathbf{w}^*\mathbf{A}\mathbf{A}\mathbf{B}\mathbf{A}\mathbf{z} = \mathbf{w}^*\mathbf{A}\mathbf{A}\mathbf{z}, \quad \mathbf{w}, \mathbf{z} \in \mathbb{C}^{q \times 1}.$$

Since this holds if and only if $\mathbf{A}^2\mathbf{B}\mathbf{A} = \mathbf{A}^2$ and since $\mathbf{A} = \mathbf{A}^+\mathbf{A}^2$, one can conclude that (iii) is equivalent to (vi). Similarly, one can find that (iv) is equivalent to (vi). The equivalence of (vi) and (vii) (resp., of (vi) and (viii)) is an easy consequence of $\mathbf{A}^+ = (\mathbf{A}^+)^2\mathbf{A}$ and $\mathbf{A} = \mathbf{A}^2\mathbf{A}^+$ (resp., of $\mathbf{A}^+ = \mathbf{A}(\mathbf{A}^+)^2$ and $\mathbf{A} = \mathbf{A}^+\mathbf{A}^2$). Besides, in view of $\mathbf{A}^+ = (\mathbf{A}^+)^2\mathbf{A}$ we obtain that (vii) holds if and only if

$$\mathbf{w}^*\mathbf{A}^+\mathbf{B}\mathbf{A}\mathbf{z} = \mathbf{w}^*\mathbf{A}^+\mathbf{A}^+\mathbf{A}\mathbf{z}, \quad \mathbf{w}, \mathbf{z} \in \mathbb{C}^{q \times 1}.$$

Thus, taking $\mathcal{R}(\mathbf{A}) = \mathcal{R}(\mathbf{A}^+)$ into account, one can conclude that (vii) is equivalent to (v). Hence, we have proven that (i)–(viii) are equivalent. Now, let the equality $\mathbf{A}^+\mathbf{x} = \mathbf{B}\mathbf{x}$ be fulfilled for each $\mathbf{x} \in \mathcal{R}(\mathbf{A})$. Suppose that $\mathbf{x} \in \mathcal{R}(\mathbf{A})$. Consequently, there is a $\mathbf{z} \in \mathbb{C}^{q \times 1}$ such that $\mathbf{A}\mathbf{z} = \mathbf{x}$ and we get

$$\mathbf{x} = \mathbf{A}\mathbf{z} = \mathbf{A}\mathbf{A}^+\mathbf{A}\mathbf{z} = \mathbf{A}\mathbf{A}^+\mathbf{x} = \mathbf{A}\mathbf{B}\mathbf{x}.$$

Therefore, if $\mathbf{A}^+\mathbf{x} = \mathbf{B}\mathbf{x}$ for each $\mathbf{x} \in \mathcal{R}(\mathbf{A})$, then (i) holds. □

Now, keeping Lemma 5.3 in mind, we point out an example which clarifies that (i) in Lemma 5.3 does not imply $\mathbf{A}^+\mathbf{x} = \mathbf{B}\mathbf{x}$ or $\mathbf{B}\mathbf{A}\mathbf{x} = \mathbf{x}$ or $\mathbf{x}^*\mathbf{A}\mathbf{x} = \mathbf{x}^*\mathbf{B}^+\mathbf{x}$ for each $\mathbf{x} \in \mathcal{R}(\mathbf{A})$ in general (not even under the claim as in Lemma 5.1 that $\mathbf{A} \geq 0_{q \times q}$ and $\mathbf{B} > 0_{q \times q}$). In fact, this example is attached to Example 5.2.

Example 5.4. Let

$$\tilde{\mathbf{A}} := \begin{pmatrix} 1 & 0 \\ 0 & 0 \end{pmatrix} \quad \text{and} \quad \tilde{\mathbf{B}} := \frac{1}{2}\begin{pmatrix} 2 & 1 \\ 1 & 2 \end{pmatrix}.$$

Then $\tilde{\mathbf{A}} \geq 0_{2 \times 2}$ and $\tilde{\mathbf{B}} > 0_{2 \times 2}$ hold, where the equality $\tilde{\mathbf{A}}\tilde{\mathbf{B}}\mathbf{x} = \mathbf{x}$ for each $\mathbf{x} \in \mathcal{R}(\tilde{\mathbf{A}})$ is satisfied. But, for each $\mathbf{x} \in \mathcal{R}(\tilde{\mathbf{A}}) \setminus \{0_{2 \times 2}\}$, the relations $\tilde{\mathbf{A}}^+\mathbf{x} \neq \tilde{\mathbf{B}}\mathbf{x}$ and $\tilde{\mathbf{B}}\tilde{\mathbf{A}}\mathbf{x} \neq \mathbf{x}$ as well as $\mathbf{x}^*\tilde{\mathbf{A}}\mathbf{x} \neq \mathbf{x}^*\tilde{\mathbf{B}}^+\mathbf{x}$ hold.

Proof. Because of Example 5.2 and (5.1) one can conclude that $\tilde{\mathbf{A}} \geq 0_{2 \times 2}$ and $\tilde{\mathbf{B}} > 0_{2 \times 2}$ hold, where the equality $\tilde{\mathbf{A}}\tilde{\mathbf{B}}\mathbf{x} = \mathbf{x}$ is satisfied for each $\mathbf{x} \in \mathcal{R}(\tilde{\mathbf{A}})$. But, for each $x \in \mathbb{C} \setminus \{0\}$, in view of $\tilde{\mathbf{A}} = \tilde{\mathbf{A}}^+$ and Example 5.2 we get

$$\tilde{\mathbf{A}}^+\begin{pmatrix} x \\ 0 \end{pmatrix} = \begin{pmatrix} x \\ 0 \end{pmatrix} \neq \frac{1}{2}\begin{pmatrix} 2x \\ x \end{pmatrix} = \tilde{\mathbf{B}}\begin{pmatrix} x \\ 0 \end{pmatrix},$$

$$\tilde{\mathbf{B}}\tilde{\mathbf{A}}\begin{pmatrix} x \\ 0 \end{pmatrix} = \frac{1}{2}\begin{pmatrix} 2x \\ x \end{pmatrix} \neq \begin{pmatrix} x \\ 0 \end{pmatrix},$$

and

$$\begin{pmatrix} x \\ 0 \end{pmatrix}^*\tilde{\mathbf{A}}\begin{pmatrix} x \\ 0 \end{pmatrix} = |x|^2 \neq \frac{4}{3}|x|^2 = \begin{pmatrix} x \\ 0 \end{pmatrix}^*\tilde{\mathbf{B}}^{-1}\begin{pmatrix} x \\ 0 \end{pmatrix}.$$

Recalling (5.1), this points up that $\tilde{\mathbf{A}}^+\mathbf{x} \neq \tilde{\mathbf{B}}\mathbf{x}$, $\tilde{\mathbf{B}}\tilde{\mathbf{A}}\mathbf{x} \neq \mathbf{x}$, and $\mathbf{x}^*\tilde{\mathbf{A}}\mathbf{x} \neq \mathbf{x}^*\tilde{\mathbf{B}}^+\mathbf{x}$ for each $\mathbf{x} \in \mathcal{R}(\tilde{\mathbf{A}}) \setminus \{0_{2 \times 2}\}$ (where $\mathcal{R}(\tilde{\mathbf{A}}) \setminus \{0_{2 \times 2}\}$ is nonempty). □

We turn now to the main result of this paper. Here, we use again the notation introduced at the beginning of Section 3 (note in particular (3.1) and (3.3)).

Theorem 5.5. *Let $(\alpha_j)_{j=1}^{\infty} \in \mathcal{T}_1$ and let $n \in \mathbb{N}$. Let $X_0, X_1, \ldots, X_n$ be a basis of the right $\mathbb{C}^{q \times q}$-module $\mathcal{R}_{\alpha,n}^{q \times q}$ and suppose that $\mathbf{G}$ is a non-singular matrix such that $\mathcal{M}[(\alpha_j)_{j=1}^n, \mathbf{G}; (X_k)_{k=0}^n] \neq \emptyset$. Furthermore, let $F \in \mathcal{M}[(\alpha_j)_{j=1}^n, \mathbf{G}; (X_k)_{k=0}^n]$. Then*

$$\mathbf{x}^* \big(F(\{u\}) \big)^+ \mathbf{x} \geq \mathbf{x}^* A_{n,u}^{(\alpha)}(u) \mathbf{x}, \quad \mathbf{x} \in \mathcal{R}\big(F(\{u\}) \big), \qquad (5.2)$$

is satisfied for each $u \in \mathbb{T}$, where equality holds in (5.2) in the case that F is a canonical solution in $\mathcal{M}[(\alpha_j)_{j=1}^n, \mathbf{G}; (X_k)_{k=0}^n]$. Moreover, if $u \in \mathbb{T}$, then

$$\big(A_{n,u}^{(\alpha)}(u) \big)^{-1} \geq F(\{u\}), \qquad (5.3)$$

where equality holds in (5.3) if and only if F coincides with the measure $F_{n,u}^{(\alpha)}$. In particular, if $u \in \mathbb{T}$, then the inequality

$$\frac{1}{\det A_{n,u}^{(\alpha)}(u)} \geq \det F(\{u\}) \qquad (5.4)$$

is satisfied, where equality holds if and only if F coincides with the measure $F_{n,u}^{(\alpha)}$.

Proof. Let $u \in \mathbb{T}$. Recall that in view of $F \in \mathcal{M}[(\alpha_j)_{j=1}^n, \mathbf{G}; (X_k)_{k=0}^n]$ and the assumption $\det \mathbf{G} \neq 0$ along with Theorem 1.1 we see that $F \in \mathcal{M}^{q,n}(\mathbb{T}, \mathfrak{B}_{\mathbb{T}})$. Furthermore, we know that (3.2) and (3.3) hold. In particular, $A_{n,u}^{(\alpha)}(u)$ is a positive Hermitian $q \times q$ matrix. If $Y_u : \mathbb{T} \to \mathbb{C}^{q \times q}$ stands for the matrix function given by $Y_u(u) = \mathbf{I}_q$ and $Y_u(z) = 0_{q \times q}$ for each $z \in \mathbb{T} \setminus \{u\}$, then we get on the one hand

$$\int_{\mathbb{T}} \big(Y(z) \big)^* F(dz) \, Y(z) \geq \int_{\mathbb{T}} \big(Y_u(z) \big)^* F(dz) \, Y_u(z) = F(\{u\})$$

for each $Y \in \mathcal{R}_{\alpha,n}^{q \times q}$ fulfilling $Y(u) = \mathbf{I}_q$. On the other hand, from [17, Theorem 22] and (3.2) one can see that there is a $\tilde{Y} \in \mathcal{R}_{\alpha,n}^{q \times q}$ fulfilling $\tilde{Y}(u) = \mathbf{I}_q$ and

$$\int_{\mathbb{T}} \big(\tilde{Y}(z) \big)^* F(dz) \, \tilde{Y}(z) = \big(A_{n,u}^{(\alpha)}(u) \big)^{-1}.$$

Thus, we get that (5.3) holds. Furthermore, from part (a) of Theorem 3.9 we already know that equality holds in (5.3) in the case of

$$F = F_{n,u}^{(\alpha)}.$$

Now, we shall show that there is no other choice of F such that that equality holds in (5.3), i.e., that $F(\{u\}) = \big(A_{n,u}^{(\alpha)}(u) \big)^{-1}$ implies also $F = F_{n,u}^{(\alpha)}$. In view of (3.13) one can realize that the setting

$$F_u := F_{n,u}^{(\alpha)} - \varepsilon_u \big(A_{n,u}^{(\alpha)}(u) \big)^{-1}$$

leads to a matrix measure belonging to $\mathcal{M}_{\geq}^q(\mathbb{T}, \mathfrak{B}_{\mathbb{T}})$. Moreover, by the already proven part of the assertion and $F_{n,u}^{(\alpha)} \in \mathcal{M}[(\alpha_j)_{j=1}^n, \mathbf{G}; (X_k)_{k=0}^n]$ one can conclude that $F = F_{n,u}^{(\alpha)}$ is the unique choice such that equality holds in (5.3) if and only if the set $\mathcal{M}[(\alpha_j)_{j=1}^n, \mathbf{G}_{X,n}^{(F_u)}; (X_k)_{k=0}^n]$ consists exactly of one element, namely F_u. Consequently, because of [19, Theorem 1] and [18, Theorem 4.4] we obtain that

$F = F_{n,u}^{(\alpha)}$ is the unique choice such that equality holds in (5.3) if and only if

$$\operatorname{rank} \mathbf{T}_n^{(F_u)} = \operatorname{rank} \mathbf{T}_{n-1}^{(F_u)}.$$

Hence, from part (d) of Theorem 3.9 we find that $F(\{u\}) = \left(A_{n,u}^{(\alpha)}(u)\right)^{-1}$ yields the identity $F = F_{n,u}^{(\alpha)}$. Based on that and (5.3), since $F(\{u\})$ is a non-negative Hermitian $q \times q$ matrix and since $\left(A_{n,u}^{(\alpha)}(u)\right)^{-1}$ is a positive Hermitian $q \times q$ matrix, an elementary result in linear algebra (see, e.g., [23, Remark 4.9]) provides that the inequality (5.4) is satisfied, where equality holds in (5.4) if and only if F coincides with the measure $F_{n,u}^{(\alpha)}$. Besides, from (5.3) along with Lemma 5.1 it follows that (5.2) holds. Finally, if F is a canonical solution in $\mathcal{M}[(\alpha_j)_{j=1}^n, \mathbf{G}; (X_k)_{k=0}^n]$, then by using Theorem 2.1 and [25, part (b) of Theorem 4.3] we get the identity

$$F(\{u\})A_{n,u}^{(\alpha)}(u)\mathbf{x} = \mathbf{x}, \quad \mathbf{x} \in \mathcal{R}\big(F(\{u\})\big).$$

Therefore, an application of Lemma 5.3 tells us that equality holds in (5.2) in the case that F is a canonical solution in $\mathcal{M}[(\alpha_j)_{j=1}^n, \mathbf{G}; (X_k)_{k=0}^n]$. $\qquad\square$

By comparing (5.2) with an inequality stated in [12, Theorem 1] regarding a moment problem for matrix measures on the real line one would expect that

$$\mathbf{x}^* \big(A_{n,u}^{(\alpha)}(u)\big)^{-1}\mathbf{x} \geq \mathbf{x}^* F(\{u\})\mathbf{x}, \quad \mathbf{x} \in \mathcal{R}\big(F(\{u\})\big), \tag{5.5}$$

for each $F \in \mathcal{M}[(\alpha_j)_{j=1}^n, \mathbf{G}; (X_k)_{k=0}^n]$ and each $u \in \mathbb{T}$, where equality holds in (5.5) if F is a canonical solution in $\mathcal{M}[(\alpha_j)_{j=1}^n, \mathbf{G}; (X_k)_{k=0}^n]$. However, the argumentation in the proof of [12, Theorem 1] contains a mistake (compare, e.g., Lemma 5.3 and Example 5.4 with the step in [12] from (1.6) to the penultimate formula on page 314). Of course (cf. part (c) of Lemma 5.1), from (5.3) it follows that (5.5) is satisfied, but equality does not need to hold in (5.5) in the case that F is a canonical solution in $\mathcal{M}[(\alpha_j)_{j=1}^n, \mathbf{G}; (X_k)_{k=0}^n]$. For instance, based on Example 4.13 (and the value of the matrix measure $F^\#$ concerning the singleton $\{z_1\}$, $\{\overline{z_1}\}$, $\{-z_1\}$, or $\{-\overline{z_1}\}$) one can get an explicit counterexample for $n = 1$ (note [25, Remark 2.5] and [18, Theorem 7.2]). Anyway, by a still easier example, the treatments at the end of this section clarify that even for $n = 0$ equality does not need to hold in (5.5) if F is a canonical solution with respect to (2.8) and (2.9).

Regarding the conception of reciprocal measures in the set $\mathcal{M}_{\geq}^q(\mathbb{T}, \mathfrak{B}_{\mathbb{T}})$ and Theorem 5.5 (see also Remark 3.12 and Proposition 3.13) we get the following.

Corollary 5.6. *Let $(\alpha_j)_{j=1}^\infty \in \mathcal{T}_1$ and let $n \in \mathbb{N}$. Let $X_0, X_1, \ldots, X_n$ be a basis of the right $\mathbb{C}^{q \times q}$-module $\mathcal{R}_{\alpha,n}^{q \times q}$ and suppose that $\mathbf{G}$ is a non-singular matrix such that $\mathcal{M}[(\alpha_j)_{j=1}^n, \mathbf{G}; (X_k)_{k=0}^n] \neq \emptyset$. Furthermore, let $F \in \mathcal{M}[(\alpha_j)_{j=1}^n, \mathbf{G}; (X_k)_{k=0}^n]$ and let $F^\#$ be the reciprocal measure corresponding to F. Then:*

(a) *For each $u \in \mathbb{T}$,*

$$\mathbf{x}^* \big(F^\#(\{u\})\big)^+\mathbf{x} \geq \mathbf{x}^* A_{n,u}^{(\alpha,\#)}(u)\mathbf{x}, \quad \mathbf{x} \in \mathcal{R}\big(F^\#(\{u\})\big),$$

where equality holds if F is a canonical solution in $\mathcal{M}[(\alpha_j)_{j=1}^n, \mathbf{G}; (X_k)_{k=0}^n]$.

(b) *If $z \in \mathbb{T}$, then*
$$\left(A_{n,z}^{(\alpha,\#)}(z)\right)^{-1} \geq F^{\#}(\{z\}),$$
where equality holds if and only if $F^{\#}$ coincides with the (special) canonical solution $(F^{\#})_{n,z}^{(\alpha)}$ which is given relating to the set $\mathcal{M}[(\alpha_j)_{j=1}^n, \mathbf{G}^{\#}; (X_k)_{k=0}^n]$ and the point z. In particular, if $z \in \mathbb{T}$, then the inequality
$$\frac{1}{\det A_{n,z}^{(\alpha,\#)}(z)} \geq \det F^{\#}(\{z\}) \tag{5.6}$$
is satisfied, where equality holds if and only if $F^{\#} = (F^{\#})_{n,z}^{(\alpha)}$.

(c) *Let $z \in \mathbb{T}$ and suppose that equality holds in (5.6). Then $F = F_{n,u}^{(\alpha)}$ for some $u \in \mathbb{T}$ if and only if one of the equivalent statements in Proposition 3.13 holds.*

(d) *Let $z \in \mathbb{T}$ and let $u = z$. If equality holds in (5.4) (resp., in (5.6)), then*
$$\left(A_{n,z}^{(\alpha,\#)}(z)\right)^{-1} > 0_{q \times q} = F^{\#}(\{z\}) \quad \left(\text{resp., } \left(A_{n,z}^{(\alpha)}(z)\right)^{-1} > 0_{q \times q} = F(\{z\})\right).$$

Proof. By [24, Remark 2.5] we know that the matrix measure F is a canonical solution in the set $\mathcal{M}[(\alpha_j)_{j=1}^n, \mathbf{G}; (X_k)_{k=0}^n]$ if and only if the matrix measure $F^{\#}$ is a canonical solution in the set $\mathcal{M}[(\alpha_j)_{j=1}^n, \mathbf{G}^{\#}; (X_k)_{k=0}^n]$. Thus, taking the definition of $\mathbf{G}^{\#}$ and [24, Remarks 6.1 and 6.2] into account, Theorem 5.5 along with Proposition 3.13 provides us with the assertion. □

Remark 5.7. Relating to Corollary 5.6 (and the conditions there), one can see that, if $u \in \mathbb{T}$ and if F is the solution in $\mathcal{M}[(\alpha_j)_{j=1}^n, \mathbf{G}; (X_k)_{k=0}^n]$ such that equality holds in (5.3), then $F^{\#}(\{u\}) \neq \left(A_{n,u}^{(\alpha,\#)}(u)\right)^{-1}$. Moreover, recalling [24, Remark 2.5] and Theorem 5.5, because of Example 4.13 it follows that the case is possible that F is the unique element in $\mathcal{M}[(\alpha_j)_{j=1}^n, \mathbf{G}; (X_k)_{k=0}^n]$ such that equality holds in (5.3) concerning each of its mass points u, but $F^{\#}(\{z\}) \neq \left(A_{n,z}^{(\alpha,\#)}(z)\right)^{-1}$ for each $z \in \mathbb{T}$.

The special case discussed in Corollary 4.10 is closely related to the scalar case $q = 1$ in some sense (cf. Remark 4.14). In particular, for that situation, the proof of the uniqueness part concerning Theorem 5.5 can be done in a more self-evident manner (cf. [22, Proposition 9.13 and Theorem 9.18]). We give now an explicit example which shows that the uniqueness with respect to equality in (5.3) for some point $u \in \mathbb{T}$ does not generally lead to the situation in Corollary 4.10.

Example 5.8. Let $\alpha_1 \in \mathbb{C} \setminus \mathbb{T}$ and suppose that X_0, X_1 is a basis of the right $\mathbb{C}^{q \times q}$-module $\mathcal{R}_{\alpha,1}^{2 \times 2}$. Furthermore, let $F := \varepsilon_1 \mathbf{A}_1 + \varepsilon_i \mathbf{A}_2 + \varepsilon_{-i} \mathbf{A}_3$, where $\mathbf{A}_1 := \mathbf{I}_2$,
$$\mathbf{A}_2 := \begin{pmatrix} 1 & 0 \\ 0 & 0 \end{pmatrix}, \quad \text{and} \quad \mathbf{A}_3 := \begin{pmatrix} 0 & 0 \\ 0 & 1 \end{pmatrix}.$$
Then $F \in \mathcal{M}_{\geq}^{2,1}(\mathbb{T}, \mathfrak{B}_{\mathbb{T}})$ and the matrix $\mathbf{G}_{X,1}^{(F)}$ is non-singular, whereas the matrices $\mathbf{A}_2$ and $\mathbf{A}_3$ are singular. Moreover, the inequality
$$\begin{pmatrix} 1 & 0 \\ 0 & 1 \end{pmatrix} \geq H(\{1\}) \tag{5.7}$$
is fulfilled for each $H \in \mathcal{M}[(\alpha_j)_{j=1}^1, \mathbf{G}_{X,1}^{(F)}; (X_k)_{k=0}^1]$, where equality holds in (5.7) if and only if $H = F$. In particular, F coincides with the measure $F_{1,1}^{(\alpha)}$ which is

defined relating to the set $\mathcal{M}[(\alpha_j)_{j=1}^1, \mathbf{G}_{X,1}^{(F)}; (X_k)_{k=0}^1]$ and $F(\{1\}) = \left(A_{1,1}^{(\alpha)}(1)\right)^{-1}$, but $F(\{\mathrm{i}\}) \neq \left(A_{1,\mathrm{i}}^{(\alpha)}(\mathrm{i})\right)^{-1}$ and $F(\{-\mathrm{i}\}) \neq \left(A_{1,-\mathrm{i}}^{(\alpha)}(-\mathrm{i})\right)^{-1}$.

Proof. Clearly, the matrices $\mathbf{A}_2$ and $\mathbf{A}_3$ are singular and equality holds in (5.7) in the case of $H = F$. Moreover, we have

$$\mathbf{c}_0^{(F)} = \begin{pmatrix} 2 & 0 \\ 0 & 2 \end{pmatrix} \quad \text{and} \quad \mathbf{c}_1^{(F)} = \begin{pmatrix} 1 - \mathrm{i} & 0 \\ 0 & 1 + \mathrm{i} \end{pmatrix}$$

which implies (cf. Example 4.12 and use, e.g., [11, Lemma 1.1.9]) that F belongs to $\mathcal{M}_{\geq}^{2,1}(\mathbb{T}, \mathcal{B}_{\mathbb{T}})$. Therefore, from Theorem 1.1 we get that the matrix $\mathbf{G}_{X,1}^{(F)}$ is non-singular. Consequently, the choice of the matrices $\mathbf{A}_1$, $\mathbf{A}_2$, and $\mathbf{A}_3$ yields in combination with Corollary 4.2 that F coincides with the canonical solution $F_{1,1}^{(\alpha)}$ in the solution set $\mathcal{M}[(\alpha_j)_{j=1}^1, \mathbf{G}_{X,1}^{(F)}; (X_k)_{k=0}^1]$. Hence, Theorem 5.5 shows that

$$F(\{1\}) = \left(A_{1,1}^{(\alpha)}(1)\right)^{-1}$$

and that (5.7) is satisfied for each $H \in \mathcal{M}[(\alpha_j)_{j=1}^1, \mathbf{G}_{X,1}^{(F)}; (X_k)_{k=0}^1]$. Finally, since $\mathbf{A}_2$ and $\mathbf{A}_3$ are singular and since $F(\{\mathrm{i}\}) = \mathbf{A}_2$ and $F(\{-\mathrm{i}\}) = \mathbf{A}_3$, we have $F(\{\mathrm{i}\}) \neq \left(A_{1,\mathrm{i}}^{(\alpha)}(\mathrm{i})\right)^{-1}$ and $F(\{-\mathrm{i}\}) \neq \left(A_{1,-\mathrm{i}}^{(\alpha)}(-\mathrm{i})\right)^{-1}$. $\square$

Let us now consider the elementary case $n = 0$ which is based on a given constant function X_0 defined on $\mathbb{C}_0$ with a non-singular $q \times q$ matrix $\mathbf{X}_0$ as value and a positive Hermitian $q \times q$ matrix $\mathbf{G}$. Suppose that $F \in \mathcal{M}_{\geq}^q(\mathbb{T}, \mathcal{B}_{\mathbb{T}})$ which fulfills (2.8). Then, similar to Theorem 5.5, one can see that

$$\mathbf{X}_0^{-*} \mathbf{G} \mathbf{X}_0^{-1} \geq F(\{u\}), \quad u \in \mathbb{T},$$

where equality holds if and only if $F = F_{0,u}^{(\alpha)}$ (see also the comments at the end of Section 3 and [22, Remark 9.16]). One can also conclude that, if $u \in \mathbb{T}$, then

$$\mathbf{x}^* \left(F(\{u\})\right)^+ \mathbf{x} \geq \mathbf{x}^* \mathbf{X}_0 \mathbf{G}^{-1} \mathbf{X}_0^* \mathbf{x}, \quad \mathbf{x} \in \mathcal{R}\left(F(\{u\})\right),$$

where equality holds if F is a canonical solution with respect to (2.8) and (2.9). But, as announced in view of (5.5), the equality $\mathbf{x}^* F(\{u\}) \mathbf{x} = \mathbf{x}^* \mathbf{X}_0^{-*} \mathbf{G} \mathbf{X}_0^{-1} \mathbf{x}$ does not hold for each $\mathbf{x} \in \mathcal{R}\left(F(\{u\})\right)$ and some $u \in \mathbb{T}$ in general if F is a canonical solution with respect to (2.8) and (2.9). This will be emphasized by the following. Let

$$\mathbf{X}_0 := \begin{pmatrix} 1 & 0 \\ 0 & 1 \end{pmatrix} \quad \text{and} \quad \mathbf{G} := \begin{pmatrix} 2 & -1 \\ -1 & 2 \end{pmatrix}.$$

So, $\mathbf{X}_0$ and $\mathbf{G}$ are positive Hermitian 2×2 matrix. In particular, $\mathbf{X}_0$ and $\mathbf{G}$ are non-singular. Furthermore, by a straightforward calculation one can check that

$$\sqrt{\mathbf{G}} = \frac{1}{2a} \begin{pmatrix} 2a^2 & -1 \\ -1 & 2a^2 \end{pmatrix}, \quad a := \sqrt{1 + \frac{\sqrt{3}}{2}}.$$

The considerations at the end of Section 2 imply that

$$F := \varepsilon_1 \sqrt{\mathbf{G}} \begin{pmatrix} 1 & 0 \\ 0 & 0 \end{pmatrix} \sqrt{\mathbf{G}} + \varepsilon_{-1} \sqrt{\mathbf{G}} \begin{pmatrix} 0 & 0 \\ 0 & 1 \end{pmatrix} \sqrt{\mathbf{G}}$$

is a canonical solution with respect to (2.8) and (2.9) in this context, where

$$F(\{1\}) = \frac{1}{4a^2} \begin{pmatrix} 4a^4 & -2a^2 \\ -2a^2 & 1 \end{pmatrix}.$$

Therefore, a vector $\mathbf{x} \in \mathbb{C}^{q \times 1}$ belongs to $\mathcal{R}(F(\{1\}))$ if and only if

$$\mathbf{x} = \begin{pmatrix} -2a^2 c \\ c \end{pmatrix}$$

for some $c \in \mathbb{C}$. If we look at such $\mathbf{x}$ with $c \in \mathbb{C} \setminus \{0\}$ we get on the one hand

$$\mathbf{x}^* F(\{1\})\mathbf{x} = |c|^2 \left(4a^6 + 2a^2 + \frac{1}{4a^2} \right)$$

and on the other hand

$$\mathbf{x}^* \mathbf{G} \mathbf{x} = |c|^2 \left(8a^4 + 4a^2 + 2 \right)$$

which yields by $a^2 = 1 + \frac{\sqrt{3}}{2}$ finally $\mathbf{x}^* \mathbf{X}_0^{-*} \mathbf{G} \mathbf{X}_0^{-1} \mathbf{x} = \mathbf{x}^* \mathbf{G} \mathbf{x} > \mathbf{x}^* F(\{1\})\mathbf{x}$.

6. Some conclusions from Theorem 5.5

The condition $(\alpha_j)_{j=1}^{\infty} \in \mathcal{T}_1$ in Theorem 5.5 is chosen, since we want to apply results on orthogonal rational matrix functions. In particular, the definition of the special measure $F_{n,u}^{(\alpha)}$ with some $u \in \mathbb{T}$ is based on such statements (see (3.4) and (3.5)). However, for the somewhat more general case that only $\alpha_1, \alpha_2, \ldots, \alpha_n \in \mathbb{C} \setminus \mathbb{T}$ is claimed, one can conclude at least the following.

Theorem 6.1. *Let $n \in \mathbb{N}$ and $\alpha_1, \alpha_2, \ldots, \alpha_n \in \mathbb{C} \setminus \mathbb{T}$. Let $X_0, X_1, \ldots, X_n$ be a basis of the right $\mathbb{C}^{q \times q}$-module $\mathcal{R}_{\alpha,n}^{q \times q}$ and suppose that $\mathbf{G}$ is a non-singular matrix such that $\mathcal{M}[(\alpha_j)_{j=1}^n, \mathbf{G}; (X_k)_{k=0}^n] \neq \emptyset$. Furthermore, let $F \in \mathcal{M}[(\alpha_j)_{j=1}^n, \mathbf{G}; (X_k)_{k=0}^n]$ and let $F^{(\alpha,n)}$ be defined by (4.4). Then the inequality*

$$\mathbf{x}^* \left(F(\{u\}) \right)^+ \mathbf{x} \geq \mathbf{x}^* A_{n,u}^{(\alpha)}(u) \mathbf{x}, \quad \mathbf{x} \in \mathcal{R}(F(\{u\})), \tag{6.1}$$

is satisfied for each $u \in \mathbb{T}$, where equality holds in (6.1) if F is a canonical solution in $\mathcal{M}[(\alpha_j)_{j=1}^n, \mathbf{G}; (X_k)_{k=0}^n]$. Moreover, if $u \in \mathbb{T}$, then

$$\left(A_{n,u}^{(\alpha)}(u) \right)^{-1} \geq F(\{u\}), \tag{6.2}$$

where equality holds in (6.2) if and only if $F^{(\alpha,n)}$ coincides with the (special) canonical solution $F_{n,u}$ which is given relating to the set $\mathcal{M}[\mathbb{T}_n^{(F^{(\alpha,n)})}]$ and the point u. In particular, if $u \in \mathbb{T}$, then the inequality

$$\frac{1}{\det A_{n,u}^{(\alpha)}(u)} \geq \det F(\{u\}) \tag{6.3}$$

is satisfied, where equality holds in (6.3) if and only if $F^{(\alpha,n)}$ coincides with $F_{n,u}$.

Proof. Our strategy is based on an application of Theorem 5.5 in the special context that the first n elements of the underlying sequence belonging to $\mathcal{T}_1$ are equal to zero. The following considerations serve as a preparation of that. Because of [17, Proposition 5] and $F \in \mathcal{M}[(\alpha_j)_{j=1}^n, \mathbf{G}; (X_k)_{k=0}^n]$ one can see that a measure $H \in \mathcal{M}_{\geq}^q(\mathbb{T}, \mathcal{B}_{\mathbb{T}})$ belongs to $\mathcal{M}[(\alpha_j)_{j=1}^n, \mathbf{G}; (X_k)_{k=0}^n]$ if and only if the measure

$H^{(\alpha,n)} : \mathfrak{B}_{\mathbb{T}} \to \mathbb{C}^{q \times q}$ given by

$$H^{(\alpha,n)}(B) := \int_B \left(\frac{1}{\pi_{\alpha,n}(z)} \mathbf{I}_q \right)^* H(\mathrm{d}z) \left(\frac{1}{\pi_{\alpha,n}(z)} \mathbf{I}_q \right)$$

belongs to $\mathcal{M}[\mathbf{T}_n^{(F^{(\alpha,n)})}]$. Obviously, for some $H \in \mathcal{M}_{\geq}^q(\mathbb{T}, \mathfrak{B}_{\mathbb{T}})$, this setting implies

$$H^{(\alpha,n)}(\{u\}) = \frac{1}{|\pi_{\alpha,n}(u)|^2} H(\{u\}), \quad u \in \mathbb{T}, \tag{6.4}$$

where $H = F$ if and only if $H^{(\alpha,n)} = F^{(\alpha,n)}$. Furthermore, by [25, Remark 2.4] we know that F is a canonical solution in $\mathcal{M}[(\alpha_j)_{j=1}^n, \mathbf{G}; (X_k)_{k=0}^n]$ is equivalent to the fact that $F^{(\alpha,n)}$ is a canonical solution with respect to the set $\mathcal{M}[\mathbf{T}_n^{(F^{(\alpha,n)})}]$. In view of Theorem 1.1 and [18, Remark 5.10] along with $\det \mathbf{G} \neq 0$ we get

$$F^{(\alpha,n)} \in \mathcal{M}_{\geq}^{q,n}(\mathbb{T}, \mathfrak{B}_{\mathbb{T}}),$$

i.e., that the block Toeplitz matrix $\mathbf{T}_n^{(F^{(\alpha,n)})}$ is non-singular. Besides, the choice $F \in \mathcal{M}[(\alpha_j)_{j=1}^n, \mathbf{G}; (X_k)_{k=0}^n]$ and (3.2) lead to

$$A_{n,u}^{(\alpha,F)} = A_{n,u}^{(\alpha)}, \quad u \in \mathbb{T}.$$

Hence, by using [17, Remarks 12 and 29] one can conclude that

$$A_{n,u}^{(\alpha)}(u) = \frac{1}{|\pi_{\alpha,n}(u)|^2} A_{n,u}(u), \quad u \in \mathbb{T},$$

where

$$A_{n,u}(u) = \left(u^0 \mathbf{I}_q, u^1 \mathbf{I}_q, \ldots, u^n \mathbf{I}_q \right) \left(\mathbf{T}_n^{(F^{(\alpha,n)})} \right)^{-1} \left(u^0 \mathbf{I}_q, u^1 \mathbf{I}_q, \ldots, u^n \mathbf{I}_q \right)^*.$$

(Note that $A_{n,u}$ is the special matrix function given via (3.1) with $w = u$ in which $\alpha_j = 0$ for each $j \in \mathbb{N}_{1,n}$ and $X_k = E_{k,q}$ for each $k \in \mathbb{N}_{0,n}$.) Thus, the assertion can be finally deduced from Theorem 5.5 and (6.4). $\qquad \square$

As an aside we remark that, with a view to Theorem 6.1, similar to the statement of Corollary 5.6 one can draw a conclusion regarding the conception of reciprocal measures in the set $\mathcal{M}_{\geq}^q(\mathbb{T}, \mathfrak{B}_{\mathbb{T}})$ for the somewhat more general case that $\alpha_1, \alpha_2, \ldots, \alpha_n \in \mathbb{C} \setminus \mathbb{T}$ is assumed instead of $(\alpha_j)_{j=1}^\infty \in \mathcal{T}_1$.

Unless otherwise indicated, in view of Problem (R), let $\mathbf{G} \in \mathbb{C}^{(n+1)q \times (n+1)q}$ and suppose that $X_0, X_1, \ldots, X_n$ is a basis of the right $\mathbb{C}^{q \times q}$-module $\mathcal{R}_{\alpha,n}^{q \times q}$ in the following, where $n \in \mathbb{N}$ and $\alpha_1, \alpha_2, \ldots, \alpha_n \in \mathbb{C} \setminus \mathbb{T}$ are arbitrary (but fixed).

With regard to Theorems 4.1 and 6.1 we introduce the following notion. Suppose that F is a canonical solution in $\mathcal{M}[(\alpha_j)_{j=1}^n, \mathbf{G}; (X_k)_{k=0}^n]$. If $u \in \mathbb{T}$, then F is called u-proper when the value $F(\{u\})$ is non-singular. We also say that F is proper if there exists some $u \in \mathbb{T}$ such that F is u-proper. Furthermore, F is called totally proper if F is proper and if the condition $F(\{z\}) \neq 0_{q \times q}$ for some $z \in \mathbb{T}$ implies that $F(\{z\})$ is non-singular.

Remark 6.2. If F is a canonical solution in $\mathcal{M}[(\alpha_j)_{j=1}^n, \mathbf{G}; (X_k)_{k=0}^n]$ which is proper, then in view of Theorem 1.1 and [18, Example 3.11] one can see that F belongs to $\mathcal{M}_{\geq}^{q,0}(\mathbb{T}, \mathfrak{B}_{\mathbb{T}})$. Furthermore, if F is a canonical solution in $\mathcal{M}[(\alpha_j)_{j=1}^n, \mathbf{G}; (X_k)_{k=0}^n]$ which is totally proper and if r is the number of mass points of F, then by [18, Theorem 6.11] one can conclude that $F \in \mathcal{M}_{\geq}^{q,r-1}(\mathbb{T}, \mathfrak{B}_{\mathbb{T}})$.

Based on Theorem 6.1, we now analyze some results of Section 4.

Proposition 6.3. *Suppose that* $\mathbf{G}$ *is a non-singular* $(n+1)q \times (n+1)q$ *matrix such that* $\mathcal{M}[(\alpha_j)_{j=1}^n, \mathbf{G}; (X_k)_{k=0}^n] \neq \emptyset$. *If* $u \in \mathbb{T}$ *and if* $F \in \mathcal{M}[(\alpha_j)_{j=1}^n, \mathbf{G}; (X_k)_{k=0}^n]$, *then the following statements are equivalent:*

(i) F *is a canonical solution in* $\mathcal{M}[(\alpha_j)_{j=1}^n, \mathbf{G}; (X_k)_{k=0}^n]$ *which is* u-*proper.*

(ii) F *is the solution in* $\mathcal{M}[(\alpha_j)_{j=1}^n, \mathbf{G}; (X_k)_{k=0}^n]$ *such that equality holds in* (6.2).

(iii) F *is the solution in* $\mathcal{M}[(\alpha_j)_{j=1}^n, \mathbf{G}; (X_k)_{k=0}^n]$ *such that equality holds in* (6.3).

(iv) *There exists a finite subset* Δ *of* $\mathbb{T}$ *such that* $F(\mathbb{T} \setminus \Delta) = 0_{q \times q}$ *holds, where* $\det F(\{u\}) \neq 0$ *and* (4.3) *are satisfied.*

(v) *There exists a finite subset* Δ *of* $\mathbb{T}$ *such that* $F(\mathbb{T} \setminus \Delta) = 0_{q \times q}$ *and* (4.3) *hold.*

Proof. By Theorem 6.1 one can immediately see that (ii) is satisfied if and only if (iii) holds. Taking (6.4) into account and that the matrix measure F is a canonical solution in $\mathcal{M}[(\alpha_j)_{j=1}^n, \mathbf{G}; (X_k)_{k=0}^n]$ if and only if the matrix measure $F^{(\alpha,n)}$ given via (4.4) is a canonical solution with respect to the set $\mathcal{M}[\mathbf{T}_n^{(F^{(\alpha,n)})}]$ (see [25, Remark 2.4]), based on Theorem 6.1 and Corollary 4.2 one can conclude that the statements (ii), (i), (iv), and (v) are equivalent as well. □

Corollary 6.4. *Suppose that* $\mathbf{G}$ *is a non-singular* $(n+1)q \times (n+1)q$ *matrix such that* $\mathcal{M}[(\alpha_j)_{j=1}^n, \mathbf{G}; (X_k)_{k=0}^n] \neq \emptyset$. *Let* $u \in \mathbb{T}$ *and let* $F \in \mathcal{M}[(\alpha_j)_{j=1}^n, \mathbf{G}; (X_k)_{k=0}^n]$ *be a canonical solution in* $\mathcal{M}[(\alpha_j)_{j=1}^n, \mathbf{G}; (X_k)_{k=0}^n]$. *Furthermore, let*

$$\widetilde{F}_u := F - \varepsilon_u F(\{u\}).$$

Then the following statements are equivalent:

(i) *The canonical solution* F *in* $\mathcal{M}[(\alpha_j)_{j=1}^n, \mathbf{G}; (X_k)_{k=0}^n]$ *is* u-*proper.*

(ii) $\widetilde{F}_u$ *is a canonical solution in the set* $\mathcal{M}[\mathbf{T}_{n-1}^{(\widetilde{F}_u)}]$.

(iii) $\operatorname{rank} \mathbf{T}_n^{(\widetilde{F}_u)} = nq$.

(iv) *The set* $\mathcal{M}[\mathbf{T}_n^{(\widetilde{F}_u)}]$ *is a singleton.*

Proof. Recalling Corollary 4.2, the assertion follows from Proposition 6.3 along with Theorem 4.1. □

Remark 6.5. Suppose that $\mathbf{G}$ is a non-singular $(n+1)q \times (n+1)q$ matrix such that $\mathcal{M}[(\alpha_j)_{j=1}^n, \mathbf{G}; (X_k)_{k=0}^n] \neq \emptyset$. Because of Proposition 6.3 and Theorem 6.1 one can see that, for each $u \in \mathbb{T}$, there exists exactly one canonical solution in $\mathcal{M}[(\alpha_j)_{j=1}^n, \mathbf{G}; (X_k)_{k=0}^n]$ which is u-proper.

Obviously, similar to Corollaries 4.3 and 4.4, by Proposition 6.3 and Theorem 6.1 one can deduce results in terms of the notion u-proper. In particular, even if Problem (R) depends on the pole location of the given rational matrix functions (i.e., on the choice of the underlying points $\alpha_1, \alpha_2, \ldots, \alpha_n \in \mathbb{C} \setminus \mathbb{T}$), in view of Theorem 6.1 one can see that the extremal solution which realizes equality in (6.2) (resp., in (6.3)) has this property in some sense independently from the concrete location of the poles. In fact, we get the following universality result.

Proposition 6.6. *Suppose that $F \in \mathcal{M}_{\geq}^{q,n}(\mathbb{T}, \mathfrak{B}_{\mathbb{T}})$. Furthermore, let $u \in \mathbb{T}$. Then the following statements are equivalent:*

(i) *F is the solution in $\mathcal{M}[\mathbf{T}_n^{(F)}]$ such that equality holds in (6.2) (resp., in (6.3)) concerning the special choice of $\alpha_j = 0$ for each $j \in \mathbb{N}_{1,n}$.*

(ii) *There are points $\alpha_1, \alpha_2, \ldots, \alpha_n \in \mathbb{C} \setminus \mathbb{T}$ and a basis $X_0, X_1, \ldots, X_n$ of the right $\mathbb{C}^{q \times q}$-module $\mathcal{R}_{\alpha,n}^{q \times q}$, where F is the solution in $\mathcal{M}[(\alpha_j)_{j=1}^n, \mathbf{G}_{X,n}^{(F)}; (X_k)_{k=0}^n]$ such that equality holds in (6.2) (resp., in (6.3)).*

(iii) *F is the solution in $\mathcal{M}[(\alpha_j)_{j=1}^n, \mathbf{G}_{X,n}^{(F)}; (X_k)_{k=0}^n]$ such that equality holds in (6.2) (resp., in (6.3)) for every choice of points $\alpha_1, \alpha_2, \ldots, \alpha_n \in \mathbb{C} \setminus \mathbb{T}$ and every basis $X_0, X_1, \ldots, X_n$ of the right $\mathbb{C}^{q \times q}$-module $\mathcal{R}_{\alpha,n}^{q \times q}$.*

Proof. Recalling Theorem 1.1, the assertion follows from Proposition 6.3. $\square$

Corollary 6.7. *Suppose that $\mathbf{G}$ is a non-singular $(n+1)q \times (n+1)q$ matrix such that $\mathcal{M}[(\alpha_j)_{j=1}^n, \mathbf{G}; (X_k)_{k=0}^n] \neq \emptyset$. Furthermore, let $F \in \mathcal{M}[(\alpha_j)_{j=1}^n, \mathbf{G}; (X_k)_{k=0}^n]$, where*

$$F(\{u_s\}) \geq H(\{u_s\}), \quad s \in \mathbb{N}_{1,\ell},$$

holds for each $H \in \mathcal{M}[(\alpha_j)_{j=1}^n, \mathbf{G}; (X_k)_{k=0}^n]$ with some integer $\ell \in \mathbb{N}$ and pairwise different points $u_1, u_2, \ldots, u_\ell \in \mathbb{T}$.

(a) *Then $\ell \leq n+1$ and F is a canonical solution in $\mathcal{M}[(\alpha_j)_{j=1}^n, \mathbf{G}; (X_k)_{k=0}^n]$ which is u_s-proper for each $s \in \mathbb{N}_{1,\ell}$.*

(b) *If $\ell = n+1$, then F admits (4.5). In particular, in this case, the canonical solution F in $\mathcal{M}[(\alpha_j)_{j=1}^n, \mathbf{G}; (X_k)_{k=0}^n]$ is totally proper.*

(c) *Suppose that $\ell \neq n+1$. Then there is some $u \in \mathbb{T} \setminus \{u_1, u_2, \ldots, u_\ell\}$ such that equality holds in (6.2) (resp., in (6.3)) if and only if*

$$\widetilde{F}(\{u\}) \geq \widetilde{H}(\{u\}), \quad \widetilde{H} \in \mathcal{M}[\mathbf{T}_{n-\ell}^{(\widetilde{F})}],$$

where $\widetilde{F}$ is defined by (4.6). In particular, for some $u \in \mathbb{T} \setminus \{u_1, u_2, \ldots, u_\ell\}$, the canonical solution F in $\mathcal{M}[(\alpha_j)_{j=1}^n, \mathbf{G}; (X_k)_{k=0}^n]$ is u-proper if and only if $\widetilde{F}$ is a canonical solution in the set $\mathcal{M}[\mathbf{T}_{n-\ell}^{(\widetilde{F})}]$ which is u-proper.

(d) *The equality $F^{\#}(\{u_s\}) = 0_{q \times q}$ holds for each $s \in \mathbb{N}_{1,\ell}$, where $F^{\#}$ stands for the reciprocal measure corresponding to F.*

Proof. Taking Theorem 6.1 into account, the assertion is a consequence of Proposition 6.6 along with Corollary 4.5. $\square$

Corollary 6.8. *Suppose that $\mathbf{G}$ is a non-singular $(n+1)q \times (n+1)q$ matrix such that $\mathcal{M}[(\alpha_j)_{j=1}^n, \mathbf{G}; (X_k)_{k=0}^n] \neq \emptyset$. Furthermore, let $F \in \mathcal{M}[(\alpha_j)_{j=1}^n, \mathbf{G}; (X_k)_{k=0}^n]$ which admits (3.17) for some $r \in \mathbb{N}_{1,n+1}$, points $u_1, u_2, \ldots, u_r \in \mathbb{T}$, and a sequence $(\mathbf{A}_s)_{s=1}^r$ of non-negative Hermitian $q \times q$ matrices. Then $r = n+1$ and F admits (4.5), where the points $u_1, u_2, \ldots, u_{n+1}$ are pairwise different. In particular, F is*

a canonical solution in $\mathcal{M}[(\alpha_j)_{j=1}^n, \mathbf{G}; (X_k)_{k=0}^n]$ which is totally proper and
$$F(\{u_s\}) \geq H(\{u_s\}), \quad s \in \mathbb{N}_{1,n+1},$$
holds for each $H \in \mathcal{M}[(\alpha_j)_{j=1}^n, \mathbf{G}; (X_k)_{k=0}^n]$, but then $F^\#(\{u_s\}) = 0_{q \times q}$ is satisfied, where $F^\#$ stands for the reciprocal measure corresponding to F.

Proof. Taking Theorem 6.1 into account, the assertion follows from Proposition 6.6 in combination with Proposition 4.7. ∎

Corollary 6.9. *Let $u_1, u_2, \ldots, u_{n+1} \in \mathbb{T}$ be pairwise different, let $(\mathbf{A}_s)_{s=1}^{n+1}$ be a sequence of positive Hermitian $q \times q$ matrices, and let F be the measure fulfilling (3.17) with $r = n+1$. Then $F \in \mathcal{M}_{\geq}^{q,n}(\mathbb{T}, \mathfrak{B}_{\mathbb{T}})$ and $\mathbf{G}_{X,n}^{(F)}$ is a non-singular matrix. Moreover, if $s \in \mathbb{N}_{1,n+1}$, then the inequality*
$$\left(A_{n,u_s}^{(\alpha,F)}(u_s) \right)^{-1} \geq H(\{u_s\}) \quad \left(resp., \quad \frac{1}{\det A_{n,u_s}^{(\alpha,F)}(u_s)} \geq \det H(\{u_s\}) \right)$$
is satisfied, for each $H \in \mathcal{M}[(\alpha_j)_{j=1}^n, \mathbf{G}_{X,n}^{(F)}; (X_k)_{k=0}^n]$, where equality holds if and only if H coincides with F, and there again $F^\#(\{u_s\}) = 0_{q \times q}$, where $F^\#$ stands for the reciprocal measure corresponding to F. In particular, F is a canonical solution in $\mathcal{M}[(\alpha_j)_{j=1}^n, \mathbf{G}; (X_k)_{k=0}^n]$ which is totally proper.

Proof. Recalling (3.2) and Theorem 6.1, the assertion follows from Proposition 6.6 in combination with Corollary 4.9. ∎

Corollary 6.10. *Suppose that $\mathbf{G}$ is a non-singular $(n+1)q \times (n+1)q$ matrix such that $\mathcal{M}[(\alpha_j)_{j=1}^n, \mathbf{G}; (X_k)_{k=0}^n] \neq \emptyset$. Furthermore, let F be a canonical solution in $\mathcal{M}[(\alpha_j)_{j=1}^n, \mathbf{G}; (X_k)_{k=0}^n]$ and let*
$$\tilde{F}_u := F - \varepsilon_u F(\{u\})$$
for some $u \in \mathbb{T}$. Then the following statements are equivalent:

(i) *The canonical solution F in $\mathcal{M}[(\alpha_j)_{j=1}^n, \mathbf{G}; (X_k)_{k=0}^n]$ is totally proper.*
(ii) *F admits (3.17) for some $r \in \mathbb{N}$, points $u_1, u_2, \ldots, u_r \in \mathbb{T}$, and a sequence $(\mathbf{A}_s)_{s=1}^r$ of positive Hermitian matrices.*
(iii) *F admits (4.5) with some pairwise different points $u_1, u_2, \ldots, u_{n+1} \in \mathbb{T}$.*
(iv) *F admits (3.17) for some $r \in \mathbb{N}_{1,n+1}$, points $u_1, u_2, \ldots, u_r \in \mathbb{T}$, and a sequence $(\mathbf{A}_s)_{s=1}^r$ of non-negative Hermitian $q \times q$ matrices.*
(v) *There is a solution $\tilde{F} \in \mathcal{M}[(\alpha_j)_{j=1}^n, \mathbf{G}; (X_k)_{k=0}^n]$ fulfilling the representation*
$$\tilde{F} = \sum_{s=1}^r \varepsilon_{u_s} \mathbf{A}_s$$
for some $r \in \mathbb{N}_{1,n+1}$, points $u_1, u_2, \ldots, u_r \in \mathbb{T}$, and a sequence $(\mathbf{A}_s)_{s=1}^r$ of non-negative Hermitian $q \times q$ matrices, where $\tilde{F}(\{u_1\}) = F(\{u_1\})$.
(vi) *For each $u \in \mathbb{T}$ with $F(\{u\}) \neq 0_{q \times q}$, the measure $\tilde{F}_u$ is a canonical solution in the set $\mathcal{M}[\mathbf{T}_{n-1}^{(\tilde{F}_u)}]$.*
(vii) *For each $u \in \mathbb{T}$ with $F(\{u\}) \neq 0_{q \times q}$, the identity $\operatorname{rank} \mathbf{T}_n^{(\tilde{F}_u)} = nq$ holds.*
(viii) *For each $u \in \mathbb{T}$ with $F(\{u\}) \neq 0_{q \times q}$, the set $\mathcal{M}[\mathbf{T}_n^{(\tilde{F}_u)}]$ is a singleton.*

Moreover, if (i) *is satisfied and if* $u \in \mathbb{T}$ *is chosen such that* $F(\{u\}) \neq 0_{q \times q}$ *holds, then equality holds in* (6.2).

Proof. Use Proposition 6.6 in combination with Corollary 4.10. □

Corollary 6.11. *Suppose that* $\mathbf{G}$ *is a non-singular* $(n+1)q \times (n+1)q$ *matrix such that* $\mathcal{M}[(\alpha_j)_{j=1}^n, \mathbf{G}; (X_k)_{k=0}^n] \neq \emptyset$. *Let* $F \in \mathcal{M}[(\alpha_j)_{j=1}^n, \mathbf{G}; (X_k)_{k=0}^n]$ *which admits* (3.17) *for some* $r \in \mathbb{N}$, *points* $u_1, u_2, \ldots, u_r \in \mathbb{T}$, *and a sequence* $(\mathbf{A}_s)_{s=1}^r$ *of nonnegative Hermitian* $q \times q$ *matrices. Furthermore, let*

$$F_{u,\mathbf{A}} := F + \varepsilon_u \mathbf{A},$$

where $u \in \mathbb{T} \setminus \{u_1, u_2, \ldots, u_r\}$ *and* $\mathbf{A} > 0_{q \times q}$. *Suppose that* $Y_0, Y_1, \ldots, Y_{n+1}$ *is a basis of the right* $\mathbb{C}^{q \times q}$*-module* $\mathcal{R}_{\alpha,n+1}^{q \times q}$. *Then* $F_{u,\mathbf{A}}$ *belongs to* $\mathcal{M}_\geq^{q,n+1}(\mathbb{T}, \mathfrak{B}_\mathbb{T})$ *and the matrix* $\mathbf{G}_{Y,n+1}^{(F_{u,\mathbf{A}})}$ *is non-singular. Moreover, the following statements are equivalent:*

(i) $\det A_{n+1,u}^{(\alpha, F_{u,\mathbf{A}})}(u) = \dfrac{1}{\det \mathbf{A}}$.

(ii) $A_{n+1,u}^{(\alpha, F_{u,\mathbf{A}})}(u) = \mathbf{A}^{-1}$.

(iii) $F_{u,\mathbf{A}}(\{u\}) \geq H(\{u\})$ *holds for each* $H \in \mathcal{M}[(\alpha_j)_{j=1}^{n+1}, \mathbf{G}_{Y,n+1}^{(F_{u,\mathbf{A}})}; (Y_k)_{k=0}^{n+1}]$.

(iv) $F_{u,\mathbf{A}}$ *is a canonical solution in* $\mathcal{M}[(\alpha_j)_{j=1}^{n+1}, \mathbf{G}_{Y,n+1}^{(F_{u,\mathbf{A}})}; (Y_k)_{k=0}^{n+1}]$ *which is* u*-proper.*

(v) $F_{u,\mathbf{A}}$ *is a canonical solution in* $\mathcal{M}[(\alpha_j)_{j=1}^{n+1}, \mathbf{G}_{Y,n+1}^{(F_{u,\mathbf{A}})}; (Y_k)_{k=0}^{n+1}]$.

(vi) F *is a canonical solution in* $\mathcal{M}[(\alpha_j)_{j=1}^n, \mathbf{G}; (X_k)_{k=0}^n]$.

Proof. Apply Theorem 6.1 and Proposition 6.6 along with Proposition 4.11. □

It is possible that the inequality in (6.2) is fulfilled for some solutions in the strong sense. In the following, some information on this situation will be given.

Proposition 6.12. *Suppose that* $\mathbf{G}$ *is a non-singular* $(n+1)q \times (n+1)q$ *matrix such that* $\mathcal{M}[(\alpha_j)_{j=1}^n, \mathbf{G}; (X_k)_{k=0}^n] \neq \emptyset$. *Furthermore, let* $F \in \mathcal{M}[(\alpha_j)_{j=1}^n, \mathbf{G}; (X_k)_{k=0}^n]$. *If* $F \in \mathcal{M}_\geq^{q,n+1}(\mathbb{T}, \mathfrak{B}_\mathbb{T})$ *and if* $u \in \mathbb{T}$, *then*

$$\left(A_{n,u}^{(\alpha)}(u)\right)^{-1} > F(\{u\}). \tag{6.5}$$

There again, if F *is a canonical solution in* $\mathcal{M}[(\alpha_j)_{j=1}^n, \mathbf{G}; (X_k)_{k=0}^n]$ *and if* $u \in \mathbb{T}$ *is chosen such that* $F(\{u\}) \neq 0_{q \times q}$, *then* (6.5) *does not hold. In particular, if* F *is a canonical solution, then there exists a* $u \in \mathbb{T}$ *such that* (6.5) *does not hold.*

Proof. Let $u \in \mathbb{T}$ and suppose that F belongs to $\mathcal{M}_\geq^{q,n+1}(\mathbb{T}, \mathfrak{B}_\mathbb{T})$. Furthermore, let $\alpha_{n+1} \in \mathbb{C} \setminus \mathbb{T}$ and let $Y_0, Y_1, \ldots, Y_{n+1}$ be a basis of the right $\mathbb{C}^{q \times q}$-module $\mathcal{R}_{\alpha,n+1}^{q \times q}$. Thus, Theorem 1.1 and $F \in \mathcal{M}_\geq^{q,n+1}(\mathbb{T}, \mathfrak{B}_\mathbb{T})$ imply that the matrix $\mathbf{G}_{Y,n+1}^{(F)}$ is non-singular. Besides, an application of [20, Corollary 4.7 and Lemma 5.1] yields

$$A_{n+1,u}^{(\alpha, F)}(u) > A_{n,u}^{(\alpha, F)}(u) > 0_{q \times q}.$$

Consequently, based on (3.2) and Theorem 6.1 with respect to the solution set $\mathcal{M}[(\alpha_j)_{j=1}^{n+1}, \mathbf{G}_{Y,n+1}^{(F)}; (Y_k)_{k=0}^{n+1}]$ we get

$$\left(A_{n,u}^{(\alpha)}(u)\right)^{-1} = \left(A_{n,u}^{(\alpha, F)}(u)\right)^{-1} > \left(A_{n+1,u}^{(\alpha, F)}(u)\right)^{-1} \geq F(\{u\}).$$

Thus, we have shown that (6.5) is fulfilled for each $u \in \mathbb{T}$ in the case that F belongs to $\mathcal{M}_{\geq}^{q,n+1}(\mathbb{T}, \mathfrak{B}_{\mathbb{T}})$. Finally, we consider the case that F is a canonical solution in $\mathcal{M}[(\alpha_j)_{j=1}^n, \mathbf{G}; (X_k)_{k=0}^n]$. Furthermore, suppose that $u \in \mathbb{T}$ is chosen such that $F(\{u\}) \neq 0_{q \times q}$ holds. Obviously (note (2.2)), there exists such a point u. So, we find some $\mathbf{x} \in \mathcal{R}\big(F(\{u\})\big) \setminus \{0_{q \times 1}\}$. From [25, Proposition 4.7] (cf. part (b) of Theorem 3.9) we obtain the identity

$$F(\{u\})A_{n,u}^{(\alpha)}(u)\mathbf{x} = \mathbf{x}. \tag{6.6}$$

Since $\mathbf{x} \neq 0_{q \times 1}$ and since the matrix $A_{n,u}^{(\alpha)}(u)$ is non-singular (cf. (3.3)), the vector $\mathbf{y} := A_{n,u}^{(\alpha)}(u)\mathbf{x}$ is not equal to $0_{q \times 1}$ as well. In view of (6.6) we also have

$$\big(A_{n,u}^{(\alpha)}(u)\big)^{-1}\mathbf{y} = \mathbf{x} = F(\{u\})A_{n,u}^{(\alpha)}(u)\mathbf{x} = F(\{u\})\mathbf{y}.$$

In particular, $\mathbf{y} \in \mathbb{C}^{q \times 1} \setminus \{0_{q \times 1}\}$ satisfies

$$\mathbf{y}^*\big(A_{n,u}^{(\alpha)}(u)\big)^{-1}\mathbf{y} = \mathbf{y}^* F(\{u\})\mathbf{y}.$$

This implies immediately that (6.5) does not hold for this case. ☐

Corollary 6.13. *Suppose that $\mathbf{G}$ is a non-singular $(n+1)q \times (n+1)q$ matrix such that $\mathcal{M}[(\alpha_j)_{j=1}^n, \mathbf{G}; (X_k)_{k=0}^n] \neq \emptyset$. Furthermore, let $z_1, z_2, \ldots, z_{n+2} \in \mathbb{T}$ be $n+2$ pairwise different points and let $F \in \mathcal{M}[(\alpha_j)_{j=1}^n, \mathbf{G}; (X_k)_{k=0}^n]$ be such that the value $F(\{z_s\})$ is non-singular for all $s \in \mathbb{N}_{1,n+2}$. Then (6.5) holds for each $u \in \mathbb{T}$.*

Proof. Since the choice of F leads in view of [18, Remark 5.12 and Theorem 6.11] to

$$F \in \mathcal{M}_{\geq}^{q,n+1}(\mathbb{T}, \mathfrak{B}_{\mathbb{T}}),$$

the assertion is a simple consequence of Proposition 6.12. ☐

The particular case studied in Corollary 6.10 is far from the general situation in a sense (see Example 5.8), but significant for the scalar case $q = 1$. In fact, with a view to Remark 4.14 one can see the following. (Note that, for $q = 1$, the differentiation between canonical solutions, canonical solutions which are proper, and canonical solutions which are totally proper is not needed.)

Remark 6.14. Let $n \in \mathbb{N}$ and $\alpha_1, \alpha_2, \ldots, \alpha_n \in \mathbb{C} \setminus \mathbb{T}$. Suppose that $X_0, X_1, \ldots, X_n$ is a basis of the linear space $\mathcal{R}_{u,n}$ and let $\mathbf{G}$ be a non-singular matrix such that $\mathcal{M}[(\alpha_j)_{j=1}^n, \mathbf{G}; (X_k)_{k=0}^n] \neq \emptyset$. Let $F \in \mathcal{M}[(\alpha_j)_{j=1}^n, \mathbf{G}; (X_k)_{k=0}^n]$ and let $F^\#$ be the reciprocal measure corresponding to F. By Theorem 6.1 and Proposition 6.6 along with Remark 4.14 one can conclude that the following statements are equivalent:

(i) F is a canonical solution in $\mathcal{M}[(\alpha_j)_{j=1}^n, \mathbf{G}; (X_k)_{k=0}^n]$.
(ii) There exists some $u \in \mathbb{T}$ such that the inequality $F(\{u\}) \geq H(\{u\})$ holds for each $H \in \mathcal{M}[(\alpha_j)_{j=1}^n, \mathbf{G}; (X_k)_{k=0}^n]$.
(iii) There exist $n+1$ pairwise different points $u_1, u_2, \ldots, u_{n+1} \in \mathbb{T}$ such that $F(\{u_s\}) \geq H(\{u_s\})$ for all $s \in \mathbb{N}_{1,n+1}$ and all $H \in \mathcal{M}[(\alpha_j)_{j=1}^n, \mathbf{G}; (X_k)_{k=0}^n]$.
(iv) F admits (3.17) for some $r \in \mathbb{N}_{1,n+1}$, points $u_1, u_2, \ldots, u_r \in \mathbb{T}$, and a sequence $(\mathbf{A}_s)_{s=1}^r$ of non-negative real numbers.

(v) $F^{\#}$ is a canonical solution in $\mathcal{M}[(\alpha_j)_{j=1}^n, \mathbf{G}^{\#}; (X_k)_{k=0}^n]$.

(vi) There exists some $z \in \mathbb{T}$ such that the inequality $F^{\#}(\{z\}) \geq H^{\#}(\{z\})$ holds
      for each $H \in \mathcal{M}[(\alpha_j)_{j=1}^n, \mathbf{G}^{\#}; (X_k)_{k=0}^n]$.

(vii) There exist $n + 1$ pairwise different points $z_1, z_2, \ldots, z_{n+1} \in \mathbb{T}$ such that
      $F^{\#}(\{u_s\}) \geq H^{\#}(\{u_s\})$ for $s \in \mathbb{N}_{1,n+1}$ and all $H \in \mathcal{M}[(\alpha_j)_{j=1}^n, \mathbf{G}^{\#}; (X_k)_{k=0}^n]$.

(viii) $F^{\#}$ admits (4.8) for some $r \in \mathbb{N}_{1,n+1}$, points $z_1, z_2, \ldots, z_r \in \mathbb{T}$, and a sequence
      $(\mathbf{A}_s^{\#})_{s=1}^r$ of non-negative real numbers.

This reasoning yields also that, if (i) holds and if $F(\{u\}) \neq 0$ for some $u \in \mathbb{T}$, then
$F(\{u\}) > H(\{u\})$ for each $H \in \mathcal{M}[(\alpha_j)_{j=1}^n, \mathbf{G}; (X_k)_{k=0}^n] \setminus \{F\}$, but $F^{\#}(\{u\}) = 0$.

In particular, for $q = 1$, Remark 6.14 shows that the weights of the reciprocal measure corresponding to a solution $F \in \mathcal{M}[(\alpha_j)_{j=1}^n, \mathbf{G}; (X_k)_{k=0}^n]$ which fulfills (6.2) with equality are extremal in this regard concerning the solution set $\mathcal{M}[(\alpha_j)_{j=1}^n, \mathbf{G}^{\#}; (X_k)_{k=0}^n]$ and associated mass points as well. Note that Example 4.13 points up that this is a special feature of the scalar case $q = 1$.

References

[1] Aronszajn, N.: *Theory of reproducing kernels*, Trans. Amer. Math. Soc. **68** (1950), 337–404.

[2] Arov, D.Z.: *Regular J-inner matrix-functions and related continuation problems*, in: Linear Operators in Function Spaces, Operator Theory: Adv. Appl. Vol. 43, Birkhäuser, Basel, 1990, pp. 63–87.

[3] Ben-Artzi, A.; Gohberg, I.: *Orthogonal polynomials over Hilbert modules*, in: Nonselfadjoint Operators and Related Topics, Operator Theory: Adv. Appl. Vol. 73, Birkhäuser, Basel, 1994, pp. 96–126.

[4] Bultheel, A.: *Inequalities in Hilbert modules of matrix-valued functions*, Proc. Amer. Math. Soc. **85** (1982), 369–372.

[5] Bultheel, A.; González-Vera, P.; Hendriksen, E.; Njåstad, O.: *Orthogonal rational functions and quadrature on the unit circle*, Numer. Algorithms **3** (1992), 105–116.

[6] Bultheel, A.; González-Vera, P.; Hendriksen, E.; Njåstad, O.: *Moment problems and orthogonal functions*, J. Comput. Appl. Math. **48** (1993), 49–68.

[7] Bultheel, A.; González-Vera, P.; Hendriksen, E.; Njåstad, O.: *A rational moment problem on the unit circle*, Methods Appl. Anal. **4** (1997), 283–310.

[8] Bultheel, A.; González-Vera, P.; Hendriksen, E.; Njåstad, O.: *Orthogonal Rational Functions*, Cambridge Monographs on Applied and Comput. Math. 5, Cambridge University Press, Cambridge 1999.

[9] Cantero, M.J.; Moral, L.; Velázquez, L.: *Measures and para-orthogonal polynomials on the unit circle*, East J. Approx. **8** (2002), 447–464.

[10] Delsarte, P.; Genin, Y.; Kamp, Y.: *Orthogonal polynomial matrices on the unit circle*, IEEE Trans. Circuits and Systems **CAS-25** (1978), 149–160.

[11] Dubovoj, V.K.; Fritzsche, B.; Kirstein, B.: *Matricial Version of the Classical Schur Problem*, Teubner-Texte zur Mathematik 129, B.G. Teubner, Stuttgart-Leipzig 1992.

[12] Duran, A.J.; Lopez-Rodriguez, P.: *N-extremal matrices of measures for an indeterminate matrix moment problem*, J. Funct. Anal. **174** (2000), 301–321.

[13] Dym, H.: *J Contractive Matrix Functions, Reproducing Kernel Hilbert Spaces and Interpolation*, CBMS Regional Conf. Ser. Math. 71, Providence, R.I. 1989.

[14] Ellis, R.L.; Gohberg, I.: *Extensions of matrix-valued inner products on modules and the inversion formula for block Toeplitz matrices*, in: Operator Theory and Analysis, Operator Theory: Adv. Appl. Vol. 122, Birkhäuser, Basel, 2001, pp. 191–227.

[15] Fritzsche, B.; Fuchs, S.; Kirstein, B.: *A Schur type matrix extension problem V*, Math. Nachr. **158** (1992), 133–159.

[16] Fritzsche, B.; Kirstein, B.: *Schwache Konvergenz nichtnegativ hermitescher Borelmaße*, Wiss. Z. KMU Leipzig, Math.-Naturw. R. **37** (1988), 375–398.

[17] Fritzsche, B.; Kirstein, B.; Lasarow, A.: *On a moment problem for rational matrix-valued functions*, Linear Algebra Appl. **372** (2003), 1–31.

[18] Fritzsche, B.; Kirstein, B.; Lasarow, A.: *On rank invariance of moment matrices of nonnegative Hermitian-valued Borel measures on the unit circle*, Math. Nachr. **263/264** (2004), 103–132.

[19] Fritzsche, B.; Kirstein, B.; Lasarow, A.: *On Hilbert modules of rational matrix-valued functions and related inverse problems*, J. Comput. Appl. Math. **179** (2005), 215–248.

[20] Fritzsche, B.; Kirstein, B.; Lasarow, A.: *Orthogonal rational matrix-valued functions on the unit circle*, Math. Nachr. **278** (2005), 525–553.

[21] Fritzsche, B.; Kirstein, B.; Lasarow, A.: *Orthogonal rational matrix-valued functions on the unit circle: Recurrence relations and a Favard-type theorem*, Math. Nachr. **279** (2006), 513–542.

[22] Fritzsche, B.; Kirstein, B.; Lasarow, A.: *On a class of extremal solutions of the nondegenerate matricial Carathéodory problem*, Analysis **27** (2007), 109–164.

[23] Fritzsche, B.; Kirstein, B.; Lasarow, A.: *On a class of extremal solutions of a moment problem for rational matrix-valued functions in the nondegenerate case I*, Math. Nachr. **283** (2010), 1706–1735.

[24] Fritzsche, B.; Kirstein, B.; Lasarow, A.: *On a class of extremal solutions of a moment problem for rational matrix-valued functions in the nondegenerate case II*, J. Comput. Appl. Math. **235** (2010), 1008–1041.

[25] Fritzsche, B.; Kirstein, B.; Lasarow, A.: *On canonical solutions of a moment problem for rational matrix-valued functions*, Operator Theory: Adv. Appl. Vol. 221, Springer Basel AG, 2012, pp. 327–376.

[26] Fritzsche, B.; Kirstein, B.; Lasarow, A.; Rahn, A.: *On reciprocal sequences of matricial Carathéodory sequences and associated matrix functions*, this volume.

[27] Geronimus, Ja.L.: *Polynomials orthogonal on a circle and their applications* (Russian), Zap. Naučno-Issled. Inst. Mat. Meh. Har'kov. Mat. Obšč. **19** (1948), 35–120.

[28] Golinskii, L.: *Quadrature formula and zeros of para-orthogonal polynomials on the unit circle*, Acta Math. Hungar. **96** (2002), 169–186.

[29] Itoh, S.: *Reproducing kernels in modules over C^*-algebras and their applications*, Bull. Kyushu Inst. Tech. Math. Natur. Sci. **37** (1990), 1–20.

[30] Jones, W.B.; Njåstad, O.; Thron, W.J.: *Moment theory, orthogonal polynomials, quadrature, and continued fractions associated with the unit circle*, Bull. London Math. Soc. **21** (1989), 113–152.

[31] Kats, I.S.: *On Hilbert spaces generated by monotone Hermitian matrix-functions* (Russian), Zap. Mat. Otd. Fiz.-Mat. Fak. i Har'kov. Mat. Obšč. **22** (1950), 95–113.

[32] Kreĭn, M.G.: *The ideas of P.L. Chebyshev and A.A. Markov in the theory of limit values of integrals and their further development* (Russian), Usp. Mat. Nauk **5** (1951), 3–66.

[33] Kreĭn, M.G.; Nudelman, A.A.: *The Markov Moment Problem and Extremal Problems*, AMS Translations Vol. 50, AMS, Providence, R.I. 1977.

[34] Lasarow, A.: *Dual pairs of orthogonal systems of rational matrix-valued functions on the unit circle*, Analysis **26** (2006), 209–244.

[35] Lasarow, A.: *On maximal weight solutions in a truncated trigonometric matrix moment problem*, Funct. Approx. Comment. Math. **43** (2010), 117–128.

[36] Rosenberg, M.: *The square integrability of matrix-valued functions with respect to a non-negative Hermitian measure*, Duke Math. J. **31** (1964), 291–298.

[37] Rosenberg, M.: *Operators as spectral integrals of operator-valued functions from the study of multivariate stationary stochastic processes*, J. Mult. Anal. **4** (1974), 166–209.

[38] Rosenberg, M.: *Spectral integrals of operator-valued functions – II. From the study of stationary processes*, J. Mult. Anal. **6** (1976), 538–571.

[39] Sakhnovich, A.L.: *On a class of extremal problems* (Russian), Izv. Akad. Nauk SSSR, Ser. Mat. **51** (1987), 436–443.

[40] Simon, B.: *Orthogonal Polynomials on the Unit Circle, Part 1: Classical Theory*, Amer. Math. Soc. Coll. Pub. 54, Providence, R.I. 2005.

[41] Simon, B.: *Rank one perturbations and the zeros of paraorthogonal polynomials on the unit circle*, J. Math. Anal. Appl. **329** (2007), 376–382.

Bernd Fritzsche, Bernd Kirstein, Andreas Lasarow
Fakultät für Mathematik und Informatik
Universität Leipzig
Postfach: 10 09 20
D-04009 Leipzig, Germany

e-mail: fritzsche@math.uni-leipzig.de
        kirstein@math.uni-leipzig.de
        lasarow@math.uni-leipzig.de

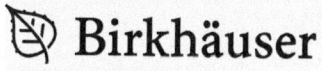 **Birkhäuser** | **www.birkhauser-science.com**

Operator Theory: Advances and Applications (OT)

This series is devoted to the publication of current research in operator theory, with particular emphasis on applications to classical analysis and the theory of integral equations, as well as to numerical analysis, mathematical physics and mathematical methods in electrical engineering.

Edited by

Joseph A. Ball (Blacksburg, VA, USA), Harry Dym (Rehovot, Israel), Marinus A. Kaashoek (Amsterdam, The Netherlands), Heinz Langer (Vienna, Austria), Christiane Tretter (Bern, Switzerland)

■ **OT 225: Sakhnovich, L. A.**, Levy Processes, Integral Equations, Statistical Physics: Connections and Interactions (2012).
ISBN 978-3-0348-0355-7

■ **OT 224: Benguria, R. / Friedman, E. / Mantoiu, M. (Eds.)**, Spectral Analysis of Quantum Hamiltonians. Spectral Days 2010 (2012).
ISBN 978-3-0348-0413-4

■ **OT 223: Jacob, B. / Zwart, H. J.**, Linear Port-Hamiltonian Systems on Infinite-dimensional Spaces (2012).
ISBN 978-3-0348-0398-4

■ **OT 222: Dym, H. / de Oliveira, M.C. / Putinar, M.** (Eds.), Mathematical Methods in Systems, Optimization, and Control. Festschrift in Honor of J. William Helton (2012).
ISBN 978-3-0348-0262-8

■ **OT 221: Arendt, W. / Ball, J.A. / Behrndt, J. / Förster, K.-H. / Mehrmann, V. / Trunk, C.** (Eds.), Spectral Theory, Mathematical System Theory, Evolution Equations, Differential and Difference Equations. IWOTA 10 (2012).
ISBN 978-3-0348-0262-8

■ **OT 220: Ball, J.A. / Curto, R.E. / Grudsky, S.M. / Helton, J.W.; Quiroga-Barranco, R. / Vasilevski, N.L.** (Eds.), Recent Progress in Operator Theory and Its Applications (2012).
ISBN 978-3-0348-0345-8

■ **OT 219: Brown, B.M. / Lang, J. / Wood, I.G.** (Eds.), Spectral Theory, Function Spaces and Inequalities. New Techniques and Recent Trends (2012).
ISBN 978-3-0348-0262-8

■ **OT 218: Dym, H. / Kaashoek, M.A. / Lancaster, P. / Langer, H. / Lerer, L.** (Eds.) A Panorama of Modern Operator Theory and Related Topics. The Israel Gohberg Memorial Volume (2012).
ISBN 978-3-0348-0220-8

■ **OT 217: Arlinskii, Y. / Belyi, S. / Tsekanovskii, E.**, Conservative Realizations of Herglotz–Nevanlinna Functions (2011).
ISBN 978-3-7643-9995-5

■ **OT 216: Ruzhansky, M. / Wirth, J.** (Eds.), Modern Aspects of the Theory of Partial Differential Equations (2011).
ISBN 978-3-0348-0068-6

■ **OT 215: Cruz-Uribe, D. / Martell, J.M. / Perez, C.**, Weights, Extrapolation and the Theory of Rubio de Francia (2011).
ISBN 978-3-0348-0071-6

■ **OT 214: Janas, J. / Kurasov, P. / Laptev, A. / Naboko, S. / Stolz, G.** (Eds.), Spectral Theory and Analysis. Operator Theory, Analysis and Mathematical Physics (OTAMP) 2008, Bedlewo, Poland (2011).
ISBN 978-3-7643-9993-1

■ **OT 213: Rodino, L. / Wong, M.W. / Zhu, H.** (Eds.), Pseudo-Differential Operators: Analysis, Applications and Computations (2011).
ISBN 978-3-0348-0048-8

■ **OT 212: Curto, R.E. / Mathieu, M.** (Eds.), Elementary Operators and Their Applications (2011).
ISBN 978-3-0348-0036-5